THE
CHARACTER
CONCEPT IN
EVOLUTIONARY
BIOLOGY

THE CHARACTER CONCEPT IN EVOLUTIONARY BIOLOGY

Edited by

Günter P. Wagner

Department of Ecology and Evolutionary Biology
Yale University
New Haven, Connecticut

ACADEMIC PRESS

A Harcourt Science and Technology Company

San Diego   San Francisco   New York   Boston   London   Sydney   Tokyo

Academic Press
A Harcourt Science and Technology Company
525 B Street, Suite 1900, San Diego, California 92101-4495, USA
http://www.academicpress.com

Academic Press
Harcourt Place, 32 Jamestown Road, London NW1 7BY, UK
http://www.academicpress.com

Library of Congress Catalog Card Number: 00-104309

International Standard Book Number: 0-12-730055-4

Printed and bound in the United Kingdom

Transferred to Digital Printing, 2011

Dedicated to my teacher and mentor
Rupert Riedl, *who, long ago, gave me*
the big picture that I am setting out to fill in.

CONTENTS

CONTRIBUTORS   xi
PREFACE   xv

Foreword   xvii
RICHARD LEWONTIN

**Characters, Units and Natural Kinds:
An Introduction   1**
GÜNTER P. WAGNER

**I HISTORICAL ROOTS OF THE
CHARACTER CONCEPT**

**1 A History of Character Concepts in
Evolutionary Biology   13**
KURT M. FRISTRUP

2 An Episode in the History of the Biological Character Concept: The Work of Oskar and Cécile Vogt 37

MANFRED DIETRICH LAUBICHLER

3 Preformationist and Epigenetic Biases in the History of the Morphological Character Concept 57

OLIVIER RIEPPEL

II NEW APPROACHES TO THE CHARACTER CONCEPT

4 Character Replication 81

V. LOUISE ROTH

5 Characters as the Units of Evolutionary Change 109

DAVID HOULE

6 Character Identification: The Role of the Organism 141

GÜNTER P. WAGNER AND MANFRED D. LAUBICHLER

7 Functional Units and Their Evolution 165

KURT SCHWENK

8 The Character Concept: Adaptationalism to Molecular Developments 199

ALEX ROSENBERG

9 The Mathematical Structure of Characters and Modularity 215

JUNHYONG KIM AND MINHYONG KIM

10 Wholes and Parts in General Systems Methodology 237

MARTIN ZWICK

III OPERATIONALIZING THE DETECTION OF CHARACTERS

11 What Is a Part? 259
DANIEL McSHEA AND EDWARD P. VENIT

12 Behavioral Characters and Historical Properties of Motor Patterns 285
PETER C. WAINWRIGHT AND JOHN P. FRIEL

13 Homology and DNA Sequence Data 303
WARD WHEELER

14 Character Polarity and the Rooting of Cladograms 319
HAROLD N. BRYANT

IV THE MECHANISTIC ARCHITECTURE OF CHARACTERS

15 The Structure of a Character and the Evolution of Patterns 343
PAUL M. BRAKEFIELD

16 Characters and Environments 363
MASSIMO PIGLIUCCI

17 The Genetic Architecture of Quantitative Traits 389
TRUDY F. C. MACKAY

18 The Genetic Architecture of Pleiotropic Relations and Differential Epistasis 411
JAMES M. CHEVERUD

19 Homologies of Process and Modular Elements of Embryonic Construction 435
SCOTT F. GILBERT AND JESSICA A. BOLKER

20 Comparative Limb Development as a Tool for Understanding the Evolutionary Diversification of Limbs in Arthropods: Challenging the Modularity Paradigm 455
LISA M. NAGY AND TERRI A. WILLIAMS

V THE EVOLUTIONARY ORIGIN OF CHARACTERS

21 Origins of Flower Morphology 493
PETER K. ENDRESS

22 Origin of Butterfly Wing Patterns 511
H. FRED NIJHOUT

23 Perspectives on the Evolutionary Origin of Tetrapod Limbs 531
JAVIER CAPDEVILA AND JUAN CARLOS IZPISÚA BELMONTE

24 Epigenetic Mechanisms of Character Origination 559
STUART A. NEWMAN AND GERD B. MÜLLER

25 Key Innovations and Radiations 581
FRIETSON GALIS

INDEX 607

CONTRIBUTORS

Numbers in parentheses indicate the pages on which the authors' contributions begin.

Bolker, Jessica A. (435) Department of Zoology, University of New Hampshire, Durham, NH 08324, *jbolker@cisunix,unh.edu*

Brakefield, Paul M. (343) Institute of Evolutionary and Ecological Sciences, University of Leiden, 2300RA Leiden, The Netherlands, *brakefield@rulsfb. leidenuiv.nl*

Bryant, Harold N. (319) Royal Saskatchewan Museum, Regina, Saskatchewan, Canada S4P 3V7, *hbryant@mach.gov.sk.ca*

Capdevila, Javier (531) The Salk Institute for Biological Studies, La Jolla, CA 92037, *jcapdevila@ems.salk.edu*

Cheverud, James M. (411) Department of Anatomy and Neurobiology, Washington University School of Medicine, St. Louis, MO 63110, *cheverud@ thalamus.wustl.edu*

Endress, Peter K. (493) Institute of Systematic Botany, University of Zürich, CH-8008, Zürich, Switzerland, *pendress@systbot.unizh.ch*

Friel, John P. (285) Department of Biological Science, Florida State University, Tallahassee, FL 32306, *friel@micromoyzon.com*

Fristrup, Kurt M. (13) Cornell Laboratory of Ornithology, Ithaca, NY 14850, *kmf6@cornell.edu*

Galis, Frietson (581) Institute for Evolutionary and Ecological Sciences, University of Leiden, 2300RA Leiden, The Netherlands, *galis@rulsfb.leidenuniv.nl*

Gilbert, Scott F. (435) Martin Biological Laboratories, Swarthmore College, Swarthmore, PA 19081, *sgiber1@cc.swarthmore.edu*

Houle, David (109) Department of Biological Sciences, Florida State University, Tallahassee, FL 32306, *dhoule@bio.fsu.edu*

Izpisúa Belmonte, Juan Carlos (531) The Salk Institute for Biological Studies, La Jolla, CA 92037, *belmonte@salk.edu*

Kim, Junhyong (215) Department of Ecology and Evolutionary Biology, Yale University, New Haven, CT 06520, *junhyong.kim@yale.edu*

Kim, Minhyong (215) Department of Mathematics, University of Arizona, Tucson, AZ 85721, *kim@math.arizona.edu*

Laubichler, Manfred D. (37, 141) Program in History of Science, Princeton University, Princeton, NJ 08542, *manfred1@Princeton.edu*

Mackay, Trudy F. C. (389) Department of Genetics, North Carolina State University, Raleigh, NC 27695, *mackay@unity.ncsu.edu*

McShea, Daniel W. (259) Department of Zoology, Duke University, Durham, NC 27708, *dmcshea@acpub.duke.edu*

Müller, Gerd B. (559) Department of Anatomy, University of Vienna, A-1090, Vienna; and Konrad Lorenz Institute for Evolution and Cognition Research, A-3422, Altenberg, Austria, *gerhard.mueller@univie.ac.edu*

Nagy, Lisa M. (455) Department of Molecular and Cellular Biology, University of Arizona, Tucson, AZ 85721, *lnagy@u.arizona.edu*

Newman, Stuart A. (559) Department of Cell Biology and Anatomy, New York Medical College, Valhalla, NY 10595, *newman@nymc.edu*

Nijhout, H. Fred (511) Department of Zoology; Evolution, Ecology and Organismal Biology Group, Duke University, Durham, NC 27708, *hfn@acpub.duke.edu*

Pigliucci, Massimo (363) Departments of Botany and of Ecology and Evolutionary Biology, University of Tennessee, Knoxville, TN 37996, *pigliucci@utk.edu*

Rieppel, Olivier (57) Department of Geology, The Field Museum, Chicago, IL 60605, *rieppel@fmppr.fmnh.org*

Rosenberg, Alex (199) Department of Philosophy, Duke University, Durham, NC 27708, *alexrose@duke.edu*

Roth, V. Louise (81) Zoology Department, EEOB Group, Duke University, Durham, NC 27708, *vlroth@apub.duke.edu*

Schwenk, Kurt (165) Department of Ecology and Evolutionary Biology, University of Connecticut, Storrs, CT 06269, *schwenk@uconnvm.uconn.edu*

Venit, Edward P. (259) Department of Zoology, Duke University, Durham, NC 27708, *epv@acpub.duke.edu*

Wagner, Günter P. (1, 141) Department of Ecology and Evolutionary Biology, Yale University, New Haven, CT 06520, *gunter.wagner@yale.edu*

Wainwright, Peter C. (285) Department of Evolution and Ecology, University of California, Davis, CA 95616, *pcwainwright@ucdavis.edu*

Wheeler, Ward (303) Division of Invertebrate Zoology, American Museum of Natural History, New York, NY 10024, *wheeler@amnh.org*

Williams, Terri A. (455) Department of Ecology and Evolutionary Biology, Yale University, New Haven, CT 06520, *terri.williams@yale.edu*

Zwick, Martin (237) Systems Science Ph.D. Program, Portland State University, Portland, OR 97207, *zwick@sysc.pdx.edu*

Schwenk, Kurt (163) Department of Ecology and Evolutionary Biology, University of Connecticut, Storrs, CT 06269, schwenk@connvm.uconn.edu

Vieth, Edward P. (359) Department of Zoology, Duke University, Durham, NC 27708, epv@acpub.duke.edu

Wagner, Günter P. (111) Department of Ecology and Evolutionary Biology, Yale University, New Haven, CT 06520, gunter.wagner@yale.edu

Wainwright, Peter C. (285) Department of Evolution and Ecology, University of California, Davis, CA 95616, pcwainwright@ucdavis.edu

Wheeler, Ward (309) Department of Invertebrate Zoology, American Museum of Natural History, New York, NY 10024, wheeler@amnh.org

Williams, Terry A. (635) Department of Ecology and Evolutionary Biology, Yale University, New Haven, CT 06520, terry.williams@yale.edu

Zelditch, Miriam L. (261) Biology, Science Ph.D. Program, Portland State University, Portland, OR 97207, zelditchm@pdx.edu

PREFACE

When the present book project was reviewed, some reviewers suggested the best thing would be for me to write a book on the character concept on my own. I did not follow this advice, and the contributions collected in this volume show I made the right decision. The problems associated with the biological character problem are so complex and multifaceted and this issue is so conceptually immature that any single author's account is doomed to be too narrow and lopsided to be of much use. I am convinced that only the interaction between the results from many model systems and techniques, the ideas of many people, and the productive criticism among colleagues will yield progress in this difficult conceptual territory.

The collection of papers united in this volume is intended to do exactly this, namely, bring together results and ideas from philosophy of science, evolutionary theory, systematics, genetics, functional morphology, and developmental biology which have implications on the way we conceptually construct and identify characters in biology. Not all authors agree on their perspective on the character concept. This is expected and good. I hope that this juxtaposition of ideas will stimulate further discussion of their merit, and ultimately stimulate the creative replacement of these ideas with better ones. If this ever happens, the present volume will have served its purpose.

I want to take this opportunity to thank all the authors, who have contributed excellent papers to this book. Their creativity makes reading the book such an intellectually rich experience. I also want to thank Richard Lewontin, who was kind enough to write a foreword in which he put our project into the proper intellectual context. My thanks also go to Chuck Crumly, who has shepherded this project through its various ontogenetic stages, and especially to Paulette Sauska, who did the most exhaustive and exhausting task of editing and formatting the manuscripts for print. Without her skill and dedication, the project would have died an agonizing death long ago.

GPW
July 2000
Yale University

FOREWORD

"Like following life through creatures you dissect, you lose it in the moment you detect." Already in the 18th century, Alexander Pope had expressed the fundamental problem in the study of living beings, a problem that continues to plague us. How are we to apply an analytic method to living objects in such a way that the phenomenon we wish to understand is not destroyed in the very process of analysis? What are the "natural" suture lines along which we can dissect the organism to understand properly its history and function? For Pope to dissect a creature meant only to cut it up with a knife, but for the modern biologist, and especially the evolutionist, the problem is conceptual. It seems that we cannot carry on our business without using characters, yet there is nothing more dangerous to the proper understanding of biological processes than that first act of characterization. How are we to recognize the "true" characters of organisms rather than imposing upon them arbitrary divisions that obscure the very processes that we seek to understand? That is the question of this book. No issue is of greater importance in the study of biology.

Modern biology is a battleground between two extreme ontological and epistemological claims about living creatures, claims that reflect the history of biological study. One takes as its model the physical sciences and their immense success in manipulating and predicting the behavior of much of the inanimate world. In that domain there are a few basic laws governing all phenomena, the laws of Newton and Einstein, of thermodynamics, of electromagnetic theory, and of nuclear forces. The objects of study either are universal elementary bits and pieces, particles, atoms, and molecules or are composed of those bits and pieces in an unproblematical way. They are Cartesian machines, clocks that can be understood by looking at the articulation of the gears and levers that are clearly recognizable as their parts. It is the extreme reductionist model that validates the various genome-sequencing projects. When we possess the complete description of the DNA sequence of an organism, we have all the information we need to understand the organism. This extreme molecular reductionism is the outgrowth of the 19th-century program to mechanize biology, to expunge the vestiges of mysticism from the study of life, and to bring that study within the domain of universal physical law. Darwinism, Mendelism, biochemistry and biochemical physiology, and Entwicklungsmechanik were its intellectual precursors.

At the other extreme is a radical holism that denies the possibility of learning the truth through analysis. The living world is a seamless whole and any perturbation of one bit of the living world may be propagated in unpredictable ways throughout the biosphere and certainly throughout the entire physiology of any individual organism. It is the whole as whole that must be studied. The belief in this holism has been greatly fortified by the discovery of chaotic regimes in fairly simple dynamical processes, showing that even minor perturbations may give rise to apparently unpredictable dynamical histories. In an attempt to make holism a science there is a movement to develop a mathematical theory of "complex" systems in the hope that complexity will have its own, irreducible laws. Modern "scientific" holism is a return to the obscurantist holism that informed biology before the middle of the mid-19th century and the romantic organicism and naturalism that opposed the development of mechanistic biology, tainted by the stench of the "dark Satanic mills."

Both extreme reductionism and holism escape the character problem. For the reductionist the characters are obvious and unproblematical. They are molecules and distinct sequences of molecular interactions. There is a signaling pathway, composed of controlling regions of DNA in certain genes and proteins coded by other genes, that determines the differentiation of a particular part of an embryo. The part is unproblematical because it is the outcome of the operation of a distinct and well-defined autonomous collection of genetic elements. If genes determine organisms, then parts of organisms are defined by the autonomous assemblages of genes that cause them. One of the ironies of the perverse history of scientific language is that, in its original sense, this reductionism is a truly organic view. The use of the word "organism" for a living being and the "organic" view of life were originally metaphors likening the body to the musical organ, a system of well-defined articulated parts that contribute to the operation of a whole. Nothing is more transparently clock-like and easy to break down into separate functional parts than a musical organ. The keyboard, the foot pedals, the bellows, each set of pipes are all clearly separate organs (in the biological sense) with easily defined relations among them. Indeed, except for the source of air all the parts can be individually disabled without interfering with the operation of others. If the vox humana stop is inoperative, nothing prevents the organist from playng the entire program with the flauto, and if the keyboard is not working, the foot pedals will still serve to play the tune. If bodies were really organic, the problem of characters would not exist and biology would be a simple science. Radical holism, on the other hand, finesses the problem of characters by denying their real causal existence, making of them merely mental constructs. Their text is Wordsworth:

> Our meddling intellect
> Mis-shapes the beauteous forms of things:—
> We murder to dissect.

Because everything is connected to everything else, the delineation of characters is necessarily arbitrary and ultimately destructive of the truth. Moreover, it leads us to a dangerous hubris about how we safely intervene in the world.

The problem for biology is that neither a radical holism nor an organic reductionism captures the actual structure of causation in the living world. It is clearly not the case that everything is connected effectively to everything, even in the relatively simple world of physical objects. Gravitational forces are everywhere, but because gravitational force is weak and falls off with the square of distance, the entire universe of objects is not locked up in an effective gravitational whole. I can feel quite confident that the motion of my body exerts no effective gravitational pull on the person who passes me in the street. It cannot be that the alteration of every part of the body of an organism has a palpable functional or developmental effect on every other part or else evolution could never have occurred. Selection on every bit of the organism would result in simultaneous pressure on every other bit, requiring the organism to be totally rebuilt in response to every selective change. Nor can every species community be totally overturned if one or another species drops out of the mix, although large perturbations may sometimes happen. Despite the popular wisdom, the death of a single butterfly in Madagascar will not have effects propagated throughout the living world or else the temporal instability of the biosphere would have long since resulted in the extinction of all life.

It is necessarily the case that the material world is subdivided into relatively small sets of objects and forces within which there are effective interactions and between which there is operational independence, what has come to be called "quasi-independence." This fact immediately predisposes us to both a conceptual and a methodological reductionism. Let us break down organisms into the smallest pieces we can and then slowly put them back together again, a bit at a time, to see which parts are, in fact, in effective interaction with each other. We can then map out the quasi-independent subsets, the natural characters that constitute living systems. No one can deny that this methodological reductionism has had great success. Everything we know about biochemistry and physiology, about basic cellular mechanisms of replication, about development, has been learned by taking things into pieces and putting them back together again. That is how we know that the differentiation of digits from a limb bud is independent of the differentiation of the external pinna of the ear or that the metabolism of alcohol as affected by the enzyme coded by the alcohol dehydrogenase gene is physiologically separable from the formation of eye pigment in *Drosophila*. That digits and ear pinna or alcohol and eye pigment biochemistry are different characters in the developmental or physiological sense need not have been the case. It simply turned out that way and we would not have been forced to reorganize our understanding of biology had things turned out differently. Methodogical reductionism seems to work, so why not pursue it as a program and let nature speak to us, delineating characters for us bit by bit? We do not do this because it is the failures of a method, not its successes, that are its test.

Organisms occupy a special part of the property space of physical objects. First, they are intermediate in size between plants and particles. Second, they are internally functionally heterogeneous, being composed of many subsystems at many levels. There are different species of molecules in the cell, spatially localized in organelles and cellular regions, and, in multicellular organisms, different cells in each organ and different organs with different functions. Third, many of the molecules and organelles in cells are present in extremely low numbers and so their reactions are not described statistically by the laws of mass action. The consequence of these properties is that organisms are the nexus of a very large number of individually weakly determining causal forms and are subject to stochastic uncertainty. What is true for one organism at one time may not be true for others at other times and in other circumstances. The obvious successes of the reductionist program are precisely in those cases where little or no difference can be seen between different organisms, like the discovery that genes homologous to the *Hox* genes of *Drosophila* are involved in anterior–posterior differentiation in an extraordinary diversity of organisms. But along the way molecular reductionism has also found an undoubted heterogeneity of causal relations from organism to organism and from circumstance to circumstance.

We might, for example, ask what the relationship is between the amino acid sequence of a protein and a functional property that would be regarded as an evolutionary character. No demonstration of the operation of natural selection is more compelling and unambiguous than the discovery by Kreitman (1983), nearly 20 years ago, that the amino acid sequence of the alcohol dehydrogenase gene in *Drosophila* was totally constrained, except at a single amino acid position, while the silent sites in the codons of the gene were 7% polymorphic. We require no assumptions about natural history or physiology to conclude that natural selection is weeding out essentially every amino acid substitution, even such a priori functionally equivalent substitutions as leucines, isoleucines, and valines. Yet this strong molecular demonstration of highly discriminating natural selection in one case turns out not to be general. Many similar studies on other genes in *Drosophila* have since shown a great diversity in the discriminatory power of natural selection on amino acid sequences.

Nor is there any clear relation between the qualitative properties of a protein and its amino acid sequence. At one extreme almost the entire amino acid sequence of a protein can be replaced while maintaining the original function. Eukaryotes, from yeast to humans, produce an enzyme, lysozyme, that breaks down bacterial cell walls. In the evolutionary divergence that has occurred in the yeast and vertebrate lines since their ancient common ancestor, virtually every amino acid in this protein has been replaced, so an alignment of their two protein or DNA sequences would not reveal any similarity. The evidence that they are descended from an original common ancestral gene comes from comparisons of evolutionarily intermediate forms which show more and more divergence of sequence in species that are more divergent. The maintenance of the function despite the replacement of the amino acids has been the result of the maintenance of the three-dimensional structure of the enzyme by the selective substitutions of just the right amino acids.

In contrast, it is possible to change the function of an enzyme by a single amino acid substitution. The sheep blow-fly, *Lucilia cuprina,* has developed resistance to organophosphate insecticides used widely to control it. R. Newcombe, P. Campbell, and their colleagues (Newcomb *et al.,* 1997) have shown that this resistance is the consequence of a single substitution of an aspartic acid for a glycine in the active site of an enzyme that ordinarily is a carboxylesterase. The mutation causes complete loss of the carboxylesterase activity and its replacement by esterase specificity. A three-dimensional modeling of the molecule indicates that the change is the result of the ability of the substituted protein to bind a water molecule close to the site of attachment of the organophosphate, which is then hydrolyzed by the water.

These examples show both the power and the weakness of methodological reductionism for the problem of characters. The method is immensely powerful in reconstructing individual stories but the very diversity that it reveals deprives us of any hope of generalization at this level. Every case will be different. As lawyers say in answer to almost any question posed in a general context, it depends on the jurisdiction.

The alternative strategy is to search for patterns in the manifest properties of a great diversity of organisms of known or inferred relationship. Of course, there is always the danger of circularity, since characters are used to infer relationship, but this tautology can be broken by using a completely different set of properties to establish the phylogeny, say molecular data, and then searching for patterns in morphology or complex behavior. Characters can then be delineated by the properties that show strong correlations or conservations across organisms. Many of the essays in this book deal with this approach to understanding evolutionary characters.

Characters defined (or detected) by the observation of correlations are thought to be given coherence by two possible constraining forces. First, there are developmental constraints. Somehow in the developmental process an increase in one dimension or property is necessarily accompanied by an increase or decrease in another. General changes in size are often invoked (bigger deer have bigger antlers) but biochemical relations may also be constraining. Despite great efforts on the part of plant breeders no one has ever succeeded in breaking the positive correlation between tar and nicotine content in tobacco or the negative relation between protein content and yield in soybeans. Second, there are functional relations. Natural selection builds characters. If the shape and size of the mammalian ear ossicles did not evolve as a unit, each fitted to the other, aural acuity would be compromised, with, presumably, some loss of fitness. As an extreme of the natural selective explanation, there may be an absolute dependence on a fit between two structures for any development at all to occur. Thus, we expect coevolution of DNA binding proteins and their binding sites in signal transduction pathways of early development. Binding protein and binding site are a single evolutionary character.

The problem with this dichotomization is that it is insufficiently evolutionary and insufficiently contingent. In the first place, developmental constraints between parts are not generally global but local in the space of genotypes. Even when a correlation appears to be as simple as one arising from total size variation, there may have been natural selective forces responsible for building it and there may even remain within species sufficient genetic variation to reverse it. An example is the length of various stiffening veins in the wings of *Drosophila*. The lengths of these veins are positively correlated within and between species and remain correlated if the flies are raised at different experimental temperatures, so shape remains constant as size varies environmentally and phylogenetically. Nevertheless, Haynes (1988) succeeded in reducing and even reversing these correlations by artificial selection in *Drosophila melanogaster,* showing that in a region of genotypic space accessible to the genus there was the possibility of a change in shape. Moreover, the unit of selectable morphology in these wings is extremely small. Weber (1992) was able to change the angle and lengths of veins in a region of only about 30 cells by selection, while leaving the rest of the wing unchanged.

Second, functional units are created by the life activities of the organism and vary from circumstance to circumstance. The loss of the last joint of my left little finger would be totally inconsequential for any function that I perform and would surely go unrecognized by natural selection, but it would be of considerable consequence indeed if any livelihood depended on playing the violin. At one moment the entire hand is a character, at another each finger is a character, and at another the hand and arm form a single functional unit. Whether particular agglomerations of bits and pieces of the organism are characters in evolution is not determined by some autonomous external prior circumstance, but by the mode of life constructed by the organism out of the available bits. The green turtle, *Chelonia mydas,* uses its hind flippers to dig a hole in the sand in which to lay its eggs, but no one who has observed that laborious and clumsy process can imagine that natural selection produced those hind appendages as an adaptation for digging. Nevertheless, it is reasonable that some alteration in those organs, chiefly used for swimming, has occurred as a compromise with their use on land. There is a constant evolutionary feedback between the characters that organisms use to make a living and their mode of employment. The environment of an organism comes into existence and changes simultaneously with changes in morphology, physiology, and behavior of organisms. Any concept of evolution that does not take the properties of the organism as both the cause and the effect of natural selection will fail to solve the problem of the character in evolution.

R. C. Lewontin

Haynes, A. (1988). Developmental constraints in the *Drosophila* wing. Ph.D. thesis, Harvard University.

Kreitman, M. (1983). Nucleotide polymorphism at the alcohol dehydrogenase locus of Drosophila melanogaster. *Nature* **304**:412–417.

Newcomb, R. D., P. M. Campbell, D. L. Ollis, E. Cheah, R. J. Russell, and J. G. Oakeshott. (1997). A single amino acid substitution converts a carboxylesterase to an organophosphorus hydrolase and confers insecticide resistance on a blowfly. *Proc. Natl. Acad. Sci. USA* **94**:7464–7468.

Weber, K. E. (1992). How small are the smallest selectable domains of form? *Genetics* **130**:345–353.

CHARACTERS, UNITS AND NATURAL KINDS: AN INTRODUCTION

GÜNTER P. WAGNER

Department of Ecology and Evolutionary Biology, Yale University, New Haven, CT 06520

INTRODUCTION

Organisms owe their special status as living beings to their material organization. There is no special force or substance that distinguishes organisms from inanimate matter; only everyday physics and chemistry prevail in the details of an organism's life. The difference between inanimate matter and organisms thus lies in the spatial and temporal organization of ordinary physical and chemical processes. The contents of the cell need to be separated from the environment, and inside higher cells compartments need to be neatly separated and the exchange of molecules among them tightly regulated to ensure the continuation of life. This implies that spatial and temporal heterogeneity is fundamentally linked to the existence of life as a material process. This intrinsic need for heterogeneity and the tendency of organisms to increase the heterogeneity of their material organization during ontogeny and phylogeny has been recognized by humans ever since our ancestors started to give names to

The Character Concept in Evolutionary Biology

things in their environment. I guess that the recognizable parts of humans, animals, and plants must have been among the first things that received names, such as hand, head, and eye, because they are of fundamental importance to the life of all peoples. This is also the origin of the character concept, i.e., the idea that organisms consist of natural units, which, in some cases, can even be identified in organisms of different species. Once this step was taken, namely to recognize that essentially the same parts can be found in different species, the idea of homology had been born. Characters are thus among the most fundamental units we use to systematize the things in our world, together with ideas about species, and different forms of matter such as air, water, and stone. Surprising then is the relative lack of attention the character concept has received (but see Gould and Lewontin, 1978; Lewontin, 1978; Ghiselin, 1984; Colless, 1985; Rosenberg, 1985; Rodrigues, 1986; Inglis, 1991; Stearns, 1992; Wagner, 1995) compared to the species concept or the concepts that form the core of the inorganic sciences. The purpose of this collection of essays is to put the character problem into the context of mechanistic biology in order to aid the development of a scientific character concept.

Only in one incarnation did the character concept receive extensive attention, namely in the context of the homology problem. The homology concept has a long but frustrating history. Only a temporary ceasefire has been reached with the introduction of cladistic methods in systematics. Cladistics gave some operational meaning to the notion of homology (Patterson, 1982). The breathtaking progress of developmental genetics, however, reignited the debate over issues that could largely be ignored in a cladistic context. For instance, what is a homologue and what accounts for its identity? The progress in developmental genetics only magnifies the problems that had already been seen by perceptive thinkers such as Spemann (1915), deBeer (1971), Riedl (1978), VanValen (1982), and others. The unsatisfactory status of the homology debate, I think, is a consequence of the fact that the debate about homology happens in a much too narrow context. The question usually asked is "How do I recognize the character in species B that corresponds to the one I know from species A?" This is an important problem for many areas of biological research, but is much too narrow compared to the scope of the problem: what are the natural unit organisms composed of? How do we recognize these units? How do we justify our choices? What is the theoretical context in which we can make these choices? Why do they exist? How did they come into existence?

In this introductory chapter I want to outline some of the theoretical background for the essays collected in this book. Why does biology need a character concept? What is the conceptual framework for addressing the character problem? What are the research problems that need to be attacked? The answers to these questions will also explain the motivation behind the present book project. I expect that a close reading of and reflection on the chapters in this book will lead to a fundamental revision of the views that I express in these pages,which is exactly the purpose of theories: inducing research that makes them obsolete.

WHY DOES BIOLOGY NEED A CHARACTER CONCEPT?

Despite the widely acknowledged functional and organizational wholeness of organisms, biology concerns itself largely with the study of often minute parts of organisms. Even so-called organismal biology mostly deals with either physical or functional parts or with abstractions of organisms such as life histories, populations, or behavioral patterns. The success of a research program depends to a large degree on the right choice of the parts or abstractions to be studied. The cell, for instance, is an example of a subsystem of higher organisms, which has all the hallmarks of a natural unit. It has a high degree of internal coherence and exhibits emergent properties that play a prominent role in the explanation of biological phenomena. On the other hand some of the intensively studied quantitative abstractions of classical physiology did not lead to generalizations because the fundamental units are often molecular or cellular rather than systemic. Hence, the identification of natural (sub)systems of organisms stands at the beginning of every research program in biology and influences its success.

The few examples of characters mentioned earlier suggest a preliminary definition of a character that could temporarily serve as a guide through the jungle of ideas and observations around the character concept:

A biological character can be thought of as a part of an organism that exhibits causal coherence to have a well-defined identity and that plays a (causal) role in some biological processes.

This (pseudo) definition is radically more ambitious than the narrow definition of a character in systematics. In systematics, characters are conceptualized as any observable difference between two groups of organisms, which can be used to "characterize" (distinguish) these groups. Our definition aims at the question why it is at all possible to individuate meaningful characters within the context of the functional integrity of the organism. What are the mechanistic conditions that make the idea of characters meaningful? A very general reason has been given in the first paragraph of this chapter, namely the need for organisms to spatially separate various processes. However, this is of course by far not sufficient because it does not provide us with an idea of how to find and individuate these characters and to formulate a research program into the nature and evolution of these units.

Of course there are parts of organisms which clearly fulfill the criteria of this definition, namely molecules and cells, and they are the subject of well-established scientific disciplines: biochemistry and cell biology. Is anything else needed? I think there is one main reason that indeed more is needed: organisms exhibit multiple levels of integration. Potentially at each level of integration units can exist that exhibit sufficient coherence and play a causal role in a process to warrant their recognition as a character. As pointed out by Robert Brandon, mechanisms can arise at any level of integration, and the assumption that mechanistic explanations can only come from smaller (molecular) units or higher units is a metaphysical commitment that has no place in science (Brandon, 1996). I suspect that the things that we call characters, defined at a certain level of integration, are the players in emergent mechanisms that occur at the respective level of integration. The existence of emergent levels of complexity with their level-specific causality seems to be generic property of complex systems (Bar-Yam, 1997).

To illustrate this point I want to recount Philip Kitchers metaphor of the reductionist madman. Kitcher uses the phenomenon of the 1:1 sex ratio in human populations to point out that the most informative explanation is the one originally given by R. A. Fisher, namely that there is frequency dependent selection for equal proportions of male and female offspring (Fisher, 1930). Only a madman (or madwoman) would in all seriousness consider a fully molecular account of this phenomenon. Such an account would need to explain how it comes that about the same number of Y chromosome and X chromosome bearing sperm are successful in fertilizing an egg and that during development the right proportion of fetus die such that in the end we have a sex ratio close to 1:1. Clearly, population genetics provides a mechanistically much more meaningful explanation than a molecular account. I take this example as an illustration for the existence of causally relevant mechanisms at higher levels of organization.

If causally relevant mechanisms exist at higher levels of organization, then it becomes interesting to ask what are the units that play a causal role in these mechanisms. I guess that most of the units that strike us as sufficiently distinct and coherent parts (i.e., characters) are those units. Otherwise it is hard to understand why these units exist at all (Wagner, 1995).

A well-developed character concept could thus help us eliminate metaphysical commitments, such as the idea that mechanistic explanations can only derive from the molecular level of description. There are of course mechanisms at the molecular level which are of fundamental importance to all aspects of life; this is not the issue. The issue is whether these molecular mechanisms are the only ones that matter in biology. This is not a metaphysical question but an empirical one.

Before I consider how other sciences define their fundamental units and what biology may learn from it, I want to discuss two biological problem areas that could benefit from a better understanding of the character problem. One is the genetic explanation of morphological variation and the other is the empirical meaning of mathematical models in biology. The problem of a genetic explanation of morphological differences between species is nicely illustrated by the debate surrounding the discovery of homologous genes involved in the development of very different light sensory organs (eyes) (Bolker and Raff, 1996; Müller and Wagner, 1996; Abouheif, 1997). Halder and collaborators (1995) have shown that the gene *Pax-6* is sufficient to initiate the development of compound eyes in the ectoderm of *Drosophila melanogaster*. What is remarkable though is the discovery that a homologous gene exists in mammals that is also necessary for the development of the "camera" eye typical of vertebrates. The eyes of mammals and flies, however, taken as anatomical characters, are certainly not homologous. Their structure is too different and mammals and flies too distantly related that either was derived from the other. Most likely both are derived from a common ancestral structure, which was neither a compound eye nor a camera eye, but maybe only a light-sensitive epithelium. The ancestral role of *Pax-6* homologues then most likely was to initiate the development of the light-sensitive epithelium, which is the phylogenetic precursor of both the camera and the compound eye. In derived species the ancestral structure became transformed into what is now either a compound or a camera eye, neither of which is the same as the ancestral structure. The role of *Pax-6* in derived species then is to determine the eyefield, i.e., the ontological precursor of complex, derived eyes. As Riedl has said so perceptively: the phylogenetic precursor of a derived character corresponds to the ontogenetic precursor of the same character in derived species (Riedl, 1978). To solve the riddle of homologous genes causing the development of nonhomologous characters requires a careful distinction between the levels of organization and the kinds of characters involved. An eye is not necessarily an eye. A clearer understanding of the biological meaning of character identity would be useful to sort out problems like these.

Another example of how a character concept might be useful in biological research is the interpretation and the empirical testing of mathematical models. The ambivalent opinion of most biologists toward mathematical models is in part caused by a structural weakness of these models (the other part is the complete lack of mathematical training). Mathematical models are an absolute necessity in many parts of biology, in particular in organismal biology, where the processes investigated are often quantitative and hard to put into a lab setting (just like in cosmology). The best examples are population biology and ecology. In Laubichler and Wagner (2000) we have argued that the models in these areas have a structural weakness because they lack a clearly defined theory of what their intended range of application is. Under exactly what conditions can we apply a life history model? What exactly are the life history characters, and how can we test whether a particular system is a legitimate instance of the model? All these questions derive from a lack of understanding of what units play a mechanistic role in a certain process. In other words, all these questions ask about what the relevant characters are for the focal mechanism.

NATURAL KINDS AND BIOLOGICAL MECHANISMS

As argued earlier, the character problem can be seen as a special case of a problem that occurs in many (all?) sciences, not only in biology, namely the question of what fundamental units play a causal role in a process. The most general form of this problem is the so-called natural kinds problem (Putnam, 1981; Keil, 1989; Boyd, 1991; Hacking, 1991; Wagner, 1996; Griffiths, 1997). In this section I want to discuss how the notion of natural kinds, as developed in other sciences, can help in guiding our thinking about the character concept.

In a seminal essay, Quine (1969) has reflected on the conceptual nature of those units which play a fundamental role in the theoretical core of a science, such as atoms and molecules. Most of the essay is dedicated to the discussion of various proposals to define natural kinds, such as similarity or statistical considerations, and he rejects all of them. Toward the end of the essay, however, and almost in passing, Quine makes a startling suggestion. He simply suggests that natural kinds are things that play a role in a law of nature. No other definition would work. In other words, natural kinds are defined by their function in interacting with other such kinds, not by their intrinsic characteristics alone. One has to appreciate that Quine's argument was formulated in the linguistic tradition of the philosophy of science and contains concepts that seem problematic from our point of view, in particular the notion of a "law of nature." I think it is fair to translate Quine's formulation into a more realistic interpretation: a natural kind is a unit that plays a mechanistic role in a process. Nothing essential has been lost in this reformulation.

A major challenge to the natural kinds concept came from Putnam (1981), who pointed out that there is no finite list of attributes that allows us to define a natural kind. His favorite example is the chemical element *Au*, gold. For instance gold could be defined among other attributes by its color. The color, however, is not a constant attribute, since evaporated gold has no color and pulverized metallic gold is black. I think a solution to this problem suggests itself from considering how chemists in fact identify and thereby operationally define chemical elements and substances. Textbooks of chemistry tell us that gold is defined as a chemical element of atom number 79. This in itself is not very useful since it only replaces one abstraction by another. What we need to find out is what gives these abstract concepts their empirical meaning. Looking back on my training as a chemical engineer I suggest that any expert does in fact identify a chemical substance or a chemical element by checking a list of attributes, apparently contra Putnam. The reason why the chemist is nevertheless not falling into the trap Putnam has pointed out is that the chemical attributes that matter in the identification of chemical elements are dispositional properties. That means that they are not just observable properties, but properties that are expected to be observed if the object is put in a particular situation. For instance, in modern chemistry a lot of chemical identification is done by spectroscopic methods. If I heat a substance containing sodium, *Na*, I expect that it will emit light with a 589-nm wavelength. The same is true of the classical chemical methods, which also test dispositional properties, such as solubility in water and acids or melting point of crystals and reactions with other chemicals. The whole system of chemical elements was erected by defining lists of dispositional properties, each of which stood for and operationally defined a chemical element. The identification of a chemical element with an atomic model was a much later achievement. What this latter conceptual step did was to summarize the lists of dispositional properties in a mechanistic model, i.e., the atomic model.

If we step back and ask what the structural features of this example are, one can recognize three distinct steps in the conceptualization of a natural kind.

1.   At the beginning stands, in accordance with Quine, the choice of a class of processes, in the case of chemistry the transformation of matter at moderate temperature levels. (Now we know that chemistry describes transformations of matter at a temperature range in which atoms are stable.)

2.   The next step is the identification of stable sets of dispositional properties. The existence of these stable sets is then taken to suggest that there are discrete units that exhibit these properties. These units are hypothesized to be the units participating in the focal processes. Of course the dispositional properties used to define these units are ones that have relevance in the focal class of processes. In chemistry these are mostly reactivities (i.e., if you combine A and B you get C and D). At this stage the natural kind concept is operationally defined but still lacks a mechanistic explanation. Note that the latter is not necessary for the scientific validity of these concepts.

3.   The last step is the formulation of a model which explains why the sets of dispositional properties that define natural kinds are stable. In chemistry this is the atomic model, which in fact does explain large chunks of empirical knowledge.

Note that there is not just one natural kinds concept but at least two historical stages: the operational definition and the theoretical model. I think this distinction between two stages of a natural kinds concept may be useful in thinking about biological characters and perhaps even other biological concepts.

Applying these ideas to the biological character problem has a number of implications. The first and perhaps most important is that any attempt to define a character has to start with defining the reference process in which this character is supposed to act as a unit. In many of the traditional uses of the character concept the implicitly assumed reference process is evolutionary change by natural selection, but this is not universally so, and many conflicts between different character definitions seem to come from a lack of agreement what the implicitly assumed reference process is (Fristrup, this volume). Examples are the process of inheritance, where Mendel discovered that genes are the units of transmission in sexually reproducing organisms. Note that in accordance with our scheme described previously that the classical definition of a gene was based on operational criteria based on crossing experiments. Only later did a model of how molecular structures can account for the Mendelian rules lead to a more mechanistic gene concept. In physiology and functional morphology the reference processes are physiological processes (see contributions by Wainwright and by Schwenk in this volume).

The second implication of the analysis of the natural kinds concept is that the defining properties of characters should be dispositional properties that are relevant to the reference process. In the case of genes this property is the segregation behavior in various crosses. In behavioral biology and physiology the relevant property can be that a sequence of actions can be triggered by a limited set of stimuli, for instance.

The third implication is that the set of dispositional properties has to be based on a causally homeostatic mechanism (Boyd, 1989; Keil, 1989; Boyd, 1991; Griffiths, 1997). In other words, the mechanisms which explain the dispositional properties have to be itself invariant in the reference process. For instance, the structure of an atom (i.e., its nucleus) does not change in chemical transformations, which only affect the structure of the electron shell. A gene is not changed by recombination and a cell keeps its integrity in (most) developmental processes.

The latter point is an important liberalization to the classical notion of natural kinds which was and is often expressed in fairly absolute terms (see chapter 7 in Griffith, 1997, for a discussion). Natural kinds are expected to have a certain degree of stability (causal homeostasis), but the stability is only relative to the interactions in the reference processes. Atoms are invariant during chemical transformations, but not in radiochemical reactions and at very high temperatures. Genes are invariant with respect to genetic transmission but not with respect to mutation. This is important, as biological characters can be quite conservative, as comparative anatomy is showing, but the conservation is of course not absolute, otherwise innovation and body plan evolution would not be possible. Hence characters are less stable than many atoms (but on what common scale?), but the difference is gradual rather than absolute.

If we try to rethink the character problem along those lines it becomes clear that this approach provides a lot of conceptual freedom to explore various options.

FROM CONCEPT TO RESEARCH PROGRAM

A character concept will only be integrated into the body of scientific thinking if and only if it somehow aids the acquisition of further scientific knowledge. A character concept that only satisfies the aesthetic predilections of theorists is ultimately doomed, regardless of how true or elegant it may be. Hence a character concept in itself cannot be the subject of a viable research program if it does not serve a wider purpose. Based on the discussion in the last section, it is clear that any character concept will be contingent on the process in which the characters are expected to play a role. This is true for all areas of biology, as hinted to earlier, but this book focuses on the character concept in evolutionary biology and thus to characters as defined relative to the process of evolution. Hence the character concept aimed at here is supposed to aid research in evolutionary biology and its associated areas such as systematics and other branches of comparative biology. It is a great advantage for the development of the evolutionary character concept that we already have a fairly sophisticated understanding of the elementary mechanisms of evolution. Any character concept thus has to fit into the existing theoretical framework of evolutionary biology, which strongly limits the range of viable candidates. A concept will have to reflect the fact that spontaneous generation of variation and sorting of variants by population dynamic processes are the elementary mechanisms of evolution. With this guideline in mind, a character concept will have to define research programs to answer at least the following three questions:

1.   What is the explanatory role of characters in evolutionary theory? I will not seriously attempt to answer this question here, but only want to point to a few directions in which answers may be found. Picking up on Lewontin's idea of characters as quasi-indepent units of evolution (Lewontin, 1978), it may turn out that characters, properly defined and delineated, are the real subject of adaptation. They are the units that interact, within the context of the organism, with the environment and "answers" to adaptive challenges. In turn, if characters are units of evolutionary change, they may be seen as providing each other with the context in which they adapt to the environment. Interactions among characters in the adaptive process would thus be an important part of the explanation of adaptive evolution. Characters with their stable properties are thus the players that form the frame and the context of the adaptive dynamics. As a consequence, the acquisition of certain characters may have long-lasting effects on the diversification of a clade, which is the intuition behind the (controversial) idea of key innovations (see Galis, this volume).

2.   What are the mechanisms underlying the causal homeostasis of characters? Assuming that indeed characters play an explanatory role in evolutionary processes, one has to assume that the relevant properties of characters are stable in these processes. Otherwise it would not be guaranteed that a reference to

different instances of a character means the same thing. This in turn raises the question of what mechanisms cause the conservation of these properties. Mutation is random, in first approximation, and selection is opportunistic and variable. Why then would any aspect of the organism be more stable than any other? The answer to this question may lead to a mechanistic understanding of character identity (homology).

3.   How do characters come into existence during the course of evolution? In the context of the evolutionary process, characters are historical individuals, such as species and clades, which have a definite beginning and potentially an end. Since characters are expected to be conservative with respect to certain properties, how did these conservative properties come into existence? This question is called the innovation problem in evolutionary biology (Müller and Wagner, 1991). Arguably this is the most challenging and exciting problem in the evolutionary biology of characters (see the contributions in Section V of this volume). Of course, the complement to the innovation problem is why and how a character may get lost in evolution, a process called reduction. This latter process is usually associated with the loss of function and the accumulation of deleterious mutations, and/or active selection against the functionless character.

I expect that these three questions, or some transformations of them, will be part of any research program that has the character concept at its center. The three areas are very similar to the three ideas that made the species concept a fertile ground for empirical and theoretical research. The modern species concept has its root in the recognition that species are the population biological units of evolutionary change (= explanatory role). They maintain their integrity through isolation mechanisms (= explanation of causal homeostasis). Finally, species originate because of the evolution of isolation mechanisms (= explanation of historical origin). It seems to me that the same conceptual outline, which made the species concept a scientific concept, may help transform the opaque notion of a character into a heuristically productive scientific concept.

LITERATURE CITED

Abouheif, E. (1997). Developmental genetics and homology: a hierarchical approach. *Trends Ecol. Evol.* 12(10):405-408.

Bar-Yam, Y. (1997). "Dynamics of Complex Systems." Addison-Wesley, Reading, MA.

Bolker, J. A., and Raff, R. A. (1996). Developmental genetics and traditional homology. *Bioessays* 18:489-494.

Boyd, R. (1989). What realism implies and what it does not. *Dialectica* 43:5-29.

Boyd, R. (1991). Realism, anti-foundationalism, and the enthusiasm for natural kinds. *Philos. Stud.* 61:127-148.

Brandon, R. N. (1996). "Concepts and Methods in Evolutionary Biology." Cambridge University Press, Cambridge.

Colless, D. H. (1985). On 'character' and related terms. *Syst. Zool.* 34:229-233.

HISTORICAL ROOTS OF THE CHARACTER CONCEPT

Scientific concepts result either from a transformation of a prescientific notion or idea or from the transformation of another scientific concept. Just as organisms are the result of a historical process of modification, concepts and theories are historical entities. This is both an advantage as well as a constraint. The trans-generational history of ideas allows the accumulation of experiences of many lifetimes and thus can lead to greater intellectual depth than the ideas of any individual might have. On the other hand, the history of a concept can constrain or at least bias the directions of thought in ways not entirely justified by empirical facts. The historicity of scientific concepts is largely irrelevant for daily scientific practice, but is highly relevant if we are dealing with the development of new concepts. One has to understand what the objectives were that led past generations to develop a concept in the way as they did. It is also useful to know which avenues of thought have been tried in the past and why they failed to make informed choices in the development of new concepts. It is for that reason that a section on the history of the character concept in this book is of more than entertainment value, and why there is more than one chapter dedicated to the history of the character concept.

Kurt Fristrup starts us off with a general historical overview of the character concept. The central theme in his account is that much of the confusion surrounding the character concept stems from the divergent research interests of systematists and evolutionary biologists. It is fascinating to read how much of the tortuous past of the character concept can be understood as resulting from the tension between these two disciplines. His narrative leads us up to the present, in which considerations of development further complicate the picture. He concludes with a warning that lack of attention to the very real differences between the traditions in different disciplines, as well as the difference between conceptual ideals and the empirical data, has hobbled biological research in the past and will do so in the future if we do not pay attention to these issues.

Manfred Laubichler brings to our attention a largely forgotten episode in the history of the character concept. He recounts the work of Oskar and Cécile Vogt, who used ideas from systematics to develop a classification of and research program about the biological basis of neurological and psychiatric diseases. It is noteworthy that this heuristic exchange between taxonomy and medicine is not just a historical fluke but was part of biomedical mainstream of the time and formative for the neurology in the first half of the 20th century. It is also fascinating to see how in this context many ideas that are discussed now have been in one way or the other already present in Vogt's theory of a character and character variation.

Olivier Rieppel reveals to us a fascinating connection between ideas in developmental biology on the one hand and systematics and evolutionary thinking on the other. The distinction between preformationist and epigenetic ideas of development is a commonplace part of the history of developmental biology. It is less appreciated, however, that a structurally similar distinction underlies the ideas about the character concept. Most obviously this is the case in attitudes toward character innovation. Is true innovation possible? A point of view that denies the possibility of innovation is clearly preformationist, while the assumption that innovations are possible has affinities to the epigenetic view of development.

1

A HISTORY OF CHARACTER CONCEPTS IN EVOLUTIONARY BIOLOGY

KURT M. FRISTRUP

Laboratory of Ornithology, Cornell University, Ithaca, NY 14850

INTRODUCTION

Biological research is vitally concerned with two issues that have no counterparts in physical and chemical research: functional significance and ancestral relationships. The identification of specific parts, qualities, or actions of individuals provides the logical basis for studying these issues. Individuals are evolutionary transients, and the enduring objects are the parts, qualities, or actions that are heritable.

Historically, the terms character and trait were used to refer to these objects, which provided the basis for comparing individuals. As emphasized by Gould and Lewontin (1979), the answer to the question "What is a trait?" is the crux of different programs for studying adaptation. Characters also formed the axes for comparative studies of historical relationships among organisms. However, adaptation or function has been dismissed as irrelevant by some comparative biologists; the sole purpose of their character analyses was to infer patterns of common ancestry among species. Species were identified and circumscribed in terms of characters, even though they were often recognized as biological individuals because of reproductive isolation.

The dual uses of characters to study of ecological "guilds" and phyletic histories (Harvey and Pagel, 1991) generated torsion over the meaning of the term. For example, some systematists referred to non-homologous features, such as the ecological category of "seed-feeder," as an "attribute" (Mickevich and Weller, 1990). This would exclude, for example, Darwin's use of the adjectives adaptive or analogical in relation to characters (Darwin, 1859, pp. 414, 427). This tension was more than semantic. Functional and historical studies sought objects with fundamentally different properties. For phylogenetic reconstruction, ideal characters arose once; their uniqueness was unambiguous. However, such characters presented serious difficulties for functional explanations. The absence of replicates disallowed discrimination among alternative hypotheses regarding their origin. The fate of a unique feature, with a single origin, is largely determined by chance, even if it conveys significant selective advantage to its bearers.

In contrast, the best characters for functional analyses arose in identical form in multiple, independent settings. Consistent association between these characters and certain functional or environmental contexts provided evidence regarding the function, origin and spread of the character. In evolutionary theory, the distinctions between unique and generalized characters interrelated with concepts of fitness. The existential view of fitness (Lewontin, 1961; Slobodkin and Rapoport, 1974) asserted that the meaningful measure was avoidance of extinction. This argument seems most appropriate for characters and lineages that arose just once. For characters whose repeated origin is assured (and sets of ecologically analogous species), the Malthusian parameter seems the more appropriate measure of fitness.

Thus, characters were used to identify organisms, to document the relationships and similarities among organisms, and as the canonical units in functional explanations. This widespread usage was accompanied by crucial variation in meaning. Aside from imprecise diction, this variation in usage manifested unresolved methodological issues in biology. To what extent could the structure and function of organisms be divided into units that had explanatory roles in biological theories? Contrarily, were such dissections artifacts? How can definitions of characters be divorced from individual or general human perceptual bias, theoretical prejudices, historical legacies, linguistic idiosyncrasies, or expediency? Contrasting positions may have stemmed from the perceptual and cognitive idiosyncrasies of investigators. What did they view, and how did they map the variety of living organisms? What representations were appropriate to concisely communicate this map, to permit others to understand and replicate their work? Rosenberg (1985a) concluded that designation of characters was not a pretheoretical matter and that identifying the facts to be explained should be jointly determined with the modes of explanation. How much familiarity with the organisms is necessary, and how well must a problem be posed, to permit designation of useful characters?

Some broad historical trends may be summarized briefly. In systematics, the kinds and numbers of characters employed were augmented as the scope of comparative analyses expanded. Limited transportation and communication restricted early comparative biology to studies local faunas. These collections could be grouped into distinct species and higher groupings, using relatively few characters. As comparative studies became intercontinental, the flood of new forms from around the globe blurred many divisions among groups, and suggested the notion of plenitude. Many more characters were employed, and groupings of species were increasingly viewed as artificial. In modern practice, the trend to more extensive character analyses has continued, although the notion of a seamless continuum of biological forms has been abandoned. A comparative study of anatomical function was established as an important discipline in Cuvier's time, and Darwin revolutionized the conceptual basis of functional studies by providing an ultimate measure of utility.

Common to all biological disciplines was a historical trend toward conceptual simplification. Characters were increasingly required to satisfy special criteria. This encapsulated complexity that practitioners viewed as distracting and less germane to their studies. Collectively, these historical processes generated increasingly specific and divergent uses of the word character. This historical perspective emphasizes the period from the 18th century to the present. The starting point was chosen in relation to the decline of Aristotelian theory in comparative biology. The discussion is organized around enduring questions regarding the designation and interpretation of characters.

DID THE TERM CHARACTER REFER TO A SPECIFIC INSTANCE OR A GENERALIZED CLASS OF THINGS?

The word character was sometimes used in a descriptive sense to refer to a part, quality, or action of one individual. The description of an individual may be more or less intelligible, evocative, or precise, but it does not constitute a hypothesis. Perceived or hypothetical similarity was implied by a second use of the word character. Here, parts, qualities, or actions associated with different individuals were designated as different instances of the same character. Similarity could refer to the function or the historical origins of the part, qualities, or actions. Similarity was operationally recognized by the situation of the part, quality, or action in the individual in relation to other characters. Where character referred to a measure of similarity, the terms character state or character value often appeared to distinguish differences in detail among subsets of the parts, qualities, or actions. In a third use, a character could be a prescription for a measurement. The prescription ensured consistency in measurement, but· did not necessarily entail a prior assumption of functional or historical relatedness.

In portions of the cladistic literature, characters were defined as nested sets of synapomorphies (Nelson and Platnick, 1981), sequences of unique historical events in the modification of a part, quality, or action, and also markers for the diversification of taxa within a clade. "A character is thus a theory, a theory that two attributes which appear different in some way are nonetheless the same (or homologous). As such, a character is not empirically observable; hence any (misguided) hope to reduce taxonomy to mere empirical observation seems futile" (Platnick, 1979, p. 543). In this view, character essentially designated descent from a particular common ancestor, and a character was labeled by the description of the corresponding feature in that ancestral taxon (Platnick, 1979; Nelson and Platnick, 1981). If birds, frogs, and mammals were said to share a character called legs, then modern snakes were said to "have" legs as well. However, if new evidence—perhaps unrelated to legs in any way—revised the position of snakes outside of tetrapods, then snakes would no longer have legs. Those parts would no longer define a portion of their ancestry. These views were inconsistent with the views of other cladists. Pogue and Mickevich (1990) reasserted the status of characters as observations. Mickevich and Weller (1990) suggested that character state trees expressed probability of transformation; probabilities only make sense if transitions among states were repeated events (states represented classes, not individuals). For a historically unique event, there was one realization and no basis for considering alternative outcomes.

Several authors have discussed the ambiguous or equivocal usage of the term character (Ghiselin, 1984; Colless, 1985; Rodrigues, 1986; Rieppel, 1991; Fristrup, 1992). Confusing use of the term often reflected imprecise diction. The author's intent may have been clear from the kind of data presented, and its treatment. More serious confusion followed when distinctions between a member and a class, or an individual and a category, were lost. Additional barriers to communication arose from specialized usage that evolved to meet the operational needs and theoretical visions of particular biological disciplines. Superficial similarities often masked these fundamental differences in usage.

DID CHARACTERS SERVE AS THE EVIDENCE USED TO RECOGNIZE AND ARRANGE TAXONOMIC GROUPS, OR DID THEY EMERGE AFTER GROUPS WERE ARRANGED BASED ON UNSPECIFIED COMPARISONS AND INTUITION?

Mayr's (1982, Chapter 4) historical treatment of systematics highlighted the latter process, despite the decisive role portrayed for characters in all schemes of classification (*ibid.*, p. 185). The early systematists, whose formal method of classification relied on logical division into groups based on a few "essential" characters, relied heavily on an informal process of inspection for their arrangements (*e.g.*, Bock, Bauhin, Tournefort, Magnol). The classifications of Cesalpino (*ibid.*, p. 160) and Linnaeus (*ibid.*, p. 179) arose from an extended process of visual inspection and comparison, which preceded formal Aristotelian analyses using characters. Mayr emphasized the surprising agreement of these pre-Darwinian systems with modern classifications and attributed it to the triumph of their intuition over flawed formal methods.

By the end of the 18th century, species were arranged based on affinity, placing the most similar species together, and subsequently designating groups ("synthesis"). However, this shift in method did not diminish the importance of informal comparison in practice. A.-L. de Jussieu's *Genera plantarum* (1789) was the first synthetic treatment of a major group that was widely accepted. He did not emphasize characters initially, and was more insistent on establishing the correct sequence of organisms using intuitive assessments (Stevens, 1994, pp. 31, 126). Subsequently, A.-P. de Candolle explicitly described undirected exploration and general comparison as the first two stages in developing the natural arrangement of species, which were followed by subordination of characters (*ibid.*, p. 81). Repeated reference to the role of "instinct" in developing classifications, and the necessity of apprenticeship (*ibid.*, p. 225) suggested that natural affinities were identified before characters were formally designated.

Despite the ascendance of hypothesis testing and objective measurement in biology, informal assessments remained an important part of the practice of delineating species and designating groups. "Hence, in determining whether a form should be ranked as a species or a variety, the opinion of naturalists having sound judgement and wide experience seems the only guide to follow" (Darwin, 1859, p. 47). "I believe that the main framework of taxonomy will always have this intuitive background. The ability to pick out those master character differences with which others are correlated is what is called 'having an eye for a species'; it is a faculty which is to a large extent innate though naturally it can be much improved by wide experience in the study of particular families or orders" (Richards, 1938, p. 97). Simpson (1961, p. 107) characterized the informal practices of classification as art.

The designation of characters and compilation of their values across groups was sometimes seen as the prerequisite to devising arrangements, but even prominent advocates of this position acknowledged the role of intuition. Adanson compiled information using a large number of characters, and his groups emerged from a composite of arrangements that these characters suggested (Mayr, 1982, p. 194; Stevens, 1994, pp. 23, 124; Sokal and Sneath, 1963, p. 16). Adanson wrote "It was by the overall view [*ensemble*] of these comparative descriptions that I perceived that plants sort themselves naturally into classes or families" (cited in Mayr, 1982, p. 194). Yet, A.-P. de Candolle called Adanson an instinctive systematist (Stevens, 1994, p. 485). In the modern practice of phenetics, which referred to Adanson as a progenitor (Sokal and

Sneath, 1963, p. 16), the critical information (unit characters) was collected prior to any analyses that developed grouping. However, they also spoke of the initial step as "preliminary choice of specimens" with "no sharp distinction between the selection of specimens and the selection of characters" (*ibid.*, p. 60).

Often it seemed that characters were designated at a later stage of comparison and analysis, perhaps more to communicate or justify arrangements than to create them. For some, this method seemed the most efficient, or the only reliable approach. "As a general rule the application of purely quantitative methods to a mixed set of animals which had not been subjected to a preliminary sorting would be a waste of time and would not lead so quickly, if at all, to the recognition of whatever distinct groups were present" (Richards, 1938, pp. 97-98).

HOW MANY CHARACTERS ARE SUFFICIENT?

The application of scholastic logic that prevailed until the end of the 18th century—the Aristotelian method of classification—defined a group by its essential character. The search for these characters, which were unique to the group and shared by all group members, was a fundamental inquiry into the nature of living things and their structural organization. These defining characters proved elusive. The Linnean classification of plants, for example, utilized relatively few characters, which were easily perceived. It had widespread and lasting impact, but its creator saw it as artificial: convenient for identification and organization, but not based on the fundamental structure and consequent affinities of plants (Stevens, 1994, p. 128; Mayr, 1982, p. 174). Buffon asserted that the search for essential characters was fated to fail, relating them to the philosopher's stone (Stevens, 1994, p. 29).

When characters were no longer viewed as defining and essential, they played a variety of conceptual roles. By the end of the 18th century, new systematic methods attempted to employ as many characters as possible, from all aspects of the organism (Buffon, Lamarck, Adanson, Jussieu). Any single character provided misleading evidence of relationships, which Adanson demonstrated by practical example (Mayr, 1982, p. 194). Darwin (1859, p. 417) echoed the importance of examining many characters, and the likelihood of errors with any one character. This thesis endured and was one of the few points of agreement among all schools of modern systematics. Studying one character in isolation could also lead to error in a functional or adaptive context. This was an important theme in Conn (1906, p. 69-93) and Gould and Lewontin's (1979) critique of vulgar Darwinism.

However, the trend to more extensive bases for comparison did not forestall the search for essential patterns in form and function. A.-L. de Jussieu clearly felt on the verge of fundamental insights regarding the structure and functional organization of living plants, and systematic arrangements that expressed these principles (Stevens, 1994, p. 39, 62,125, and also A. de Candolle, *ibid.*, p. 248). With the shift to evolutionary explanations, the concept of "essential characters" was transmuted into "evolutionary novelties." These were widely cited as evidence that refuted Darwin's theory of natural selection, which relied upon random differences between parents and offspring that were small in effect. Even after evolution became widely accepted, saltational explanations for the origins of evolutionary novelties persisted (Goldschmidt, 1940). The refutation of these theories was so thoroughly pursued that Mayr wrote: "There has been such and extreme emphasis on the gradual nature of all evolutionary change that the problem of the origin of evolutionary novelties has largely been neglected" (Mayr, 1976, p. 89). However, evolutionary novelties and "key innovations" (Hunter, 1998; Thomson, 1992) abide as critical phenomena for studies of adaptation and phyletic patterns.

The use of more extensive and detailed descriptions of organisms in classification changed the relationship between characters and higher taxa. The distributions of characters were no longer necessarily coincident with the circumscriptions of these groups. Some characters might be common to all members of the group, but also be found in other species that were excluded. Some characters might be unique to the group, but not possessed by all its members. Thus, the demand for more extensive and varied evidence regarding relationships entailed an explicit acceptance of discordant evidence, and what were called polythetic taxa. After noting the existence of fossil intermediates that blurred the apparent distinctiveness of contemporary groups, Romer (1966, p. 4) described systematic practice in terms that echo Lamarck, De Jussieu, and their peers almost two centuries earlier. "We cannot always give definitions which will hold true of all members of a group; we can merely cite the characters of typical members and build the groups about them."

If, as Simpson (1961, p. 71) asserted, "it is an axiom of modern taxonomy that the variety of data should be pushed as far as possible to the limits of practicability," then biologists confronted a very real problem in digesting this information and visualizing its patterns. Adanson's practice of working with large numbers of characters had limited impact in the 18th century, in part because his methods were viewed as cumbersome or impractical (Sokal and Sneath, 1963, p. 16; Mayr, 1982, pp. 194-5; Stevens, 1994, p. 23). Computers eased the computational burden, but the number of potential characters has been considered indeterminate (Mayr, 1942, p. 18; Rosenberg, 1985a, p. 185). Sokal and Sneath (1963, p. 115) estimated that on the order of 100 independent characters should be used for their methods. This large number of measurements imposed formidable sample size requirements to ensure statistical reliability. For example, in any analysis that quantified interrelationships among pairs of variables, the desired sample size of individuals from each population scaled with square of the number of characters. After emphasizing the necessity of characterizing variation within species, Stebbins (1950, pp. 4, 13) explicitly recognized the problem of excessive dimensionality.

Essential characters aside, abbreviated sets of characters were utilized to create keys for the identification of specimens. In this context, important characters were easily distinguished on each specimen in a diverse group, and different character scores divided the group and speed the diagnosis of new specimens. Mayr (1942, p. 19) presented related criteria: useful characters were easily visible and had low variability within populations. The logical principles and mathematical considerations for the construction of efficient identification keys were clearly expressed by Lamarck 200 years ago (Stevens, 1994, p. 22).

Clearly, the selection of characters was limited. Any feature of an organism could serve, but economy was crucial in selecting the ensemble of characters. Stebbins' (1950, p. 13) solution to this problem was intuition: "On the basis of preliminary exploration the investigator must decide what parts of this pattern are likely to provide the most significant information on the evolution of the group concerned, and he must then select the methods which will enable him to obtain this information as efficiently as possible." As noted by many authors (e.g., Thiele, 1993; Pogue and Mickevich, 1990), the difficulty with informal designation of characters was that the principles, assumptions, or predispositions that guided the selection were often obscure by the time the final results were published.

WHAT IS THE SIGNIFICANCE OF CORRELATION AMONG CHARACTERS?

One of Cuvier's lasting contributions was the principle of functional correlation of organs and parts—organic integration. He accented this principle by asserting that a whole animal could be inferred by examining a few parts (Mayr, 1982, p. 460). Mayr asserted that prior to Cuvier, taxonomists acted as if each character were independent of every other character (*ibid.*, p. 183). In an adaptive research paradigm, these correlations appeared as constraints. Lewontin and Gould (1979) follow Seilacher (1970) in classifying constraints as architectural, developmental, and phyletic.

Explicit recognition of these correlations among parts in systematics caused researchers to consider what constituted independent evidence for relationships. If a compact set of characters were to provide a comprehensive picture of organisms, each character had to provide independent information. In this context, however, independence had many meanings. Sokal and Sneath (1963, p. 66) distinguished between logically and empirically correlated characters. Some of their examples of logical correlations were the presence of hemoglobin and redness of blood (if the latter were strictly a consequence of the former), or an alternative measurement of the same dimension. Logically correlated characters were always pared down to one representative, both to make best use of a limited character set and to avoid multiplying the influence of one piece of information relative to all others.

More complex issues arose with empirical correlations, and the logical-empirical dichotomy seemed artificial given the grade of issues it encompassed. Hecht and Edwards (1977) argued that a suite of characters whose modification could be tied to a single developmental mechanism should be treated as one character. Mayr (1982, p. 235) argued that a suite of morphological changes that could be tied to a single functional shift (a new food source, a new courtship process) should be treated as a single character. Simpson (1961, p. 88) stated that perfectly or highly associated characters should not be considered independent lines of evidence, but as one item. Rosenberg (1985a, p. 184) asserted that all properties that covary perfectly or very closely among all pairs of organisms should be treated as single properties, "to circumvent any bias toward the choice of taxonomically insignificant properties imported from presystematic, common sense descriptions of organisms." However, Sokal and Sneath (1963, p. 68) noted, as an extreme example, the characters that differentiate birds and mammals. They argue that it would be nonsensical to treat these as one character, because it was inconceivable that all of them were based on a single gene.

Thus, the independence of characters has been evaluated in terms of the mechanism of inheritance and development, the range of possible character values they could generate, or in terms of the selective process that could cause the observed changed in character values. Simpson and Rosenberg pared down a set of highly correlated characters regardless of the potential explanation.

In cladistic analysis, independence was established by the existence of branch points that documented the temporal separation of historical events. Thus, when characters were viewed as nested sets of synapomorphies, the pattern of "nesting" and the independence of these sets followed from the groups of taxa they delimited (Pogue and Mickevich, 1990).

WHAT DETERMINES THE IMPORTANCE OR SIGNIFICANCE OF A CHARACTER?

A wide variety of criteria were proposed for subordinating some characters to others, or for giving characters unequal weight, in the study of systematic relationships. The binary weighting imposed by straightforward numerical techniques could be tempered to produce a more graded scale of evidence. *A priori* weighting of characters was elemental to Aristotelian theories of relationships (Mayr, 1982, p. 186; Sokal and Sneath, 1962, p. 16), and subordination of characters was an important part of Jussieu's and Cuvier's contributions. It remained an important subject in modern systematic discourse. Such weightings imply that characters differ in the reliability or significance of the information they convey regarding phylogeny (Thiele, 1993).

Adanson and his followers broke with the Aristotelian tradition, and rejected the *a priori* weighting of characters (Stevens, 1994, pp. 23, 35; Sokal and Sneath, 1963, p. 118; but Mayr, 1982, p. 194 dissents). Modern pheneticists (Sokal and Sneath, 1963, p. 118) and some cladists (*e.g.* Farris, 1990) adopt this position. Lamarck gave characters weight in proportion to their frequency of occurrence across all plants. He interpreted this pattern to indicate the general importance of these characters, and the universality of their functions, in the life of plants (Stevens 1994, p. 18). A.-L. de Jussieu conceived "an analytic hierarchy of characters that was connected with their importance in the life of the organism (*ibid.*, pp. 33,126)". He distinguished between external features and characters proper, which were internal features that were critical elements of the organization of plants (*ibid.*, pp. 30-31).

In ranking characters, Jussieu sought a system applicable to all plants, although he recognized exceptions and developed weighting within the context of families (*ibid.*: 35, 38-39). Jussieu also maintained that a dependent character, whose appearance was conditional on another character, could only be as important as the character on which it depended. Cuvier's subordination of characters derived from the principle of the correlation of parts, and in practice from the constancy of characters across a broad range of taxa (Mayr, 1982, p. 183). The existence of exceptional cases for characters that otherwise seem quite conservative was noted by Darwin (1859, p. 426), and was incorporated into the synthetic view of evolution (*e.g.* Mayr, 1942, p. 21).

Darwin explicitly recognized that his theory confronted a conceptual difficulty posed by prevailing systematic practice. Natural selection required variability in order to act, but systematists focused on characters that did not vary. "Important characters" were practically defined by invariance (Darwin, 1859, p. 46). Thus, Darwin identified conflicting biological criteria for characters, which led some biologists to exclude evidence that would be significant for others. Darwin did not restrict his use of the term character to features satisfying his specific requirements.

Although Darwin agreed with the systematic practice of giving high weight to characters that nearly uniform and common to a great number of forms, and not common to others (*ibid.*, p. 418), he explicitly challenged the primacy of physiological function as an indicator of systematic value (contra Cuvier). "That the mere physiological importance of an organ does not determine its classificatory value, is almost shown by the one fact, that in allied groups, in which the same organ, as we have every reason to suppose, has nearly the same physiological value, its classificatory value is widely different" (Darwin, 1859, p. 415).

Following the distinction between analogous and homologous characters stressed by Richard Owen and others, Darwin clearly enunciated the problems that strong natural selection would pose for inferring propinquity of descent. "It might have been thought (and was in ancient times thought) that those parts of the structure that determined the habits of life, and the general place of each being in nature, would be of very high importance in classification. Nothing can be more false." "It may even be given as a general rule, that the less any part of the organisation is concerned with special habits, the more important it becomes for classification" (Darwin, 1859, p. 413). Many systematists sustained this view (Mayr, 1942, p. 21; Wheeler 1986), although Simpson (1961, p. 89) dissented, asserting that the adaptive value of features could not be fully understood. Kirsch (1982) connected categories of biological causation and levels of physiological organization. He suggested that the most valuable phylogenetic characters would be those in which the observed patterns of variation are least influenced by natural selection. His increasing grade of adaptive influence was biochemical, metabolic, ontogenetic, and anatomical characters.

Darwin (1859, p. 449) developed his argument to explain the reliability of embryonic characters as indicators of common descent. The early stages in life will generally be less adapted to special conditions, except for those stages that actively fend for themselves. Many recent authors also attach greater weight to embryonic characters in systematic analyses (Simpson, 1961, p. 88; Hecht and Edwards, 1977). Similarly, Darwin attached high weight to rudimentary organs because they also appeared to be shielded from natural selection (Darwin, 1859, p. 450).

Darwin also attached weight to correlated characters, "when not explained by bond of connexion" (Darwin, 1859, p. 418). This view is accepted by a wide range of systematists (Mayr, 1942, p. 21; Simpson, 1961, p. 88; Farris 1969). A related view attaches high weight to "complex" or "highly integrated" characters (Simpson, 1961, p. 89; Hecht and Edwards 1977; Wheeler 1988). One conceptual justification for these views held that they reflected broader changes in the genome or the developmental architecture of the organism. An alternative justification cited more confident, consistent character identification because complex characters offered many points of comparison, and resemblance could be assessed in minute detail (Simpson, 1961, p. 88, Neff 1986).

CAN A CHARACTER BE JUST LIKE A MEASUREMENT?

A historical trend toward increasingly quantitative treatments of characters contrasted with enduring questions regarding the suitability and effectiveness of quantitative characters for biological analyses. Buffon asserted that all parts of natural history were too complicated for useful employment of mathematics (Mayr, 1982, p. 41). Mayr asserted that although quantification and other mathematical approaches have a high explanatory value in the physical sciences and much of functional biology, the contributions to systematics and much of evolutionary biology were very minor. For some, quantification was simply impractical in cases of interest: "Often those characters which are most easy to measure have no biological significance, while those for which measurement is most needed are least susceptible to it" (Robson and Richards, 1936, p. 10).

In cladistics, where complex computerized analyses became a mainstay, the received view questioned whether quantitative characters could provide valid phylogenetic information (Pimentel and Riggins, 1987; Cranston and Humphries 1988). Pimentel and Riggin (1987), who described themselves as advocates of morphometrics, nonetheless excluded these data from cladistic analyses on the same terms that they rejected other quantitative features.

More inclusive treatments resorted to quantitative characters when they were the only means to obtain sufficient data for resolving an analysis (Chapill, 1989). Thiele (1993) specified three possible grounds for the frequent dismissal of quantitative characters in cladistics. Branching diagrams were not suited to displaying these data. Quantitative data were too noisy to recover a useful phyletic signal. Quantitative characters required more complex treatment (e.g., Archie, 1985; Baum, 1988; Rae, 1998). The problem of increased complexity also applied to categorical characters that showed variation within populations: "Cladistic characters are features of taxa, and comprise frequency distributions of attribute values over individuals of a taxon" (Thiele, 1993).

Recognition of intrapopulation variability was a conceptual foundation of natural selection, and mensural confirmation of this phenomenon was a major theme of research at the close of the 19th and the beginning of the 20th centuries (e.g., Bateson, 1894; MacLeod, 1926; Robson and Richards, 1936). Once again, some interesting characters eluded quantification. "In short, in regard to the study of birds where observations have been most carefully made, there is not a single part of the body that is measurable, and therefore subject to exact tabulation, that is not widely variable. Other features that are not measurable are more difficult to tabulate, such as differences in habit, in general shape of body, in strength, etc." (Conn, 1906, p. 107).

Nevertheless, avoidance of characters that expressed intrapopulation variability was a governing simplification in systematic practice for a very long time. Mayr (1942, p. 22) acknowledged the artifice of treating systematic characters as invariant within taxa. Stevens (1987) and Baum (1988) emphasized the arbitrary and subjective aspects of characterizing living forms with categorical descriptions. Gift and Stevens (1997) documented the idiosyncratic nature of character perception. The critical issue was the extent to which expedient procedures reshape conceptual views. The potential for this pragmatic systematic practice to create a bias has been long recognized. Darwin (1859, p. 45) observed that most systematists were far from pleased at finding variation in important characters. Richards (1938, p. 98) noted "But there is a tendency not only to rely too much on the intuitive method but to ignore or to underestimate characters on which it cannot conveniently be exercised."

The controversial status of quantitative characterizations also reflects different perceptions of characters and evolutionary processes. Those who emphasized the continuity of microevolutionary processes may have viewed categorical distinctions as entirely arbitrary: "All structures, no matter how unique they may appear in isolated evolutionary stages, have evolved from other structures by means of gradual alterations." "Thus, the distinction by different names between structures of common ancestry is perforce arbitrary" (Smith 1960, p. 11). Those who accepted more punctuated or saltational modes of the origin and fixation of new features may have found categorical labels appropriate on the proper temporal or descriptive scales. Lastly, those who envisioned characters as markers of historical events may have rejected quantitative considerations as irrelevant in their determinations.

The cladists' rejection of quantitative characters may have arisen from other considerations. It may have reflected what Hull (1970) called "a blanket distaste on the part of some taxonomists for mathematical techniques as such and, in particular, for the pheneticist's attempt to quantify taxonomic judgement." Felsenstein (1988) suggested that many systematists were relatively innumerate, that they rejected the statistical methodologies that were widely accepted throughout science. However, his remark might have applied equally to any biologist who worked with complex morphological, physiological, or behavioral features and who saw insurmountable barriers to describing them with simple measurements. Such biologists may have been uncomfortable with multivariate statistical methods that seemed difficult to use or interpret. They may also have been dubious that any such approach would yield consistent or intelligible results. Proponents of landmark-based morphometrics (Bookstein, 1991) might hope to offer resolutions for these dilemmas.

Biological analyses that utilized suites of quantitative characters would require more sophisticated methods of description as well as analysis. Multivariate statistical methods may be required in an exploratory phase, to devise a relatively concise representation of the data for later analysis, and to determine the most defensible models and assumptions for later analyses. Rosenberg (1985a) asserted that this complexity, and the failure to achieve general results analogous to the periodic table of the elements in chemistry, promoted the declining influence of phenetic theory and methods in systematics.

All of these questions bore equal force in analyses of proximate function and adaptation. The emerging consensus that adaptation could not be properly be studied without reference to historical patterns of relationship (e.g., Dobson, 1985; Coddington, 1988) intensified the traction of these questions in functional research.

ARE CHARACTERS A METHODOLOGICAL ARTIFACT OR DO THEY REPRESENT FUNDAMENTAL ELEMENTS OF LIVING ORGANIZATION?

In the Aristotelian view, essential characters defined groups of living organisms. The rejection of this approach to classification did not end the search for fundamental principles of anatomical organization. However, the succeeding systematic methods admitted a broader range of evidence to evaluate relationships among organisms (Mayr, 1982; Stevens, 1994), raising new issues about the nature of that evidence. Among the leading architects of a synthetic approach to classification, there was substantial discrepancy in their views of characters. For Lamarck, Stevens (1994, p. 18) asserted: "Thus although the order as a whole (the arrangement) was that of nature, because it was the naturalist who decided which characters should be emphasized, groupings circumscribed using those (or any other) characters—the classification itself— must be alien to nature." From his procedures, it seems Adanson regarded characters as convenient measurements. However, Jussieu's views differed and exhibited some compatibility with prior essentialist themes. "Among the characters that plants provide, there are some essential, general and invariable, which, it appears, must serve as the basis of the order that we seek. They are not arbitrary, but based on observation, and are not to be obtained except by proceeding from the particular to the general" (Jussieu. 1778, p. 221, translated in Stevens, 1994, p. 29). "That the characters used in the delimitation of major groupings of both plants and animals were similar suggested to Jussieu that some fundamental understanding of how organisms were built up or compounded was within his grasp; in both groups, it was the first-formed organ that was important" (Stevens, 1994, p. 39).

Jussieu's work influenced Cuvier, who asserted the primacy of function in developing general principles of anatomical coordination. Cuvier also stressed an important role for interaction among characters: "The separate parts of every being must ... possess a mutual adaptation; there are therefore, certain peculiarities of conformation which exclude others, and some again which necessitate the existence of others. When we know any given peculiarities to exist in a particular being we may calculate what can and cannot exist in conjunction with them. The most obvious, marked, and predominant of these, those which exercise the greatest influence of the totality of such a being, are denominated its *important* or *leading characters*; others of minor considerations are termed *subordinate*" (Cuvier, 1827; quoted in Panchen, 1992, p. 14).

Neglect of interactions among characters or functions was one of the important criticisms that Gould and Lewontin (1979) directed at the assumption of universal adaptation. They also critiqued the arbitrary partitioning of the organism into characters. They advocated greater attention to ontogenetic factors, a position echoed elsewhere (*e.g.*, Alberch *et al.*, 1979; Wagner 1989). For example, they criticized the study of the human chin as an adaptation, because its structure resulted from the joint action of the alveolar and mandibular growth fields. The importance of identifying developmental pathways was generally accepted in systematic practice, with Hecht and Edwards' (1977) analysis of salamander paedomorphosis being a prominent example. Alberch and Gale (1983, 1985) demonstrated that patterns of digit reduction and loss among related species of modern Anura (or Urodela) were remarkably congruent with the pattern of digit reduction and loss that can be experimentally induced by compounds that inhibit mitosis. The congruence of experimental and systematic patterns within these groups gained significance because the match obtained despite the difference between the groups: the thumb was the digit first affected in frogs, while in salamanders it was a postaxial digit.

However, if a chin were just like wing, where an important function was clear and the consequent fluid dynamic parameterizations were well known, would Gould and Lewontin's critique of this character have seemed as forceful? The most concise descriptions of the wing's properties as an airfoil would undoubtedly incorporate the consequences of multiple developmental fields, which would not invalidate the more concise functional characterization as a basis for studying adaptation to flight. This functional focus also would imply a different basis for delimiting the anatomical extent of the character. This scenario is not necessarily at odds with Gould and Lewontin's analysis. Agreement might be reached in this example by emphasizing inability of the compact functional characterization to simultaneously serve as an accurate and concise representation of heritable variation in wing shape. Returning to the chin, perhaps the questionable function of this part, combined with its compound developmental basis, makes it an invalid or useless character.

The roles of anatomical coordination and developmental programs in relation to adaptation were clearly acknowledged by Darwin and his immediate successors. "For the crude belief that living beings are plastic conglomerates of miscellaneous attributes, and that order of form or Symmetry have been impressed upon this medley by Selection alone; and that by Variation any of these attributes may be subtracted or any other attribute added in indefinite proportion, is a fancy which the Study of Variation does not support" (Bateson, 1894, p. 80). "On the one hand, we find those who are so thoroughly convinced of the universality of the principle of natural selection that they insist that all specific characters are useful, however useless they may seem. It is beyond question that they are led to this belief in the utility of all characters, not from observation, but simply from their belief in the sufficiency of the law of natural selection" (Conn, 1906, pp. 81-82).

Although their critique of pan-adaptational research programs identified the arbitrary division of organisms into characters as a problem, Gould and Lewontin (1979) did not dismiss characterization altogether, and the practices of their paleontological and genetic research repudiated a commitment to radical holism (Rosenberg 1985a). Two theoretical issues stress the value of characterization. Regarding the use of optimization, Rosenberg (1985b, p. 238), likens the theory of natural selection to Newtonian or quantum physics, where extremal principles provide powerful tools. An extremal formulation intensified the concentration on characters: "The use of such principles in physical science nevertheless does show that the dynamical structure of physical systems can be formulated so as to make focal the effect of constituent elements and subsidiary processes upon certain global properties of the system taken as a whole" (Nagel, 1984, p. 327). Focus on characters also arose from the need to designate the persistent currency of evolutionary change and historical relatedness because individuals were transient and unique. Recognizing organisms as units—understanding what constitutes reproduction and death—was usually straightforward (Gould, 1995), but the critical step of designating characters lacked equivalent perceptual or conceptual resolution.

Thus, the designation of a more enduring basis for comparison—characterization—began with substantial latitude for the expression of intent. A completely operational approach to characterization seemed unlikely to succeed. In criticizing phenetic methods, Rosenberg (1985b, pp. 184-185) agreed with Richards (1938, pp. 97-98) that indiscriminant designation of a large number of characters—uninformed by prior knowledge—could not be expected to group taxa in any useful way. To recast this criticism in statistical terms, even if important information were captured in the data, latent biological patterns would be obscured by the unnecessarily high dimensionality of subsequent analysis. The cost of estimating excess parameters, to quantify the interrelationships among these characters, would be reduction in degrees of freedom necessary to sustain inference. However, Gift and Stevens (1997) challenged this view following their experimental test of subjective consistency in delimiting character states.

Robson and Richards (1936) proposed a "radical characterization" of evolutionary explanatory models. "To suggest that the character is the most fundamental unit is to open the door to all kinds of complications, chief among which is that the limits of characters are usually very hard to define; but the suggestion has a particular value from our point of view. Evolution is essentially a matter of character-changes. Individuals are bundles of characters which have each a history of their own, and the divergent groups manifest a progressive accumulation of character-divergences. It is a matter of more than academic or formal interest to keep the individual character before our minds throughout this discussion (cf. lineages, p. 65) and to remember that the individual may be resolved into its constituent elements ('structural units' – Swinnerton, 1921, p. 358). The organism has its peculiar autonomy and 'wholeness,' but each of its structural units has an individual history of change which, though related to the needs of the whole organism, can be treated as a separate evolutionary episode."

Wagner (1995) offered a more formal hypothesis that structural homologues—special kinds of characters—play elemental roles as the building blocks of organisms. Here, building blocks connoted structural elements that had both clear functional roles and were replicated with relatively high fidelity, like genes. He suggested that this mode of organization conferred evolutionary advantages. He asserted that organisms with building block organization would exhibit greater capacity for rapid change and breadth of diversification than organisms in which parent-offspring variation was expressed independently in all possible features. Wagner's reasoning follows developments in computer science, where genetic algorithms have seen wide usage for the solution of optimization problems with large numbers of parameters and convoluted utility functions. Genetic algorithms that fostered the bundling of parameters into "schema" (Holland, 1992) converged more rapidly and reliably to globally optimal values than completely stochastic algorithms or locally efficient algorithms based on gradient searches. The genetic algorithms did not completely specify the structure of the schema, and the emergence of schema was strictly a function of relatively neutral rules for "selection," "mating," "crossing over," and the like. One explanation for this advantage was the dramatic reduction in the dimensionality of the search space, which allowed a unit of search effort to span a greater range. This advantage may compensate for the loss of freedom to alter all parameters simultaneously. In many computer optimizations, and perhaps in biotic evolution as well, the structures and effects of schema changed in the course of an extended optimization.

Wagner's concept harmonized with the views of Alberch *et al.* (1979) regarding the significance of developmental processes. Moy-Thomas (1938), among others, anticipated their views: "The tendency among morphologists at the present time is to believe that the fundamental problems of their science lie more in determining the processes underlying the organization of animals rather than in providing further evidence for the now generally accepted theory of evolution by the production of more and more complete phylogenies. The morphologist should consider the developmental processes that have contributed to the formation of the adult, and not be content merely to compare the resulting forms without reference to these."

In contrast, characters are clear methodological artifacts in the practice of phenetics, where a character was treated as a prescription for a measurement. There was little distinction between the terms character and variable. Phenetics emphasized collecting measurements for as many different features of organisms as possible, to seek an adequate representation of the entire genome and a balanced perspective of overall similarity among organisms. Although this systematic practice could, in principle, have organized the selection of measurements based on perceived structural elements, the discipline seemed to favor relatively undirected collection of data. Important regularities of form were instead sought in the analysis and reduction of these initial measurements (Thompson, 1917; Sneath and Sokal, 1973, pp. 157-161). Some of the cladistic literature also treated characters as methodological artifacts. "The definition of characters relates in two ways to the methods and biases that influence character selection. First, a particular approach or technology may limit the number of characters that come into the view of the systematist. And second, the extent of the detailed comparison may influence how precisely two or more characters can be recognized" (Wheeler, 1986).

HOW CLOSE TO EMPIRICAL OBSERVATIONS, AND HOW FREE FROM THEORETICAL ASSUMPTION, SHOULD CHARACTERS BE?

Francis Bacon's ideas, and the perceived speculative excesses of *Naturphilosophen,* led to a profoundly atheoretical sentiment among most 19th century systematists (Stevens 1994, pp. 221-225). "Systematics was seen as a science of observation. Theory should not intrude between the observer and nature; systematic data themselves were obtained from direct observation of plants" (Stevens, 1994, p. 224). Similar sentiments were expressed by modern systematists, *e.g.*: "The goal of systematics should be to reflect observations in nature as closely as possible" (Pogue and Mickevich, 1990).

However, even "the great literalist" Cuvier, who argued incessantly that science should move from observation to theory (Gould, 1977, pp. 21, 59), complained of a regrettable divorce of rationalism from "practical" botany (Stevens, 1994, p. 223). The romantic appeal of "pure" observation and the impossibility of achieving this state were vividly expressed by Bateson (1894, p. *vii*). "In the old time the facts of Nature were beautiful in themselves, and needed not the rouge of speculation to quicken their charm, but that was long ago, before Modern Science was born. Besides this, to avoid the taint of theory in morphology is impossible, however much it may be wished. The whole science is riddled with theory. Not a specimen can be described without the use of a terminology coloured by theory, implying the acceptance of some one or other theory of homologies." Many modern authors would argue that facts, or perceptions of facts, never exist outside theory (*e.g.* Kuhn 1962; Gould, 1977, p. 38; Rosenberg, 1985, pp. 182-187).

Theoretical requirements or a practical need to for data compression sometimes required considerable analysis or screening of simple observations to produce characters. Thompson's (1917) evocative illustrations of shape deformations suggested that morphological transformations involving a grid of landmarks (the simple observations) could be concisely expressed by relatively few deformation parameters. In morphometrics (Bookstein 1991), landmark-based distance data were reduced to produce compact parameterizations with analogous properties. The characters resulting from these kinds of analyses could be viewed as artificial, or as documentation of fundamental patterns that revealed the building blocks of organisms.

Among cladistic examples, Neff (1986) distinguished several steps in the designation of characters, and Pogue and Mickevich (1990) contrasted "synthetic" (combining or dividing observed conditions) and "reflective" (in relation to cladogram) definitions. Pimentel and Riggins (1987) required that character states were ordered, and Platnick (1979) asserted that characters represent explicit hypotheses of common historical origin. These requirements clearly excluded "raw" observations as valid characters. The goals of these requirements or analyses seemed to be the production of data that eased the manufacture of congruent phyletic pictures of relationship. Wheeler (1986) noted that analytical errors in character designation were common in practice: convergence and parallelism often misled.

TO WHAT EXTENT DO PREVAILING CHARACTER DEFINITIONS REFLECT ATTEMPTS TO COMPARTMENTALIZE COMPLEXITY AND ESTABLISH A SIMPLE CONCEPTUAL FRAMEWORK?

The varied usage of "character" recalls the caution that definitions are arbitrary, involving subtle considerations of expedience (Lotka, 1956, p. 3). This varied usage spanned a range of concepts: facts, inferences, and hypotheses. Divergence grew from efforts to simplify the relationship between characters and the goals of biological research. For example, once comparative observations were reduced to a matrix of binary data (characters by taxa), a toolbox of accepted algorithms was available to objectively produce the cladogram. The problem of apparent homoplasy in characters could be eliminated by further conditions. For example, "Were we to correctly delimit all of the characters in the data set there would be no conflict and a single, uncontested hypothesis of relationships (cladogram) would emerge from the analysis" (Wheeler, 1986). Note that the exclusion of quantitative characters dismisses several issues: the difficulty of translating them into more convenient binary codes, the issue of within-taxon variation, and the fundamental problem of specifying rules of descent with modification for mensural features.

Thus, by investing characters with properties that could not be established by simple inspection of specimens, or ignoring other properties, the process of inferring the pattern of phylogeny was greatly simplified. The appeal of simplicity in cladistics was further evidenced by the popularity of three taxon statements as guiding examples (and a basis for algorithms), and the defense of parsimony as the only relevant principle for cladogram construction. Hull (1970) emphasized the aesthetically satisfying aspects of Hennig's principles, which were straightforward and exceptionless. Felsenstein (1983, 1988) asserted that many systematists preferred the principal of parsimony because it appeared to avoid explicit models of evolutionary change and because it had an aura of certainty that a statistical framework could not provide.

The modern reliance on genes as canonical units of evolutionary theory manifested the appeal of simplicity in models of adaptation. Williams (1966) and Dawkins (1976) popularized the primacy of genes as the units of selection, emphasizing the fidelity and direct copying of these heritable factors. This approach eliminated the need to consider heritability, and simplified the description of variability. The cost, however, was significant. The representation of natural selection was limited to methodological artifacts, in the form of genic selection coefficients (Sober and Lewontin 1984). Like some depictions of cladistic characters, selfish genes excluded quantitative variation from view. Such simplified approaches divert attention from the critical problem of determining an evolutionarily significant basis for dissecting the developmental trajectory of the phenotype into parts.

CONCLUSION

Despite wide acknowledgement that scientific impressions were a function of the methods used (*e.g.* Stebbins, 1950, p. 7), substantial differences in character concepts obstructed the exchange of ideas among evolutionary biologists. This difficulty was exacerbated because the ideal unit or measure of evolutionary process for a systematist was diametrically opposed to the ideal for an adaptationist. Systematists sought indicators of unique historical events, free from functional or selective pressures that caused confusing parallel or convergent similarity. Adaptationists sought functionally identical units that originated repeatedly, so historical contingency could be ignored, or controlled for. Data that emerged from simple comparisons of organisms were not ideal in either sense, and both idealizations would have to be abandoned to refocus on the study of variation (Bateson 1894), emergent patterns of development (Alberch *et al.*1979) or the building block hypothesis (Wagner 1995).

Limited communication between evolutionary theory and systematics has a long history. Darwinism transformed the notion of a "natural system" from expressing the continuity of living things (affinity, *scala naturae*, plenitude) to expressing propinquity of descent. However, *The Origin* had little impact on the practice of systematics: "Systematists will be able to pursue their labours as at present; but they will not be incessantly haunted by the shadowy doubt whether this or that form be in essence a species" (Darwin, 1859, p. 484). Dobzhansky (1942) lamented a related lack of dialog and expressed hope of reconciliation. "A correlation of this sort has been necessary for some time; even in the recent past there existed a notorious lack of mutual comprehension between the systematists on one hand and the representatives of the experimental biological disciplines on the other. That this lack of mutual comprehension was due in part to unfamiliarity with each other's factual materials and methods, and in part to a sheer misunderstanding of the respective points of view, was felt by many systematists as well as by experimentalists. But it remained for a systematist of Dr. Mayr's caliber, possessing a wide familiarity with and a perfect grasp of the apparently conflicting disciplines, to demonstrate conclusively that this conflict is spurious."

Characterization has always been the *sine qua non* for comparative biology and evolutionary theory. However, yesterday's compelling metaphors and character definitions may seem quixotic from our modern perspective. Many historical examples (*e. g.*, essentialism, *Naturphilosophie*, and quinarian classification) illustrate the costs of hobbling biological research with aesthetic principles. How do these historical observations bear on selfish genes and cladistics? In their extreme forms, these modern paradigms rest upon atomic functional and historical characters respectively, abstractions that cannot be observed. It is unquestionable that these paradigms have stimulated significant advances in comparative and evolutionary biology, but it also clear that puritans of each faith have broadened divergence in character concepts. In addition to obscuring issues that unify evolutionary biology, these specialized character concepts mask critical methodological issues. Even the specialist, who may feel quite secure in a narrowly circumscribed discipline, should attend to the distinctions between ideal information and the empirical data at hand.

LITERATURE CITED

Alberch, P., Gould, S. J., Oster, G. F., and Wake, D. B. (1979). Size and shape in ontogeny and phylogeny. *Paleobiology* **5**:296-317.

Alberch, P., and Gale, E. A. (1983). Size dependence during development of the amphibian foot: colchicine-induced digital loss and reduction. *J. Embryol. Exp. Morph.* **76**:177-197.

Alberch, P., and Gale, E. A. (1985). A developmental analysis of an evolutionary trend: digital reduction in amphibians. *Evolution* **39**:8-23.

Archie, J. W. (1985). Methods for coding variable morphological features for numerical taxonomic analysis. *Syst. Zool.* **34**:326-345.

Bateson, W. (1894). "Materials for the Study of Variation." Macmillan and Co., London.

Baum, B. R. (1988). A simple procedure for establishing discrete characters from measurement data, applicable to cladistics. *Taxon* **37**:63-70.

Bookstein, F. L. (1991). "Morphometric Tools for Landmark Data: Geometry and Biology." Cambridge University Press, Cambridge.

Chappill, J. A. (1989). Quantitative characters in phylogenetic analysis. *Cladistics* **5**:217-234.

Coddington, J. A. (1988) Cladistic tests of adaptational hypotheses. *Cladistics* **4**:3-22

Colless, D. H. (1985). On 'character' and related terms. *Syst. Zool.* **34**:229-233.

Conn, H. W. (1906). "The Method of Evolution." C. Putnam & Sons, New York.

Cranston, P. S., and Humphries, C. J. (1988). Cladistics and computers – a chironomid conundrum. *Cladistics* **4**:72-92.

Darwin, C. (1859). "On the Origin of Species." 1979 facsimilie of the 1st ed., Harvard University Press, Cambridge, MA.

Dawkins, R. (1976). "The Selfish Gene." Oxford University Press, Oxford.

Dobson, F. S. (1985) The use of phylogeny in behavior and ecology. *Evolution* **39**:1384-1388.

Dobzhansky, T. (1942). Introduction. *In* "Systematics and the Origin of Species" (E. Mayr, ed.). Columbia University Press, New York.

Farris, J. S. (1990). Phenetics in camouflage. *Cladistics* **6**:91-100.

Felsenstein, J. (1983). Methods for inferring phylogenies: a statistical view. *In* "Numerical Taxonomy" (J. Felsenstein, ed.), pp. 313-334. Nato ASI series G number 1. Springer-Verlag, Berlin.

Felsenstein, J. (1988). The detection of phylogeny. *In* "Prospects in Systematics" (D. Hawksworth, ed.), Clarendon Press, Oxford.

Fristrup, K. (1992). Character: current usages. *In* "Keywords in Evolutionary Biology" (E. F. Keller and E. A. Lloyd, eds.), pp. 45-51. Harvard University Press, Cambridge, MA.

Ghiselin, M. (1984). 'Definition,' 'character' and other equivocal terms. *Syst. Zool.* **33**:104-110.

Gift, N., and Stevens, P. F. (1997). Vagaries in the delimitation of character states in quantitative variation - an experimental study. *Syst. Biol.* **46**:112-125.

Goldschmidt, R. (1940). "The Material Basis of Evolution." Yale University Press, New Haven, CT.

Gould, S. J. (1995). The Darwinian body. *N. Jb. Geol. Paläont. Abh.* **195**:267-278.

Gould, S. J., and Lewontin, R. C. (1979). The spandrels of San Marco and the Panglossian paradigm. *Proc. R. Soc. Lond. B* **205:**581-598.

Harvey, P. H., and Pagel, M. D. (1991). "The Comparative Method in Evolutionary Biology." Oxford University Press, Oxford.

Hecht, M. K., and Edwards, J. L. (1977). The methodology of phylogenetic inference above the species level. *In* "Major Patterns in Vertebrate Evolution" (M. K. Hecht, P. C. Goody, and B. M. Hecht, eds.), pp. 3-51. Plenum Press, New York.

Holland, J. H. (1992). "Adaptation in Natural and Artificial Systems." MIT Press, Cambridge, MA.

Hull, D. L. (1970). Contemporary systematic philosophies. *Annu. Rev. Ecol. Syst.* **1:**19-51.

Hunter, J. P. (1998). Key innovations and the ecology of macroevolution. *Trends Ecol. Evol.* **13:**31-36.

Jussieu, A.-L. de. (1778). Exposition d'un nouvel order des plantes adopté dans les démonstrations du Jardin Royal. *Mém. Math. Phys. Adad. Roy. Sci.* (Paris) (1773):214-240.

Jussieu, A.-L. de. (1789). "*Genera plantarum.*" Hérissant and Barrois, Paris.

Kirsch, J. A. W. (1982) The builder and the bricks: notes toward a philosophy of characters. *In* "Carnivorous Marsupials" (M. Archer, ed.), pp. 587-594. Royal Zoological Society of New South Wales, Sydney, Australia.

Lewontin, R. C. (1961). Evolution and the theory of games. *J. Theor. Biol.* **1:**382-403.

Lotka, A. J.(1956). "Elements of Mathematical Biology." Dover Publications, New York.

MacLeod, J. (1926). "The Quantitative Method in Biology." Longmans, Green & Co. Ltd., London.

Mayr, E. (1942). "Systematics and the Origin of Species." Columbia University Press, New York.

Mayr, E. (1976). "Evolution and the Diversity of Life." Belknap Press, Cambridge, MA.

Mayr, E. (1982). "The Growth of Biological Thought." Belknap Press, Cambridge, MA.

Mickevich, M. F., and Weller, S. J. (1990). Evolutionary character analysis: tracing character change on a cladogram. *Cladistics* **6:**137-170.

Moy-Thomas, J. A. (1938). The problem of the evolution of the dermal bones in fishes. *In* "Evolution" (G. R. de Beer, ed.), pp. 305-320. Clarendon Press, Oxford.

Nagel, E. (1984). The structure of teleological explanations. *In* "Conceptual Issues in Evolutionary Biology" (E. Sober, ed.), pp.319-346. MIT Press, Cambridge, MA.

Neff, N. A. (1986). A rational basis for *a priori* character weighting. *Syst. Zool.* **35:**110-123.

Nelson, G., and Platnick, N. (1981). "Systematics and Biogeography: Cladistics and Vicariance." Columbia University Press, New York.

Panchen, A. L. (1992). "Classification, Evolution, and the Nature of Biology." Cambridge University Press, Cambridge.

Pimentel, R. A., and Riggins, R. (1987). The nature of cladistic data. *Cladistics* **3:**201-209.

Platnick, N. (1979). Philosophy and the transformation of cladistics. *Syst. Zool.* **128:** 537-547.

Pogue, M. G., and Mickevich, M. F. (1990). Character definitions and character state delineation: the bete noire of phylogenetic inference. *Cladistics* **6:**319-361.

Rae, T. C. (1998). The logical basis for the use of continuous characters in phylogenetic systematics. *Cladistics* **14:**221-228.

Richards, O. W. (1938). The formation of species. *In* "Evolution." (G. R. de Beer, ed.), pp. 95-110. Clarendon Press, Oxford.

Rieppel, O (1991). Things, taxa, and relationships. *Cladistics* **7:**93-100.

Robson, G. C., and Richards, O. W. (1936). "The Variation of Animals in Nature." Longmans, Green and Co., London.

Rodrigues, P. D. (1986). On the term character. *Syst. Zool.* **35:**140-141.

Romer, A. S. (1966). "Vertebrate Paleontology." 3rd ed. Univ. of Chicago Press, Chicago.

Rosenberg, A. (1985a). Adaptationalist imperatives and Panglossian paradigms. *In* "Sociobiology and Epistemology" (J. H. Fetzer, ed.), pp. 161-180. D. Reidel Pub. Co., Dordrecht.

Rosenberg, A. (1985b). "The Structure of Biological Science." Cambridge University Press, Cambridge.

Seilacher, A. (1970). Arbeitskonzept zur Konstruktions – Morphologie. *Lethaia* **3:**393-396.

Simpson, G. G. (1961). "Principles of Animal Taxonomy." Columbia University Press, New York.

Slobodkin, L. B., and Rapoport, A. (1974). An optimal strategy of evolution. *Q. Rev. Biol.* **49**:181-200.

Smith, H. M. (1960). "The Evolution of Chordate Structure." Holt, Rinehart and Winston, Inc., New York.

Sneath, P. H. A. and Sokal, R. R. (1973). "Numerical Taxonomy." W. H. Freeman, San Francisco.

Sober, E., and Lewontin, R. C. (1984). Artifact, cause, and genic selection. *In* "Conceptual Issues in Evolutionary Biology" (E. Sober, ed.), pp. 210-231. MIT Press, Cambridge, MA.

Sokal, R. R., and Sneath, P. H. A. (1963). "Principles of Numerical Taxonomy." W. H. Freeman, San Francisco.

Stebbins, G. L. (1950). "Variation and evolution in plants." Columbia University Press, New York.

Stevens, P. F. (1987). Pattern and process: phylogenetic reconstruction in botany. *In* "Biological Metaphor and Cladistic Classification" (H. M. Hoenigswald and L. F. Wiener, eds.), pp. 155-179. University of Pennsylvania Press, Philadelphia.

Swinnerton, H. H. (1921). The use of graphs in palaeontology. *Geol. Mag.* **58**:357-364.

Thiele, K. (1993). The holy grail of the perfect character: the cladistic treatment of morphometric data. *Cladistics* **9**:275-304.

Thompson, D. W. (1917). "On Growth and Form." Cambridge University Press, Cambridge.

Thomson, K. S. (1992). Macroevolution - the morphological problem. *Amer. Zool.* **32**: 106-112.

Wagner, G. P. (1989). The biological homology concept. *Annu. Rev. Ecol. Syst.* **20**:51-69.

Wagner, G. P. (1995). The biological role of homologues: a building block hypothesis. *N. Jb. Geol. Paläont. Abh.* **195**:279-288.

Wheeler, Q. D. (1986). Character weighting and cladistic analysis. *Syst. Zool.* **35**:102-109.

Williams, G. C. (1966). "Adaptation and Natural Selection." Princeton University Press, Princeton, NJ.

AN EPISODE IN THE HISTORY OF THE BIOLOGICAL CHARACTER CONCEPT: THE WORK OF OSKAR AND CÉCILE VOGT

MANFRED DIETRICH LAUBICHLER

Program in History of Science, Princeton University, Princeton, NJ 08542

INTRODUCTION

The question of what constitutes a relevant biological character and how such characters can be identified is a perennial problem that lies at the heart of many biological disciplines. To study the assumptions that at any given time informed ideas about the nature of biological characters is to reconstruct a complex web of negotiations between theoretical commitments and experimental practices. On the one hand, theoretical ideas, such as the notion of adaptation by means of natural selection, the form-function complex, the principle of recapitulation, or patterns of Mendelian inheritance, guide the interest of researchers, while on the other hand, experimental practices and technological advances, such as histological staining techniques, methods of electrophysiological stimulation, or statistical methods of pedigree analysis, both facilitate and limit the implementation of these theoretical ideas in empirical studies. In this paper I will focus on one particular episode in the history of the biological character concept, the theoretical synthesis of Oskar (1870-1959) and Cécile (1875-1962)

Vogt. Their synthesis is one of the best illustrations of the confluence of different theoretical assumptions and experimental practices that is constitutive of every interpretation of biological characters.

Oskar and Cécile Vogt's main interest was the identification and characterization of distinct regions in the Neo-cortex by both functional and structural criteria. Together with their collaborators they eventually identified more than 200 different areas, some of them still recognized today (Vogt, O., 1926a; Vogt and Vogt, 1937; Brodmann, 1909; Nolte, 1993). While it is not at all surprising that the study of an organ as complex as the Neo-cortex requires a variety of different research methodologies (see, e.g., Breidbach, 1996, and Hagner, 1997, for recent historical treatments of the subject), Oskar and Cécile Vogt's approach differs from others in that they actively integrated separate research traditions in a systematic as well as a programmatic way. In their search for distinct areas of the Neo-cortex they used anatomical and histological analysis and electrical stimulation of the Neo-cortex of both monkeys and humans, the study of brain pathologies, and the analysis of pedigrees of patients with psychiatric conditions. In addition, Oskar Vogt also employed experimental *Drosophila* genetics and the study of geographic variation (he owned one of the world's largest collections of bumble-bees) as model systems for the analysis of biological characters and their variations within the Neo-cortex.

The development of their theory of biological characters took place over a period of 40 years and culminated in the publication of their two-volume magnum opus *Sitz und Wesen der Krankheiten im Lichte der topistischen Hirnforschung und des Variirens der Tiere* ("On the location and nature of diseases in the light of topistic brain research and the variation of animals," Vogt and Vogt, 1937, 1938.) Here I will present the logical structure of this idiosyncratic synthesis and evaluate its role as an exemplary contribution to theories of biological organization and, more generally, to theoretical biology. While the neurobiological contributions and the sociopolitical context of Oskar and Cécile Vogt's careers have been the subject of some recent studies (Hagner, 1994; Richter, 1976a, b; Satzinger, 1998), the theoretical dimension of their work has not been explored in similar detail. But it is only in the context of their theories of biological organization that one can understand how the neuroanatomist Oskar Vogt came to define the concepts of penetrance and expressivity, concepts that have been central in medical genetics up to the present day. Furthermore, if one looks at Oskar Vogt's remarkable career, one that spanned more than six decades and was more or less devoted to a single research problem—to localize the material correlates of higher brain functions—one cannot appreciate his consistency in the face of the many revolutionary developments in all biological sciences without a thorough understanding of his theoretical commitments. The fact that Oskar Vogt remained focused on a single theoretical issue also helps us to comprehend how he could incorporate such diverse fields as neuroanatomy, histology, neurophysiology, *Drosophila* genetics, clinical studies, human genetics, and neuropharmacology within a single research institute devoted solely to the investigation of his ideas.

It is the thesis of this paper that Oskar and Cécile Vogt's theory of biological organization and their ideas on biological characters were characteristic of a more conceptually oriented theoretical biology that was prominent in the first decades of the 20th century and, furthermore, can serve as an important model for similar discussions today. To illustrate my thesis I will begin by sketching the development of Oskar and Cécile Vogt's research program in the context of some of the scientific developments of their times. Next, I will summarize the main tenets of their theoretical synthesis, focusing especially on their conception of biological characters. In the last section I will reflect on what we can learn from their theoretical synthesis for our attempts to develop a character concept within modern biology. Before I start I will introduce a methodological distinction between a context of structural identification and a context of functional individualization and argue for their respective roles in any analysis of the biological character concept.

METHODOLOGICAL PROLOGUE: CONTEXT OF STRUCTURAL IDENTIFICATION AND CONTEXT OF FUNCTIONAL INDIVIDUALIZATION

If we analyze the seemingly straightforward proposition, x is a biological character C, we will soon find that this statement has two elements. For x to be a biological *character* C, there has to exist a set of criteria that enable us to identify x as a biological *object* O that is distinct from other such objects. In addition there has to be to a specific biological *role* F of x that would justify the original assessment that x is a biological character C (Laubichler, 1997a, b, 1999; but see also Millikan, 1984; Wright, 1973; Cummins, 1975; Allen, Bekoff, and Lauder, 1998). In other words, any designation of a biological character requires us to first identify a biological object and second to attribute a specific functional role to this previously identified object. Biological characters are not just found objects waiting to be picked up in nature; rather they are conceptual abstractions and as such can only be identified within the well-defined context of a specific biological theory. The context of a specific theory defines the functionally relevant biological role of an object and it also suggests ways in which these objects can be identified or measured. Here I refer to these two distinct steps in the assessment of biological characters as context of structural identification and context of functional individualization. The two are of course connected and in practice they quite often overlap, but they are logically distinct and to keep them conceptually separate pays off both in developing an operationally useful notion of biological characters for modern biology and in analyzing historical episodes.

An example will help illustrate these ideas. For the sake of consistency I will get ahead of myself and use the problem of distinct areas in the Neo-cortex. It is well known that two specific areas of the Neo-cortex, Broca's area and Wernicke's area, are connected with language abilities. Let us now analyze how Broca's area came to be known as the biological character for the nonfluent, motor, or expressive aspects of language production. It all started when expressive aphasia, or the difficulty to produce either written or spoken language while still being able to comprehend it, was recognized as a distinct pathology by Broca in 1861. He then proceeded to identify this pathology with lesions in what we now call the opercular and triangular parts of the inferior frontal gyrus or Broca's area. Continuing research on this complex pathology over the years has led to major revisions in our understanding of the neuronal mechanisms of language production, but for our purposes these details do not matter since the same principles of identification and individualization of a biological character apply to all of them. First, a biological role such as the motor program connected with language productions is identified. In Broca's case this biological role was identified by its associated pathology, expressive aphasia. Next, this biological role (or its pathology) is correlated with a specific structure, a lesion specific to the left side of the brain in the first patient and a more specific location in the inferior frontal gyrus in the second patient that Broca studied (Broca, 1861, 1865). Further research then added specificity to both the biological role (pathology) and the anatomical location (Broca's area). What all these studies have in common is that the analysis of a biological character such as Broca's area is always based on two separate approaches: (1) the identification of a biological object by means of a measurement procedure (anatomical characterization, histological analysis, a compound mapping of different lesions with the same symptoms, MRI imaging, etc.) and (2) the correlation of an object identified by these means with a specific biological role (function). As mentioned earlier, I refer to the first step as the context of structural identification and to the second step as the context of functional individualization of a biological character. In research practice structural identification and functional individualization are often connected in a process of mutual reinforcement. For example, further anatomical specificity often leads to a clearer clinical diagnosis (biological role) while, by the same token, a more refined diagnosis can also trigger further advances in localization.

When interpreting historical case studies, such as the one of Oskar and Cécile Vogt's research program and its connection to their ideas on biological organization and the biological character concept, the distinction between a context of structural identification and a context of functional individualization will prove very useful indeed. Particularly, since both of these approaches are usually connected with quite distinct research methodologies. Let us therefore proceed with the analysis of Oskar and Cécile Vogt's research program.

HISTORICAL CONTEXT: THE DEVELOPMENT OF THE SCIENTIFIC CAREERS OF OSKAR AND CÉCILE VOGT

Oskar Vogt was born in 1870 in Husum, Prussia, into a family of Lutheran ministers. Following his early interests in natural history and philosophy, he studied philosophy and medicine first in Kiel (1888-1890) and then in Jena. Among his teachers were such eminent scientists as Ernst Haeckel (zoology), Max Fürbringer (anatomy), and Otto Binswanger (psychiatry). Vogt's academic *Wanderjahre* led him to Zürich in 1894, where he studied with August Forel, known today primarily for his interest in scientific hypnosis and the social life of insects. Vogt's intellectual partnership with Forel would later lead to the joint editorship of two scientific journals. After his stay in Zürich, Vogt moved to Leipzig (1894-1895), where he worked with the highly respected anatomist and psychiatrist Paul Flechsig, whose ideas to employ the phenomenon of myelogenesis for the localization of distinct regions in the cortex were a major new development in the century-old quest to map the brain. Flechsig argued that one can identify functionally integrated regions in the cortex based on their pattern of simultaneous myelogenesis. Based on this method, he was able to identify the already known senso-motoric regions in the cortex and also several additional regions that he called associative centers. In Flechsig's view the senso-motoric centers were composed of projection fibers (*Projektionsfasern*), whereas association fibers (*Assoziationsfasern*) connected the associative centers both to the senso-motoric centers and among themselves. Vogt, who after some major disagreements with Flechsig, left Leipzig for Paris in order to gain more knowledge of the clinical side of neurology, would soon challenge Flechsig's myelogenetic criterion for the identification of distinct regions of the brain by demonstrating that Flechsig's associative centers also contained projection fibers (Vogt, O., 1897).

In Paris, Vogt studied with the eminent French neuroscientists Jules and Augusta Dejerine. He also met his future wife, Cécile Mugnier, a medical student who worked with the neuroanatomist Pierre Marie. In 1897 he left Paris for Berlin, where he set up his own research facilities (funded by his own private medical praxis and supported by grants from the Krupp Family). Cécile Mugnier joined him the next year and thus started a private and professional partnership that would last for almost 60 years.

Over the next six decades Oskar Vogt displayed exceptional organizational skills in the pursuit of his scientific vision. He was the editor of three scientific journals and a series of monographs and he founded or helped found five research institutes that were all more or less devoted to his scientific agenda. In 1898 he opened the *Neurologische Zentralstation,* which in 1902 was incorporated into the University of Berlin as the *Neurologisches Laboratorium.* Also in 1902, Oskar Vogt and August Forel started the new series of their joint journal. The change in the title—from the old *Journal für Hypnotismus, Psychotherapie sowie andere psychophysiologische und psychopathologosche Forschungen* (Journal for Hypnosis, Psychotherapy and Related Psychophysiological and Psychopathological Research) to the new *Journal für*

Psychologie und Neurologie, Mitteilungen aus dem Gesamtgebiet der Anatomie, Physiologie und Pathologie des Zentralnervensystems, sowie der medizinischen Psychologie (Journal for Psychology and Neurology: Communications from all areas of the Anatomy, Physiology and Pathology of the Central Nervous System and of Medical Psychology)—already indicates the central issues of Oskar Vogt's research program. An even clearer expression of this program can be found in the editorial "*Psychologie, Neurophysiologie und Neuroanatomie* (Psychology, Neurophysiology, and Neuroanatomy)" that inaugurates the new journal (Vogt, O., 1902, but see also Vogt, O., 1897). Here Oskar Vogt states his convictions that (i) every psychological phenomenon has a material correlate in the form of a vital process within the brain, (ii) the anatomical structures are closely correlated with their respective functions and, accordingly, neuroanatomy and neurophysiology will have to be integrated to form what he refers to as neurobiology (the anatomy and physiology of the nervous system), and (iii) in certain cases it should be possible, as knowledge about the brain increases, to infer the function of a specific part of the brain from its structure and vice versa. This last statement, in its recognition of the close connection between form and function, is an early indication of how Vogt would later conceptualize the problem of biological characters.

Another important dimension of Oskar and Cécile Vogt's research project was the inclusion of pathology. Within the parameters of their research program the study of pathologies was a logical extension of their search for the material basis of brain functions. From the proposed connection between the structure of a biological character and its function, it follows that any variation in structure will lead to a correlated variation in function and, correspondingly, that any observed variation in function will be accompanied by a variation in structure. In their view the difference between so-called normal variations and pathologies is just a quantitative one.[1] It was one of the explicit goals of the newly founded journal to overcome the institutionally imposed separation between neurology and brain pathology. However, a prerequisite for any comprehensive study that would correlate variation in structure with variations in function was the identification of those individual regions in the brain that could be correlated with specific functions.

The project of mapping the brain was, of course, not new. In modern times these attempts date back at least to the phrenology of Franz Josef Gall (Breidbach, 1996; Hagner, 1995, 1997; Harrington, 1987). Throughout the 19th century the idea that there is a material basis for specific behavioral and intellectual capacities went through several modifications and by the end of the century a variety of brain maps had been proposed. We have already encountered Flechsig's proposal that specific regions can be identified based on their pattern of myelogenesis. Oskar Vogt had criticized Flechsig's conclusions in an early paper (Vogt, O., 1897) but he too needed an operational criterion that

---

[1] See also George Canguilhem's "The Normal and the Pathological" (Canguilhem, 1991) for some further reflections on the historical importance of these distinctions.

would allow him to localize specific regions. His idea was to focus on brain architectonics. This choice was based on the assumption that the structural differences that are important for the correlated functional differences are those that "Oskar Vogt identified as architectonic differences" (Vogt and Vogt, 1926 p. 1190). Brain architectonics was different from histology, another potential method of identifying structural differences, in that while the latter is the study of the fine structure of the elements that characterize specific regions of the brain, the former is concerned with the analysis of the number, size, and form of the nerve cells within the laminae of the cortex. In brain architectonics, special attention is paid to the specific arrangement of these elements in various regions. Individual areas are then identified by local differences in composition, number, and kind of the structural elements that are identifiable in specific preparations (see also Hagner, 1995.) Following this method, which requires the analysis of whole series of microscopic preparations and their three-dimensional reconstruction, the Vogts and their collaborator, Korbinian Brodmann, were able to identify up to 200 separate regions in the cortex.

While the research program in brain architectonics would give the Vogts a technique to identify structurally defined regions, it did not in itself provide an answer to the question of whether the regions so identified could be linked with a corresponding function. To establish these links, the Vogts employed the technique of electrical stimulation of specific regions of the cortex. This work was done primarily with monkeys (Vogt and Vogt, 1907, 1919a, 1926). Eventually they brought the results of both lines of investigation together and produced maps that were characterized by both structural and functional criteria (see especially Vogt and Vogt, 1937, for a synthesis of their results). It should be noted, however, that the Vogts did not work in isolation. Word War I had been a somewhat macabre bonanza for brain research due to the high number of brain injuries (see also Harrington, 1996, for further details).[2] Of the many neuroscientists confronted with these often highly localized lesions, it was especially Otfried Foerster who began to map the human brain (Foerster, 1923, 1925, 1926). After completing his electrophysiological mapping of the human cortex, Foerster's proposal of a functional map was in amazing agreement with the one produced by the Vogts that was based on known cytoarchitectonic fields combined with functional assignments that were derived from their studies with monkeys. For the Vogts this was a triumph not only of their method but also of their ideas of using animal models in the study of the human brain.

At the heart of any consideration of brain pathologies lies the question of the classification of psychiatric diseases. In Germany this problem was linked to the doyén of clinical psychiatry, Emil Kraepelin, still recognized today for his

---

[2] It is interesting to note that other researchers drew rather different scientific conclusions from all the cases of head injuries that were available for study after WWI. While the Vogts and Otfried Foerster were able to support their claims of the specific localization of higher brain functions, other such as Kurt Goldstein were more impressed by the ability of the brain to compensate for the effects of specific lesions, thereby developing a more holistic conception of the brain (see also Harrington 1996, Goldstein 1995[1934]).

various attempts at psychiatric classification. Kraepelin was somewhat skeptical as to whether any classificatory system based on clear distinctions between different entities could accommodate all the intermediate expressions of diseases that were to be expected, since diseases were manifestations of living processes and thereby subject to the same rules of variability. Acknowledging this fundamental limitation, Kraepelin nevertheless developed a highly influential classification of psychiatric diseases. In the introduction to the second volume of the 7th edition of his influential textbook of psychiatry, he argued that an ideal system should be based on three independent lines of evidence: pathological anatomy, symptomatology, and etiology of the disease (Kraepelin, 1904). However, due to a lack of available data in the other two areas, Kraepelin's classificatory system was primarily based on symptoms or clinical signs. The Vogts took on his challenge and contributed significantly to the knowledge of the pathological anatomy and etiology of certain diseases (Vogt and Vogt, 1919b, 1920). They also based their own theoretical system on Kreapelin's conviction that any accurate classification that is based on true knowledge of any one of these three sources of evidence would in essence agree with any such system that could be obtained by using any of the others.

Psychiatric diseases were of course not the only area where questions of classification featured prominently. Zoological and botanical systematics also gained importance at the turn of the century, particularly regarding questions of evolution and speciation. Here it was especially the problem of variation that was at the center of attention. Geographical variation had been put forward as an important argument in favor of a Darwinian picture of gradual evolution by means of natural selection, but the extent of gradual and continuous variation in nature was thoroughly contested (Bateson, 1894). Similarly there was widespread disagreement about the causes of variation, as is evidenced by the highly charged debate between Biometricians and Mendelians (Provine, 1971). Furthermore, the connection between variation and its causes and the problem of speciation was also the subject of intense discussions. Oskar Vogt was very interested in all these issues. He had been collecting bumble-bees ever since his early youth and by the time he finished publishing his ideas on the species questions in 1911, his collection contained more than 75,000 specimens (Vogt, O., 1909, 1911). Vogt's study of the geographic variation of bumblebees and his approach to the problem of speciation already show some of the elements of his theory of biological organization that would later become the cornerstone of his theoretical synthesis. He argued, for instance, for the importance of inner organization in constraining the possibilities of variation (Vogt, O., 1909), a phenomenon that he discovered by arranging all known variations of a character such as color patterns in a systematic way. Based on this arrangement he could then establish the sequence of transitions from one character state to the next. With reference to the discussions of his period he interpreted this as evidence for Eimer's theory of orthogenesis. Another noteworthy feature of these papers is his discussion of geographic variation. Vogt observed that major differences in coloration are connected with regional varieties. Drawing on his extensive collection, he also recognized regional convergence in coloration between different species and regional gradation of coloration within different species. Since he could not establish any selective reason for these differences he argued that environmental factors must cause this form of variation. Vogt also advocated a separation of a physiological (i.e., biological) species concept from morphologically recognizable forms and flirted with the possibility of the inheritance of environmentally induced geographic variations.

This last point brings us to the final aspect of the Vogts research agenda that I want to introduce here, the question of inheritance. The problem of variation obviously has a strong connection to the questions of genetics and development. Genetics, the study of patterns of inheritance, and developmental biology both deal with causes of variation. In turn, the existence of natural and artificially induced variations proved to be essential to progress in both disciplines. Oskar Vogt was keenly aware of the importance of genetics and developmental biology for his own research agenda. As we will see later, Oskar and Cécile Vogt's theory of biological organization and their attempt to derive criteria for the identification and individualization of biological characters and their variations was based on a synthetic integration of both development and genetics. Here I want to focus on the historical context of their ideas. Studies of the inheritance of pathological conditions, especially of mental and psychological deficiencies, were a major concern of the eugenics movement (see, e.g., Adams, 1990; Kevles, 1985; Paul, 1995). The Vogts were aware of these trends, but their own research was only marginally important to the eugenics movement. Nevertheless, they often phrased their results in the language of "improvement," for instance, when they argued that "the practical goals of brain research are the care and selective enhancement (*Höherzüchtung*) of the human brain (Vogt and Vogt, 1929, p. 438). Even though their involvement with the ideology of "improvement" was largely theoretical, it was based on specific ideas about the mechanisms of inheritance.

As Jonathan Harwood points out in his magisterial study of the German genetics community (Harwood, 1993) there were at least two different schools of genetics within Germany. The *"pragmatics,"* who were eager to adopt a Morgan-style genetics with its emphasis on transmission, and the *"comprehensives,"* who preferred to study genetics in the context of evolution and development. Harwood argues that these scientific preferences in genetics can, to some extent, also be correlated with more general "styles of thought" that include the breadth of biological knowledge, ideas about what counts as science (*Wissenschaft*), and the socioeconomic and political position of the respective researchers. Within this classification of geneticists, Oskar Vogt is somewhat of an oddity. While his ideas on the mechanism of inheritance are most certainly "comprehensive"—his framework is thoroughly developmental and he emphasized the role of extra-nuclear and environmental factors even to the point where he considered the possibility of the inheritance of acquired characters and the existence of so-called *Dauermodifikationen*—his insistence on a strict correlation of structural and functional variations and his materialistic belief in the localized and material basis of higher brain functions puts him closer to the "pragmatists" sensu Harwood. In any case, what was most central to Vogt's ideas on inheritance was his synthetic outlook. He combined a Morgan-type Mendelism with a developmental view of gene action. Consequently, he was critically involved in attempts to complexify the framework of classical genetics in order to accommodate more complicated biological characters and more ambiguous patterns of inheritance.

I will explain some of the details of Vogt's contributions to genetics later. Here I will conclude this brief account of Oskar and Cécile Vogt's careers by describing their success in institution building. I have already mentioned that Oskar Vogt's *Neurologische Zentralstation* became the *Neurologisches Laboratorium* of the University of Berlin in 1902. In 1915 the Kaiser Wilhelm Institute for Brain Research was formally incorporated and the *Neurologisches Laboratorium* was gradually merged with this new institution. Generous new facilities, including a clinic, were built in Berlin Buch and the institute moved in 1931. Previously, a department of experimental genetics headed by Nicolai Timofeeff-Ressovskij had been established in 1925. The purpose of this department was to pursue *Drosophila* genetics as a model system for the study of the inheritance of higher brain functions and their pathologies. Oskar Vogt hired Timofeeff-Ressovskij when he was in Moscow to help establish an Institute for Brain Research in the Soviet capital. One of the main goals of this institute was to find the material basis of exceptional mental abilities by way of the study of Lenin's brain. The Nazis forced Oskar Vogt into retirement, but the Vogts continued their research in a small private research institute in the Black Forest built in 1937, again with funds from the Krupp family. During WWII *the Journal for Psychology and Neurology* ceased publication in 1942, and in 1954 Oskar Vogt founded his last scientific publication, the *Journal für Hirnforschung* (Journal for Brain Research). Oskar Vogt died in 1959; Cécile Vogt 3 years later in 1962.

THE BIOLOGICAL CHARACTER CONCEPT IN THE THEORETICAL SYSTEM OF OSKAR AND CÉCILE VOGT

In the previous section I briefly sketched the development of Oskar and Cécile Vogt's careers. In this section I will focus on the logical reconstruction of those parts of their theoretical system that are relevant to the problem of the biological character concept. As I have indicated earlier, the Vogts published a systematic presentation of their theoretical arguments and a summary of the main results of their empirical studies in a two-part article in the *Journal for Psychology and Neurology* (Vogt and Vogt, 1937, 1938). The scope of these two papers is enormous (together they are 375 pages long) and I can only give a concise account of their principal ideas here.

The Vogts conception of biological characters is based on several crucial assumptions. First there is the postulate that all functional differences are based on structural differences and the related assumption that there is no qualitative difference between pathological variation and natural variation. The latter point is especially important for it provides the justification for their methodological principle that natural variation can be a model system for the study of pathological variation and vice versa. The postulate that there is a correlation between functional and structural variation also led them to a dual research strategy where structural differences that were identified by the study of brain architectonics had to be correlated with a functional role and functional differences that were most prominently identified by their pathological manifestations in the form of psychiatric diseases had to be correlated with a structural modification in a relevant region of the brain. Here we can clearly see how the Vogts two-step research strategy reflects the distinction between a context of structural identification and a context of functional individualization. Their strategy included the use of techniques of structural (anatomical, histological, and architectonic) and functional (physiological, pathological, and genetic) analysis. The results they obtained by using these various research methodologies were then interpreted within the conceptual framework that postulated a correlation between structural and functional variation, even though direct evidence for this link could not always be found immediately.

As I have already indicated earlier, the Vogts interpretation of biological characters is closely linked to the problem of the classification of (mainly) psychiatric diseases. The issue of functional individualization finds it counterpoint in the problem of psychiatric diagnosis, and the issue of structural identification is essentially the problem of neurological localization. The Vogts were keenly aware of this connection. In a paper of 1926 on the importance of zoological systematics for psychiatry (Vogt, O., 1926b, see also Vogt, O., 1925)—incidentally the paper that contains definitions of the genetic concepts of expressivity and penetrance (see later)—Oskar Vogt concluded that their research program with its emphasis on the structural and functional localization of specific regions in the cortex and the use of insect genetics as a model system for the study of variation in the brain had already converged with Kraepelin's ideal system for the classification of psychiatric diseases based on pathological anatomy, symptomatology, and etiology of diseases.

A second postulate that was crucial for the Vogts theory of biological organization was the idea that all biological characters are in essence dynamic and not static in nature. Every individualized biological character follows a specific ontogenetic trajectory throughout the lifecycle of the organism, a process they termed *Bioklise*. This postulate implied that the origin and modification of each character was caused by specific formative processes that acted throughout the lifecycle of the organism. A biological character was then defined as a *topistic unit* (topos, Greek for place). In the case of the central nervous system, a topistic unit was any combination of parts (neurons) that reacted in a coordinated way to a formative stimulus and that was characterized by a common functional and physicochemical specificity. Within this developmental perspective of biological characters the Vogts also assumed that these topistic units reacted in a coordinated way to any disturbance or variation in the formative processes that led to the establishment and maintenance of these characters. Changes among the formative factors then led to correlated changes in the expression of characters. The Vogts observed that these changes in character states did not occur at random but that there existed a specific sequence of transitions in the expressions of characters. We have already seen that Oskar Vogt discovered such a pattern in the geographic variation of the coloration of bumble-bees (Vogt, O., 1909, 1911). In order to explain these observations the Vogts argued that the reason for these regular, as opposed to random, changes in the transitions between character states lay in the organizational properties of the organism. In their view, a biological organization was a dynamic unit characterized by the constant interaction of many formative factors (such as genes, cells, and other somatic and environmental factors) and the rules of these interactions constrained the variation in the expression of character states which are the endproduct of all these formative processes. The Vogts referred to this specific sequence of character states as *Eunomie* and they collected many examples of such *eunomic* series both from brain pathologies and from the geographic variations of insects.

The third postulate that characterized the Vogts theory of biological organization is the least controversial. It simply assumed that all biological characters are inherited. This assumption only becomes an issue when we remember that the Vogts were studying the material basis of higher brain functions and that they were studying these questions in the first half of this century in Germany. Here I do not have the space to explore the potential problems that lie at the intersection of science, politics, and ideology in regard to these issues. I want to focus instead on the logical implications of this third postulate for the Vogts theory of biological characters. When discussing inheritance the Vogts were no reductionists and neither did they equate inheritance with genetic determinism. They did not even see genes as privileged factors in explanations of inheritance. In their view, genetic factors participated in the formative processes of biological characters but they did not believe that genes were the only factors that mattered. Instead, they emphasized the role of interactions between all of the formative factors (genetic, somatic, and environmental). The main role of genetic factors was, in their opinion, to contribute to the stability of certain characters throughout the generations and not to cause the formation of any biological character. However, even for the purpose of generating stability, genetic factors were not solely responsible, the rest of the soma also contributed. In short, the Vogts view of the causal role of genes in development was quite similar to that of many German developmental biologists and developmental geneticists (e.g., Goldschmidt, 1927, 1938; Kühn 1934; Kühn and Henke, 1929, 1932, 1936; see also Gilbert, 1991; Haraway, 1976; Harwood 1993).

The reason why the Vogts emphasized genetic factors at all can be found in the methodological consequences of such a move. The complexity of the brain prevented them from testing experimentally any of their ideas about formative processes, pathological deviations, and patterns of inheritance. They could not manipulate the brain directly and had to rely on whatever clinical material came their way, but if, as they claimed, these processes were fundamentally the same in all biological organizations, then studying these processes in insect model systems should provide them with insights that would also apply to the more complex cases. As we have seen, the Vogts relied on model systems throughout their career. For them, naturally occurring variation became a stand-in for pathological conditions, biological systematics became the model for the classification of psychiatric diseases, and *Drosophila* genetics would illuminate the patterns of inheritance of pathological conditions in the brain.

The adoption of *Drosophila* genetics as a model system for the study of inheritance of psychiatric diseases could, however, not be accomplished in a straightforward manner. The main problem was that psychiatric diseases are complex characters and that in the mid-1920s patterns of inheritance for complex characters could not easily be accounted for within the boundaries of classical genetics. Oskar Vogt was aware of these problems when he went to Moscow in 1925 to help establish the Institute for Brain Research in the Soviet capital. There he learned of studies by Nicolai Timofeeff-Ressovskij and D. D. Romaschoff on the phenotypic expressions of genetic variations in *Drosophila funebris* (Timofeeff-Ressovskij, 1925; Romaschoff, 1925). They had observed that the presence of a specific allele in pure lines of flies does not always lead to the phenotypic expression of its associated trait and that the degree of phenotypic manifestations of these traits can also vary between these lines. Oskar Vogt was very interested in these results, for they provided empirical support for his ideas about biological organization and they offered him a model system in which he could study these phenomena. This was particularly important since similar phenomena—varying degrees of expression of the phenotypic trait of known or suspected genes—were also commonly observed in psychiatric diseases (see also Patzig, 1933; Vogt and Vogt, 1937, 1938). Vogt arranged for the publications of these results in the *Journal of Psychology and Neurology* and hired Timofeeff to establish a division of animal genetics at the KWI for Brain Research.

Vogt also incorporated this empirical evidence in his theoretical system. In 1925 he published a programmatic article, *Psychiatrisch wichtige Tatsachen der zoologisch-botanischen Systematik* (important facts for psychiatry derived from zoological and botanical systematics) in the *Zeitschrift für die gesamte Neurologie und Psychologie* (Journal for General Neurology and Psychology). In this paper Vogt developed the first fully elaborated synthesis of his studies on geographical variation in bumble-bees and associated questions of zoological classification with both his and his wife's work on brain pathology and localization that was specifically mediated by an inclusion of genetics. In order to accomplish this integration, Vogt had to develop three new concepts within the framework of classical or Mendelian genetics. Based on the experimental results by Timofeeff and Romaschoff, he defined the concepts of expressivity, penetrance, and specificity that have since then become a cornerstone of medical genetics up to the present day. Expressivity refers to the degree a phenotypic character of a certain gene (allele) is manifested in the individuals of a lineage, penetrance is defined as the percentage of individuals in a lineage with a certain genetic variation (*Genenvariation*) that show the phenotypic manifestation of the associated character, and specificity is considered an indication of the degree to which the phenotypic manifestations of a genetic variation in the individuals of a lineage resemble each other.

These three concepts (expressivity, penetrance, and specificity) allowed Oskar Vogt to connect the study of the expression of well-defined phenotypic manifestations of genetic variations in insects with the study of pathological variations in the brain. In his view the mechanistic causes for the phenomena described by expressivity, penetrance, and specificity could be found in the formative processes that create and maintain biological organizations. Interactions with other formative factors preclude the identical expression of phenotypic characters in all cases that share a specific genetic factor (genetic variation). Furthermore, the expression of phenotypic variation is constrained by the rules of interactions between these various formative factors, as is evidenced by the prevalence of eunomic series of variations. Disease, as a pathological variation, can therefore be interpreted as a specific step in an eunomic series of variation. Its occurrence has to be understood in the context of the formative processes that lead to the expression of any biological character. The explanation of a pathology thus lies in its etiology, or in the formative process that created it. Insofar as the analysis of variations in insects illuminates the nature of the formative processes that create them it also contributes to the understanding of disease. As Kreapelin suggested in 1904, since every classification of psychiatric diseases should ultimately be based on their etiology, the biological insights gained from the study of insect genetics are also relevant to this rather elusive problem. It is therefore understandable why the study of insect genetics would come to occupy such a prominent position in the Kaiser Wilhelm Institute for Brain Research. The genetics group headed by Timofeeff-Ressovskij soon became one of the centers for genetic research within Germany and, due to the collaboration between Timofeeff, Zimmer, and Delbrück on the world famous *Dreimännerarbeit* (Timofeeff, Zimmer, and Delbrück, 1935), one of the cradles of molecular biology.

Oskar Vogt's foray into classical genetics was no accident. As we have seen, he had consistently explored areas of biology that were even marginally relevant to his main research program, the identification of the material basis for higher brain functions. The theoretical synthesis that he developed in collaboration with his wife over a period of 40 years contains elements that were derived from many different areas of biological investigation.

After this brief analysis of the conceptual structure of Oskar and Cécile Vogt's theory of biological organization, let me begin to conclude with a concise description of their conception of biological characters. This description will take the form of a summary of what has been said so far. I base this reconstruction primarily on the Vogts own synthesis of their lifelong scientific research (Vogt and Vogt, 1937, 1938).

The basic element of Oskar and Cécile Vogt's theory of biological organization is the *topistic unit*. This already indicates that their theory is centered around the localization of individual structures or biological objects. Every such structure is the outcome of a *formative process* and, throughout the lifecycle of the organism, is either modified or maintained by one or more *vital processes*. Biological structures therefore exist only as part of a continuous process. The developmental processes that lead to the formation of biological structures or topistic units are characterized by the interaction of various *formative factors*. These formative factors or causes of development can be divided into *exogenous* and *endogenous* factors. Among the latter the Vogts distinguished between the *genome* and the soma (*somatom*). They shared the view of many German biologists of the 1910s and 1920s that in the development of a structural unit, somatic interactions are more important than strictly genetic factors (see, e.g., Harwood, 1993; Sapp, 1987). Consequently, they also emphasized the role of somatic inheritance, a fact that they expressed in the distinction between the *genotype* (the sum of all genes) and the *idiotype* (the sum of all hereditary factors.) They also defined the *phenotype* somewhat idiosyncratically as the structure of an individual (characterized by the sum of the genotype and the idiotype) at any specific point of time in its development as opposed to the *constitution* of an individual, which they defined as the sum of all potential modifications of this present phenotype (a notion that is somewhat similar to Schmalhausen's norm of reaction). In this system the clue to the formation of any structure or biological object is found in the interactions between all participating factors. The interactions between these factors can be of a physical or a chemical nature and in theory one could distinguish between *acting* and *reacting* factors. (In practice, however, this distinction might be difficult to establish.) Also, in the course of development, newly differentiated structures or characters can become formative or vital factors in the differentiation and maintenance of other topistic units. In this theory of biological organization *morphogenesis* is the result of a tightly integrated process. Individual biological characters can only be identified and individualized in the context of dynamic processes, either as the result of a developmental or stabilizing process or as a factor in another such process.

It should also be noted that even though the Vogts views on inheritance and development can be placed within the range of ideas discussed in Germany during the 1920s and 1930s, there seems to have been little interaction between them and the rest of the German scientific establishment (see also Satzinger, 1998). The Vogts also preferred to define concepts within their own theoretical system, which makes it difficult to compare their definitions with the ones used by other biologists. Despite these qualifications a few connections can easily be established. Their notion of the gene is one of a dynamic object that is defined by its effect. As such, it is in line with similar views of the gene that can be found in Kühn's developmental and Goldschmidt's physiological conception of the gene. The Vogts conception of the gene is one of a material object but not restricted to the chromosomes. Regarding their conception of development as an interactive process that involves a variety of physical and chemical factors, this notion bears a certain resemblance to widely discussed ideas such as the morphogenetic field (see, e.g., Weiss, 1939). The Vogts were convinced of the ultimately chemical nature of this process and even used a chemical criteria of specificity to localize areas in the brain that reacted in an identical way to certain environmental effects, such as drugs or poisons.

I have mentioned earlier that a biological character or a topistic unit is a dynamic object that is actively engaged in either its further transformation or in its stabilization. The Vogts defined the normal process of the lifelong development of a biological character as *Bioklise* and, correspondingly, the process that leads to the formation of a pathological condition as *Pathoklise*. A change in any of the formative factors can cause a variation in a biological character. This factor then becomes a variation factor. Different individual variation factors can lead to the same phenotypic variation and the same variation factor can have different phenotypic effects in different contexts. In many cases, such as the coloration of bumblebees, the spots on beetles, the veins on *Drosophila* wings, or the degeneration (lesions) of localized areas in the cortex, the expression of phenotypic variations follows a specific pattern (see Vogt and Vogt, 1937, 1938, for a summary of these results). The Vogts defined this directed sequence of characters states as *Eunomie*. In their opinion, such eunomic series of character states were a quite frequent phenomenon that found its explanation in the rules of interactions of formative and vital factors.

To summarize, three aspects of the Vogts conception of biological characters stand out: (i) the integrative developmental perspective, (ii) their attention to different aspects of the biological character concept, such as variation, pathology, comparison, causation, constraint, and functional dependency, and (iii) their reliance on model systems and the mediation of genetics. This last issue deserves some further attention. The use of model systems in biology has gained prominence over the last few decades and has raised some important conceptual questions. Foremost among them is how one can assure that the insights gained from the study of a model system can be generalized. Two ways suggest themselves immediately: (i) one can develop a generalized theory based on the results of the analysis of model system and (ii) one can construct an argument based on homology. In this view, generality is achieved by arguing that because of a similarity between different instances (often based on the notion of common descent but not necessarily so), the insights gained in one case can be applied to another.

Within their research program, the Vogts employed both strategies. This comparative and generalized outlook helps us to understand the peculiarities of many of their ideas. Consider, for instance, their concepts of expressivity, penetrance, and specificity. The logical theoretical context for these concepts would be the theory of quantitative genetics. Even though these concepts helped address a theoretical problem in quantitative genetics, technical problems soon emerged that seriously diminished their importance. This was, however, not an issue for the Vogts. For them, these concepts derived their meaning from within their own theoretical framework. Therefore, if we want to understand the role and eventual fate of all these concepts, including the Vogts notion of biological characters, we have to analyze them in the theoretical context within which they were developed in the first place.

CONCLUSION

In this paper I have introduced Oskar and Cécile Vogt's ideas on the biological character concept and their theory of biological organization in the context of their research program and the biological discussions of their times. Here, by way of conclusion, I want to raise two more issues: (i) the relevance of the Vogts approach to the problem of biological characters for current research and (ii) the role of conceptual analysis in theoretical biology.

The question of what is a biological character and what are the criteria for its identification and individualization is an important issue in modern (evolutionary) biology, as is made clear in the present volume. This set of papers also demonstrate that the problem of biological characters is not confined to a single biological discipline and furthermore that we cannot continue to treat the issue of defining biological characters separately from the advances made in other fields. Important questions that depend on a clear conception of what we mean by "biological character" are the issue of functional kinds in biology, the problem of understanding variation, the relation between molecular and morphological data, the individualization of parts in biological systems, the problems of adaptation and selection, the question of building blocks in development, the homology problem and its conceptual twin the novelty problem, etc. All these problems call for an integrative approach. As Walter Elsasser put it in his *Reflections of a Theory of Organisms*, progress will come when we "improve the cognitive frame of reference into which the individual data of biological science can be organized" (Elsasser, 1998[1987], page xxvi). The conceptual synthesis of Oskar and Cécile Vogt can hereby serve as a model on more than one level. The problems they studied and the conceptual solutions they developed still apply to be found in many of the questions I have just listed. For this alone they deserve our attention. Their theory of biological organization that identifies a biological character as a dynamic structural object, produced and maintained by a complicated set of physical and chemical interactions, and individualizes it by its functional role in the context of a specific biological process is one of the more productive conceptions of biological characters currently available. The Vogts relevance, however, does not stop there. In a time of far-ranging reorganization of biological research, the structure of the Kaiser Wilhelm Institute for Brain Research can serve as a blueprint for how one can organize different biological disciplines around a central biological problem.

Finally, I want to briefly argue for the role of conceptual analysis in biology. Ever since theoretical biology gained prominence in the Anglo-American world after WWII its focus has mainly been on mathematical biology. On one level this has been a success story; today everybody accepts the importance of mathematical techniques and models for biological research. On the conceptual front, however, theoretical biology has not been as successful (the emergence of a whole new discipline of philosophy of biology notwithstanding). This was not always the case. Between 1900 and WWII theoretical biology was mostly conceptual rather than mathematical. The theoretical synthesis of Oskar and Cécile Vogt can be seen as a prime example of this kind of theoretical biology. Today we are at a crossroad and we have the opportunity to bring together both incarnations of theoretical biology. As Walter Elsasser argued, comparing the physical with the biological sciences, "..the character of biology as compared to physical science must appear in *conceptual innovations* rather than mathematical ones "(Elsasser, 1998[1987], p. 11). The mathematics needed is already there; it is concepts we miss. Oskar and Cécile Vogt realized that long ago.

ACKNOWLEDGMENTS

I thank the following individuals and audiences for their critical input: Michael Hagner, Larry Holmes, Hans-Jörg Rheinberger, and Sahotra Sarkar were always available for discussions and the questions raised by the audiences at the 1998 Joint Atlantic Seminar in the History of Biology in Baltimore and the Program Seminar of the Princeton Program in the History of Science were very helpful indeed. Gerry Geison, Scott Gilbert, and Günter Wagner read the whole manuscript and contributed significantly to its improvement. I also acknowledge the financial support of the Max Planck Institute for the History of Science.

LITERATURE CITED

Adams, M. (ed.) (1990). "The Wellborn Science." Oxford University Press, Oxford.

Allen, C., Bekoff, M., and Lauder, G. (eds.) (1998). "Nature's Purposes." MIT Press, Cambridge, MA.

Bateson, W. (1894). "Materials for the Study of Variation." Macmillan, London.

Breidbach, O. (1996). "Die Materialisierung des Ichs." Suhrkamp, Frankfurt/Main.

Broca, P. (1861). Perte de la parole, ramollissement chronique et destruction partille du lobe anterieur gauche du cerveau. *Bulletins de la Société d' Anthropologie* 2:235-238.

Broca, P. (1865). Du siege de la faculté du langage articulé dans l'hemisphére gauche du cerveau. *Bulletins de la Société d' Anthropologie* 6:377-393.

Brodmann, K. (1909). "Vergleichende Lokalisationslehre der Großhirnrinde in ihren Prinzipien dargestellt aufgrund des Zellenbaues." Barth, Leipzig.

Canguilhem, G. (1991). "The Normal and the Pathological." Zone Books, New York.

Cummins, R. (1975). Functional Analysis. *Journal of Philosophy* 72:741-765.

Elsasser, W. (1998). "Reflections on a Theory of the Organism." Johns Hopkins Univeristy Press, Baltimore.

Foerster, O. (1923). Die Topik der Hirnrinde und ihre Bedeutung für die Motilität. *Deutsche Zeitschrift für Nervenheilkunde* **77**:124-139.

Foerster, O. (1925). Zur Pathogenese und chirurgischen Behandlung der Epilepsie. *Zentralblatt für Chirurgie* **25**:531-556.

Foerster, O. (1926). Zur Pathogenese des epileptischen Krampfanfalls. *Deutsche Zeitschrift für Nervenheilkunde* **94**:15-56.

Gilbert, S. (ed.) (1991). "A Conceptual History of Modern Embryology." Plenum, New York.

Goldschmidt, R. (1927). "Physiologische Theorie der Vererbung." Springer, Berlin.

Goldschmidt, R. (1938). "Physiological Genetics." McGraw-Hill, New York.

Goldstein, K. (1995). "The Organism." Zone Books, New York.

Hagner, M. (1994). Lokalization, Funktion, Cytoarchitektonik. *In* "Objekte, Differenzen und Konjunkturen: Experimentalsysteme im Historischen Kontext" (M. Hagner, H.-J. Rheinberger, and B. Wahrig-Schmidt, eds.), pp.121-150. Akademie Verlag, Berlin.

Hagner, M. (1997). "Homo cerebralis. Der Wandel vom Seelenorgan zum Gehirn." Berlin Verlag, Berlin.

Haraway, D. (1976). "Crystals, Fabrics, and Fields." Yale University Press, New Haven

Harrington, A. (1987). "Medicine, Mind, and the Double Brain." Princeton University Press, Princeton.

Harrington, A. (1996). "Reenchanted Science. Holism in German Culture from Wilhelm II to Hitler." Princeton University Press, Princeton.

Harwood, J. (1993). "Styles of Scientific Thought." University of Chicago Press, Chicago.

Kevles, D. (1985). "In the Name of Eugenics." Knopf, New York.

Kraepelin, E. (1904). "Psychiatrie. Ein Lehrbuch für Studierende und Ärzte," 7th ed. Barth, Leipzig.

Kühn, A. (1934). "Grundriss der Vererbungslehre." Quelle and Meyer, Heidelberg.

Kühn, A., and Henke, K. (1929). Genetisch und entwicklungsphysiologische Untersuchungen an der Mehlmotte Ephistia kühnielle Zeller, I-VII. *Abhandlungen der Gesellschaft der Wissenschaften zu Göttingen, Math.-Phys. Klasse, N.F. 15*, no. 1:1-121.

Kühn, A., and Henke, K. (1932). Genetisch und entwicklungsphysiologische Untersuchungen an der Mehlmotte Ephistia kühnielle Zeller, VIII-XII. *Abhandlungen der Gesellschaft der Wissenschaften zu Göttingen, Math.-Phys. Klasse, N.F. 15*, no. 1:127-219.

Kühn, A., and Henke, K. (1936). Genetisch und entwicklungsphysiologische Untersuchungen an der Mehlmotte Ephistia kühnielle Zeller, XIII-XIV. *Abhandlungen der Gesellschaft der Wissenschaften zu Göttingen, Math.-Phys. Klasse, N.F. 15*, no. 1:225-272.

Laubichler, M. D. (1997a). "Identifying Units of Selection." Ph.D. dissertation, Yale University.

Laubichler, M. D. (1997b). The Nature of Biological Concepts. *European Journal for Semiotic Studies* **9**:251-277.

Laubichler, M.D. (1999). A semiotic perspective on biological objects and biological functions. *Semiotica* **127**:415-432.

Millikan, R. (1984). "Language, Thought and Other Biological Categories." MIT Press, Cambridge, MA.

Nolte, J. (1993). "The Human Brain: An Introduction to Its Functional Anatomy," 3rd ed. Mosby Year Book, St. Louis.

Patzig, B. (1933). Die Bedutung der schwachen Gene in der menschlichen Pathologie, insbesondere bei der Verebung striärer Erkrankungen. *Naturwissenschaften* **21**:410-413.

Paul, D. (1995). "Controlling Human Heredity." Humanities Press, Atlantic Highlands.

Provine, W. (1971). "The Origins of theoretical Population Genetics." University of Chicago Press, Chicago.

Richter, J. (1976a). Oskar Vogt, der Begründer des Moskauer Staatsinstituts für Hirnforschung. *Psychiatrie, Neurologie und medizinische Psychologie* **28**:385-395.

Richter, J. (1976b). Oskar Vogt und die Gründung des Berliner Kaiser-Wilhelm Institutes für Hirnforschung unter den Bedingungen imperialistischer Wissenschaftspolitik. *Psychiatrie, Neurologie und medizinische Psychologie* **28**:449-457.

Romaschoff, D. D. (1925). Über die Variabilität in der Manifestierung eines erblichen Merkmals bei Drosophila funebris. *Journal für Psychologie und Neurologie* 31:298-304.

Sapp, J. (1987). "Beyond the Gene." Oxford University Press, Oxford.

Satzinger, H. (1998). "Die Geschichte der genetisch orientierten Hirnforschung von Cécile und Oskar Vogt." Deutscher Apotheker Verlag, Stuttgart.

Timofeeff-Ressovskij, N. W. (1925). Über den Einfluß des Genotypus auf das phänotypische Auftreten eines einzelnen Gens. *Journal für Psychologie und Neurologie* 31:305-310.

Timofeeff-Ressovskij, N. W., Zimmer, K. G., and Delbrück, M. (1935). Über die Natur der Genmutation und der Genstruktur. *Abhandlungen der Gesellschaft der Wissenschaften zu Göttingen, Math.-Phys. Klasse, N.F. 15*, no. **VI**:190-245.

Vogt, O. (1897). Flechsig's Associationscentrenlehre, ihre Anhänger und Gegner. *Zeitschrift für Hypnotismus* 5:347-361.

Vogt, O. (1902). Psychologie, Neurophysiologie und Neuroanatomie. *Zeitschrift für Psychologie und Neurologie* 1:1-3.

Vogt, O. (1909). Studien über das Artproblem. 1. Mitteilung. Über das Variiren der Hummeln. 1. Teil. *Sitzungsberichte der Gesellschaft Naturforschender Freunde zu Berlin* 1:28-84.

Vogt, O. (1911). Studien über das Artproblem. 2. Mitteilung. Über das Variiren der Hummeln. 2. Teil. *Sitzungsberichte der Gesellschaft Naturforschender Freunde zu Berlin* 3:31-74.

Vogt, O. (1925). Psychiatrische Krankheitseinheiten im Lichte der Genetik. *Zeitschrift für die gesamte Neurologie und Psychiatrie* 100:26-34.

Vogt, O. (1926a). Die physiologische Bedeutung der architektonischen Rindenfelderung und–schichtung der menschlichen Großhirnhemisphäre. *Arch. Psychiatr. Nervenkr.* 76:649-651.

Vogt, O. (1926b). Psychiatrisch wichtige Tatsachen der zoologisch-botanischen Systematik. *Zeitschrift für die gesamte Neurologie und Psychiatrie* 101:805-832.

Vogt, C., and Vogt, O. (1907). Zur Kenntnis der elektrisch eregbaren Hirnrindengebiete bei den Säugetieren. *Zeitschrift für Psychologie und Neurologie* 8:277-456.

Vogt, C., and Vogt, O. (1919a). Allgemeinere Ergebnisse unserer Hirnforschung. *Zeitschrift für Psychologie und Neurologie* 25:277-461.

Vogt, C., and Vogt, O. (1919b). Zur Kenntnis der pathologischen Veränderungen des Striatum und des Palladium und zur Pathophysiologie der dabei auftretenden Krankheitserscheinungen. *Sitzungsberichte der Heidelberger Akademie der Wissenschaften. Math. Natwissen. Klasse. Abt B* 14:1-56.

Vogt, C., and Vogt, O. (1920). Zur Lehre der Erkrankungen des striären Systems. *Zeitschrift für Psychologie und Neurologie* 25:631-846.

Vogt, C., and Vogt, O. (1926).Die vergleichend-architektonische und die vergleichend-reizphysiologische Felderung der Großhirnrinde unter besonderer Berücksichtigung der menschlichen. *Naturwissenschaften* 14:1190-1194.

Vogt, C., and Vogt, O. (1929). Hirnforschung und Genetik. *Zeitschrift für Psychologie und Neurologie* 39:438-446.

Vogt, C., and Vogt, O. (1937). Sitz und Wesen der Krankheiten im Lichte der topistischen Hirnforschung und des Variieres der Tiere. Erster Teil. Befunde der topistischen Hirnforschung als Beitrag zur Lehre vom Krankheitssitz. *Journal für Psychologie und Neurologie* 47:237-457.

Vogt, C., and Vogt, O. (1938). Sitz und Wesen der Krankheiten im Lichte der topistischen Hirnforschung und des Variieres der Tiere. Zweiter Teil, 1. Hälfte. Zur Einführung in das Variieren der Tiere. Die Erscheinungsseiten der Variation. *Journal für Psychologie und Neurologie* 48:169-324.

Weiss, P. (1939). "Principles of Development." Holt, New York.

Wright, L. (1973). Functions. *Philosophical Review* 82:139-168.

3

PREFORMATIONIST AND EPIGENETIC BIASES IN THE HISTORY OF THE MORPHOLOGICAL CHARACTER CONCEPT

OLIVIER RIEPPEL

Department of Geology, The Field Museum, Chicago, IL 60605

INTRODUCTION

The notion of what a "character" is will vary with the perspective from which a character is being analyzed. An embryologist will have a different notion of a character than a systematist, and an ecologist may look at "characters" in a somewhat different way than a functional anatomist. Common to all these diverse perspectives is the understanding, however, that the notion of a character is by no means trivial. What is being construed as a character will always depend on the research objective for which the character will be used, but at the same time it will always be understood that any character remains, to some degree, an abstract notion, as it requires the individualization, or abstraction, of traits from a complex organic whole. A character is as much a product of observation as it is of conceptual thinking or, as Etienne Serres put it (1827, p. 49): "Organized matter is the field of the anatomist, philosophy provides the tools with which to work this field. The first tool of the anatomist is

The Character Concept in Evolutionary Biology

observation, his second tool is abstraction." The identification of any character presupposes a comparison of different organisms in the attempt to discover similarities versus differences. Comparison and distinction are correlative principles of comparative anatomy. As every organism is itself determined by the sum of its internal relations, every identification of a character requires the conceptual decomposition of an organism into parts that become the object of comparison and distinction (Rieppel, 1988a).

There are, therefore, two fundamental ways to conceptualize the organism in comparative biology: once as a developmentally and functionally integrated complex whole or alternatively as a composite of parts that can be analyzed and compared in isolation from one another. The method of systematic biology requires decomposition of the organisms into parts, or "characters," which can then be used in the reconstruction of phylogenetic relationships. Developmental biologists or functional anatomists may take issue with the ways systematists decompose the complex organic whole because this procedure may disrupt what they recognize as vital internal relations determining the organism. Indeed, the nature of "characters" used in systematics is an object of continuous debate because of the element of abstraction that underlies the decomposition of the organic whole. Conversely, an attempt can be made to use information on the developmental generation of the organic whole as a guide for the decomposition of the organism in search for "characters." An investigation of the history of ideas about the generation of animals reveals that developmental biologists were themselves conceptualizing the organic whole in different and mutually exclusive ways, which have become known as epigenetic versus preformationist (Roger, 1971; Rieppel, 1986a). It is the purpose of this chapter to review the history of the epigenetic versus preformationist approach to embryogenesis and the ways in which these alternate "ways of seeing" bear on the nature of "characters."

ARISTOTLE AND THE PRINCIPLE OF CONTINUITY

The mystery of creative processes of nature has always attracted the human intellect, philosophers and biologists alike. How do species originate, how do species propagate, how does metamorphosis work? Such questions, it was believed, would be answered with a deeper understanding of the nature of animal reproduction. Everybody could agree that plant and animal life was ordered into species, entities of nature whose form (or appearance) was reproduced throughout the cycle of generations. As the organism develops, from its conception to its death, its material appearance is subject to continuous change, yet never did this change appear to transcend the species-specific principle of form. For Aristotle, change meant movement, and the notion of movement presupposed a first unmoved mover. Under such premises, the cycle of reproduction of organisms appeared as much to be part of the intricate fabric of nature as did the cyclical movement of celestial bodies.

Subject to continuous movement, i.e., change in Aristotle's terms, the planets nevertheless travel on immutable, seemingly eternal trajectories on the basis of laws of nature which had to emanate from the first unmoved mover. Aristotle's universe had been called one of "dynamic permanence" (Regnéll, 1967): nature is subject to continuous motion, but this movement, or change, is governed by eternal laws which constrain movement, or change, to immutable cycles of reappearance. The sun slowly and gradually changes its appearance throughout the day, yet it repeats that cycle of change on a daily basis according to the eternal laws which govern its movement. In a similar sense, the organisms change in their appearance from conception to death, yet that same trajectory of change is eternally repeated throughout the cycle of generations according to the eternal laws which govern the process of reproduction. In Aristotle's universe, an element of "being" correlates with an element of "becoming." The form of an organism is continuously "becoming" throughout life, but the species-specific "principle of form" which underlies that change remains the same forever.

In his theory of (animal) reproduction, Aristotle considered the male seminal fluid, itself derived from the blood, the active component of generation (Balss, 1943). The principle (or "knowledge") of form was believed by Aristotle to be inherent in the male seminal fluid, as was the potential for movement. As such, the species-specific form of an organism was preformed in the male seminal fluid, but only as a potential ("*in potentia*"), not in an actual, i.e., material sense! The species-specific form would become actualized, through embryogenesis, in the receptive female substance (the menstrual blood according to Aristotle) which would be set into motion and worked upon by the male seminal fluid via the *aura seminalis*. Embryogenesis as such is a matter of "becoming," but the species-specific form which is being actualized through development is a principle of "being."

Aristotle studied the development of cephalopods, fishes, and the chicken, and it is these latter studies that became most influential in the history of biology. With his observations Aristotle was able to show that the embryo is a new development, something that develops "de novo" and is not just a combination of preexisting or preformed parts or particles. No trace of the embryo can be seen in a chicken egg until after 3 days and 3 nights of incubation. After that time, there appears a little red dot which "jumps and moves," which was recognized by Aristotle as the first appearance of the heart, the motor of subsequent development. Two meandering blood vessels (arteries) could eventually be seen to grow from the heart into the membranes surrounding the yolk. After 10 days, Aristotle was able to recognize the head with its large eyes, while the rest of the embryonic body remained relatively inconspicuous. Implicit in Aristotle's account is the idea that the embryo is a new development and that its organs develop one after the other, as well as one out of the other, just as blood vessels grow out of the heart. This model of development eventually became known as *epigenesis*: one organ formed becomes the material cause of the next one to develop. The process results in an integrated organic whole determined by the sum of its internal relations.

With his theories on generation, Aristotle opposed the tradition of atomistic philosophy as founded by Democritus and revised by Epicurus, and its explanation of animal generation. According to atomists, the universe is built from immutable and indestructible atoms that move freely through empty space. All material appearances are temporary aggregates or composites of atoms, including plants and animals. In the case of animal reproduction, atomists again postulated a male and female seminal fluid. According to their theory, every part of the parent body was represented by equivalent, i.e., preformed but miniaturized atoms in its seminal fluid. Generation would consist of a mixture of the male and female seminal fluids in the maternal body and would result in the aggregation of the atoms which form the fetus. This theory explained the similarities which the offspring shares with both parents, while errors in the combination of atoms would explain the production of malformations. Movement of the atoms, and their aggregation to form organisms, would be governed by the properties of matter, and by chance. Atomism did not appeal to a first unmoved mover, nor to an underlying "principle of form" governing embryogenesis. Atomistic models of generation have become known as preformationist because the parts of the embryo are preformed as atoms in the parental seminal fluids (Roger, 1971).

A deeper issue in the debate over atomism versus epigenesis was the principle of continuity and its significance for scientific explanation. As a first principle, the principle of continuity precedes observation as a guide for the conceptual organization of the world. "Continuity and discontinuity, I would argue, are not constructs of nature, but constructs of the human mind used to interpret nature" (Mendelsohn, 1980, p. 107). In atomistic philosophy, the immutable and indestructible atoms were believed to be separated by empty space, a void which allowed unconstrained movement of those atoms. Aristotle abhorred the idea of a void in nature and stressed the continuity of phenomena: "Nature proceeds little by little from things lifeless to animal life in such a way that it is impossible to determine the exact line of demarcation, nor on which side thereof an intermediate form should lie" (Aristotle, quoted from Mendelsohn, 1980, p. 81). The available writings from Aristotle do not fully explore the consequences of continuity, but they certainly laid the foundation for the idea that all products of nature, inanimate and animate, were ordered along a continuously graded series of form ascending to ever higher levels of perfection (the Great Chain of Being: Lovejoy, 1936), and that the gradual individual development of animals recapitulates the graded series of "lower" forms of life (Rieppel, 1988a). For Aristotle, continuity was the ordering principle for a world in which every movement, every change, had its immediate cause as well as its meaning and its goal.

WILLIAM HARVEY, SUCCESSOR OF ARISTOTLE

Ex ovo omnia are the famous words which appear on the frontispiece of William Harvey's (1578-1657) treatise on animal generation, again based mainly on the study of the development of the chicken (Harvey, 1651 [1981]). Harvey differed from Aristotle with his opinion that it is blood which appears first, then the heart develops, which in a next step starts pulsating. Otherwise, however, he remained true to his mentor with his view that "... it is certain that the chick is built by *epigenesis*, or the addition of parts *budding out* from one another ..." (Harvey, 1651 [1981, p. 240; emphasis added]). The embryo again is a new formation, not initially present in the egg. Embryogenesis, according to Harvey, was a continuous process of growth, budding, subdivision, and differentiation. The constituent organs of an embryo develop successively, one out of the other, and *individualize* through differentiation: "... the first [part] to exist is the genital particle by virtue of which all the remaining parts do later arise as from their first original ... at the same time that part *divides up* and forms all the other parts in their due order ..." (Harvey, 1651 [1981, p. 240; emphasis added]). And again: "These homogenous parts, I repeat, are not made from heterogeneous or dissimilar elements united together, but they arise by way of generation from similar or homogenous material, and are *differentiated* and made dissimilar" (Harvey, 1651 [1981, p. 207; emphasis added]).

Harvey's comment about "dissimilar elements united together" opposed claims by René DesCartes (1596-1650) and Pierre Gassendi (1592-1655) that ultimately discontinuous particles of matter underlie all apparent continuity of phenomena (Mendelsohn, 1980). In his *Syntagma philosophicum* (published posthumously, 1658), Gassendi specified: "... we infer in every seed the presence of a variety ... of true rudiments ... of such *parts* as require only to be changed in position and to be so coordinated in order to be fashioned into *members* of the animal ..." (Adelmann, 1966, Vol. II, p. 805; emphasis added). These parts or particles "unite ... because of their innate motions, that like particles are held together with the like" (Adelmann, 1966, Vol. II, p. 806). The notion of attractive forces holding like and like together demonstrates Hippocrates' influence in Gassendi's writing (Roger, 1971).

In matters of embryology, it was Marcello Malpighi (1628-1694) who raised his voice against Harvey (Rieppel, 1994). Malpighi's position has been interpreted in controversial ways (Adelmann, 1966), but by his contemporaries he was understood as a preformationist who held the view that organisms develop by the juxtaposition of preformed parts or rudiments (Roe, 1981; Bowler, 1971). Describing the development of the heart in the chicken, Malpighi noted how the ventricles are joined together and how at a later date several arterial tubules are joined to the heart like fingers to a hand (Adelmann, 1966, Vol. II, p. 973). In a similar way, the brain was described to develop by the union of formerly separate parts or vesicles (Adelmann, 1966, Vol. II, p. 967; see also Rieppel, 1994). Given the perspective that embryos form by the juxtaposition of preformed parts, Malpighi found it "indeed superfluous to inquire whether the heart is formed before the brain, or the blood before the heart ..." (Adelmann, 1966, Vol. II, p. 867). Since the parts of the embryo were preformed, they had to be present (in the egg) even if they remained invisible to the human eye: "... we are forced to await the manifestation of the parts as they successively come to view" (Adelmann, 1966, Vol. II, p. 937).

As stated earlier, each organism is determined by the sum of its internal relations. For Harvey, these relations guide the epigenetic development of the organism through which becomes actualized (i.e., materialized) what potentially exists in the egg. Thus, Harvey stated that "it [the blastoderm] is the principal part, as being that in which all other parts exist potentially and from whence they later arise, each in its due order" (Harvey, 1651 [1981, p. 274]). One organ formed becomes the material cause of the next one to develop in a graded and continuous sequence of growth, budding, subdivision and differentiation subject to a "formative faculty" (Harvey, 1651 [1981, p. 284]). As in Aristotle, a vitalistic principle inherent in the organism determines internal relations and therewith the epigenetic developmental process.

By contrast, a purely mechanical conceptualization of preformation would require that internal relations of the developing organism are a consequence of, rather than a prerequisite for, the juxtaposition of parts. However, this is where atomists such as Pierre Gassendi departed from the ancient tradition by proposing "this awesome combination of audacious speculation incorporating the atomism of Democritus and Epicurus, the ensouled semen of Aristoteles, a dash of Church fathers, a supçon of Cartesianism, tinged with both mechanism and vitalism, permeated with preformationism, and ending with the humble acknowledgment of the futility of all his theories ..." (Adelmann, 1966, Vol. II, p. 777). According to Gassendi, ensouled atoms are shed by both sexes into their respective seminal fluids. As these are mixed, the atoms combine instantaneously to form the functionally integrated whole under the guidance of their soul: "if the parts of an animal were produced one after the other those which were formed first would then impede the fashioning of others ..." (Adelmann, 1966, Vol. II, p. 811). The organs do not enclose empty space through which atoms may freely move during their successive combination! Malpighi's writings are less easily interpreted, but he, too, believed the essence of internal relations, the "deeper origins" of the organismal form, to preexist in terms of an obscure latent organization (Adelmann, 1966, Vol. II, p. 869). Malpighi eschews the problem of successive versus instantaneous combination of parts by restricting his theorizing to the statement that the parts "successively come to view," a statement which does not preclude their possible primeval and simultaneous production (Adelmann, 1966, Vol. II, p. 886).

One of the major problems left unsolved by epigenetic theories of generation was the functional correlation of parts. Stating explicitly that the organs of the integrated whole would develop one after the other, epigenetic theories were unable to answer the question of the functional correlation of parts. How could the heart start to beat prior to the development of the brain, when the heartbeat requires nervous stimulation in order to occur? Atomistic theories of generation solved this problem by an instantaneous juxtaposition of parts. Finally, in the last analysis, both camps would resort to some metaphysical, i.e., vitalistic "principle of form" which would guide either epigenetic development, or the combination of parts, to form a living whole.

THE DEVELOPMENTAL AND FUNCTIONAL CORRELATION OF PARTS

If all parts of organisms are developmentally as well as functionally correlated with one another, the organism becomes completely determined by its internal relations. Under this presumption, preexistence is the only possible hypothesis to explain the problem of generation. Development constitutes first and foremost growth, concluded Charles Bonnet (1720-1793), but growth presupposes nutrition, nutrition presupposes a circulatory system, a circulatory system presupposes the existence of a heart, the existence of the heart presupposes the existence of the brain, and so on: "Arteries presuppose veins, the ones and the others presuppose nerves, the nerves presuppose the brain, the latter presupposes a heart, and all of these presuppose a multitude of other organs" (Bonnet, 1764, p. 154). There is no room for epigenesis, for the successive development of the parts of a developmentally and functionally integrated whole. Nor is there room for the successive juxtaposition of preformed parts. The embryo must preexist in its entirety, either in the male spermatozoon (animalculism), an idea which Gottfried Wilhelm Leibniz (1646-1716) adapted from Antoni van Leeuwenhoek (1632-1723) and Nicolaas Hartsoeker (1656-1725), or in the female egg (ovulism), as Bonnet concluded after he had discovered parthenogenesis in aphids (Bonnet, 1745). Since all organs of an organism can only function in concert, they all must exist during the initial phases of development, even if they remain hidden from the human observer because of their small size and initial translucency. In this theory, the process of development is reduced to the simple growth and unfolding (*évolution* in Bonnet's terms) of the preexistent embryo. Whereas the doctrine of preexistence explains development of the germ, it does not explain its generation through sexual reproduction. In this respect, Bonnet resorted to a doctrine of encapsulation (*emboîtement* in his terms), which postulated that the female of every species created *ab initio* contained within its eggs all the preexistent germs of all the generations the species was destined to produce, all those generations being encapsulated one within the other. In analogy to the continuum that can be divided into an infinite number of ever smaller segments, Bonnet maintained that there are no limits to the almightiness of God who may create things as small as he pleases (for a closer analysis of Bonnet's theories and the influence of Leibniz, see Rieppel, 1988b). Bonnet proposed again an Aristotelian world of "dynamic permanence": The encapsulated germs are "being," their actualization, their "becoming," is ensured through the cycle of generation.

Other than attacking atomistic (preformationist) conceptualizations of life as being rooted in atheism, Bonnet raised important biological arguments against the idea that embryos could form by the juxtaposition of preformed particles. Atomists of Bonnet's time, such as Pierre Louis Moreau de Maupertuis (1698-1759) or Denis Diderot (1713-1784), repeatedly referred back to the appositional growth of crystals as an analogy to the development of embryos (Roger, 1971). Bonnet looked into studies of the growth of mollusk shells as a biological system, which might be comparable to a crystal, and correctly noted that on the basis of that analogy, embryonic development would be reduced to appositional growth ("accroîssement par juxtaposition": Bonnet, 1764, pp. 61-62). This may work for exoskeletons such as mollusk shells, but it would not seem to work so easily for endoskeletal elements like bones. Bonnet believed that Albrecht von Haller (1708-1777) had convincingly shown that bone grows differently. Studies by Haller on the development of the chicken had led him to believe that bones are preexistent structures, small and translucent at first (Haller, 1758). Later during growth, Haller noted the deposition of an "earthy" (i.e., mineral) substance inside the gelatinous element, a pattern of growth which Bonnet called intussuzeptional ("accroîssement par intussuszeption"). The difference between the exoskeletal mollusk shell and the endoskeletal bone is that the latter is vascularized, and it is through the blood vessels that the "earthy substance" gets into the bone. The conclusion which Bonnet drew from Haller's observations was that the deposition of nutritive molecules into the preexisting germ is the mechanism driving its "évolution," its unfolding. On a more general level, intussuzeptional growth of animal tissue is an important concept relating to epigenesis as well, as it is the only mechanism by which one organ formed can become the material cause of the next one to develop in a process of growth, budding, subdivision, and individualization!

The doctrine of preexistence is rooted in the belief of a profound and all-pervading functional correlation of the organs of the integrated organic whole. On August 11, 1770, Charles Bonnet wrote to Albrecht von Haller: "... it is enough to show you a foot or a hand for you to envisage the whole ..." (Sonntag, 1983, p. 890). According to Bonnet, each organism forms in its entirety a singular, tightly integrated system whose parts are closely and reciprocally related to each other and together strive toward the performance of the same function, which is the life of an organism perfectly adapted to its environment (for further comments on the notion of perfect adaptation see Ospovat, 1981). Through its preexistence due to the functional correlation of its parts, the organism becomes fully determined.

THE DISCOVERY OF *HYDRA VIRIDIS* AND THE TRIUMPH OF MATERIALISM

It appears like an irony of history that it was Bonnet's cousin Abraham Trembley (1710-1784) who, in 1740, reported his discovery of *Hydra viridis*. Here was an animal which was of green color, sessile, and capable of reproduction through budding, just like a plant. Yet the animal contracted if stitched with a needle (subject to *irritabilité* which Albrecht von Haller had declared an essential property of animal fibers), it would capture prey with its tentacles, and it could move around (in somersaults) just like an animal. Trembley had problems in classifying this strange organism either as a plant or as an animal. Given the background knowledge and instrumentalization of his time, Trembley was unable to discover that the green color of *Hydra viridis* results from a cellular symbiosis with green algae, and thus decided to perform what could be called an "Aristotelian test" on the creature (Rieppel, 1988a). Aristotle had noted different capabilities of regeneration in plants and animals. If a twig is cut off from a tree, not only will the tree regrow the twig, but the twig will regrow a tree! If, on the other hand, a leg is cut off from a salamander, the salamander will regenerate the leg, but the severed leg will not regrow a complete salamander! So Trembley cut the polyp in two, and both parts regenerated the whole animal. He cut the polyp into four, eight, and more pieces, but each time each piece would regrow an entire polyp. Trembley could take the top half of one animal and graft it on to the bottom part of another individual! Evidently, organisms were formed by the juxtaposition of parts.

Following Trembley's experiments, a controversy erupted, under the leadership of René-Antoine Ferchault de Réaumur (1683-1757), as to the role of the soul in the regeneration of *Hydra*. The brain was believed to be the seat of the soul, and the brain of *Hydra* was believed to be located in the top part of the creature. If an individual was sectioned horizontally, there was no problem to explain where the regenerating head part would have gotten its soul from. The difficulty was explaining where the regenerating foot part would get a soul from. The heretical answer to that question was delivered by Julien Offray de La Mettrie (1709-1751) in his book *L' Homme Machine* published in 1747: either the soul does not exist or it has to be coextensive with matter! *Hydra* eventually became the paradigm on which several authors based their atomistic conceptualization of animal generation, among which Georges Buffon (1701-1788) perhaps ranks as the best known one.

In his *Histoire des Animaux*, published in 1749, Buffon referred back to *Hydra* and concluded: "An individual is nothing but an agglomeration of *germs*, or little individuals of the same kind." Like Maupertuis, Buffon compared organismal growth to the growth of crystals: if "salt and other minerals are composed of parts, which not only resemble each other but also the composite whole, then organisms are composed of other organisms whose ultimate parts again resemble each other and consist of organized matter. All these considerations lead to the conclusion that infinitely many small, organized particles exist in nature which are endowed with life, and which aggregate to form all these organized beings large enough so we can see them with our eyes." Buffon termed these particles "organic molecules" (*molécules organiques*), and believed these to cycle eternally through nature, comparable to the eternal trajectories of planets. A dead plant or animal decomposes, its organic molecules become part of the topsoil. From there they enter plants through assimilation, and thus become part of the foodchain, until they disintegrate again to become part of the topsoil. In the spring of 1748, Buffon had conducted experiments on spontaneous generation in collaboration with the Irish priest John Turberville Needham (1713-1781), who visited Paris at the time. These experiments seemed to prove Buffon's theory of organic molecules. Dead plant or animal material was cooked in a flask which then was left standing for a week or two. Sure enough, the decomposition of the dead organic material seemed in the meantime to have given birth to spontaneously generated, minuscule yet moving organisms. These experiments resulted in a heated debate and generated elegant tests conducted by the Italian Priest Lazzaro Spallanzani (1729-1799), but it was left to Louis Pasteur (1822-1895) to deal the death blow of the doctrine of spontaneous generation. A broth of plant (tea) or animal (meat) matter is called an *infusion* in French; the animals supposedly spontaneously generating in these broths became known as infusorians.

Having what he considered experimental proof for spontaneous generation, Buffon went on to explain the generation of animals through sexual reproduction. As all atomists, or preformationists before him, Buffon postulated a male and female seminal fluid. Albrecht von Haller shook his head in distress, as he believed to have convincingly shown that a female seminal fluid did not exist (Haller, 1752). According to Buffon, the parent organism would acquire organic molecules through nutrition. These organic molecules would be assimilated to the organism during growth. At the time of maturity, when growth slowed down or stopped altogether, any surplus of organic molecules would be shed by the parent organism into the seminal fluid, which at this time would first start to form. Before the molecules reached the seminal fluid, however, they would circulate through the parent body, and in that process become imprinted, each one with an organ-specific trait of the parent body. The imprinting of the organic molecules would, according to Buffon, be mediated by "internal molds" (*moule intérieur*), a notion which notoriously escaped clear understanding, and which Buffon might have adopted from Louis Bourguet (1678-1742) (Roger, 1971; Rieppel, 1987). Following the lead of Aristotle, it was again Haller who raised the question of why mutilations were not heritable.

Buffon believed that sexual reproduction would result in the mixture of the two parental seminal fluids. Whereas spontaneous generation would result from the accidental combination of organic molecules, these would be endowed with a "penetrating power" (*force pénétrante*) in sexual reproduction. The endowment of organic molecules with a penetrating power certainly represents a vitalistic component of Buffon's theory of animal generation, but Buffon himself compared this force to the attractive forces which had been invoked by Newton in his explanation of gravity. In case of man, the total amount of seminal fluid would suffice to produce two offspring. Normally, only one fetus would form, while the rest of the organic molecules would contribute to the formation of extraembryonic membranes and the placenta. Prior mixture of the seminal fluids would nevertheless explain the similarities which the offspring shares with both parents.

Buffon's atomistic theory of preformation certainly had a big advantage over the doctrine of preexistence in the explanation of the variation of the offspring, of the recombination of "characters" of both parents in the offspring, of the phenomena of hybridization, and of the formation of "monsters." However, as other atomists, Buffon still had to come to grips with the functional correlation of parts. Buffon cited *Hydra* as an example for the fact that organisms exist that do not form an integrated whole. For organisms of higher complexity, however, Buffon maintained that they represent developmentally and functionally integrated wholes, the parts of which cannot function in isolation. Looking back on Harvey's discussion of Aristotle's and his own views as to whether the heart or the blood is formed first, Buffon concluded that both sides were wrong: "Everything forms at the same time." The aggregation of the organic molecules to form a fetus "lasts perhaps as long as a blink of the eye." The dual vision of an organized being manifests itself again in Buffon's writing: the notion of the developmentally and functionally integrated whole contrasts with the decomposition of this whole into its constituent elements.

Nevertheless, Buffon's atomism allowed him to speculate about species transformation, a process he called "degeneration" (*dégénération*) of species. Although his reasoning with respect to this issue remains ambivalent to some degree, he did believe that the nature of the organic molecules changes with changing environmental conditions. In his *De la Dégénération des Animaux*, published in 1766, Buffon referred to the different appearance of human beings in different parts of the world, living in different climatic conditions. Although they all belong to the same species, they nevertheless illustrate the effect of the climate on the organic molecules and the way they are assimilated and propagated by the parent body. If the earth is subject to continuous cooling, as Buffon believed, then the climate will change during earth history, and with it the nature of the organic molecules and the animals they form. Modern elephants therefore could represent degenerated mammoths, a hypothesis which was later destroyed by Georges Cuvier in his famous lecture of 1796.

DARWINISM AND ATOMISM

"I have read Buffon: whole pages are laughably like mine," Charles Darwin (1809-1882) wrote to Thomas Huxley (Darwin, 1887, p. 45). Like Buffon, Darwin adopted an atomistic perspective of life, and his later theory of pangenesis was traced back to his early exposure to chemistry as a teenager (Schweber, 1985). With his theory of pangenesis, Darwin attempted to solve the problem of heritable variation, which would then become subject to natural selection. Through his exposure to animal breeders, Darwin was well versed in matters of heredity, but at his time there was no satisfactory theory available to causally explain the phenomena of inheritance. Chemistry was a reputable science firmly rooted in empiricism, yet based on elements, or atoms, too small for the human eye to perceive. Like Buffon, Darwin eventually postulated invisible particles to account for observed inheritance of variability. Sloan (1985, p. 85) records Darwin's interest in Chemistry upon his enrollment at Edinburgh, and quotes from T.C. Hope's lecturenotes: "Almost all animal matters have certain / "characters" in common ... they all consist of the same chemical / elements, which are few in number; and that / the diversity among them arises from a difference / in the proportions of these elements, or in the / manner, the atoms of these Elements are associated / and grouped together..." A second and perhaps even greater influence on Darwin's early exposure to atomism was his research on invertebrates, initially following the lead of Robert Grant (Sloan, 1985).

Upon arrival in Edinburgh, Darwin joined the Plinian Society where Grant held a leading position for those interested in research on, and discussion of, invertebrate marine biology. Grant's particular interest was the status of "zoophytes," colonial invertebrates. Grant introduced Darwin to the study of "polyzoa" or bryozoans, who then discovered the development of ciliate planctotrophic larvae in one form, Flustra, which Darwin referred to as "ova." From this and similar observations, Grant "drew the startling conclusion that all of these forms [colonial invertebrates] reproduced by a body directly analogous to the infusoria [see Needham's experiments above], and that these same infusorial 'monads' formed the elementary units, not only of the zoophytes, but of all organic tissues" (Sloan, 1985, p. 83). By the time he left for the voyage with the Beagle, Darwin was also aware of the controversy triggered by Robert Brown, who discovered the spontaneous molecular motion referred to as "Brownian motion." In a paper published in 1828, Brown concluded from the observation of moving plant pollen that the capacity of motion was inherent in all particular matter of organic and inorganic bodies, and that the motile molecules of organic matter probably were the *molécules organiques* of Buffon and Needham (Sloan, 1985).

In his theory of pangenesis, Darwin postulated, like Buffon before him, a male and female seminal fluid which derived from invisibly small *gemmules* cruising through the parent body. Collectively, these gemmules represent all parts of the parent organism in miniature. Darwin explained the supposed inheritance of acquired characteristics by the circulation of these gemmules through the parent body before they became concentrated in the seminal fluid. Mixture of these seminal fluids in the course of sexual reproduction resulted in the aggregation of the gemmules to form offspring. Again, the offspring would for this reason combine both maternal and paternal characteristics, an important source of variation.

The formation of all organisms from a mixture of parental gemmules established an important material continuity throughout the diversity of living forms. In addition, it provided the basis for an understanding of *fundamental variation*. Darwin's view of life was built on the understanding that "... natural selection is daily and hourly scrutinizing, throughout the world, every variation, even the slightest; rejecting that what is bad, preserving and adding up all that is good..." (Darwin, 1859, p. 84). To be malleable by natural selection, every organism must be subject to ever so slight variation in all of its constituent elements, and the smaller the constituent elements are hypothesized to be, the more easily could slow, continuous, and gradual change take place. "From echinoderms to Englishmen, all had arisen through the lawful redistribution of living matter in response to an orderly changing geological environment. This was rank materialism, and Charles knew it" (Moore, 1985, p. 452).

Of course, Darwin recognized the problems which his atomistic and materialistic approach created in the understanding of the evolution of complex and tightly integrated structures. In the *Origin* he wavers: "To suppose that the eye, with all its inimitable contrivances ... could have been formed by natural selection, seems, I freely confess, absurd in the highest possible degree," but then he asserts: "the difficulty of believing that a perfect and complex eye could be formed by natural selection, though insuperable by our imagination, can hardly be considered real" (Darwin, 1859, p.187). Pressed by one of the reviewers of his book, Darwin admitted: "The eye to this day gives me a cold shudder, but when I think of the fine known gradations, my reason tells me to conquer the cold shudder" (Darwin, letter to Asa Gray, 8 or 9 February 1860: Burkhardt *et al.*, 1993, p. 75). In the end, however, Darwin supported the supremacy of fundamental variation and natural selection and referred to the limited powers of human cognition in view of difficulties faced by his theory: "So even with the Eye, as numerous fine gradations can be shown to exist, the perfecting this wondrous organ by Nat. Selection I must look at as a difficulty to our imagination and not to our reason" (Darwin, letter to Asa Gray, 24 February 1860: Burkhardt *et al.*, 1993, p. 107).

DARWINISM: PROS AND CONS

Darwin's theory of generation established a one to one relationship between heritable units and the characters of organisms. As with Buffon, these heritable units were not immutable and eternal, but potentially subject to change: "in variations caused by the direct action of changed conditions ... the tissues of the body ... are directly affected ... and consequently throw off modified gemmules, which are transmitted with their newly acquired peculiarities to the offspring" (Darwin, 1868, pp. 394-395). Variation, the raw material for selection, is generated by recombination and change in the continuous redistribution of living matter which establishes the "hidden bond" between organisms of different appearance: "We can understand why a species or a group of species may depart, in several of its most important characteristics, from its allies, and yet be safely classified with them. This may be safely done ... as long as a sufficient *number of characters* ... betrays the hidden bond of community of descent" (Darwin, 1859, p. 426; emphasis added). The number of characters corresponds to the number of shared gemmules and is evidence for common descent. The corollary of this hypothesis is that everything that exists in nature is nothing but a modification of what existed before. When examining the "community of plan" of the vertebrate skull, Thomas H. Huxley (1825-1895) was not deterred by differences of appearance but was determined to reveal the "hidden bond" of common descent: "The biological science of the last half-century is honourably distinguished from that of preceding epochs, by the constantly increasing prominence of the idea, that a community of plan is discernible amidst the manifold diversities of organic structure. That there is nothing really aberrant in nature; that the most widely different organisms are connected by a hidden bond; that an apparently new and isolated structure will prove, when its characters are thoroughly sifted, to be only a modification of something which existed before ...: (Huxley, 1859, p. 382). Evolution creates nothing new, but only modifies what already existed.

Darwinism was strongly opposed by the last great epigenesist, Karl Ernst von Baer (1792-1876). Studying once again the development of the chicken, von Baer (1828) likewise recognized a "community of plan," but at a different level than Darwin. Von Baer saw the hierarchical classification of animals reflected in their ontogeny: the development of the embryo relates to the type of organization in such a way as if it passed through the animal kingdom classified according to the French *méthode analytique*, i.e., according to Cuvier's method of dichotomization (von Baer, 1828). Darwin (1859, p. 449), by contrast, found the embryo to represent "the animal in its less modified state; and in so far it reveals the structure of its progenitor." For Darwin, embryology provided a clue to evolutionary transformations linking the ancestor to its descendant; for von Baer, embryology was a process of progressive divergence, separation, and individuation of organisms. To von Baer, embryology is not a key to the evolutionary transformation of structures which existed before, but a process of growth, budding, subdivision, and differentiation which creates new structures that were not in existence before and which individualize "types" of organization. For Darwin, evolution shapes ontogeny; for von Baer, ontogeny creates hierarchy. Being opposed to the idea of species transformation, von Baer found the hierarchy of the natural classification of organisms to reflect the logic of the plan of creation, which also determines the goal (*causa finalis*) of ontogeny, i.e., the creation of a new individual (von Baer, 1886). This paradigmatic difference between Darwin and von Baer is best illustrated with an example, such as the homology of the tetrapod stapes with the hyomandibula of fishes (Rieppel, 1993). Recognition of this homology is usually attributed to Reichert (1837), although working under the von Baerian paradigm, Reichert compared the tetrapod stapes to the hyomandibula of fishes not in terms of descent with modification, but in terms of topological equivalence. Studying the embryology of the splanchnocranium throughout the "series" of vertebrates, Reichert (1837) first introduced the distinction of visceral arches (the first two arches) from the succeeding branchial arches. He recognized that the dorsal half of the first visceral arch metamorphoses to become the primary upper jaw, whereas its ventral half differentiates to become the primary lower jaw. The dorsal half of the second visceral arch becomes involved in jaw suspension in fishes (hyomandibula), whereas in tetrapods it metamorphoses to become an ear ossicle (stapes). In no way did Reichert imply an evolutionary transformation of the fish hyomandibula into the tetrapod stapes, i.e., the development of a new structure by modification of something that existed before. Rather, he concluded: "I want to stress that after their early anlage, the *existence* of visceral arches in lung-breathing vertebrates *comes to an end*" (Reichert, 1837, p. 142; emphasis added). To Reichert, the visceral arches had a developmental identity as embryonic elements of both "gill-breathing vertebrae" and "lung-breathing vertebrae." They are part of the vertebrate type. During subsequent development, these visceral arches metamorphose, or differentiate, whereby they lose their prior identity and assume a new one, this individualizing the type of "lung-breathing vertebrae." This is a far cry from Huxley's protocol, who introduced the term "hyomandibular" bone to refer to the ossified dorsal part of the visceral arch in fishes and then set out to trace its evolutionary transformations "by the interpolation of transitional gradations of structure" (Huxley, 1859, p. 382).

DISCUSSION AND CONCLUSIONS

The atomistic conception of nature, and of organisms in particular, was variably subjected to vitalistic overtones, but ultimately found its way back to its origins rooted in materialism. As such, it lent itself easier to the explanation of generation and, ultimately, species transformation on the basis of natural causes alone. The atomistic background persisted in Mendelian genetics, in population genetics, and ultimately persists in the Synthesis (Mayr, 1982). Organisms continue to be conceptualized as the sum of their parts, i.e., as a sum of "characters," and a match is sought between these characters and the underlying units of inheritance. As discussed earlier, atomism is historically rooted in preformationism, and its tradition continues to assume historical and material continuity between parts of different organisms that are believed to be homologous, i.e., inherited from a common ancestor. Variation and natural selection modify organismal components that already exist in a continuous process of change. Fundamental variation, i.e., the potential of variation in every one of its traits or parts, requires the decomposition of the organic whole into as many constituent elements as possible, with the potential for modification or transformation of all of these elements. As such, the atomistic background is incompatible with the notion of a tight developmental and/or functional integration of the organismic whole and continues to face problems in the explanation of the evolutionary origin and modification of complex systems. In recent time, dissatisfaction with the atomistic tradition has triggered an interest in the stability of structures and the difficulty of their change (Gould, 1982; Wake and Roth, 1989). Developmental integration results in developmental constraints and "impossible morphologies" (Holder, 1983) which denies organisms the capacity of fundamental variation: "We need to see the organism as an integrated whole, the product of a developmental program and constrained by developmental and functional interactions. Conversely, hierarchical developmental integration coupled with epigenetic influences can result in the creation of evolutionary novelties (Müller and Wagner, 1991). In evolution, selection may decide the winner of a given game but development non-randomly defines the players" (Alberch, 1980: 665).

The epigenetic conceptualization of generation and, ultimately, of species transformation, had and still has to overcome its own problems in order to establish itself as a modern research program. Generation is not viewed as a juxtaposition of performed parts or their underlying heritable units, but as a process of growth, budding, subdivision, and differentiation for which intussuzeptional growth is a necessary prerequisite. The embryo is not preexistent in the egg, but the product of a genuine developmental process. There is no inheritance of historical and material identity of parts, there only is the inheritance of an ancestral developmental program. Evolution again is no longer a matter of recombination and modification of constituent elements of an organic structure or their underlying heritable units, but rather results from a change of developmental pathways (Oster and Alberch, 1982). Developmental programs are not considered to be capable of fundamental variation; instead, modification of the developmental program is believed to be constrained by internal relations which determine the organismic whole. Internal relations result from developmental and/or functional integration of tissues and, ultimately, organs, and these internal relations are thought to unfold and establish themselves successively in the course of the developmental process. They are no longer explained in terms of obscure vitalistic concepts, but as consequences of geometrical, physical, and chemical properties of tissues forming biological structures. As a result, evolution is no longer restricted to the stepwise modification of existing structures, thereby determining ontogeny. Instead, ontogeny is viewed as a creative process, capable of generating genuinely new structures through a sequence of growth, subdivision, and differentiation, which results in the individuation of taxa and, ultimately, in phylogeny (Müller and Wagner, 1991). The taxic approach to ontogeny and phylogeny (Rieppel, 1993) treats those innovations as homologies, at their appropriate level of inclusiveness. These homologies share no historical and material identity with adult ancestral structures. Such transformational relations must remain an abstract construct (Rieppel, 1994). Instead, evolutionary novelties may share with the putative ancestor the initial developmental pathways resulting in similar embryonic rudiments until divergent differentiation initiates individualization.

LITERATURE CITED

Adelmann, H. B. (1966). "Marcello Malpighi and the Evolution of Embryology," Vols. I - IV. Cornell University Press, Ithaca, NY.

Alberch, P. (1980). Ontogenies and morphological diversification. *Am. Zool.* **20**:653-667.

Balss, H., ed. (1943). "Aristoteles Biologische Schriften." Verlag Ernst Heimeran, München.

Barel, C. D. N. 1984 (1985). Form-relations in the context of constructional morphology: the eye and suspensorium of lacustrine Cichlidae (Pisces, Teleostei). *Neth. J. Zool.* **34**:439-502.

Bonnet, Ch. (1745). "Traîté d' Insectologie; ou observations sur les Pucerons," Première Partie. Durand Librairie, Paris.

Bonnet, Ch. (1764). "Contemplation de la Nature," 2 vols. Marc-Michel Ray, Amsterdam.

Bowler, P. (1971). Preformation and pre-existence in the seventeenth century: a brief analysis. *J. Hist. Biol.* **4**:221-244.

Burkhardt, F., Porter, D. M., Browne, J., and Richmond, M. (1993). "The Correspondence of Charles Darwin," Vol. 8, 1860. Cambridge University Press, Cambridge.

Darwin, Ch. (1859). "On the Origin of Species." John Murray, London.

Darwin, Ch. (1868). "The Variation of Animals and Plants under Domestication." John Murray, London.

Darwin, Fr. (1887). "The Life and Letters of Charles Darwin," 3 Bd., 3. Aufl. John Murray, London.

de Pinna, M. C. C. (1991). Concepts and tests of homology in the cladistic paradigm. *Cladistics* **7**:367-394.

Gould, S. J. (1982). Darwinism and the expansion of evolutionary theory. *Science* **216**:380-387.

Haller, A.v. (1752). Vorrede über des Herrn von Buffon Lehre von der Erzeugung. *In* "Sammlung kleiner hallerischer Schriften," 2. Aufl. , pp. 81-117. Verlag Emanuel Haller, Bern [1772].

Haller, A.v. (1758). "Sur la Formation du Coeur dans le Poulet; sur l'Oeil; sur la Structure du Jaune &c. Premier Mémoire." Marc-Michel Bousquet, Lausanne.

Harvey, W. 1651 (1981). "Disputations Touching the Generation of Animals." Blackwell, London. (Translated with an introduction and notes by G. Whitteridge.)

Hawkins, J. A., Hughes, C. E., and Scotland, R. W. (1997). Primary homology assessment, "characters," and character states. *Cladistics* **13**:275-283.

Holder, N. (1983). Developmental constraints and the evolution of vertebrate digit patterns. *J. Theor. Biol.* **104**:451-471.

Huxley, Th. (1859). The Croonian Lecture. - "On the Theory of the Vertebrate Skull." *Proc. R. Soc. Lond.* **9**:381-457.

Kluge, A. G. (1991). Boine snake phylogeny and research cycles. *Misc. Publ. Mus. Zool. Univ. Mich.* **178**:1-58.

Liem, K. F. (1981). A phyletic study of the Lake Tanganyika cichlid genera *Asprotilapia, Ectodus, Lestradea, Cunningtonia, Ophthalmochromis,* and *Ophthalmotilapia. Bull. Mus. Comp. Zool.* **149**:191-214.

Lovejoy, A. O. (1936). "The Great Chain of Being." Harvard University Press, Cambridge, MA.

Lyell, Ch. (1830-1833). "Principles of Geology, being an Attempt to Explain the Former Changes of the Earth's Surface, by Reference to Causes now in Action," 3. vols. John Murray, London.

Mayr, E. (1942). "Systematics and the Origin of Species." Columbia University Press, New York.

Mayr, E. (1982). "The Growth of Biological Thought." The Belknap Press at Harvard University Press, Cambridge MA.

Mendelsohn, E. (1980). The continuous and discrete in the history of science. In "Constancy and Change in Human Development" (O. G. Brim and J. Kagan, eds.), pp. 75-112. Harvard University Press, Cambridge, MA.

Moore, J. R. (1985). Darwin of Down: the evolutionist as Squarson-naturalist. In "The Darwinian Heritage" (D. Kohn, Hrsg.), pp. 435-481. Princeton University Press, Princeon, NJ.

Müller, G. B., and Wagner, G. P. (1991). Novelty in evolution: restructuring the concept. Annu. Rev. Ecol. Sys. 22:229-256.

Ospovat, D. (1981). "The Development of Darwin's Theory: Natural History, Natural Theology and Natural Selection," pp. 1838-1859. Cambridge University Press, Cambridge.

Oster, G., and Alberch, P. (1982). Evolution and bifurcation of developmental programs. Evolution 36:444-459.

Pogue, M. G., and Mickevich, M. F. (1990). Character definitions and character state delineation: bête noire of phylogenetic inference. Cladistics 6:319-361.

Regnéll, H. (1967). "Ancient Views on the Nature of Life." C. W. K. Gleerup, Lund.

Reichert, C. (1837). Über die Visceralbogen der Wirbelthiere im allgemeinen und deren Metamorphosen bei den Vögeln und Säugethieren. Arch. Anat. Physiol. Wissenschaftlichen Med. 1837:120-222.

Rieppel, O. (1986a). Atomism, epigenesis, preformation and pre-existence: a clarification of terms and consequences. Biol. J. Linn. Soc. 28:331-341.

Rieppel, O. (1986b). Der Artbegriff im Werk des Genfer Naturphilosophen Charles Bonnet (1720-1793). Gesnerus 43:205-212.

Rieppel, O. (1987). "Organization" in the Lettres Philosophiques of Louis Bourguet compared to the writings of Charles Bonnet. Gesnerus 44:125-132.

Rieppel, O. (1988a). "Fundamentals of Comparative Biology." Birkhäuser Verlag, Basel.

Rieppel, O. (1988b). The reception of Leibniz's Philsophy in the writings of Charles Bonnet (1720-1793). J Hist. Biol. 21:119-145.

Rieppel, O. (1993). The conceptual relationship of ontogeny, phylogeny, and classification. The taxic approach. Evol. Biol. 27: 1-32.

Rieppel, O. (1994). Homology, topology, and typology: the history of modern debates. In "Homology, the Hierarchical Basis of Comparative Biology" (B. K. Hall, ed.), pp. 63-100. Academic Press, San Diego.

Roe, S. A. (1981). "Matter, Life and Generation: Eighteenth-Century Embryology and the Haller-Wolff Debate." Cambridge University Press, Cambridge.

Roger, (1971). "Les Sciences de la Vie dans la Pensée Française du XVIII Siècle, 2nd ed. Armand Colin, Paris.

Schweber, S. S. (1985). The wider British context in Darwin's theorizing. In "The Darwinian Heritage" (D. Kohn, Hrsg.), pp. 35-69. Princeton University Press, Princeon, NJ.

Serres, E. (1827). Recherches d'anatomie transcendante, sur les lois de l'organogénie à l'anatomie pathologique. Ann. Sci. Nat. 11:47-70.

Sloan, P. R. (1985). Darwin's invertebrate program, 1826-1836. In "The Darwinian Heritage" (D. Kohn, Hrsg.), pp. 71-120. Princeton University Press, Princeon, NJ.

Smith, A. B. (1994). "Systematics and the Fossil Record: Documenting Evolutionary Patterns." Blackwell Scientific Publications, London.

Sonntag, O. (1983). "The Correspondence between Albrecht von Haller and Charles Bonnet." Huber Verlag, Bern.

Von Baer, K. E. (1828). "Über Entwickelungsgeschichte der Thiere. Beobachtung und Reflexion," Vol. 1. Gebr. Bornträger, Königsberg.

Von Baer, K. E. (1886). Über den Zweck in den Vorgängen der Natur. *In* "Reden gehalten in wissenschaftlichen Versammlungen und kleinere Aufsätze vermischten Inhalts. Zweiter Theil. Studien aus dem Gebiete der Naturwissenschaften," 2nd. ed. Friedrich Vieweg & Sohn, Braunschweig.

Wake, D. B., and Roth, G. (eds.) (1989). "Complex Organismal Functions: Integration and Evolution in Verterates." John Wiley & Sons, Chichester.

NEW APPROACHES TO THE CHARACTER CONCEPT

Picture a 19th century landscape in southern Germany or Austria with a road leading to a town. On that road two farmers are on their way to the market, one leading a cow to be sold, the other without. After some miles of silent travel the farmer without a cow to sell says to his companion: "Do you see this toad here at the wayside? If I eat that toad, would you give me the cow?" "Sure – " said the other farmer and, fair enough, the farmer eats the toad and gets the cow. A few miles down the road the pair sees another toad and the scene is repeated, only now the farmer who originally had the cow is eating the toad and gets back his cow. Another few miles down the road, the farmer starts to think and finally asks: "Why did we eat that toad?" – Well, what a sensible question, and one that a reader may ask, after reading the section on the character concept. In other words, what can one expect to gain from an investigation into the meaning of a concept; an exercise sometimes denounced as a "semantic issue," meaning an issue only about words and not about scientific contend? This is a legitimate question and one that needs to be taken seriously. There are, however, many examples from the recent history of biology which convince us that conceptual questions, properly handled and solved, are far from "semantic." For example, regardless of whether Mayr's species concept will turn out to be the final word, the fact is that the biological species concept was instrumental in forging a highly productive research program into the nature and the origin of species. This was the case because the biological species concept made clear what the critical issue is to understand speciation. It says that we need to understand the evolution of isolation mechanisms. Nowadays, as this is textbook knowledge, it is hard to imagine how illuminating this insight was. Only a proper species concept could clarify what it is we need to look for to understand the origin of species. I committed myself to working on the character concept because of this example. Only a well-developed character concept will allow us to define a research program into the origin of characters, i.e., find a solution to the problem of evolutionary innovation. We need to know what a character *is* (assuming we agree on what "IS" is, Clinton '99) in order to understand which mechanisms can cause its origin.

In developing a scientific concept one faces a double constraint. On the one hand the concept has to fit the empirical facts it is supposed to capture, as for instance patterns of variation and evolution in the case of the character concept. On the other hand the new concept also has to connect to the relevant existing knowledge. Otherwise the concept would fail to inform the rest of biology and would thus at most be of limited value. In addition it will fail to be informed by already existing knowledge, which would limit its theoretical depth. This double constraint makes the development of scientific concepts particularly challenging. This is the challenge that the authors of this section have taken on from a variety of viewpoints; each forging a tie between the character concept and another field of knowledge.

Louise Roth approaches the character concept from the theory of natural selection. In a seminal paper from 1991 she noted a fundamental similarity between the notion of a replicator, at least at the formal level, but not with regard to the ontological assumptions, and a character. This is a radically new attempt to construct a character concept entirely from a process notion, that of natural selection.

David Houle approaches the character problem from the standpoint of abstraction. The enormous complexity of any organism forces us to introduce abstractions in order to obtain some understanding of what is going on. The simplified picture obtained by abstraction is one where characters summarize the complexity of molecular detail in its most salient features, salient for the process we are interested in. Houle examines two traditional approaches to character abstraction, quantitative genetics and life history theory, and finds serious practical problems in their application. He finally proposes the idea of phenomics, i.e., exploiting genomic information to find out which genes matter to the evolution of important phenotypes.

Wagner and **Laubichler** similarly argue that characters are abstractions obtained from a mental decomposition of organisms. This implies that the character concept is secondary to the organism concept, just as many concepts from molecular biology are secondary to the cell concept. To illustrate this approach a mathematical method for character decomposition is proposed based on Lewontin's notion of quasi-independence. It is found that criteria for the identification of dynamically independent characters derive from the symmetry properties of the dynamical equations describing the process of interest (in this case natural selection).

Kurt Schwenk critically evaluates and extends yet another tradition that has sought to solve the problem of character abstraction, i.e., the idea of functional units in morphology. He shows that there are at least three concepts in use in functional morphology: structural units, mechanical units, and, more recently, evolutionarily stable configurations of characters. He notes that the functional unit concept may be able to overcome the schism between atomistic and structuralist notions of the organism, as they are found in the gray area between unit characters and Baupläne.

Alex Rosenberg approaches the character problem from the point of view of taxonomy arguing for a radical abandonment of functionally individuated characters at the phenotypic level. His alternative is to reduce the description of the organism to that of structural and regulatory genes, the smallest functional units. He argues that this level of description will lead to a taxonomy that approaches the predictive value of the periodic system of chemical elements.

WARNING: the next two chapters by Kim and Kim, as well as the one by Zwick, use very abstract mathematical language. The reader does not need much specialized mathematical knowledge to appreciate the message in these papers, but it is essential to be used to highly abstract arguments and be willing to invest the energy necessary to understand these arguments.

Junhyong Kim and **Minhyong Kim** take a radically fundamental approach to the character problem. No ontological commitment is made other than organisms exist and attributes can be discerned and that organisms vary with respect to these attributes. They then ask under what condition sets of attributes assume the properties one expects of a proper character, like having demographically replaceable "character states." Based on their construction they argue that the origin of modular characters is statistically unlikely and therefore in need of a causal explanation. The reason is that modular characters are "measure zero" set of states (which means possible but with zero a priori probability) in the space of all possible distributions of attributes.

Martin Zwick introduces us to a tradition that originated outside biology but which nevertheless is highly relevant to the character problem. The character problem can be seen as a special case of the part-whole distinction, or the identification of natural subsystems. Natural subsystem decomposition received considerable attention in the area of general systems theory and provides a general framework for discussing the character problem. Since this chapter may be less accessible to some readers than others, I want to summarize a few observations from this chapter relevant to the character concept. Just as characters in biology, which are thought of as quasi-independent parts (Lewontin, 1978), subsystems are thought of as quasi-independent static or dynamic units. The challenge is to find a subsystem decomposition in which the subsystems represent as much of the constraints on the system as possible but otherwise is as simple as possible, i.e. has as few restrictions as possible on the combination of states of different subsystems. Hence the constraints on the possible states of the system are "captured" in the constraints on the possible states of the subsystems. Applying this idea to characters leads to the conclusions that characters are not only quasi-independent parts of the organisms, but also characterized by constraints on their possible states. This is identical to the idea that character identity resides in the developmental constraints on their variation (Wagner, 1986).

Lewontin, R. C. (1978). Adaptation. *Am. Sci.* **239**:156-169.
Roth, V. L. (1991). Homology and hierarchies: problems solved and unresolved. *J. Evol. Biol.* **4**:167-194.
Wagner, G. P. (1986). The systems approach: an interface between development and population genetic aspects of evolution. *In* "Patterns and Processes in the History of Life" (D. M. Raup and D. Jablonski, eds.), pp. 149-165. Springer-Verlag, Berlin.

4

CHARACTER REPLICATION

V. LOUISE ROTH

Zoology Department, EEOB Group, Duke University, Durham, NC 27708

INTRODUCTION

Organisms have observable features—characters—at multiple levels of organization. This in itself is unremarkable, as structural hierarchy is characteristic of all matter: Just as any object may be composed of parts, materials, molecules, atoms, or subatomic particles, living organisms can be examined at multiple magnifications, at which different features become prominent, such as organs, cells, and organelles.

As the level of magnification changes, so do the observable attributes; so, for example, an object that appears homogeneous at one level may be heterogeneous at another: a uniform sheet of tissue may comprise many distinguishable cells. Accompanying the changes in structural level and in observable attributes are changes in the processes in which structures participate. What we call fertilization at the level of gametic cells (eggs and sperm), for example, involves diffusion and bonding of molecules at another.

A central, defining quality of living things is their ability to make copies of themselves, i.e., to replicate or reproduce. Especially important to evolutionary biology are those features that are reproduced faithfully from generation to generation. Such characters are termed "homologous," or in the context of population genealogies, "identical by descent," and by the faithfulness of their replication they serve as historical markers of lineages.

It is the faithful replication of characters that is the basis of heredity. The stability provided by heredity—wherein traits recur through successive generations—is essential in microevolutionary processes such as natural selection or drift. If a character becomes altered in a manner that is faithfully replicated through subsequent generations, and if it also increases in frequency until it becomes fixed in a population, it may demarcate a clade and thereby provide macroevolutionary information useful for both taxonomy and phylogenetic reconstruction.

One of the most remarkable characteristics of organisms is their ability to produce copies that correspond at multiple levels of organization, and to do so without a process of replication occurring directly within each structure that is replicated. When a multicellular organism reproduces, for example, new individual organisms are generated that have similar characteristics, including structurally similar cells and molecules, but this replication occurs without each constituent itself directly undergoing one-for-one duplication. Some characters appear to undergo replication without even maintaining structural correspondences at lower levels of organization. The feature of gross morphology of the eye known as a lens in vertebrates, for example, has been replicated through diverse lineages for hundreds of millions of years; yet different proteins have been incorporated into this structure in different taxa (Piatigorsky and Wistow, 1991). Most characters, moreover, are not manifest continuously throughout the life cycle; yet copies appear faithfully and maintain their integrity when they are expressed in successive generations. The subjects of this essay are (i) the implications that the hierarchical organization of living things has for the process of replication, (ii) the indirect nature of replication for most characters, and (iii) the questions this hierarchical organization and the indirectness of replication raise about the origin of novelty and about development and evolution.

BASICS: TYPES OF OBSERVABLE FEATURES OR CHARACTERS

Characters can be delineated in various ways. Depending upon the context, the kinds of observable features that interest us may be (a) a biological structure or part: studies of phenotypic evolution may trace the persistence, loss, or transformation of, for example, a specific mitochondrial protein, a nucleotide position within the gene coding for that protein, or an appendage used in flight; (b) an attribute—aspect or feature—of those parts, such as color, body temperature, or arrangement into a particular architecture or bauplan; (c) the capacity to participate in a process: e.g., sexuality, social behavior, or the expression of a particular biosynthetic or regulatory pathway or network.

The process of replication may apply to any of these types of characters (whether or not the mechanism for such replication is clear): it is common practice to speak of the heritability or homology of structures, of attributes, and of capacities. Even so, as scientists we naturally expect replication to have a material basis, so we expect the replication of all types of characters—not just biological structures but also attributes and processes—to be manifest as the transmission or replication of material and its physical structure. For the sake of concreteness, my discussion will focus on characters as objects, but it should be understood that similar, if more extended, arguments apply to characters that are attributes or capacities. For example, in a feather, heritable transmission of the attribute of being blue may be traced to the manufacture of a pigment or to a set of proteins arrayed into a reflection grating of a certain gauge (Nassau, 1983; Giradella, 1984). Heritable transmission of the capacity to participate in a process such as self-fertilization or photosynthesis can be considered in terms of the organs, organelles, and proteins of appropriate structure it requires, although it remains legitimate to speak of photosynthesis itself as a character. Understanding the material basis for the formation and transmission of complex or intangible characters presents some of the most challenging problems currently confronting biology, including, for example, the analysis of the biological bases for behavior and the explication of all aspects of development. (The difficulty of these problems also underlies some of the hottest controversies involving the misuse of biological information in formulating social policy, but that is another topic.)

RELATIONSHIPS AMONG COPIES IN THE MECHANICS OF REPLICATION

In this paper I use the words "copying," "replication," and "reproduction" more or less interchangeably.[1] Figures 1-3 illustrate different types of copying processes in a formal way: Figure 1 shows how the relationships among copies can affect the introduction of novelty. Figure 2 illustrates some processes of replication that may not directly involve an object in producing copies of itself. Figure 3 shows some relatively simple mechanisms of information transfer that may occur during replication: progeneration (to be defined later) and the use of a template.

Two ways in which copies can be related to one another are illustrated in Fig. 1. Copies can be reiterated parallel products of replication of a single "original" or source (Fig. 1B), or they can occur in series, as a succession of sequential products (Fig. 1A). Only in copies related as lineal descendants (Fig. 1A) can novel traits that are introduced be propagated (Fig. 1C).

---

[1] See, however, Griesemer (2000, in press) for a different, useful, and highly original perspective in which more specific definitions of these terms, and usage different from that followed here, are proposed.

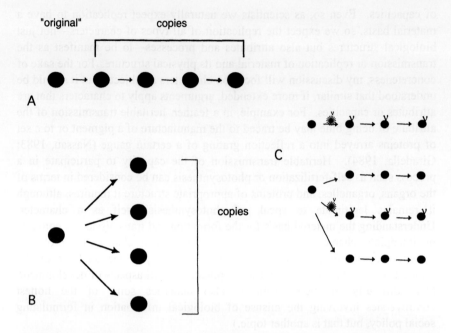

FIGURE 1 How a novel trait is introduced depends on where the copies themselves stand in the causal sequence of replication – whether the path is linear, as in A, or branching, as in B (see also Fig. 2). Each arrow represents a bout of replication. Copies of an original are produced in both A and B. In A the copies are produced through a succession of generations; in B through reiteration within a single generation. C. A new trait (v), shown arising at the starburst, is propagated only among copies related as lineal descendants (i.e., those linked by arrows of replication).

Multiple copies can also be made in ways that do not involve an original (Fig. 2). The simplest mechanisms of this type are autocatalytic chemical cycles: The process itself can be considered a character that is replicated with each turn, or the character of interest may be a product, yielded with each iteration of the process. Novelty is introduced by any persistent (lasting) change in the reiterating process.

Alternatively, previously produced copies may play an important role in the process. Griesemer (2000, in press) has pointed out that all cases of biological reproduction involve what he calls progeneration, or material overlap between an original and its offspring (Fig. 3A). Replication of DNA is semiconservative: both daughter molecules incorporate a strand from their parent. The same is true of chromosomes. Membrane-bound organelles reproduce by splitting in two. Cellular contents are distributed at cell division. New multicellar organisms develop from single cells that were produced by their parents or as ramets (Harper, 1977) that may arise from even larger fragments.

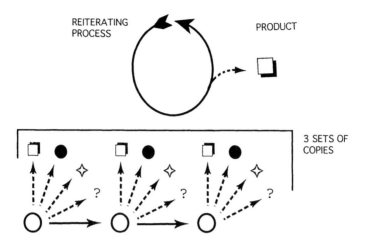

FIGURE 2 "Replication" without an original. Production of multiple copies can occur through the reiteration of a process, shown in this diagram with arrows. Copies may or may not themselves participate in the process. In simplest form, such a mechanism does not involve a copy further in subsequent replication.

Where progeneration is involved, novel traits in the parental copy or original can be directly incorporated into the offspring. Thus those parts of the offspring in Fig. 3A shown in solid outline relate to the parent as do the copies of Fig. 1A to one another. However, direct transmission of material between parent and offspring does not create new copies until the missing parts are reconstituted: In Griesemer's terms (Griesemer, 2000, in press), biological reproduction consists of progeneration and development, with development being the set of processes that must occur before the copy is itself capable of reproduction.

A character is said to be replicated only if its numbers are thereby increased ("[my daughter] did not inherit my blue eyes; I still have them"; Cartmill, 1994, p.118) and if the resulting copies bear some sort of one-to-one correspondence (which will be discussed later) to one another. Some mechanism must be employed to construct or reconstruct any information that is missing once the parental material is distributed.

Use of a template is one mechanism that assures such correspondence or fidelity among copies. "Thus each daughter cell inherits from its mother a complete set of specialized cell membranes. This inheritance is essential because a cell could not make such membranes *de novo,*" as one popular cell biology textbook (Alberts *et al.*, 1994, p. 560) states it.[2] Growth and formation of additional membranes are to some extent self-organizing and are guided by the structure of an existing membrane.

---

[2] A functional consequence of this continuity of membranes is the maintenance of chemical gradients across them. Particular differences between the milieux of inside and outside are also thereby maintained, without having to be reconstituted repeatedly (G. P. Wagner, pers. comm.)

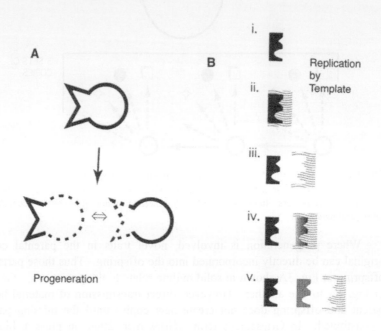

FIGURE 3 (A)   A copying process involving transfer of material from parent to offspring, or "progeneration" (sensu Griesemer, 2000, in press).   Portions of the parent that are transmitted directly to the offspring are shown in solid outline; missing parts, shown by dotted outline, must be reconstituted.   (B). The sequence i - v shows use of a template (hashed) to produce a copy (filled) in one-to-one correspondence to an original (black).

Successive copies that are produced with a single reusable template (Fig. 4, v-vi) will be related to each other as in Fig. 1B, by a mechanism that can be represented as in Fig. 2A.   If the template is remolded by the copy or formed anew on it, however (Fig. 4, v-ix'), the situation is like Fig. 1A.

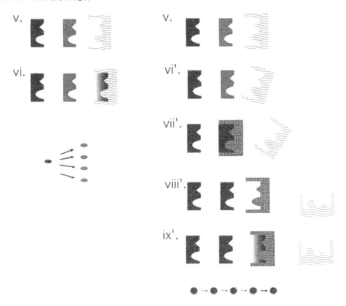

FIGURE 4 Continuation of the copying process shown in Fig. 3B. If the template is reused and unaltered by the previous copying process (vi), successive copies will be reiteratively related (as in Fig.1B). If a new template is formed off the copy, or if the old template is remolded by the copy during the copying process (vi' - ix'), copies will be lineal descendants (Fig. 1A).

THE INDIRECTNESS OF REPLICATION

DNA replication is the classic example of the biological use of templates, its fidelity readily understood as the result of base-pair matching. The one-to-one correspondence between matching purines and pyrimidines in nucleic acid replication may be what caused Dawkins (1976) initially to consider genes to be the only true replicators, and to consider all other features of the organism components of a "vehicle," evolved through selection by virtue of their roles in enhancing the replication of genes. Importantly, however, the process of DNA replication requires not just a molecule with the ability to act as a template, but also the materials and conditions provided in a living cell (or in a functioning thermal cycler, or in the primal soups, sandwiches, or surfaces involved in the origin of life). The success and fidelity of replication depend upon these auxiliary factors. Conversely, novelty is introduced, and the fidelity of replication is reduced, not only by changes in the template, but also by changes in anything that affects the copy-production process. An outline of the processes that introduce novelty or maintain fidelity in replication for various levels of biological organization is shown in Table I (reproduced from Roth, 1994). A key point is that, as simple in principle as information transfer is between complementary strands of nucleic acids, even DNA replication (in the present day at least) is not simple, self-contained, or independently self-catalyzing.

TABLE 1   Some entities that form lineages and that in appropriate circumstances can serve as special replicators[a]

Entities whose replication forms lineages	Intervening processes (in which variation may be introduced)[b]	Sources of maintenance of fidelity[c]
DNA	DNA replication (uncoiling, initiation, polymerase binding, ligasing, etc.)	use of a parental DNA template in a semi conservative process, DNA repair mechanisms
cells	processes occurring at any phase of the cell cycle	mechanisms that assure accurate replication and segregation of organelles and redistribution of cytoplasm (using, e.g., DNA repair enzymes, spindle microtubules, etc.)
organisms	development	regulatory processes of development (involving induction, hormonal control, growth factors, etc.)
populations	anagenesis (continuation of population through generations) cladogenesis (splitting into more than one population lineage)	mate-recognition systems (may keep population intact or contribute to its propensity to speciate)

[a] The list is not exhaustive.
[b] Processes that intervene, between bouts of replication, and that are involved in the material manifestation of the information that is replicated.   If the lineage-forming entity is a special replicator, the variation is potentially heritable.
[c] Mechanisms for maintaining fidelity (these might be viewed as developmental constraints); ultimately stabilizing selection, as an overarching process, will also contribute to maintaining these sources of fidelity.

Life presumably originated with replicating molecules, but in the present day even the processes of molecular replication are fairly elaborate. To take an example mentioned earlier, the structure of an existing cell membrane provides a necessary scaffolding for growth and addition of a new membrane. However, the lipid and protein constituents of the new membrane are constructed not from existing lipids or proteins in that membrane, but instead through metabolic processes taking place elsewhere in the cell. Copies of membrane molecules are not direct copies of one another; they are reiterated products of a machinery that itself is produced, maintained, and reproduced quite separately from those products.

For many characteristics replication occurs not through the direct transfer of portions of the original, and not through direct interaction between original and replica. Instead, copies are generated through repetition of a mechanism (Fig. 2). Information about a character is embodied in, or evoked by, something other than the character itself; that is, it is coded (Griesemer, 2000, in press; Maynard Smith and Szathmáry, 1995). Where coding is involved, an exemplar of a character is unnecessary for its replication, as long as material that confers an ability to produce it is preserved or transmitted, along with the machinery to effect translation. Some of this machinery may also be transmitted in coded form, and must itself undergo construction; this is the case increasingly across the spectrum of increasing complexity of organismal development, from replicating compartments of molecules to reproducing organisms of complicated structure. As a consequence, with highly encoded development, objects as compact as single spores or zygotes can give rise to entire multicellular organisms with different cell types organized into body parts and a body plan.

Replication that employs a template, with its alternating production of part and counterpart, may be seen as indirect in the sense that the counterpart serves as an intermediary in the process (Fig. 3B). Part way through the replication process, the original can disappear or cease to exist without affecting subsequent production or expression of the copy. (Yet because the original initially plays an influential role in such replication – standing, in the chain of production, in linear series with subsequent copies, changes to the original can propagate to its copies.) By comparison, however, the replication of some characters of the gross phenotypes of multicellular organisms is stunningly indirect.

I will digress here for a moment to comment on the distinction made by Dawkins (1982, p.83) between "germ-line" and "dead-end" replicators: "a germ-line replicator is potentially the ancestor of an indefinitely long line of descendant replicators. A gene in a gamete [as opposed to a somatic cell] is a germ-line replicator. So is a gene in one of the germ-line [as opposed to somatic] cells of the body, a direct mitotic ancestor of a gamete."

Only when it appears in germ-line replicators is a novelty propagated to other generations, with potentially long-term impact on evolution, and this is the source of their significance. However, this distinction does not on its own allow a full account of the manner in which novelty is introduced during replication.

Dawkins stressed that he applied the distinction of germ-line replicator to entities that (through replication) were potentially immortal; in practice, a germ-line replicator could fail, becoming terminal, without losing its status as germ-line. Emphasizing potential rather than actual roles presumably allowed Dawkins to generalize and make statements about the types of things capable of participating in processes such as natural selection. My specific interest here, however, is in the process of copying, and in locating sources of novelty and how they are transmitted; this requires a closer look at precisely where and when changes are introduced.

If we focus on where novelty arises in a copying process we can apply something like the germ-line/dead-end distinction to points discussed earlier in this section: The relationship shown in Fig. 1A is that of a germ-line replicator. In Fig. 2, the reiterating cycle can be seen as germ-line, and the product is dead-end if it is not itself involved in further cycles; the hollow circles linked by solid arrows are germ-line, while objects at the ends of dashed arrows are dead ends. In Fig. 3, the parts drawn with solid outlines, which are distributed from parent to offspring through progeneration, are germ-line; the reconstituted parts (shown dashed) may also serve in the germ-line, during the next bout of progeneration. Often, a code will be germ-line; its associated characters are not. A reusable template (Figs. 3 and 4) is germ-line, as was the copy on which it was originally molded. However, after the process of molding, the original becomes dead-end. If the template is remolded at each step (Fig. 4, vi'-ix'), the previous template also becomes dead-end. Thus, with respect to the introduction of novelty into descendants, it is not the objects themselves that carry the property of being dead-end or germ-line, but rather where they stand at any particular time in the causal chain of processes. Anything involved in a copying process—whether it is conventionally viewed as a germ-line entity or not—can introduce novel features into the copies, and any durable change in the process of replication can be a source of novelty.

Such durable change is often the result of a change in another structure that assists in the copying process and which itself has a distinct codical basis (germ line). So, for instance, we can attribute a lasting (heritable) change in the structure of a cell membrane to many sources, including the code for the amino acids incorporated in its proteins, the codes for polypeptides in the enzymes involved in assembly of those proteins or any other participating molecules, the codes for the polypeptides in the enzymes and proteins involved in assembling other enzymes and in genetic regulatory mechanisms, the conformation of preexisting membrane in which additional molecules must insert or attach (and which itself is affected by its constituent molecules), and so forth. A process of copying that appears at one level of description to be linear is on closer examination revealed to contain branches, anastomoses, reticulations, and feedback loops.

Coding provides an explanation for some apparent paradoxes: A lineage is by definition continuous, yet the characters we use to infer lineal continuity are expressed discontinuously. A character that has disappeared can nevertheless be copied. When a trait due to a recessive allele disappears for multiple generations, its later reappearance is attributed to persistence (through replication) of the code embodied in its DNA. The expression (or not) of a trait can be understood in terms of the regulation of gene expression – the state of a gene as reflected in its various chromatin conformations being what Jablonka and Lamb (1995) refer to as a "gene's phenotype." Jablonka and Lamb identify three sorts of cell memory systems other than nucleic-acid base sequences that they refer to as "epigenetic inheritance systems," by which variations are inherited through multiple generations of cells: self-perpetuating metabolic patterns, inherited structures, and chromatin marking. Such systems "were probably pre-requisites for the evolution of complex multicellular organisms" because they "maintain the determined state of cell lineages and allow different lineages to retain their specificity in spite of cell turnover" (Jablonka and Lamb, 1995, p. 199).

But multicellular organisms also have observable features at a scale greater than that of a single cell and its contents.

HIERARCHY

Buss (1987) and Maynard Smith and Szathmáry (1995) have identified as major evolutionary transitions the origin of "larger wholes" that are capable of reproduction. These larger wholes—chromosomes, eukaryotic cells, sexual populations, and multicellular organisms—incorporate entities previously capable of independent replication, which now replicate only as part of the larger whole. Thus polymerized nucleic acids, which presumably once replicated independently, can replicate only as part of a linked set known as a chromosome. Mitochondria and chloroplasts, which are descended from free-living prokaryotes, can only replicate with and within the eukaryotic cells that are their hosts. Asexual organisms form lineages of individuals, but today most eukaryotes can replicate only as part of a sexual population. Cells of multicellular organisms do not live freely or reproduce independently in the same manner as their protistan ancestors. How can a character formed by the participation of multiple cells be replicated?

Some features of gross morphology develop as mosaics, organized simply as aggregates of their constituent cells. Cell lineages can be identified in nematodes and other aschelminthes that give rise to specific organs, and in the extreme, development of the body plan emerges in a rigidly determined sequence of cell divisions, through progressive differentiation of independent cell lineages.

There are other situations, however (even in nematodes; Goldstein, 1992), in which mechanisms that stabilize form have evolved at a scale larger than single cells and their lineages. Here, morphogenesis is not a matter of lineages of cells differentiating autonomously, reproducing a character in successive generations through a point-for-point recapitulation. Rather, characters of gross morphology are contingent on cellular interactions and on the mechanical properties and spatial relationships of groups of cells. The components of a single character are not "monophyletic" with respect to lineages of cells. A cell's fate is influenced by other, separately differentiated ("non-monophyletic") groups of cells that provide feedback in the form of chemical or electrical cues or the transmission of mechanical forces. In the development of the eyes in many vertebrates, for example, the proper functional, spatial relationship between lens (which is derived from epidermal ectoderm) and retina (derived from the optic cup, an outgrowth of neurectoderm) is assured by a parsimonious mechanism: the former is induced by the latter. Thus irrespective of what particular cells are involved or from what cell lineages they arose, a lens normally forms in the epidermis only in proximity to a presumptive retina.

Moreover, there are many examples, now widely cited, of homologous characters whose phylogenetic continuity of expression at a level of gross morphological organization is not matched by similarities in other domains: Patterns of gene expression are not reliable indicators of homology in gross morphology; characters in two taxa may be homologous despite differences in their embryological precursors, patterning mechanisms, or sequences of development (de Beer, 1971; Sander, 1983; Roth, 1988; Wagner, 1989a; Abouheif et al., 1997). How can a character show phylogenetic continuity and be reproduced with fidelity through the generations, despite fundamental changes in the processes that generate it?

To describe the phylogenetic decoupling of phenotypic characters from their associated patterns of gene expression and development, I have borrowed language used in discussions of levels of selection (Roth, 1991, 1994); in particular, the idea of "screening off." Brandon (1990) used this notion (introduced by Salmon, 1971) to identify for any specified example the level—gene, organism, group, species, avatar, or clade—at which the process of natural selection is occurring. If A and B are putative causes (at two different levels) of effect E, we say that A screens off B from E, if A renders B statistically irrelevant with respect to outcome E, but not vice versa. As Brandon (1990, p. 84) argued, in standard cases of organismic selection (his example was directional selection for height of a plant), "manipulating the phenotype without changing any aspect of the genotype can affect reproductive success," whereas "tampering with the genotype without changing any aspect of the phenotype cannot." Differential reproduction of tall plants "is best explained in terms of differences in organismic phenotypes, because phenotypes screen off both genotypes and genes from the reproductive success of organisms" (Brandon, 1990, p. 85).

In place of the effect "reproductive success of organisms," consider instead the effect "faithful replication through history" or "faithful copying." From the evidence given by de Beer (1971) and others (in sources listed earlier), it appears that for some sets of homologous characters, the phenotype's utility and its ability to develop through multiple alternative pathways (together these serve as "A" in the notation just described) screen off the particular genetic and developmental processes involved in its production ("B") from the process of its faithful replication through generations of history ("E").

An example will make this statement concrete: take gastrulation. Possession of a gut (cavity or tube) is a synapomorphy shared (or possessed historically, but lost) by all Eumetazoa, and the homology of this character within the Chordata is undoubted. Yet textbook accounts of gastrulation (e.g., Balinsky, 1970; Gilbert, 1997) typically require separate descriptions for each of several model systems (zebrafish, *Xenopus*, chick, mouse) and even "the study of amphibian gastrulation has been complicated by the fact that there is no single way that all amphibians gastrulate. Different species employ different means toward the same goal" (Gilbert, 1997, p. 221). Gastrulation may begin with epiboly, invagination, or delamination and the migration of individual cells; a stage in which cells that form the notochord are integrated into the endoderm of the primitive gut may (in mouse) or may not (in chick) exist. Clearly, although copies of an observable feature called the gut have been manifest faithfully generation after generation, through divergent lineages over the course of a half-billion years, there have been novel features introduced into the mechanism for producing it. What sort of mechanism could maintain the form of a character when the very process that generates it is altered? Textbooks suggest that the divergent modes of gastrulation are associated with differences in the volume of yolk in the egg. Perhaps it is no coincidence that vocabulary drawn from discussions of natural selection has proved useful for discussion of homology: perhaps selection, acting on phenotypes, is what has ensured their stable replication on a phylogenetic time scale, irrespective of what mechanical impediments (such as a large volume of yolk) are present or what developmental mechanisms or materials are used.

This is a very natural conclusion from one perspective: why shouldn't pathways and mechanisms of development evolve, just like any other character? And why should not their evolutionary preservation or maintenance be dissociated, on occasion—through piracy of gene products, changes in inductive stimuli, introduction of physical barriers, what have you—from the preservation or maintenance of their ultimate products? It should not be surprising that mechanisms of development appear "opportunistic," in the sense of incorporating (with that incorporation becoming fixed in a population though selection or drift) novel features that do not disrupt the building of a character of gross phenotype. Regardless of other changes, the ultimate product (say, an internal, epithelium-lined tube that opens to the outside at both ends and is used in digestion and nutrient absorption) will of course be maintained by selection (that is, differential reproduction of the organisms that possess it), if it is useful.

From another perspective, however—one in which genes and the genome are seen as encoding information about characters—this conclusion is an unusual one. The implication of the argument from screening off is that, for the long-term replication of a character, the utility and developmental robustness of the phenotype may be causally more relevant to its faithful replication than the particular genes or developmental mechanisms that produce it, because selection acting on that phenotype is what maintains it. But if homology is a manifestation of continuity of information (Van Valen, 1982), we must ask where that information resides. Clearly, genealogical continuity—that is, continuity at the level of organismal reproduction, through generations and lineages of organisms—is essential, whether or not the same genes remain involved. Intriguingly, it appears that on some macroevolutionary time scales, it is genealogical but not genetic continuity, combined with consistency in the selective environment, and robustness (i.e., existence of a multiplicity of routes to the same end) of development, that has allowed some characters to be reliably replicated.

Let me clarify this statement about the roles of genetics, genealogy, and selection. Overall historical continuity in developmental mechanics is still required. By overall continuity I mean that while stepwise substitution of some elements of the process occurs, correspondence remains for other elements. Imagine a developmental mechanism that generates character X, with elements $y, e, s \rightarrow X$ (the arrow signifies "yields"). For an evolutionary transformation sequence of $y, e, s \rightarrow X$; $y, e, o \rightarrow X$; $n, e, o \rightarrow X$; $n, o \rightarrow X$, considering X in the first $(y, e, s \rightarrow X)$ and last $(n, o \rightarrow X)$ instance to be homologous would be justified, even though nothing (other than their ability ultimately to produce the same end product) remains common to their developmental mechanisms. Absent the fact that the intermediate stages $(y, e, o \rightarrow X$; $n, e, o \rightarrow X)$ existed, however, a conclusion of homology would be unwarranted.[3]

Picture, for example, an ancestral predatory animal with claws on its digits. By some unlikely scenario it gives rise spontaneously to a descendant with teeth in their place. Claws and teeth are both hard sharp body parts, so one could hypothesize that a character homologous at this level of description has been maintained. Although (1) the digital teeth and claws are linked genealogically by virtue of appearing in ancestral and descendant organisms and (2) they may be selected to serve the same function (say, manual capture of prey), they would not be homologous characters because there is no phylogenetic continuity of development in an immediate and wholesale substitution of a tooth for a claw. A related line of reasoning (Nilsson, 1996; Bolker and Raff, 1996; Abouheif et al., 1997) causes one to reject the homology of fly and mouse eyes even if there

---

[3] McKittrick (1995: 5) cites the story of the Tin Woodsman in "The Wizard of Oz", who "began life as a flesh and blood human, and successive accidents with his axe led to successive replacement of human parts with tin parts." The Tin Woodsman "was still a historical individual at the end of this ordeal, however, even though his parts were different."

is homology between the genetic switches used to initiate the development of each (Quiring *et al.*, 1994; Zuker, 1994; Halder *et al.*, 1995).

The implication of the notation I have used is that **n** and **o** are not modifications of **y**, **e**, or **s**, but entirely different elements. In a real biological example, introduction of **n** and **o** into the process would presumably be possible because their genetic precursors existed elsewhere in the same organisms, although they were not previously involved in producing **X**; hence the importance of genealogical continuity at the level of the whole organism.

TREES

The conclusion that selection plus genealogical (but not necessarily genetic) continuity provide the biological basis of homology has important implications for the practice of phylogeny reconstruction, and in particular, for the utility of gross phenotypic characters and the phylogenetic information they contain. Continuity of expression is absent for gross phenotypic characters; nor do they form treelike lineages.   Yet gross phenotypic characters have been extraordinarily useful for phylogenetic inference.   It is instructive to look closely at patterns of replication in entities that form treelike lineages, in order to examine their relationship to replicating characters that do not.

The usual procedure in phylogenetic reconstruction is to sample one or more individual organisms in different taxa for characters whose distribution is expected to be phylogenetically informative.   For cladistic parsimony methods, phylogenetic information is provided by characters whose presence in sister taxa is hypothesized to be due to their inheritance from a common ancestor, rather than a separate origin in each lineage (Fig. 5).   With likelihood methods, explicit models of character-state change are employed to specify probabilities of character persistence or change, so, with sufficiently rapid rates or high probabilities of character-state change relative to branching events, the scenario shown at the right in Fig. 5 may be more likely than the one inferred in the middle of Fig. 5.   In either case, characters are used as markers of lineages and clades and are assumed to provide evidence of evolutionary events (the origin or change of a trait) in the shared histories of divergent lineages.

The logic underlying the kind of inference represented in Fig. 5 is straightforward enough, but of course the relationship between lineages and the entities that are actually compared in a systematic study are not correctly represented by such diagrams.   The trees reconstructed in most phylogenetic work are intended to represent histories of taxa.   Any point on one of the trees shown in Fig. 5 corresponds to an entire population.   The characters mapped onto or used to construct the trees, however, are features observed in individual organisms.

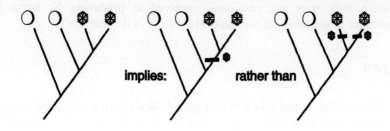

FIGURE 5   "Tree-reasoning" using cladistic parsimony methods. With different assumptions about the rate or probability of character-state change relative to branching frequency, the scenario shown at the right may be the preferred interpretation.

I suggested earlier (Roth, 1991) that the appropriate population-level characters for which the scoring of organismal traits serves as a surrogate are not the presence or absence of a trait in individual organisms, but rather the fixation or complete loss of the character in the population as a whole. It is well known that gene or character trees are not necessarily good reflections of the history of the taxa that contain them, and molecular systematists especially are conscious that even different parts of a single genome may have experienced different phylogenetic histories (Pamilo and Nei, 1988; reviews in Doyle, 1997; Maddison, 1997). It is of interest to look more closely at lineages of entities that replicate at different levels of organization.

We can start with the history of populations or species as represented by a tree (Fig. 5) – a set of linear segments and nodes, representing replication of the population through time, and splitting of the population in two, respectively. "Tree-reasoning"—as shown in Fig. 5, as represented by Patterson's (1981) criteria for homology, and as applied in Hennigian phylogenetic inference— may apply to objects with this type of topology, which I've called "special replicators" (Roth, 1994), but it does not necessarily apply to others. In a tree, a novelty introduced at one point is propagated to points distal (all descendants) until it is superceded by another novelty; but the propagation of new characters in other kinds of networks is more complex.

Bringing the species tree under higher magnification, we see a population of individual organisms making up each point in a lineage segment. For sexual populations, reproduction at this next level (of individual organisms) is more complex than a linear series, as each offspring is the product of two parents, not one; a parent can pair with more than one other individual in producing different offspring; and individuals are linked together in a meshwork ("tokogeny" in Hennig's 1966 terminology) through the generations, within the boundaries defined by the species-level lineages. The propagation of novel characters through such a meshwork is much less predictable than through a tree, for three reasons:

1. Connections not only diverge, they anastomose.
2. Each link between parental individual and offspring represents (if we discount maternal effects and consider only the nuclear genome) 50%, not all, of the information available from the parent.
3. Which half of the parental information an offspring receives is a matter of chance, and the rules governing the division of information in two (during meiosis) to which the 50% probability is applied are intricate. It is not simply a random sampling of half of the base-pair sites contained in a parental cell's (nuclear) genome, because the genome is physically structured: base-pair sites are themselves paired, with sampling occurring between the two members of each pair. Yet because there is linkage (on chromosomes) and because degrees of linkage vary among sites (because of crossing-over), sites are not sampled independently, and the joint probabilities of sampling particular sets of members of the pairs in the next generation will vary.

As Maddison (1997, pp. 523-524) commented, "Sexual reproduction and recombination within populations may appear to but actually do not cause genetic history to be reticulating." This is an accurate statement if "genetic history" is read to mean "the history of linked sets of nucleotides." But organismal genealogies are reticulate, and in this context sex causes the genetic history of phenotypic characters to be reticulate also, where genetic influences on a phenotype involve interacting loci or groups of sites that recombine.

For linked nucleotides, as Maddison further suggested, sex "break[s] up the genomic history into many small pieces, each of which has a strictly treelike pattern of descent." It may improve (though not simplify) visualization to think instead of the history of a genome not as "broken up" by sex but rather as a bundle or skein of what are many trees (perhaps one for every nucleotide site) to begin with.

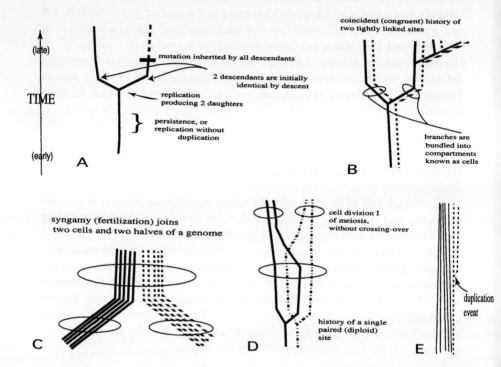

FIGURE 6 Trees representing the histories of individual nucleotide sites: persistence, replication, and the grouping (or bundling or packaging together) of nucleotide sites with other sites, within cells and organisms. A. Persistence (or replication without duplication) of a site, followed by a branching event: replication that produces two daughter lineages. A mutant form serves as a cladistic marker of shared evolutionary history. B. After replication produces branching, branches are separately bundled together into compartments known as cells, which are grouped into even larger wholes known as multicellular organisms. Except in events of polyploidy, branching immediately precedes cell division, when branches are subsequently bundled into different cells. Meiosis involves an elaborate sorting and segregation of branches into separate bundles (D). E. The history of the genome of a single cell showing an event of duplication at a single site.

A branch in the tree is an event of replication that produces more than one descendant (Fig. 6A). Where the genome is replicated altogether in its entirety, through cell lineages within an organism, we can imagine a bundle of trees (each tree representing the history of a single nucleotide site), all branching congruently (segments, nodes, and the numbers and placement of branches coincide, running parallel and tracking together throughout much of their length) – their congruence reflecting their shared history (Fig. 6B).

A nucleotide tree also branches with gene replication at meiosis, though here the association of branches among different trees is complex, following the probabilistic rules described in #3:

First, all trees branch. Keep in mind that within this metaphor, each tree represents the history of a single site. In diploid organisms, where each tree is paired with another, this initial branching process yields two parallel sets of bifurcations, identifiable as four distinct branches (Fig. 6D). Of these four, branches joined at a node represent sites duplicated with the most recent chromosomal replication; branches not joined in this way are corresponding sites on different members of a pair of chromosomes. With meiotic cell divisions, the large bundle of trees representing the duplicated genome of an individual organism is twice peeled apart longitudinally into two (picture a stalk of celery, whose individual strands represent segments of single trees). Cell division I of meiosis bundles together half (i.e., two) of the branches of each pair of trees, typically keeping together branches joined at the most recent node; but if crossing-over occurs, individual strands are traded across bundles, crossing over to join strands of the opposite tree to become braided into a different bundle. Cell division II of meiosis subdivides the two bundles just created into two again, this time separating not only members of a pair of trees, but also carefully segregating each of the two strands joined as branches at the most recent node. Thus after meiosis II each resulting bundle (cell; gamete) contains single strands corresponding to each site; other members of the bundle are segments of trees representing other sites that together constitute a haploid genome.

Fertilization (Fig. 6C) brings two such bundles of tree segments together, to share future history. With subsequent cell cleavage and development into a multicellular organism these tree segments replicate by branching in concert through lineages of cells (as described earlier, immediately preceding the description of meiosis) for an organismal generation. Among groups of trees representing linked nucleotide sites, history through many generations (of bundling, branching, rebundling, and braiding) will be shared, so segments and nodes will coincide, running parallel and bundled into cells together for much of their length (Fig. 6B). Events of within-organism gene duplication (paralogy), which have had an important role in the history of life (Ohno, 1970), can be represented by a set of adventitious branching events affecting a subset of the trees in the genome (Fig. 6E). This creates additional tree branches that are incorporated into the same bundle.

All this is basic biology well understood by any biologically informed reader, but I labor through this extended metaphor in order to highlight some points that are easily overlooked. The metaphor serves to make vivid the fact that (1) the history of individual nucleotide sites of replicating DNA both within and between organisms, sexual and asexual, is indeed strictly treelike; and (2) the notion of "shared history" requires attention both to the structure of single trees and to their grouping into cells and organisms. Coalescence diagrams (e.g., Hudson, 1990, Figs. 3, 4; Hein, 1993, Figs. 2-4) often fail to represent as distinct the difference between a branching event (a replication producing two daughters) and bundling (segregation of two distinct parts of a single genome into different cells or chromosomes), although this difference is obviously recognized and well understood by the authors who model the processes of recombination and coalescence. Where the history of recombining loci is conveyed in such published diagrams, it may not be represented by the topology and grouping of sets of trees, but instead may be shown as a network of individual segments of identical width, accompanied at each segment and node by another small diagram showing the resulting linkage relationships of relevant alleles. Accordingly, because replication and repackaging (or rebundling, if we use the visual metaphor of tree segments) are not clearly distinguished (both are drawn as branching events), it is easy to conflate or ignore the different roles they play in (a) the origin and propagation of novelty (through lineages) and (b) allowing interactions (dominance, epistasis) among loci within the common environment of a cell.

With regard to replication and repackaging, the one often accompanies the other: replication (shown as a branching event in my scheme) commonly occurs just before the two branches become separated into different cells. However, it is one thing to share a history as lineal descendants—where the source of novelty is mutation, transmitted through subsequent generations by replication (Fig. 1A)—and quite another to share time together temporarily in the same cell or organism—where physiological interactions occur and create characters at a higher level of organization, and where birth, death, or failure to contribute to subsequent generations apply to the entire assemblage as a unit.

Discussions of the relationships between "gene trees" and "species trees" in phylogenetic reconstruction often overlook the organizational units of cells and organisms, and compare the congruence of trees without overt reference to the temporal series of entities (individual organisms) through which they must pass. But because the fate of a gene is tied for a time to that of other genes in the same organism, a full account of the relationship between "gene trees" and "species trees" must also make explicit reference to how genes are packaged into cells and organisms. It is a whole organism, spore, or gamete that lives or dies. Interactions among loci within a cell or organism affect the ability of the alleles at all loci to persist and replicate into subsequent generations: despite recombination, in standard organismal reproduction alleles persist and are transmitted only within the context of a diploid or haploid (or polyploid) genome, with successful persistence and replication depending upon organismal phenotypes. Thus understanding the history of a particular gene or sets of genes calls upon an understanding of the relationship between the replication of genes and the transmission of phenotypic characters through the generations (in a word – development, including the history of an organism throughout its life cycle). Tracing this is a difficult undertaking (Doyle, 1996), requiring as a first step the elaboration of appropriate visual and conceptual representations to superimpose organisms and their phenotypic characters on gene trees. As Doyle (1997:547) pointed out, "genes and morphological characters belong to different organizational rather than hierarchical levels: genotype and phenotype."

It is helpful to reiterate what is meant here metaphorically by a single tree. I suggest a single nucleotide site because it is indivisible, and its duplication is easily represented by a branch, its deletion by truncation, transposition by a rebundling together with a different set of trees, and modification by a heritable mark that propagates distally. Concerted evolution is a challenge to the metaphor, but is perhaps best conceived as simultaneous and identical modification of a subset of bundled loci. Loci that are linked of course travel as a bundle together and branch simultaneously for a very long time – so long that for a while (a portion of their length) they may look like a single thick tree. Such thicker trees—what Doyle (1995) refers to as c-genes—have a duration (a length) that is inversely related to the number of individual nucleotides they comprise (their thickness) (Williams, 1966). It is recombination that eventually separates and rebundles some of the component strands of these thick, composite trees, pulling the strands out in groups, and exchanging them for other groups while braiding them into other bundles.

As detailed earlier, the rules that govern distribution, combination, and recombination of individual-site trees are intricate, and they have two consequences: *variation* among individual organisms in the particular sets of gene trees that pass through them, and a *precise allocation* of numbers and kinds of trees distributed into individual organisms each generation, so that ordinarily each organism includes a full complement of sites in its genome. The precision assures that, despite variation, the necessary complement of genetic material is available to each organism to influence the expression of characters that are observed at a level of gross phenotype. The reliability of the process of meiosis thus plays a central role in the capability of characters of gross phenotype to be replicated.

In his "building block hypothesis," Wagner (1995) noted that any given genetic locus or any given character (each separately at its own respective level) may (1) serve a unique functional role; (2) exist as variants that compete with one another for representation in a population (e.g., alleles compete at loci, and likewise characters find alternative expression as different character states); and (3) retain its own unique contribution to organismal performance when put in combination with other such entities (i.e., other loci or additional characters).

By virtue of these properties, Wagner argues, loci and characters are "building blocks" because their performance can be optimized individually by selection among their variants, and evolving in this fashion ultimately contributes to enhanced performance of the system as a whole.

In this framework (of loci serving as building blocks), meiosis would have an important bookkeeping function within a genome, alleles being the entities that compete for representation at a locus, and meiosis maintaining (in the short term, at least) the number of loci constant. For phenotypic characters, self-regulatory mechanisms that maintain the integrity and distinctness of character modules during development presumably play an analogous role – though they are more complicated and at present less well characterized than meiosis.

Novel nucleotides (or entire alleles) typically enter a population by substituting for a previously existing allele (Wagner's criterion #2), and they succeed in entering the population only if their introduction does not substantially disrupt organismal function (criteria #1 and #3). These are properties of interchangeability and combinability, and because of them individual alleles and variants of characters commonly show patterns of inheritance that are haphazard with respect to the phylogenetic history of an entire lineage. However, variants may more reliably track the history of population-level lineages if they become fixed in the population and, in doing so, acquire a unique functional role.

For sexual taxa, the structural levels at which clearly treelike lineages are observed are (a) lineages of diverging, reproductively isolated populations; (b) lineages of asexual organisms; (c) lineages of cells (and their subcellular components); and (d) linked sets of nucleotide sites. As has been noted earlier, gross phenotypic characters exhibit neither continuity of expression nor treelike lineages, and genealogies of sexual organisms are reticulate, not treelike. Thus the success of the long tradition of applying "tree-reasoning" (as illustrated in Fig. 5) to phenotypic characters of individual organisms in inferring the evolutionary history of population/species lineages is impressive, and exceeds any reasonable expectation. The measure of this success is the striking consistency and congruence of phylogenetic histories that are inferred from different sets of data, widely publicized exceptions notwithstanding. Despite the attention that conflicts on the placement of particular branches have received in specific cases ("The guinea pig is/is not a rodent": *cf* Graur, Hide, and Li, 1991; Luckett and Hartenberger, 1993; "The platypus is not a rodent": Kirsch and Mayer, 1998), it is on the whole fair to say that there is more pattern than chaos in the distributions of character states, given the huge number of combinatorial alternatives.

What accounts for this success? The foregoing arguments suggest that two processes, meiosis and natural selection, are important. The precise allocation of genetic material that takes place in meiosis sustains the close relationships between the histories of genes and the histories of phenotypic characters, in the short term. Natural selection maintains the stability of characters through the long term, permitting a match between the histories of characters and the histories of population-lineages.

New variants can spread within, but (by definition) not between reproductively isolated population-lineages, so the history of populations constrains the spread of variants that arise at lower levels. In practice, characters are observed in individual organisms that are assumed to represent the population as a whole. This assumption may not be warranted because maintenance of ancestral polymorphisms through splitting events in the history of a lineage has the potential to confound cladistic conclusions. However, it is in a fairly restricted set of circumstances that this sort of process will cause character-state distributions to be positively misleading (Tajima, 1983; Pamilo and Nei, 1988; Neigel and Avise, 1986; Roth, 1991), and incorrect inferences can be reduced by the use of multiple independent yet congruently varying characters. It must further be acknowledged that not just any arbitrarily-chosen character will serve in phylogenetic reconstruction. For reasons already described, the characters used for inferring phylogenies are typically *not* those observed to behave in Mendelian fashion within the populations whose history is being inferred.

As it happens, the process of sexual recombination has both integrative and disruptive effects, both of which can be used to advantage in the choosing of characters and interpretion of their distributions: On the one hand, the separation, shuffling, and merging of variant nucleotides among individuals serve to integrate a sexual population into a lineage at a higher level by linking organisms together in a network of gene exchange. Genetic recombination also quickly reveals which characters are too labile (with respect to this shuffling) to be used in phylogenetic analysis.

Studies of deep phylogeny have been (well) informed by nucleotide or amino acid sequences that change very slowly on a macroevolutionary time scale (constrained to infrequent substitution and rapid fixation, presumably, by selection). However, the morphological characters that are most revealing of ancient divergences—such as for metazoans, overall body plans—typically are homologous at the level of the gross phenotype, which is maintained through history, but not necessarily at the level of the particular combination of alleles required for its expression, which can change.

The production of characters of gross phenotype involves networks and nonlinear interactions of gene products, and properties affecting processes at the level of gross or histological anatomy (such as, for instance, the quantity or the positional relations of groups of cells competent to respond to induction) that "screen off" the effects of single alleles. Smith and Schneider (1998), for example, point to important flaws of argument and anatomical interpretation in several recent studies of gene knockout experiments and double null mutations in mice that recent workers had concluded were atavisms, or evolutionary reversals of the mammalian to a reptilian condition. For analyses of the morphogenesis of complex characters, "...a gene knockout or misexpression experiment explains less about gene function per se, and more about the conditions and materials necessary to maintain the normal course of embryonic events" (Schneider, 1998, p.140).

Schneider (1998) has also examined the effects of gene disruptions in mice on the morphological organization of the sphenoid, a region of the skull many of whose anatomical relations have remained largely unchanged for hundreds of millions of years. Here he found that abnormalities resulting from gene disruptions do not segregate according to modules of skull organization defined by comparative anatomy and embryology (in particular, the division between prechordal and chordal regions of the head; or between dermato-, neuro-, and viscerocranial elements). This finding suggests more integration than segregation among components in their reliance on particular gene products for normal development. It also points to an important role for what he termed "secondary" effects – disturbances propagated to structures neighboring those in which a gene is expressed.

One scheme for partitioning the head skeleton into modules, that based on the embryonic source of skeletal tissue, *was* concordant with the pattern of abnormalities resulting from genetic perturbations: "However, while many genes affect both neural crest and mesodermal derivatives, there appears to be a suite of genes that may play a role only in the neural crest part of the head" (Schneider, 1998, p.152), whereas none were observed that affected only mesoderm. These observations, as he pointed out, are consistent with the hypothesis of Gans and Northcutt (1983; see also Northcutt and Gans, 1983) that the portion of the vertebrate head derived from neural crest (and the use of neural crest to form skeletal elements, which by contrast in the vertebrate trunk and in protochordates are formed from mesoderm) was the major innovation in the evolutionary transition between protochordates and vertebrates. The relationship between morphological homologues (sensu Wagner, 1989b) and particular patterns of gene expression, however, are often more complex.

RECAPITULATION

Replication of characters occurs through a combination of direct transfer of material, and a reconstitution of the features that are not transmitted in this way. In reconstituting missing features (those not transmitted directly from a parent or original), the fidelity of copies may be assured either by a reiteration of the same process that produced the original, or by a process of information transfer that uses a template or a coding process. Mechanisms of replication in modern living organisms are extraordinarily complicated, and a change in any of the participants involved in the copying process can introduce novelty. Hence Dawkins's distinction between "replicators" and "vehicles," and between "germ-line" and "dead-end" replicators, does not allow a complete account of the nonlinear processes that are involved in making copies through the generations.

When a novel feature is introduced, its propagation is easily traced through a linear or treelike pattern of replication, but neither complex phenotypic characters nor sexually reproducing organisms replicate in such fashion. Phenotypic characters are expressed intermittently (after disappearing for portions of a life cycle they reappear in subsequent generations); they are "polyphyletic" with respect to cell lineages; and their phylogenetic continuity may be partially independent of any continuity at the level of particular genes, or any repetition of particular patterns of gene expression. However, complex phenotypic characters provide (what high levels of congruence in their taxonomic distributions suggests is) reliable information about phylogenetic history at the level of population and species lineages. Analyses of the levels at which natural selection operates have suggested that phenotypes "screen off" genotypes with respect to selection; it appears that properties of phenotypes also screen off genotypes with respect to their own (the phenotypes') propagation (replication) through large numbers of generations in evolutionary history, most likely because natural selection maintains the features that are conserved.

The assembling of genes, which do form treelike lineages, within cells and multicellular organisms allows interactions at higher levels of organization to occur. The intricate pattern of meiosis and syngamy permits genes to be reassociated (despite recombination) in reliable patterns and thereby assists in the reliable production and reproduction of characters through generations of these higher levels (cells and organisms). A synthetic understanding that integrates transmission genetics with the developmental basis of morphogenesis remains a major challenge and a frontier in the understanding of how complex phenotypic characters replicate and evolve.

ACKNOWLEDGMENTS

I thank Robert Brandon, Jim Griesemer, Dan McShea, John Mercer, and Günter Wagner for comments and stimulating discussion and NSF DEB97-26855 for support of tree-constructing activities.

LITERATURE CITED

Abouheif, E., Akam, M., Dickinson, W. J., Holland, P. W. H., Meyer, A., Patel, N. H., Raff, R. A., Roth, V. L., and Wray, G. A. (1997). Homology and developmental genes. *Trends Genet.* **13**:432-433.
Alberts, B., Bray, D., Lewis, J., Raff, M., Roberts, K., and Watson, J. D. (1994). "Molecular Biology of the Cell," 3rd ed. Garland Publishing, New York.
Balinsky, B.I. (1970). "An Introduction to Embryology," 3rd ed. W. B. Saunders, Philadelphia.
Bolker, J. A., and Raff, R. A. (1996). Developmental genetics and traditional homology. *Bioessays* **18**:489-494.
Brandon, R. N. (1990). "Adaptation and Environment." Princeton University Press, Princeton.
Buss, L. W. (1987). "The Evolution of Individuality." Princeton University Press, Princeton.

Cartmill, M. (1983). A critique of homology as a morphological concept. *Am. J. Phys. Anthropol.* **94:**115-123.

Dawkins, R. (1976). "The Selfish Gene." Oxford University Press, Oxford.

Dawkins, R. (1982). "The Extended Phenotype." Oxford University Press, Oxford.

De Beer, G. R. (1971). "Homology, an Unsolved Problem." Oxford University Press, Oxford.

Doyle, J. J. (1995). The irrelevance of allele tree topologies for species delimitation, and a non-topological alternative. *Syst. Bot.* **20:**574-588.

Doyle, J. J. (1996). Homoplasy connections and disconnections: genes and species, molecules and morphology. *In* "Homoplasy: The Recurrence of Similarity in Evolution" (M. J. Sanderson and L. Hufford, eds.), pp. 37-66. Academic Press, San Diego.

Doyle, J. J. (1997). Trees within trees: genes and species, molecules and morphology. *Syst. Biol.* **46:**537-553.

Gans, C., and Northcutt, R. G. (1983), Neural crest and the origin of vertebrates: a new head. *Science* **220:**268-274.

Ghiradella, H. (1984). Structure of iridescent lepidopteran scales: variations on several themes. *Ann. Entomol. Soc. Am.* **77:**637-645.

Gilbert, S. F. (1997). "Developmental Biology," 5th ed. Sinauer Associates, Sunderland, MA.

Goldstein, B. (1992). Induction of gut in *Caenorhabditis elegans* embryos. *Nature* **357:**255-257.

Graur, D., Hide, W. A., and Li, W.-H. (1991). Is the guinea-pig a rodent? *Nature* **351:**649-652.

Griesemer, J. (2000). Reproduction and the reduction of genetics. *In* "The Concept of the Gene in Development and Evolution" (P. Buerton, R. Falk, and H-J. Rheinberger, eds.). pp. 240-285. Cambridge University Press, Cambridge.

Griesemer, J. (in press). Development, culture, and the units of inheritance. Philosophy of Science (Proceedings).

Halder, G., Callaerts, P., and Gehring, W. (1995). Induction of ectopic eyes by targeted expression of the *eyeless* gene in *Drosophila. Science* **267:**1788-1792.

Harper, J. L. (1997). "The Population Biology of Plants." Academic Press, London.

Hein, J. (1993). A heuristic method to reconstruct the history of sequences subject to recombination. *J. Mol. Evol.* **36:**396-405.

Hennig, W. (1966). "Phylogenetic Systematics." University of Illinois Press, Urbana.

Hudson, R. R. (1990). Gene genealogies and the coalescent process. *Oxf. Surv. Evol. Biol.* **7:**1-44.

Jablonka, E., and Lamb, M. (1995). "Epigenetic Inheritance and Evolution." Oxford University Press, Oxford.

Kirsch, J. A. W., and Mayer, G. C. (1998). The platypus is not a rodent: DNA hybridization, amniote phylogeny, and the palimpsest theory. *Phil. Trans. Roy. Soc., B* **353:**1221-1237.

Luckett, W. P., and Hartenberger, J.-L. (1993). Monophyly or polyphyly of the order Rodentia: possible conflict between morphological and molecular interpretations. *J. Mamm. Evol.* **1:**127-147.

Maddison, W. P. (1997). Gene trees in species trees. *Syst. Biol.* **46:**523-536.

Maynard Smith, J., and Szathmáry, E. (1995). "The Major Transitions in Evolution." W. H. Freeman, Oxford.

McKittrick, M. C. (1994). On homology and the ontological relationship of parts. *Syst. Biol.* **43:**1-10.

Nassau, K. (1983). "The Physics and Chemistry of Color." John Wiley and Sons, New York.

Neigel, J. E., and Avise, J. C. (1986). Phylogenetic relationships of mitochondrial DNA under various demographic models of speciation. *In* "Evolutionary Processes and Theory" (E. Nevo and S. Karlin, eds.), pp. 515-534. Academic Press, New York.

Nilsson, D.-E. (1996). Eye ancestry: Old genes for new eyes. *Curr. Biol.* **6:**39-42.

Northcutt, G., and Gans, C. (1983). The genesis of neural crest and epidermal placodes: a reinterpretation of vertebrate origins. *Q. Rev. Biol.* **58:**1-28.

Ohno, S. (1970). "Evolution by Gene Duplication." Springer-Verlag, New York.

Pamilo, P., and Nei, M. (1988). Relationships between gene trees and species trees. *Mol. Biol. Evol.* **5**: 568-583.

Patterson, C. (1982). Morphological characters and homology. *In* "Problems of Phylogenetic Reconstruction" (K. A. Joysey and A. E. Friday, eds.), pp. 21-74. Academic Press, London.

Piatigorsky, J., and Wistow, J. (1991). The recruitment of crystallins: new functions precede gene duplication. *Science* **252**:1078-1079.

Quiring, R., Walldorf, U., Kloter, U., and Gehring, W. (1994). Homology of the *eyeless* gene of *Drosophila* to the *small eye* gene in mice and *aniridia* in humans. *Science* **265**:785-789.

Roth, V. L. (1988). The biological basis of homology. *In* "Ontogeny and Systematics" (C. J. Humphries, ed.), pp. 1-26. Columbia University Press, New York.

Roth, V. L. (1991). Homology and hierarchies: problems solved and unresolved. *J. Evol. Biol.* **4**:167-194.

Roth, V. L. (1994). Within and between organisms: replicators, lineages, and homologues. *In* "Homology: the Hierarchical Basis of Comparative Biology" (B. K. Hall, ed.), pp. 301-337. Academic Press, San Diego.

Salmon, W. C. (1971). "Statistical Explanation and Statistical Relevance." University of Pittsburgh Press, Pittsburgh.

Sander, K. (1983). The evolution of patterning mechanisms: gleanings from insect embryogenesis and spermatogenesis. *In* "Development and Evolution" (B.C. Goodwin, N. Holder, and C.C. Wylie, eds.), pp. 137-159. Cambridge University Press, Cambridge.

Schneider, R. A. (1998). "An Experimental Analysis of Sphenoid Skeletogenesis in Birds and Mammals." Ph.D. dissertation, Duke University.

Smith, K. K., and Schneider, R. A. (1998). Have gene knockouts caused evolutionary reversals in the mammalian first arch? *Bioessays* **20**:245-255.

Tajima, F. (1983). Evolutionary relationships of DNA sequences in finite populations. *Genetics* **105**:437-460.

Van Valen, L. (1982). Homology and causes. *J. Morphol.* **173**:305-312.

Wagner, G. P. (1989a). The origin of morphological characters and the biological basis of homology. *Evolution* **43**:1157-1171.

Wagner, G. P. (1989b). The biological homology concept. *Annu. Rev. Ecol. Syst.* **20**:51-69.

Wagner, G. P. (1995). The biological role of homologues: a building block hypothesis. *N. Jb. Geol. Paläont., Abh.* **195**:279-288.

Williams, G. C. (1966). "Adaptation and Natural Selection." Princeton University Press, Princeton.

Zuker, C. S. (1994). On the evolution of eyes: would you like it simple or compound? *Science* **265**:742-743.

5

CHARACTERS AS THE UNITS OF EVOLUTIONARY CHANGE

DAVID HOULE

Department of Biological Sciences, Florida State University, Tallahassee, FL 32306

INTRODUCTION

As evolutionary biologists, we would like to understand the history and predict the course of evolution. This is a difficult task, as the complexity of biological entities is staggeringly high. The phenotype of an organism can be described as a collection of many traits — limb length, bristle number, enzyme activity, hormone levels, etc. Furthermore, each of these traits will change through the life of an individual, making the task of describing the phenotype one of infinite complexity. Clearly, if we need to consider an infinite number of phenotypes to understand evolution, we can make no progress.

Turning to the genetic basis of the phenotype seems hardly better. The genotype of a multicellular organism consists of 10^4 or more genes. Although the number of genes is at least finite, it is small improvement to argue that if only we knew the states of 10,000 genes in a population could we understand evolution. Two major complicating features of the genetic system are that it is polygenic and pleiotropic. Pleiotropy describes the fact that each gene influences many phenotypic traits; it tells us that the fitness of genotypes at each locus will be difficult to predict, as it will depend on its effects on many phenotypes. Polygenic

The Character Concept in Evolutionary Biology

inheritance refers to the fact that many genes influence each phenotypic trait; it tells us that the genetic basis of each phenotype may be difficult to decipher.

To understand evolution, we need to reduce the level of complexity of the problem at both phenotypic and genotypic levels to one that is both sufficiently precise for our purposes, yet simple enough to understand and test. This then is the character problem I will consider: what are the entities which need to be studied in order to provide a description of the evolutionary process? I will call these entities evolutionary characters. This paper is concerned with the nature of these evolutionary characters. Why do they matter? How do we identify what they are? How many do we need to study to decipher the riddles which fascinate us?

I will explore these ideas through the lens of life history theory. Life history theory is a natural point to begin a search for evolutionary characters because it is a widely accepted method for reducing the phenotypic complexity of the evolutionary problem. A life history is most economically defined as the age-specific schedule of reproduction and mortality of a population of individuals. Reproduction and mortality are the events which influence fitness, so a description of a life history defines its fitness.

I begin by describing the two whole-organism approaches which have been employed to identify the nature of evolutionary characters underlying life histories. These are quantitative genetics, which is a data-driven method for summing up the relationship between genotypes and phenotypes (Lande, 1982). It requires no *a priori* assumptions about the nature of evolutionary characters. The second method in wide use, which I link to optimality models (Parker and Maynard Smith, 1990), is to make an assumption about the nature of the evolutionary characters that underlie the traits of interest. These assumptions are then tested either by comparing the predictions of models which assume the nature of the characters against observational data, or by experimental tests of their existence. Both types of studies are implicitly directed toward an understanding of the *functional architecture,* the set of pathways which connect genotypes to phenotypes. Genomic studies are currently directed toward the same goal. We can anticipate that merging of genomic and phenotypic approaches to this problem will be come increasingly fruitful; I suggest the term phenomics for this area of inquiry.

FUNCTIONAL ARCHITECTURE AND EVOLUTIONARY CHARACTERS

Many evolutionary biologists have pointed out the importance of a more thorough understanding of the pathways between genotypes and phenotypes (Lewontin, 1974; Wright, 1977; Schlichting and Pigliucci, 1998), what I have called the functional architecture (Houle, 1991). I start with the intuition that we can begin to understand the tangled mass of genetic relationships with three well-known facts. First, each gene has a very specific set of functions in the organism. Second, these functions are often organized into pathways, which themselves also carry out specific functions, such as production of metabolites, internal signaling, or gathering information about the external world. Third, these pathways are coordinately regulated, for example through the actions of hormones.

These elementary facts from development, biochemistry, and physiology suggest several things. First, rather than paying attention to all the genes in an organism, we may be able consider evolution of the pathways, rather than all of the genes that underlie them, at an enormous simplification in the dimensionality of the system. In our terms, the evolutionary characters we are seeking are the properties of the functional pathways themselves. Second, the pleiotropic effects of variation in a pathway will be restricted to phenotypes to which it is functionally connected. We need not consider the effects of each pathway on all of the arbitrarily large number of phenotypes. Finally, the hierarchical nature of some functional architecture suggests that some pathways may be more important than others, to the point where perhaps focusing on a few simple pathways may capture the essential nature of some evolutionary transitions.

This intuition about the organization of biological complexity is represented in Fig. 1. At the top, in the genome domain, we have the DNA sequence and the proteins and RNA molecules that the genome directs the synthesis of. In the second domain, that of the proteome (Kahn, 1995), each of the proteins is recognized for its specific biological function, and these functions are organized into the pathways which perform more general functions, such as aspects of metabolism and development. The domain at the bottom, labeled the phenome, is the organization and regulation of the pathways into the functional architecture, which lays out the relationships between biological function and fitness. In order to understand evolution we must ultimately understand the phenome as well as the genome and the proteome.

This sort of reasoning is the implicit biological justification for many models that consider the optimization of only one or a few evolutionary characters. There are now a fair number of architectural models of evolution (Riska, 1986; Slatkin, 1987; Wagner, 1989; Houle, 1991; de Jong and van Noordwijk, 1992). Wagner's (1989) model is particularly helpful for understanding the implications of this way of thinking. Wagner assumed that there are a set of loci which each determine one "physiological" property of the organism. These physiological variables are then "mapped" onto the phenotypes with a "developmental" function which gives the effects of each physiological state on each of the phenotypic traits. Wagner restricted his analysis to a linear form of this model where the developmental function could be represented as a matrix of linear coefficients, the B matrix, which represent the effects of each physiological variable on each of the phenotypes, with all the other physiological variables held constant. While this linear form is mathematically convenient, Wagner noted that in general the developmental function need not be linear and can involve interactions between the physiological variables. In addition, I suggest that we can relax the assumption that each physiological variable is controlled by only one locus, and consider it to be a function of all of the genes in a particular pathway. All of these genes that interact to determine one physiological property, such as the flux of material through the pathway, can be treated together when we are concerned with the evolution of particular phenotypic traits, such as life histories.

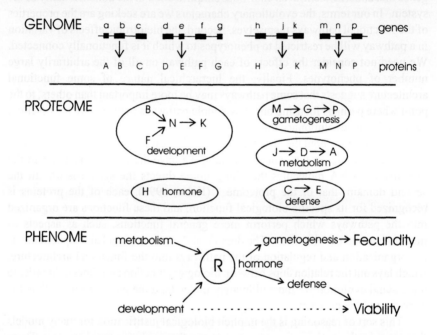

FIGURE 1 Three different representations of biological complexity: the genome, the proteome, and the phenome. The genome is the DNA sequence, and the molecules which it directs the synthesis of. Genomics is the study of the DNA sequence. The proteome consists of all the interacting biomolecules created by the genome. Proteomics is the study of the functions of these proteins (and nucleic acids) and the networks of relationships among these molecules. The phenome is the relationship between the pathways which make up the proteome and the phenotype, and especially fitness. In this representation, the morphology of the organism is dictated by development. This, in combination with basic metabolism, determines the amount of resources (R) which the organism can acquire. These resources are then spent to enhance fitness by increasing both fecundity (here represented as a function of gametogenesis) and viability (represented as some costly defensive function). The relative amount of resource allocated to these two functions is assumed to be regulated by a hormone. In addition to affecting resource acquisition, morphology and development may influence other aspects of the phenome, as shown by the dotted line between development and viability.

A very simple example of this type of model is the Y model of a life history, shown in Fig. 2 (van Noordwijk and de Jong, 1986; Houle, 1991; de Jong and van Noordwijk, 1992). It assumes that the organism acquires some resources, R, from the environment, which it can then allocate to either of two traits, z_1 or z_2. For example, we can assume that z_1 represents somatic function enhancing survival, while z_2 represents reproductive functions. The proportion of all resources allocated to z_2 is P, which leaves fraction 1-P for z_1. In general, the level of expression of z_1 and z_2 may be nonlinear functions of resources allocated. Let us further assume that the values of R and P are determined by separate functional pathways, each containing independent sets of genes. R and P are then evolutionary characters, as they are assumed to be capable of responding to selection independently. However, the biologist studying this life history can observe survival and reproduction, but knows neither the actual allocation hierarchy nor the gain functions which express the relationship between resources spent and the life history traits.

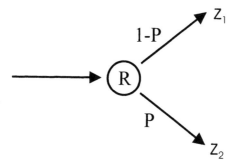

FIGURE 2   The Y model of a pair of life history traits. The organism acquires resources, R, and allocates proportion P of them to trait z_2, leaving proportion 1-P for trait z_1.

The bottom panel of Fig. 1 represents a somewhat more realistic view of the Y model. Here the three arrows of Fig. 2 are given biological names, and allocation is assumed to be regulated by a hormone. In addition, Fig. 1 shows the possibility that the simple Y model may not be sufficient to describe the real situation. For example, it is usually true that the same morphological form which is necessary for resource acquisition will also influence survival directly, as shown by the dotted line. For example, plant leaves carry out photosynthesis and gas exchange, which directly affect acquisition; while the resistance of those same leaves to herbivory can also directly influence survival.

In summary, I propose that evolutionary characters are functional pathways. This suggests that evolutionary characters are free to take on values which are, to a large degree, independent of those of other evolutionary characters, although the pathways will interact with each other to determine fitness. Identification of these characters would allow a description of a population sufficient both to predict its evolutionary dynamics and the eventual equilibrium state. The question then is: How best can we go about identifying these evolutionary characters?

LIFE HISTORY THEORY

Life history theory concerns the evolution of the schedule of reproduction and mortality through the life of an organism (Roff, 1992; Stearns, 1992; Charlesworth, 1994). The appeal of life history theory is that when one has specified the life history of a population of genetically identical organisms, then one has also specified that genotype's fitness. Every real population of organisms is a collection of genotypes which may differ in their age-specific schedule of birth and death. If we could describe the life history of every genotype in a population, we could predict its evolutionary trajectory. Many persistent problems in evolutionary biology would be soluble if it were easy to measure the fitness of a genotype.

Sadly this is not the case. There are a few organisms where it is convenient to measure something close to fitness under special circumstances – clonal growth in microorganisms (Paquin and Adams, 1983), seed output of asexual plants (Stratton, 1992), or the competitive ability of Drosophila genotypes (Fowler *et al.,* 1997). However, the special circumstances necessary to obtain these relatively comprehensive measures compromise their generality. In the vast majority of organisms, we can only capture fragmentary glimpses of their fitness through the trees of their normal environment.

For most organisms some parts of the life cycle are readily studied. Therefore, we can fill in some parts of the schedule of age-specific mortality and reproduction, the life table, of most organisms. Life history theory provides a framework with which to interpret these fragmentary pieces of fitness. For example, the fitness of a life history is usually acutely sensitive to the age at which reproduction commences because of its correlation with generation time and the size and fecundity of adult organisms. This justifies the study of organisms where timing of reproduction and the size of breeding individuals are known, even when much of the rest of the life history is obscure.

The origins of the study of the evolution of life histories lie in demography. The necessary pieces of information for making demographic predictions are the rate at which new offspring are produced and the rate at which individuals die. The most common representation of a life history is that of the discrete life table (Charlesworth, 1994). To make a life table, the potential life cycle of an organism is split up into a series of stages defined by a measurable variable such as age, developmental stage, or size. For each of these stages, we measure the probabilities of survival through that stage, transformation to other stages, and the reproductive rate of individuals in that stage. Demography and fitness follow from these parameters. The life table representation suggests a finite set of parameters to be estimated – at most the square of the number of stages defined. The choice of the nature and number of stages is subject to conflicting goals of those which are practical to measure, and those adequate to capture the current state of the population we want to represent.

Clearly a life table is only an approximation of a real life history. One important aspect of this approximation is the assumption that the set of transition probabilities can accurately capture the fate of individuals. For example, transition probabilities may depend not just on the current state of an individual in the model, but also on that individual's history, which would be expected to affect the state of variables, such as energy reserves, which are outside the purview of the model. Another major approximation is the implicit assumption that the parameters of a life table will not change. This will be violated by changes in the environment, including density and frequency of genotypes in the population.

One way to proceed is simply to add more parameters to this discrete model, including information about the environment, subdividing the stages more finely, etc. Practically speaking, this is not a very promising solution, as our ability to estimate parameters is soon outstripped by such complexity. In the limit, a life history will have infinitely many such parameters. An attractive alternative is to postulate a continuous version of a life table, where survival and reproductive rates are continuous functions of variables such as age or size and the state of the environment. Perhaps relatively simple equations with a small number of parameters can sum up a life history where a life table is a poor approximation to reality. Roff (1992) reviews many such models. The goal of any life history representation is the same: we want sufficient complexity that the state of the population is captured, without requiring us to estimate an unrealistic number of parameters.

While the demographically focused approach can adequately describe the state of the population, it is clearly not adequate for understanding the evolutionary reasons for that state, nor for making evolutionary predictions. To see this, one has only to consider what sort of life history would be predicted based on the just-described representations. Fitness will clearly be maximized by increasing survivorship and reproduction at all ages. This should lead to the evolution of a single population of "Darwinian demons," which live forever, and produce an infinite number of offspring. Since Darwinian demons do not exist, there must be some set of factors which prevent this state. These are the evolutionary constraints on life histories.

CONSTRAINTS, TRADE-OFFS, AND CHARACTERS

I define a constraint as something which prevents a population from evolving to a naturally selected optimum. One kind of constraint would be the existence of some absolute limit to a life history trait. For example, many sources of mortality are completely beyond the capacity of evolution to alleviate, so it is unrealistic to imagine perfect survivorship. If assuming a limit to each life table parameter made it possible to predict life histories, there would be no need to understand the nature of evolutionary characters, as we would be assuming that each character can be optimized independently. In reality, we need to know what the evolutionary characters are to understand the constraints on the joint distribution of sets of life history traits. These multivariate constraints arise due to the nature of the evolutionary characters that underlie them and the pattern of pleiotropic effects that are necessary consequences of changes to the characters. For example, in the Y model in Fig. 2, there will be a negative relationship between the degree to which an individual commits resources to survival versus reproduction, if all other things are equal. This idea that there are limits on the joint expression of life history traits is usually referred to as the trade-off problem. Trade-offs provide the backbone around which life histories must evolve, and so deciphering their nature has long been an important goal of experimental work in life history evolution (Reznick, 1985; Bell and Koufopanou, 1986; Partridge and Harvey, 1988; Sinervo and Basolo, 1996). Four of the eight chapters in Stearns (1992) book on life history evolution concern either the concept of trade-offs or the joint evolution of particular pairs of traits which are likely to be traded off against one another, such as age and size at maturity.

There are two basic sorts of reasons for trade-offs among life history parameters. The first is resource limitation, as in the Y model. Second, a single phenotype must serve all the needs of the organism. For example, a morphology, behavior, or physiology which maximizes survival may not maximize reproduction (Schluter *et al.,* 1991). In either case, trade-offs result from the necessary pleiotropic effects of the evolutionary characters.

Constraints may be operative over a range of time scales. It is sometimes useful to regard any factor which prevents evolution from increasing fitness at the maximum rate as a constraint (Clark, 1987). This is a quantitative constraint. At the other extreme, constraints may be regarded as those factors which completely prevent the attainment of some optimal state (Houle, 1991), such as the Darwinian demon. This is an absolute constraint. These two kinds of constraints form the end points of a continuum connected by variance in the environment. If the selective environment of a population stays constant, then the population will overcome any quantitative constraints, and the equilibrium achieved will be only a function of the absolute constraints. However, when the environment is changing, and therefore the selected optimum is not constant, a quantitative constraint can dictate whether the population ever reaches an optimum state, how far behind the optimum the population will lag (Maynard Smith, 1976), and even whether the population can evolve fast enough to avoid extinction (Lynch and Lande, 1992; Gomulkiewicz and Holt, 1995).

The relationship of quantitative and absolute constraints to the underlying evolutionary characters is quite different. Absolute constraints imply that the biological system is incapable of evolving some combination of phenotypic traits. If we conceive of the phenotype of an organism as defining a multidimensional space, with each axis representing the value of a particular life history trait, then at equilibrium it will be impossible to proceed in certain directions in that space. At this equilibrium point, there would be fewer evolutionary characters than there are phenotypic dimensions to the organism. For example, in the Y model, evolution is expected to maximize R, the rate of acquisition of resources rather quickly, after which time only shifts in P, the allocation of resources between somatic and reproductive functions, will be possible. In Wagner's B model, we would find that the number of physiological characters capable of responding to selection is less than the number of traits we are considering. An absolute constraint can therefore in theory be detected by determining whether the dimensionality of the evolutionary characters is less than the dimensionality of one's description of the phenotype. I discuss how one might do this in more detail later.

To detect quantitative constraints, we must examine both the amount and nature of the genetic variation for each trait. The amount of genetic variation will set an upper limit to the rate of evolution of each trait. The nature of the pleiotropic effects of each of the evolutionary characters which underlie a trait will determine the correlated response to selection. If selection on one trait drags other traits away from their optima, progress toward the overall multivariate optimum will be slowed. To predict a quantitative constraint we need to identify evolutionary characters and measure the variance caused by each of them. This is a more difficult task than in detecting an absolute constraint, as it is easier to identify an absence of variation in some dimensions than to understand the nature of all the variation which is present.

QUANTITATIVE GENETICS

One potential solution to the character problem is to use quantitative genetics to search them out (Lande, 1982). The quantitative genetic formulation of the evolutionary process uses information both on natural selection and on the phenotypic and genetic covariances among traits to make a numerical prediction of the future course of evolution. The primary advantage of this approach is that it requires no *a priori* assumptions about the nature of the evolutionary characters. An obvious disadvantage is that it is often difficult to obtain the necessary estimates of genetic relatedness of individuals while they are in a natural environment. Marker based techniques promise to alleviate such difficulties in the future (Ritland and Ritland, 1996).

In the quantitative genetic model, if we construct a vector of mean trait values, $\bar{z}$, the change in this vector due to a single generation of selection, $\Delta\bar{z}$, is predicted to be

$$\Delta\bar{z} = GP^{-1}S \tag{1}$$

where S is the vector of covariances between each trait and fitness; G is the genetic and P the phenotypic variance-covariance matrix of the traits (Lande, 1979). This equation is known as the breeder's equation, from its usefulness in artificial selection programs. The variance and covariance matrices contain information both on how variable the traits are, in the variances (found along the matrix diagonal), and how the expression of each trait is related to that of the other traits, the covariances (the off-diagonal elements). Recall that correlations are just covariances standardized so that they lie between -1 and +1. Thus covariances summarize the linear relationships between pairs of traits.

A given amount of selection, S, affects the traits z through the filter of phenotypic and genotypic covariances. The covariance between z_1 and fitness (S_1) reflects both directional forces directly on z_1, but also selection on all traits phenotypically correlated with z_1 (Lande and Arnold, 1983). If z_1 and z_2 are positively correlated and z_1 is being selected to increase, then z_2 will also show a correlation with fitness, even if it is not directly selected. As usual with correlational data, the observed covariance with fitness (S) does not imply anything about the cause of the covariance. However, the term $\beta = P^{-1}S$ gives the standardized regression of each trait on fitness, i.e., the strength of the covariance between the trait and fitness, holding all other traits in the analysis constant. A difficult challenge in quantitative genetic studies of natural selection is to include all selected traits which are correlated with the traits of interest (Lande and Arnold, 1983; Mitchell-Olds and Shaw, 1987). Inclusion of traits capturing an adequate representation of the life history, and therefore fitness, in the study alleviates this problem, as the correlated effects on fitness through the focal traits will be captured.

There is a second step to the filtering process, which is determining the genetically mediated effects of selection on phenotypes of offspring. One part of this is straightforward: for a given value of β_i, the larger the value of G_{ii}, the genetic variance in z_i, the larger the response to selection will be. In addition to this, selected alleles that influence one trait will likely have effects on other traits as well. For example, if we assume that there is also a positive genetic covariance between z_1 and z_2 ($G_{12} > 0$), then a genetic response in z_1 will lead to a correlated response in z_2. The same is true for alleles influencing every trait in the organism.

Selection on one trait may affect a wide variety of traits in ways which would be unpredictable without information on the variances and covariances of all selected traits. Thus multiplication of the selection gradient by **G** in equation 1 sums up all these direct and indirect genetic responses to give the net effect of selection on **z**.

In principle, the quantitative genetic formulation of life history evolution is an empiricist's dream come true. It provides a blueprint for future research by enumerating precisely a set of quantities (**S**, **G**, and **P**) which would predict evolutionary responses. In addition, this approach places difficult interpretational issues aside, such as why natural selection favors particular combinations of phenotypes or why some traits are more highly correlated than others. One apparently does not have to understand how the system works to predict or even to shape its behavior. Furthermore it has been suggested that the **G** matrix can reveal both absolute and quantitative constraints on evolution (Lande, 1979; Cheverud, 1984). Absolute constraints are indicated by a low degree of dimensionality. Quantitative constraints can be quantified by the directions of maximum variance and covariance of **G**.

Part of the attraction of quantitative genetics is that it could provide an empirically sufficient dynamic model, which would hold over periods of many generations, perhaps long enough to include the differentiation of species, genera, or higher taxa (Lande, 1979, 1980, 1982; Arnold, 1981b). If so, it would allow both retrospective analyses of the history of selection and predictions of the future course of evolution under novel environments. Unfortunately, it is now clear that quantitative genetics is unlikely to be suitable for such a task (Mitchell-Olds and Rutledge, 1986; Turelli, 1988; Roff, 1997). There are a number of conceptual, statistical, and practical pitfalls which lead to this conclusion. Chief among these is the requirement that the **G** matrix must remain constant over the relevant evolutionary time frame. This requires a host of restrictive subsidiary assumptions that selection be weak and constant in form, the number of loci influencing each trait be large, the average effects of alleles be small, and that genotype-environment interactions are not large enough to be troublesome (Lande, 1979, 1982; Mitchell-Olds and Rutledge, 1986; Turelli, 1988). Some of these assumptions are particularly problematic for life history traits. For example it is difficult to see how selection on life history traits can be weak, as life history determines fitness, and there are many examples of alleles with large effects on life histories (Roff, 1986; Templeton *et al.*, 1993).

Despite this consensus, one recent study suggests that the primary axis along which populations evolve is similar to the major axis of variation in **G** in several species (Schluter, 1996). However, Schluter notes these observation can be explained either by **G** playing a dominant role in constraining evolution or as natural selection reshaping **G** to fit the prevailing selective environment.

Detecting Constraints With Quantitative Genetics

In response to these problems, practitioners are now suggesting quantitative genetics be employed to detect whether a population is at a local fitness optimum, and to supply hypotheses about why a population might deviate from the optimum if it is not. This sort of goal is much easier to justify, as there is ample evidence that the quantitative genetic approach is capable of rather accurate predictions in the short term (Falconer, 1989; Roff, 1997). This more limited program is most clearly laid out by Kirkpatrick (Kirkpatrick *et al.,* 1990; Gomulkiewicz and Kirkpatrick, 1992; Kirkpatrick and Lofsvold, 1992), who suggests a three step process. First, the selection gradient is estimated to determine whether a population is already at a local optimum. At an optimum, all of the selection gradients will equal 0. If this condition is not met, then selection favors a different combination of phenotypes than the population mean. In this case, the second step is to estimate whether the **G** matrix has variation to allow response to selection in all directions, or reflects some genetic constraint which would prevent progress in one or more directions. If the matrix lacks variation in some dimensions, examination of the pattern of variation would provide hypotheses about the nature of the constraint present. Finally, it would be necessary to compare the nature of the apparent constraint with the selection gradients which remain non-zero, to see whether they correspond.

The first step of this program has rarely been attempted for life history traits. An exception is a small study of size and timing of reproduction in *Arabidopsis* (Mitchell-Olds, 1996). Instead, those studying life histories have depended on having an *a priori* demographic model of the life history with which to infer selection gradients. If the demographic model is in fact correct, this is an important shortcut to the Kirkpatrick program. A formidable set of difficulties arises in the estimation of the selection gradients (Lande and Arnold, 1983; Mitchell-Olds and Shaw, 1987). However, without experimental tests of the demographic model's validity, there still seems to be substantial opportunity for error in specifying it, for example, if biologically important features of life history have been ignored, such as time lags or density dependence.

If the selection gradients suggest that the population is not at a local optimum, then the population is either not at equilibrium or is genetically constrained from reaching the optimum. The quantitative genetic model predicts that these constraints will leave their mark in **G** (Lande, 1979, 1982). In mathematical terms, a matrix which is lacking variation in some combinations of traits is *singular* and will have a determinant which is 0. Detection of these singularities is the second step in the Kirkpatrick program. A singularity would indicate that the **G** matrix lacks variation to allow evolution in one or more directions in phenotype space. Even if we were not confident that **G** would remain constant in the long term, a singular **G** would nevertheless be satisfying evidence for a genetic constraint at the present time.

A nonsingular **G** matrix suggests that there are at least as many evolutionary characters as there are traits in the matrix. Furthermore, it suggests that the evolutionary characters have a pattern of effects on the measured traits such that a response is possible in each trait, while holding the others constant. Note that this relationship need not be a simple one. For the Y model, there are just as many evolutionary characters (R, and P) as there are life history traits (z_1 and z_2), but no life history trait is a function of only one evolutionary character. If both evolutionary characters have genetic variance, then the **G** matrix will not be singular, and the system is predicted to be able to evolve to any phenotype.

In the Y model, assuming that the population is not perturbed by directional pressures other than selection, the population is predicted to evolve until R reaches a constrained maximum value R_{opt}, where natural selection has fixed alleles which have positive effects on R, and those which reduce R have been eliminated. At that point, R will have no genetic variance and effectively ceases to exist as an evolutionary character. Conversely, the selection on the allocation would be less intense, as the passing of resources between different functions, each of which increases fitness, would be subject to less intense selection (Lande, 1982). This example is a special case of more general reasoning which considers only variation due to trade-offs, while variation in quantities whose value is maximized by natural selection is assumed to be absent. This assumption is crucial to the use of quantitative genetics as a means of detecting absolute constraints.

There is good reason to doubt this assumption. In addition to the variation due to trade-offs, we expect the **G** matrix will contain two other kinds of variation. The first arises from mutation and gene flow, which often perturb populations away from their selected equilibrium states. In these cases, the equilibrium achieved is a function both of these perturbations and of the resulting selection pressures. A large number of experiments have shown that mutation reduces fitness (Mukai *et al.*, 1972; Simmons and Crow, 1977; Shabalina *et al.*, 1997; Keightley and Ohnishi, 1998) and that life history traits seem to receive a substantially larger input of mutational variance than morphological traits (Houle *et al.*, 1996; Houle, 1998). The large amount of variation in fitness correlates implies that such directional perturbations are also important (Houle, 1992; Burt, 1995). Theoretical work has shown that the consequences of such perturbations can have important effects on the equilibrium life history (Charlesworth, 1990) and the form of the variance-covariance matrix (Houle, 1991).

To take this into account, we need to modify the breeder's equation (equation 1) to account for the vector of perturbations, **D**

$$\Delta \bar{z} = GP^{-1}S + D \qquad (2)$$

At equilibrium, the perturbations would be precisely compensated for by the response to selection, so that

$$-\mathbf{D} = \mathbf{GP}^{-1}\mathbf{S} \qquad (3)$$

Earlier, I made the assumption that $R=R_{opt}$ and that therefore we might not expect to see variation in R at all. However, either mutation or gene flow may reduce R below R_{opt} through the introduction of deleterious variation. These introduced alleles could, for example, reduce the ability to catch prey or photosynthesize at peak efficiency. This then generates additional variance in all of the life history and provides the opportunity for an additional response to selection to recover this new loss in fitness.

The second form of variation in the **G** matrix, other than that due to trade-offs, is experimental error. The large standard errors of quantitative genetic parameters are legendary. Indeed, it is rare to see a correlation matrix for more than a few traits where all of the correlations fall in the range -1 to +1. This poses statistical problems for analyses of **G** (Hayes and Hill, 1981). More generally, the existence of a constraint can never be conclusively demonstrated, as one can never prove a complete absence of variance in a particular direction.

There is substantial older literature on the detection of constraints which was uninformed by these issues. The goal of many experiments was to find simple evidence for singular **G** matrices, such as a lack of variation in particular traits, or a perfect negative correlation between traits assumed to be subject to trade-offs. Single-trait studies have almost always found a good deal of genetic variance in life histories (Lewontin, 1974; Mousseau and Roff, 1987; Roff and Mousseau, 1987), although it can be particularly hard to see against the background of large residual variances that such traits carry (Houle, 1992). On the other hand, a few studies have failed to detect genetic variance for fitness correlates in ecologically important situations (Bradshaw, 1991; Futuyma *et al.*, 1995).

The detection of genetic constraints through examination of genetic correlations has a much more ambiguous experimental history, in part because the expectation is unclear. If two traits vary only due to a perfect trade-off between them, then we expect a perfect negative correlation (Charnov, 1989), as in the Y model. Unfortunately, constraints are likely to involve more than two traits (Pease and Bull, 1988; Charlesworth, 1990), Even making the assumption that all of the variation is due to trade-offs, in the multitrait case, there is no longer any simple criterion for determining whether **G** values reflect constraints, unless the form of the trade-offs is precisely known (Charlesworth, 1990). Charlesworth (1990) presents a numerical example of a constrained five trait life table where at equilibrium there are five negative correlations, three positive correlations, and two zero correlations. It is not surprising then that the experimental evidence shows a wide diversity of correlations among life history traits, with only about 40% of the estimates being less than 0 (Roff, 1996). Very few estimates fall near -1 or +1, where an absolute pairwise constraint would be indicated. The preponderance of positive correlations may suggest that directional perturbations are an important source of covariance among life history traits.

Identification Of Evolutionary Characters

A somewhat more promising way to infer the existence of evolutionary characters or constraints from **G** matrices is to examine the structure of the entire matrix through the related multivariate techniques of factor analysis (Gale and Eaves, 1972; Arnold, 1981a) or principal components analysis (Kirkpatrick *et al.,* 1990). Both techniques seek to summarize the entire pattern of variation and covariation expressed in a covariance matrix like **G**, rather than focusing on only a few elements at a time. In principle, each can be used to test the dimensionality of the matrix and the existence of absolute constraints, but they are most useful to suggest hypotheses about the nature of the evolutionary characters which underlie the most important axes of variation in the data, i.e., quantitative constraints.

Let's assume that we have measured p traits, so the **G** matrix is of dimension p $x\, p$. Principal components analysis decomposes the matrix into a series of linear combinations of the original traits, called principle components or eigenvectors. Associated with each eigenvector is an eigenvalue whose magnitude is related to the amount of variation in the data in the direction of the eigenvector. In effect, this analysis may be thought of as beginning with the question: In which direction relative to the multivariate mean is there the most variation? The direction is the first eigenvector, and the magnitude of the variance is related to the corresponding first eigenvalue. Once this largest direction of variation is removed, the same question is asked of the residual values, subject to the additional constraint that the next direction chosen must be at right angles to (*orthogonal* to) the first eigenvector. This proceeds for p steps, with the ith eigenvector subject to the constraint that it must be orthogonal to the previous i-1 eigenvectors. The result is an analysis that captures the full structure of the original **G** matrix. This is a major disadvantage of principal components, as it means that the error variance of **G** will also be included in all of the eigenvectors, biasing both their direction and the amount of variance they explain.

In contrast, factor analysis begins with the assumption that there are less than p unobserved factors, which account for the majority of the correlation structure in **G** (Johnson and Wichern, 1982). As in principal components, a factor is a linear combination of the measured traits. Factor analysis explicitly assumes that some of the apparent 'structure' in **G** is due to error variance. Therefore, the first step in a factor analysis is not to try to explain the maximum amount of variance, but to sequester what are sometimes referred to as "unique" or "specific" factors which apply only to a single trait in the analysis. Only after this step does the analysis proceed in a manner analogous to principal components to ask in which direction would a hypothesized set of factors best explain the remaining data. The underlying assumption is a functional one, that causal, or "common"' factors exist which can explain the covariance among traits. This is reminiscent of my assumptions about the nature of evolutionary characters above.

Factor analysis is not a single technique, like principal components, but a complex family of techniques meant to uncover the underlying correlation-causing factors. Versions of factor analysis exist that counter many of the difficulties of principal components, such as the requirement that the principal components be orthogonal to one another, or that the first component be chosen by the narrow criterion of maximizing the explained variance. Among this arsenal of techniques are maximum likelihood models for testing the minimum number of common factors necessary to explain the observed G matrix. If one has an *a priori* notion of what the underlying evolutionary characters are, then related techniques such as path analysis or structural equation models can be used to validate them (Crespi and Bookstein, 1989).

In either analysis, the result is a set of linear combinations of the original variables, along with an estimate of how much variation is associated with each combination. These linear combinations can then be used as the basis for other analyses, such as the detection of selection on the presumed factors. Kirkpatrick *et al.* (1990) proposed that such techniques could be used to test the dimensionality of the underlying G matrix by constructing confidence limits on each of the eigenvalues in a principal components analysis. The number of those whose confidence limits do not overlap 0 is the inferred dimensionality of the system. Maximum likelihood factor analysis was devised to provides tests of similar hypotheses concerning the minimal number of factors required to explain the data structure (Johnson and Wichern, 1982, pp. 415-423). While these techniques are in theory promising, the large experimental errors in determining G suggest that inferences about the least variable dimensions of the genetic system are likely to be unreliable. Furthermore, the results of such analyses have never been subjected to experimental tests of their reliability.

Selection Gradients And Tradeoffs

A very different and relatively unknown method for inferring the existence of constraints is to examine the selection gradients for metric traits with respect to several life history traits (Schluter *et al.*, 1991). If selection of conflicting direction is observed, then this suggests that the metric trait (or a trait it is correlated with) affects fitness through its effects on the life history traits, but that no single value of the metric trait is optimal in all circumstances. For example, Schluter and Smith (1986) observed that beak length in song sparrows was positively correlated with overwinter survival, but negatively correlated with female reproductive success. This suggests that a trade-off between fecundity and survivorship must exist such that they cannot simultaneously be maximized. The developmental pathway which leads to the phenotype then is the cause of the trade-off.

Selection Experiments

The goal of evolutionary quantitative genetics is prediction of the outcome of selection. An attractive alternative to the estimation of quantitative genetic parameters, with all the attendant problems, is simply to either apply artificial selection in an interesting direction and observe the response or set up conditions where the action of natural selection is constrained to a known direction. This latter sort of experiment has recently been reviewed under the name laboratory evolution by Rose *et al.* (1996), who carefully distinguish between artificial selection and natural selection in the laboratory. This distinction is not as important in our context; the critical issue is that the selection gradient be known.

The first question to be asked with selection experiments is whether the population responds to selection at all. For life histories, the answer seems to depend on the size of the populations used. In many small experiments no response is observed (Lints *et al.*, 1979), but when the experiments are later repeated with larger populations, spectacular responses result (Zwaan *et al.*, 1995a, b; Rose *et al.*, 1996; Chippindale *et al.*, 1997). This contrast suggests that artificial selection is a problematic source of data on evolutionary characters because of the small population sizes usually involved. These latter studies and many others like them, suggest that many life history traits have substantial genetic variation, as indicated by the direct measurement of variation. Furthermore, this variation is available for natural selection to bring about a reshaping of the overall life history.

Once a response is observed, the pattern of responses to selection may provide an indication of the nature of the pleiotropic effects of the selected variation. For example, Chippindale *et al.* (1996) selected for increased starvation resistance in *Drosophila melanogaster* and observed a correlated increase in lipid storage, accompanied by decreased larval survival and growth rate. A reasonable hypothesis from these results is that the underlying evolutionary character which has responded to selection is really lipid metabolism and that the changes in lipid metabolism have costs seen in growth rates and viability.

Selection experiments have the advantage that they compound the effects of selection over many generations, and potentially in very large populations. The within population genetic variance is converted to variation among populations. This increases the range of phenotypes, especially if both high and low selected lines are included, making it far easier to, in effect, detect the variance in the original population. An important disadvantage is that only a single selection gradient can be applied to each population. Information will only be gained about the variation in this one direction in phenotype space. In contrast, a study of within population variance can be extended to a much wider range of traits simultaneously, and thus may be more suitable for exploratory purposes.

OPTIMALITY STUDIES

The quantitative genetic approaches to the study of life history traits all focus, directly or indirectly, on evolutionary dynamics, i.e., the response to selection. They require few *a priori* assumptions about the nature of the underlying evolutionary characters. The alternative "optimality" approach begins with assumptions about the nature of the evolutionary characters, then tests those assumptions with a combination of observations or experiments. I call this the optimality approach because the implications of the assumptions are usually worked out by constructing a model which predicts the optimum phenotype, given the assumptions. Since we do not wish our models to predict the evolution of nonexistent Darwinian demons, it is clear that we must immediately make an assumption about the nature of the constraints and trade-offs which limit the range of possible phenotypes (Partridge and Harvey, 1988; Parker and Maynard Smith, 1990). In doing so, we essentially reify these hypothetical factors into evolutionary characters. In addition, we assume that all of the other potential evolutionary characters are of lesser importance. The optimality approach therefore assumes that absolute constraints are important, while quantitative constraints are not. Once the functional architecture and the relationships of the phenotypes to fitness are assumed, it is a conceptually straightforward task to find what the optimal phenotype would be, although it is frequently very difficult to find an analytical solution.

A good set of examples of optimality models are the reproductive effort models, which consider the proportion of available resources which should be allocated to reproduction (Roff, 1992, Chapter 8; Stearns, 1992, Chapter 8; Charlesworth, 1994, pp. 213-223). The Y model is a very simple example of a reproductive effort model. Williams (1966) first proposed this way of looking at life histories. He suggested that both fecundity and survival would be positive functions of reproductive effort. In the terms of this paper, Williams implies that reproductive effort is an evolutionary character with antagonistic effects on mortality and fecundity. This insight has inspired a great deal of theoretical and experimental work, which either assumes the existence of a "cost of reproduction" or tests for the existence of such costs.

It is important to bear in mind that there may be many potential explanations of an evolutionary pattern; the problem will often be to distinguish among many formally correct models. While it is easy to accept William's basic insight that there is a cost to reproduction, there are many potential ways in which such costs could be manifested. For example, Sibly and Calow (1986, pp. 66-71) discuss a simple life history model that imagines that the life cycle is divided into a juvenile period prior to first breeding and an adult period in which reproduction takes place periodically. Even in this simple life cycle, fecundity per reproductive bout can be correlated with adult survival rate, with interbreeding interval, juvenile survival of the reproducing individual, the time to maturity of the breeding individual, the quality of the offspring, and juvenile survival of the offspring through levels of parental care. Similarly, shortening the adult interbreeding interval may also increase lifetime fecundity, but at a cost in adult survival or offspring quality of survivorship. Each of these possible trade-offs is a different assumption about the nature of the evolutionary character reproductive effort.

The result of this is that we can come up with many possible explanations for any pattern. For example, the goal of reproductive effort models is often to explain the evolution of reproductive lifespan, in particular the existence of semelparous (annual, or breed once) and iterparous (perennial, or breed multiple times) life histories. Models have shown us that we can explain differences among populations due to differences in mean levels of extrinsic sources of mortality (Charnov and Schaffer, 1973), to the variance in mortality or fecundity (Orzack and Tuljapurkar, 1989), to differences in the relationship between reproductive effort and effective fecundity, and between somatic effort and survival (reviewed by Charlesworth, 1994, pp. 213-223), as well as, no doubt, other possibilities we have not thought of yet. This reinforces that observational or experimental tests are an essential component of optimality explanations.

Testing Optimality Models

Models may be tested either through their predictions or through their assumptions. As noted earlier, the predictions of several models may fit the same data, so models whose assumptions are well tested should be preferred. Since the nature of evolutionary characters is usually key assumptions, the identification of these characters is of crucial importance to the optimality approach. For the reproductive effort models, the key assumption is that mortality and reproduction are in some way inversely related, and therefore a huge amount of effort has been devoted to detecting "costs of reproduction." There has been considerable debate on the best ways to make these tests (Reznick, 1985, 1992; Bell and Koufopanou, 1986; Partridge and Harvey, 1988; Leroi et al., 1994b; Sinervo and Basolo, 1996). The three attractive options are experimental manipulations, selection experiments, and quantitative genetics.

Ultimately, we are interested in the evolutionary potential of populations, which argues that quantitative genetic and selection based approaches to these questions are the most useful (Reznick, 1985, 1992). However, as outlined earlier, the pitfalls of the quantitative genetic approach approach are many. These difficulties are not so great when a particular relationship, such as between reproduction and mortality, is of a priori interest, in contrast to the exploratory role I emphasized earlier. If we have a good measure of mortality and of reproduction, and a high negative correlation is observed, we can be fairly confident that we have found a trade-off. Selection experiments can be a powerful way to test for correlated responses predicted from assumptions about the nature of the evolutionary characters in short-lived model organisms (Rose et al., 1996), as discussed previously.

In the experimental approach, the experimenter directly alters a life history trait. The extent and nature of correlated effects on other traits are then used to infer something about the nature of the evolutionary characters underlying the traits (Partridge and Harvey, 1988; Sinervo and Basolo, 1996). Experimental manipulations allow a wider variety of organisms to be investigated, often with less effort. It also can generate a range of variation far greater than that amenable to an observational genetic study. However, the nature of the manipulation needs to be carefully considered, as the observed correlations are not necessarily of evolutionary relevance (Reznick, 1985).

Sinervo and de Nardo (1996) distinguish between three categories of manipulations with varying relevance. First, the environment of an individual can be altered to affect the life history. However, the resulting plastic response may be different from what an evolutionary response would be, especially if the goal of the manipulation is to extend the range of observed variation beyond that in the population. For example, the response of *Drosophila melanogaster* lines to manipulations of food availability bears little resemblance to their evolutionary response to food limitation (Leroi *et al.,* 1994a).

Second, one can manipulate the value of a trait directly. For example, adding or removing eggs from nests as a means of manipulating reproduction is readily accomplished in birds, and so has been carried out numerous times (Gustafsson and Sutherland, 1988). However, direct manipulations do not necessarily capture all the evolutionary costs of fecundity. The eggs themselves may be metabolically costly to produce (Winkler, 1985), so parents with reduced clutches still pay some costs of offspring they no longer care for, while those with enhanced clutches escape these costs. A more subtle problem is the level of parental care offered to an altered clutch. The assumption of the manipulator is that parent birds would respond with a level of parental care appropriate to the clutch size it finds itself caring for, but as with purely environmental manipulations, the plastic response to altered clutch size may not be appropriate, especially at extreme clutch sizes (Gustafsson and Sutherland, 1988), indicating costs that could be alleviated over evolutionary time, or displacing the costs away from the evolutionarily relevant phenoypes.

The ideal manipulation is to alter the evolutionary character itself in the same way that evolution would do, hence the relevance of selection experiments. However, this precise a manipulation requires that the evolutionary character is already known from independent evidence, and so must be used in conjunction with comparative or genetic approaches. An outstanding example of this type of study is the ongoing work of Zera on wing-polymorphic crickets (Zera and Denno, 1997). Many insect species have wing polymorphisms, where short-winged flightless and long-winged flight-capable individuals are found, often in the same populations. Compared to wingless individuals, winged individuals have longer time to reproductive maturity and lower fecundity overall. This is consistent with the idea that flight capability is costly and competes for resources with reproduction; in other words, the polymorphism captures a classic lifehistory trade-off.

Direct evidence for this trade-off has come from controlled studies demonstrating the two wing morphs in two *Gryllus* cricket species consume and assimilate similar amounts of nutrients, while wingless individuals accumulate about 50% less wing muscle mass and 50% less lipid, an important flight fuel in this group, resulting in whole-organism respiration rates significantly lower than that of winged individuals (Mole and Zera, 1993; Zera *et al.*, 1994, 1998). These energetic and material savings are sufficient to account for the 50% larger ovarian mass of short-winged individuals.

The proximal basis for the switch between wing morphs has long been supposed to be juvenile hormone (JH) levels (Roff, 1986), and there is now reasonably convincing evidence that this is true for *G. rubens* (Zera and Denno, 1997). While JH biosynthesis itself does not differ between morphs, juvenile hormone esterase, an enzyme that degrades JH in order to trigger molt to an adult form in insects, does. Remarkably, direct application of JH to *Gryllus assimilis* individuals, a species of cricket which is monomorphic for long wings, resulted in short-winged individuals extremely similar to naturally occurring short-winged individuals in other *Gryllus* species (Zera *et al.*, 1998). This similarity extended to a host of correlated features: flight muscle mass, enzyme activities, muscle respiration, ovarian mass, and lipid and triglyceride levels. This strongly suggests that JH levels are the key element of an evolutionary character which affects all these aspects of the life history.

It is useful to contrast these studies of JH in wing polymorphisms, where there is a rich set of background information, with the technically more sophisticated studies of mice genetically engineered to express rat growth hormone (Kajiura and Rollo, 1996; Rollo *et al.*, 1997). We would like to be able to manipulate one aspect of the life history in an evolutionarily relevant manner to decipher the "costs of growth." The fact that the creation of transgenic lines yields genetic changes in growth makes them an attractive model for evolutionary changes. However, in the mouse there is no indication from comparative or other evidence that this is an evolutionarily relevant manipulation.

Transformed mice are constitutively 50% larger than control mice, and yet they consume less food than control mice when body size is controlled for (Kajiura and Rollo, 1996). There are also pleiotropic effects on reproduction, which is postponed until 25% later in life, and ultimately yields only one-third the number of offspring per female relative to controls (Rollo *et al.*, 1997). There are many possible interpretations of these results. This manipulation may reveal necessary costs of increasing growth rates or size, but this is difficult to accept given that other rodents have much larger body sizes. Stepping down a level of generality, it may be telling us about the necessary costs of one particular way of becoming large: increased expression of growth hormone. This is the interpretation favored by Rollo and his colleagues, who attribute the maladaptive aspects of this manipulation to energy stress. They suggest that the mice do not appropriately relieve this energy stress through ad libitum feeding because feeding is regulated to meet growth and protein needs and is insensitive to energy demands per se (Webster, 1993). Given this scenario, the evolutionary implications of the observed costs of growth are not so clear. For example, a population of mice subject to increased energy stress could be expected to evolve a new and more appropriate feeding strategy, which might relieve the observed costs of growth. There might also be other mechanisms for increased size which are less evolutionarily costly, such as postponing reproduction without affecting early growth rates. A final interpretation is that the costs observed are due to peculiarities of the manipulation itself. As Rollo *et al.* (1997) point out, the normal circadian rhythm of GH expression is suppressed in the manipulated mice, and the level of other hormones is altered due to regulatory interactions. We need a more complete understanding of the functional basis of growth to determine what the implications of this manipulation are.

NEW FUNCTIONAL ALTERNATIVES

As evolutionary biologists, we have long been concerned with the evolutionary character problem by other names – in particular as constraints or trade-offs. We have been hopeful that the evolutionary character problem could be solved by whole organism approaches, such as those I have discussed previously. While these approaches can be informative, they have many pitfalls which can only be avoided by thorough, multidisciplinary approaches, such as those being taken with wing-polymporphic insects (Zera and Denno, 1997). This suggests the desirability of exploiting technical advances which will simplify the process of indentifying the functional wiring underlying evolution.

Most efforts by evolutionary biologists to understand evolutionary characters have been directed at the important but crude question of whether there is a pleiotropic relationship between a pair of traits. However, theory makes clear that it is not simply the existence of a trade-off that is important to validate a particular evolutionary scenario, but the precise quantitative nature of the trade-off. For example, current modeling efforts on the reproductive effort problem seek to incorporate growth and age and size at maturity into the reproductive effort model in a more general way (Sibly *et al.*, 1985; Kozlowski, 1992; Bernardo, 1993). All such models assume that age and size at maturity are in effect one trait, with a strong positive relationship between them, while reproductive fitness is enhanced both by being early and by being large. Two different growth curves have been used to model deterministic growers. One set of models takes a von Bertalanffy growth equation as an assumption (Roff, 1984; Stearns and Koella, 1986; Berrigan and Koella, 1994, reviewed by Day and Taylor, 1997). The von Bertalanffy model assumes that there is a maximum possible size that an organism can achieve and that growth rates diminish as this asymptotic size is approached. An alternative is to model growth as a power function of mass (Roff, 1983; Kozlowski and Weigert, 1987; Kozlowski and Weiner, 1996). Here there is no maximum size, so additional growth always results in larger size. Cessation of growth is assumed to be due to the shunting of resources to reproduction, away from growth. In each case, there is a tight, trade-off-driven relationship between age and size at maturity, dictated by the growth model. Only the shape of this relationship differs between the models. Day and Taylor (1997) showed that even with such similar models, the effect of a change in growth rate on the optimal age at maturity differs between them. When juvenile and adult mortality are held constant, the von Bertalanffy-based models predict that an increase in growth rate will usually lead to a decrease in age at maturity, as an individual will simply approach its maximum size more rapidly. The power function based models predict that a fast growing individual should always mature later, as it would be more willing to pay the mortality costs of additional growth.

As evolutionary biologists we need to know the precise nature of the evolutionary characters which underlie life history transitions. To get this information, we will have to depend more heavily on studies of gene function and interrelationships. Genomic analyses ultimately hold the promise of a complete understanding of the functional wiring of organisms. Quantitative trait locus (QTL) mapping holds the promise of identifying which loci, out of all the ones in the functional wiring, actually vary within and among populations. Explicitly evolutionary studies will ultimately be necessary to tell us which of the loci that vary are important in effecting evolutionary transitions.

Genomics

A fundamentally new tool which will ultimately prove the key to developing the detailed wiring diagram underlying potential life history transitions are the complete descriptions of gene sequences which the various genome projects have produced (e.g., Fleischmann *et al.*, 1995; Clayton *et al.*, 1997; *C. elegans* Sequencing Consortium, 1998). Just how much we have to understand is emphasized by the high proportion of protein coding sequences found which have no known function. For example, 56% of the over 6400 proteins coded for by the yeast genome had no known function at the time the sequence was completed (Clayton *et al.*, 1997). Substantial efforts are now being directed at understanding the functions of these unknown proteins (James, 1997). Evolutionary studies that make use of whole genomes are in their infancy, but already yield some intriguing results relevant to our study of evolutionary characters. For example, comparisons of the genome complements of pathogenic and nonpathogenic species suggest hypotheses about which pathways are necessary for a pathogenic life history (Huynen and Bork, 1998). In the longer term, application of comparative genomic methods to suites of taxa differing in life histories will prove a powerful source of hypotheses concerning the characters responsible for evolutionary transitions.

A complementary approach that nibbles at the edges of this ignorance is mutagen screening. Such screens have been used in model systems to identify large numbers of genes with particular classes of mutant phenotypes. For example, mutant screens in the nematode worm *Caenorhabditis elegans* have led to the discovery of at least eight genes which are capable of affecting mortality rates (Hekimi *et al.*, 1998). Consideration of what we have learned from these mutations highlights both the possibilities and the limitations of this approach. These genes seem to fall into two categories with independent effects on mortality (Lakowski and Hekimi, 1996), and so identify two evolutionary characters which can affect life span. One set of genes affects the entry of worms into a resting stage, known as a dauer larva. Life span of a dauer larva can be up to 6 months, compared to a normal life span of 15 days. The genes identified seem to be part of the signaling pathway that turns on the dauer phenotype when conditions deteriorate. The existence of these genes is predictable from the fact that it has an inducible resting stage. However, when the constitutively dauer genotypes are placed at a temperature too high for induction of the morphological dauer larva phenotype, life span is still lengthened considerably. This has revealed unexpected complexity to the basis of the phenotype which would otherwise have been difficult to find. Otherwise, these loci are not very informative, since they do not help identify the physiological basis of the alterations to life span.

The second class of genes affecting mortality rates are the clock genes, so called because they seem to affect a host of traits that have a temporal component (Hekimi *et al.*, 1998). The effects on life span seem to be highly correlated with effects on development rate. The genes act maternally, suggesting that they are involved in setting the rate of living of worms early in life and that this rate is difficult to change later. One of the clock genes has been cloned and shows extensive homology with a yeast gene which is involved in regulating the switch from growth on glucose to growth on nonfermentable carbon sources. The implication is that, in the worm, the clock genes are involved in activating a fundamental energetic pathway, with mutants depriving worms of the energy the pathway supplies, which has the pleiotropic effect of slowing the rate of living and decreasing the mortality rate. This suggests that the clock genotype is a classic example of an antagonistically pleiotropic one, where the fitness advantages of increased metabolism on fitness early in life more than compensates for the gain in fitness late in life.

Possession of the entire sequence of the *C. elegans* genome will catalyze the identification of the missing elements of these known pathways. What is known is tantalizing rather than conclusive, but clearly reveals the power of the genomic approach to the identification of evolutionary characters.

QTL Mapping

Quantitative trait locus mapping seeks to find the genetic location of segregating variation, usually in artificially constructed populations. To do the mapping, the population is first constructed by crossing stocks that differ in the phenotype of interest. In addition, the two stocks are characterized for genetic markers at previously mapped marker loci. The QTL mapping is then carried out by looking for correlations between the markers and the phenotype in the descendants of the original cross. The technique is beginning to be widely applied to life history traits by evolutionary biologists (Mitchell-Olds, 1995) and plant (Stuber, 1995) and animal breeders (Haley, 1995).

While ultimately the goal is to identify the specific loci responsible for the mapped variation, this is rarely possible today using QTL mapping alone, given the large numbers of unknown loci in the genome and the small numbers of markers employed in QTL mapping (Mackay and Fry, 1996). A related problem is that a single QTL containing region may harbor several loci affecting the trait. Successes in identifying the specific loci involved have only come by comparing QTL maps with previously mapped loci which are known to influence the trait of interest, which are called candidate genes (Doebley *et al.,* 1995, 1997; Long 1995; Mackay and Fry, 1996). Presumably the information necessary to resolve these issues will increase in model organisms as the genomic data increase.

Even without knowledge of the function of a QTL, its identification can delineate the existence of evolutionary characters. For example, Mitchell-Olds (1996) mapped two genes controlling flowering time in *Arabidopsis thaliana.* Both gene regions also had correlated effects on plant size, such that late flowering plants were large. This reflects a classic time vs. size trade-off, as postulated in many life history theories. Interestingly, the strains crossed to initiate mapping did not vary in flowering time, but each carried one early and one late allele whose effects approximately canceled out.

A second example of the usefulness of QTL mapping for the identification of evolutionary characters are maps of the differences between primitive cultivars of maize and its wild progenitor teosinte (Doebley and Stec, 1993; Doebley *et al.,* 1995). The two strains are extremely different in large numbers of traits, so different that they were originally classified as different genera (Iltis, 1983). Most of the morphological differences reflect increased reproductive effort and harvestability in the cultivar, both of great practical value to agriculturalists. The list of major changes includes at least a dozen traits in the ear and the overall architecture of the plant. Remarkably, only five regions of the genome seem to control the lion's share of the differences in all of these traits. In some cases, the correlated effects of a single locus seem sensibly related to each other. For example, the seed of teosinte is protected by a hard outer glume, covering the opening of the thick cupule in which the seed sits. A mutation at a single locus, *tga1*, withdraws the protective cupule and softens the glume to the edible form found in maize (Dorweiler *et al.,* 1993). However, two other loci jointly determine a bewildering variety of seemingly unrelated traits, including the conversion of the long lateral branches of teosinte to the short ones of maize, the conversion of the terminal inflorescence on the lateral branches from male to female, an increase in the number of seeds/ear, rearrangement of the seeds along the ear, increase in seed size, and for increasing the tendency of ears to stay intact during handling (Doebley *et al.,* 1995). One of these, *tb1*, has been identified and cloned (Doebley *et al.,*

1995, 1997). All of this evidence suggests the presence of a functional pathway whose nature could never have been predicted *a priori*.

The conservation of gene order in the grasses enables the comparison of QTL positions among maize, rice, and sorghum, and remarkably these two gene regions also have effects on mass per seed in crosses between wild and cultivated strains in all three species (Paterson *et al.*, 1995). Correspondence of regions influencing other traits, such as day-length response is also apparent, suggesting that the same evolutionary characters have been involved in independent domestication events.

PHENOMICS

Very soon, we will have a catalog of all of the genes in a wide sample of model multicellular organisms. This has already led to a shift in attention away from genomics, to the decipherment of the biological roles of the proteins a genome is capable of constructing. The set of proteins in an organism has been dubbed its proteome (Kahn, 1995). Understanding the proteome will occupy more reductionist scientists such as biochemists and developmental biologists for some time. Within a few years, or decades at the most, we can anticipate that the proteome too will be deciphered, i.e., we will know what all the genes do and which biochemical or developmental pathways they are organized into.

There is another domain which we need to understand, and that is the potential effects of variation in the proteome on the phenotype (see Fig. 1). This is the same task I have suggested we need to pursue to promote our understanding of evolution, the decipherment of evolutionary characters. I suggest the term *phenomics* for this task, to suggest that this is ultimately as important as the genome projects themselves. The task of phenomics is to understand the implications of functional architecture for biology. Some pathways are likely to be important for evolution, while others rarely are. The ones that underlie major evolutionary transitions, respond to phenotypic selection most readily, or cause the most genetic load should receive the most attention from all sorts of biologists. Other pathways may be ignored as a first approximation. These are crucial tasks which evolutionary biologists should embrace as one for which their conceptual training makes them uniquely suited.

CONCLUSION

In order to understand evolution we must simplify the enormous phenotypic and genetic complexity of organisms to an understandable, yet hopefully predictive, level. Life history theory provides a convenient summary of phenotypic complexity, sufficient to approximate fitness of individuals. I argue that the pathways connecting genotype and phenotype are the units that change during evolution, and that properties of these pathways can be used to summarize the immense complexity of the genetic system. I term such units evolutionary characters. Evolutionary characters matter because they determine both the opportunities for evolutionary change and the constraints on evolution. I review two traditional whole-organism approaches to identifying the nature of evolutionary characters: quantitative genetics and model building and testing. Quantitative genetic methods are valuable because of their exploratory nature, but have serious practical obstacles to their use. Selection experiments are a powerful genetic technique for exploring issues of constraint and opportunity in the limited range of species where they may be applied to large populations. Model building and testing is also powerful and informative in well-understood systems. These approaches are now being supplemented by comprehensive genomic approaches to the identification of characters. Evolutionary biologists should exploit this new information to build a comprehensive understanding of the characters which underlie evolutionary transitions. This process of understanding the evolutionary implications of functional architecture is the capstone of genomic studies. I suggest that it should be referred to as phenomics to emphasize its importance. Evolutionary biologists should embrace the study of the phenome as a task they are uniquely positioned to undertake.

ACKNOWLEDGEMENTS

Thanks to Günter Wagner, Tom Miller, Alice Winn, Locke Rowe, and Ellie Larsen for comments on previous drafts.

LITERATURE CITED

Arnold, S. J. (1981a). Behavioral variation in natural populations. I. Phenotypic, genetic and environmental correlations between chemoreceptive responses to prey in the garter snake, *Thamnophis elegans*. *Evolution* **35**:489-509.

Arnold, S. J. (1981b). Behavioral variation in natural populations. II. The inheritance of a feeding response in crosses between geographic races of the garter snake, *Thamnophis elegans*. *Evolution* **35**:510-515.

Bell, G., and Koufopanou., V. (1986). The cost of reproduction. *Oxf. Surv. Evol. Biol.* **3**:83-131.

Bernardo, J. (1993). Determinants of maturation in animals. *Trends Ecol. Evol.* **8**:166-173.

Berrigan, D., and Koella, J. C. (1994). The evolution of reaction norms: simple models for age and size at maturity. *J. Evol. Biol.* **7**:549-566.

Bradshaw, A. D. (1991). The Croonian Lecture, 1991. Genostasis and the limits to evolution. *Philos. Trans. R. Soc. London, Ser. B* **333**:289-305.

Burt, A. (1995). Perspective: the evolution of fitness. *Evolution* **49**:1-8.

C. elegans Sequencing Consortium. (1998). Genome sequence of the nematode *C. elegans:* a platform for investigating biology. *Science* **282**:2012-2018.

Charlesworth, B. (1990). Optimization models, quantitative genetics, and mutation. *Evolution* **44**:520-538.

Charlesworth, B. (1994). "Evolution in Age-Structured Populations," 2nd ed. Cambridge University Press, Cambridge.

Charnov, E. L. (1989). Phenotypic evolution under Fisher's fundamental theorem of natural selection. *Heredity* **62**:113-116.

Charnov, E. L., and Schaffer, W. M. (1973). Life history consequences of natural selection: Cole's results revisited. *Am. Nat.* **107**:791-793.

Cheverud, J. M. (1984). Quantitative genetics and developmental constraints on evolution by selection. *J. Theor. Biol.* **110**:155-171.

Chippindale, A. K., Chu, T. J. F., and Rose, M. R. (1996). Complex trade-offs and the evolution of starvation resistance in *Drosophila melanogaster. Evolution* **50**:753-766.

Chippindale, A. K., Alipaz, J. A., Chen, H.-W., and Rose, M. R. (1997). Experimental evolution of accelerated development in *Drosophila melanogaster.* I. Developmental speed and larval survival. *Evolution* **51**:1536-1551.

Clark, A. G. (1987). Genetic correlations: the quantitative genetics of evolutionary constraints. *In* "Genetic Constraints on Adaptive Evolution" (V. Loeschke, ed.), pp. 25-45. Springer-Verlag, Berlin.

Clayton, R. A., White, O., Ketchum, K. A., and Venter, J. C. (1997). The first genome from the third domain of life. *Nature* **387**:459-462.

Crespi, B. J., and Bookstein, F. L. (1989). A path-analytic model for the measurement of selection on morphology. *Evolution* **43**:18-28.

Day, T., and Taylor, P. D. (1997). Von Bertalanffy's growth equation should not be used to model age and size at maturity. *Am. Nat.* **149**:381-393.

de Jong, G., and van Noordwijk, A. J. (1992). Acquisition and allocation of resources: genetic (co)variances, selection, and life histories. *Am. Nat.* **139**:749-770.

Doebley, J., and Stec, A. (1993). Inheritance of morphological differences between maize and teosinte: comparison of results for two F_2 populations. *Genetics* **134**:559-570.

Doebley, J., Stec, A., and Gustus, C. (1995). *teosinte branched1* and the origin of maize: evidence for epistasis and the evolution of dominance. *Genetics* **141**:333-346.

Doebley, J., Stec, A., and Hubbard, L. (1997). The evolution of apical dominance in maize. *Nature* **386**:485-488.

Dorweiler, J., Stec, A., Kermicle, J., and Doebley, J. (1993). *Tesosinte glume architecture1:* a genetic locus controlling a key step in maize evolution. *Science* **262**:233-235.

Falconer, D. S. (1989). "Introduction to Quantitative Genetics," 3rd ed. Longman Scientific and Technical, Harlow.

Fleischmann, R. D., Adams, M. D., White, O., Clayton, R. A., Kirkness, E. F., Kerlavage, A. R., Bult, C. J., Tomb, J.-F., Dougherty, E. F., Merrick, J. M., McKenney, K., Sutton, G., FitzHugh, W., Fields, C., Gocayne, J. D., Scott, J., Shirley, R., Liu, L.-I., Glodek, A., Kelley, J. M., Weidman, J. F., Phillips, C. A., Spriggs, T., Hedblom, E., Cotton, M. D., Utterback, T. R., Hanna, M. C., Nguyen, D. T., Saudek, D. M., Brandom, R. C., Fine, L. D., Fritchman, J. L., Fuhrmann, J. L., Geoghagen, N. S. M., Gnehm, C. L., McDonald, L. A., Small, K. V., Fraser, C. M., Smith, H. O., and Venter, J. C. (1995). Whole-genome random sequencing and assembly of *Haemophilus influenzae* Rd. *Science* **269**:496-512.

Fowler, K., Semple, C., Barton, N. H., and Partridge L. (1997). Genetic variation for total fitness in *Drosophila melanogaster. Proc. R. Soc. Lond. Ser. B* **264**:191-199.

Futuyma, D. J., Keese, M. C., and Funk, D. J., (1995). Genetic constraints on macroevolution: the evolution of host affiliation in the leaf beetle genus *Ophraella. Evolution* **49**:797-809.

Gale, J. S., and Eaves L. J. (1972). Variation in wild populations of *Papaver dubium* V. the application of factor analysis to the study of variation. *Heredity* **29**:135-149.

Gomulkiewicz, R., and Holt, R. D. (1995). When does evolution by natural selection prevent extinction? *Evolution* **49**:201-207.

Gomulkiewicz, R., and Kirkpatrick, M. (1992). Quantitative genetics and the evolution of reaction norms. *Evolution* **46**:390-411.

Gustafsson, L., and Sutherland, W. J., (1988). The costs of reproduction in the collared flycatcher *Ficedula albicollis. Nature* **335**:813-815.

Haley, C. S. (1995). Livestock QTLs – bringing home the bacon? *Trends Genet.* **11**:488-492.

Hayes, J. F., and Hill, W. G. (1981). Modification of estimates of parameters in the construction of genetic selection indices ('bending'). *Biometrics* **37**:483-493.

Hekimi, S., Lakowksi, B., Barnes,T. M., and Ewbank, J. J. (1998). Molecular genetics of life span in Elegans: how much does it teach us? *Trends Genet.* **14**:14-20.

Houle, D. (1991). Genetic covariance of fitness correlates: what genetic correlations are made of and why it matters. *Evolution* **45**:630-648.

Houle, D. (1992). Comparing evolvability and variability of quantitative traits. *Genetics* **130**:195-204.

Houle, D. (1998). How should we explain variance in the genetic variance of traits? *Genetica* **102/103**:241-253.

Houle, D., Morikawa, B., and Lynch, M. (1996). Comparing mutational variabilities. *Genetics* **143**:1467-1483.

Huynen, M. A., and Bork P., (1998). Measuring genome evolution. *Proc. Natl. Acad. Sci. USA* **95**:5849-5856.

Iltis, H. H. (1983). From teosinte to maize: the catastrophic sexual transmutation. *Science* **222**:886-894.

James, P. (1997). Protein identification in the post-genomes era: the rapid rise of proteomics. *Q. Rev. Biophys.* **30**:279-331.

Johnson, R. A., and Wichern, D. W. (1982). Applied Multivariate Statistical Analysis. Prentice-Hall, Englewood Cliffs, NJ.

Kahn, P. (1995). From genome to proteome: looking at a cell's proteins. *Science* **270**:369-370.

Kajiura, L. J., and Rollo, C. D. (1996). The ontogeny of resource allocation in giant transgenic rat growth hormone in mice. *Can. J. Zool.* **74**:492-507.

Keightley, P. D., and Ohnishi, O. (1998). EMS-induced polygenic mutation rates for nine quantitative characters in *Drosophila melanogaster*. *Genetics* **148**:753-766.

Kirkpatrick, M., and Lofsvold, D. (1992). Measuring selection and constraint in the evolution of growth. *Evolution* **46**:954-971.

Kirkpatrick, M., Lofsvold, D., and Bulmer, M. (1990). Analysis of the inheritance, selection and evolution of growth trajectories. *Genetics* **124**:979-993.

Kozlowski, J. (1992). Optimal allocation of resources to growth and reproduction: implications for age and size at maturity. *Trends Ecol. Evol.* **7**:15-19.

Kozlowski, J., and Weigert, R. G. (1987). Optimal age and size at maturity in annuals and perennials with determinate growth. *Evol. Ecol.* **1**:231-244.

Kozlowski, J., and Weiner, J. (1996). Interspecific allometries are byproducts of body size optimization. *Am. Nat.* **149**:352-380.

Lakowski, B., and Hekimi, S. (1996). Determination of life-span in *Caenorhabditis elegans* by four clock genes. *Science* **272**:1010-1013.

Lande, R. (1979). Quantitative genetic analysis of multivariate evolution applied to brain:body size allometry. *Evolution* **33**:402-416.

Lande, R. (1980). Genetic variation and phenotypic evolution during allopatric speciation. *Am. Nat.* **116**:463-479.

Lande, R. (1982). A quantitative genetic theory of life history evolution. *Ecology* **63**:607-615.

Lande, R., and Arnold, S. J. (1983). The measurement of selection on correlated characters. *Evolution* **37**:1210-1226.

Leroi, A. M., Kim, S. B., and Rose, M. R. (1994a). The evolution of phentoypic life-history trade-offs: an experimental study using *Drosophila melanogaster*. *Am. Nat.* **144**:661-676.

Leroi, A. M., Rose, M. R., and Lauder, G. V. (1994b). What does the comparative method reveal about adaptation? *Am. Nat.*. **143**:381-402.

Lewontin, R. C. (1974). The Genetic Basis of Evolutionary Change. Columbia University Press, New York.

Lints, F. A., Stoll, J., Grunway, G., and Lints, C. V. (1979). An attempt to select for increased longevity in *Drosophila melanogaster*. *Gerontology* **25**:192-204.

Long, A. D., Mullaney, S. L., Reid, L. A., Fry, J. D., Langley, C. H., and Mackay, T. F. C. (1995). High resolution mapping of genetic factors affecting abdominal bristle number in *Drosophila melanogaster*. *Genetics* **139**:1273-1291.

Lynch, M., and Lande, R. (1992). Evolution and extinction in response to environmental change *In* "Biotic Interactions and Global Change" (P. M. Kareiva, J. G. Kingsolver, and R. B. Huey, eds.), pp. 234-250. Sinauer Associates, Sunderland, MA.

Mackay, T. F. C., and Fry, J. D. (1996). Polygenic mutation in *Drosophila melanogaster*: genetic interactions between selection lines and candidate quantitative trait loci. *Genetics* **144**:671-688.

Maynard Smith, J. (1976). A comment on the Red Queen. *Am. Nat.* **110**:325-330.

Mitchell-Olds, T. (1995). The molecular basis of quantitative genetic variation in natural populations. *Trends Ecol. Evol.* **10**:324-328.

Mitchell-Olds, T. (1996). Pleiotropy causes long-term genetic constraints on life-history evolution in *Brassica rapa. Evolution* **50**:1849-1858.

Mitchell-Olds, T., and Rutledge, J. J. (1986). Quantitative genetics in natural plant populations: a review of the theory. *Am. Nat.* **127**:379-402.

Mitchell-Olds, T., and Shaw, R. G. (1987). Regression analysis of natural selection: statistical inference and biological interpretation. *Evolution* **41**:1149-1161.

Mole, S., and Zera, A. J. (1993). Differential allocation of resources underlies the dispersal-reproduction trade-off in the wing polymorphic cricket, *Gryllus rubens. Oecologia* **93**:121-127.

Mousseau, T. A., and Roff, D. A. (1987). Natural selection and the heritability of fitness components. *Heredity* **59**:181-197.

Mukai, T., Chigusa, S. I., Mettler, L. E., and Crow, J. F. (1972). Mutation rate and dominance of genes affecting viability in *Drosophila melanogaster. Genetics* **72**:335-355.

Orzack, S. H., and Tuljapurkar, S. (1989). Population dynamics in variable environments. VII. The demography and evolution of iteroparity. *Am. Nat.* **133**:901-923.

Paquin, C. E., and Adams, J. (1983). Relative fitness can decrease in evolving asexual populations of *S. cerevisiae. Nature* **306**:368-371.

Parker, G. A., and Maynard Smith, J. (1990). Optimality theory in evolutionary biology. *Nature* **348**:27-33.

Partridge, L., and Harvey, P. H. (1988). The ecological context of life history evolution. *Science* **241**:1449-1455.

Paterson, A. H., Lin, Y.-R., Li , Z., Schertz, K. F., Doebley, J. F., Pinson, S. R. M., Liu, S.-C., Stansel, J. W., and Irvine, J. E. (1995). Convergent domestication of cereal crops by independent mutations at corresponding genetic loci. *Science* **269**:1714-1718.

Pease, C. M., and Bull, J. J. (1988). A critique of methods for measuring life-history tradeoffs. *J. Evol. Biol.* **1**:293-303.

Reznick, D. (1985). Costs of reproduction: an evaluation of the empirical evidence. *Oikos* **44**:257-267.

Reznick, D. (1992). Measuring the costs of reproduction. *Trends Ecol. Evol.* **7**:42-45.

Riska, B. (1986). Some models for development, growth, and morphometric correlation. *Evolution* **40**:1303-1311.

Ritland, K., and Ritland, C. (1996). Inferences about quantitative inheritance based on natural population structure in the yellow monkey-flower, *Mimulus guttatus. Evolution* **50**:1074-1082.

Roff, D. A. (1983). An allocation model of growth and reproduction in fish. *Can. J. Fish. Aquat. Sci.* **40**:1395-1404.

Roff, D. A. (1984). The evolution of life history parameters in teleosts. *Can. J. Fish. Aquat. Sci.* **41**:989-1000.

Roff, D. A. (1986). The evolution of wing dimorphism in insects. *Evolution* **40**:1009-1020.

Roff, D. A. (1992). "The Evolution of Life Histories: Theory and Analysis." Chapman and Hall, New York.

Roff, D. A. (1996). The evolution of genetic correlations: an analysis of patterns. *Evolution* **50**:1392-1403.

Roff, D. A. (1997). "Evolutionary Quantitative Genetics." Chapman and Hall, New York.

Roff, D. A., and Mousseau, T. A. (1987). Quantitative genetics and fitness: lessons from *Drosophila. Heredity* **58**:103-118.

Rollo, C. D., Rintoul, J., and Kajiura, L. J. (1997). Lifetime reproduction of giant transgenic mice: the energy stress paradigm. *Can. J. Zool.* **75**:1336-1345.

Rose, M. R., Nusbaum, T. J., and Chippindale, A. K. (1996). Laboratory evolution: the experimental wonderland and the Cheshire cat syndrome. *In* "Adaptation" (M. R. Rose and G. V. Lauder, eds.), pp. 221-241. Academic Press, San Diego.

Schlichting, C. D., and Pigliucci. M. (1998). "Phenotypic Evolution: A Reaction Norm Perspective." Sinauer, Sunderland, MA.

Schluter, D. (1996). Adaptive radiation along genetic lines of least resistance. *Evolution* 50:1766-1774.

Schluter, D., and Smith. J. N. M. (1986). Natural selection on beak and body size in the song sparrow. *Evolution* 40:221-231.

Schluter, D., Price, T. D., and Rowe, L. (1991). Conflicting selection pressures and life history trade-offs. *Proc. R. Soc. Lond, Ser. B* 246:11-17.

Shabalina, S. A., Yamploksy, L. Y., and Kondrashov, A. S. (1997). Rapid decline of fitness in panmictic populations of *Drosophila melanogaster. Proc. Natl. Acad. Sci. USA* 94:13034-13039.

Sibly, R. M., Calow, P., and Nichols, N. (1985). Are patterns of growth adaptive? *J. Theor. Biol.* 112:553-574.

Simmons, M. J., and Crow, J. F. (1977). Mutations affecting fitness in Drosophila populations. *Annu. Rev. Genet.* 11:49-78.

Sinervo, B., and Basolo, A. L. (1996). Testing adaptations using phenotypic manipulations, *In* "Adaptation" (M. R. Rose and G. V. Lauder, eds.), pp. 149-185. Academic Press, San Diego.

Sinervo, B., and DeNardo, D. F. (1996). Costs of reproduction in the wild: path analysis of natural selection and experimental tests of causation. *Evolution* 50:1299-1313.

Slatkin, M. (1987). Quantitative genetics of heterochrony. *Evolution* 41:799-811.

Stearns, S. C. (1992). "The Evolution of Life Histories." Oxford University Press, Oxford.

Stearns, S. C., and Koella, J. C. (1986). The evolution of phenotypic plasticity in life-history traits: predictions of reaction norms for age and size at maturity. *Evolution* 40:893-913.

Stratton, D. A. (1992). Life-cycle components of selection in *Erigeron annuus.* I. Phenotypic selection. *Evolution* 46:92-106.

Stuber, C. W. (1995). Mapping and manipulating quantitative traits in maize. *Trends Genet.* 11:477-481.

Templeton, A. R., Hollocher, H., and Johnston, J. S. (1993). The molecular through ecological genetics of *abnormal abdomen* in *Drosophila mercatorum.* V. Female phenotypic expression on natural genetic backgrounds and in natural environments. *Genetics* 134:475-485.

Turelli, M. (1988). Phenotypic evolution, constant covariances and the maintenance of additive variance. *Evolution* 42:1342-1347.

van Noordwijk, A. J., and de Jong, G. (1986). Acquisition and allocation of resources: their influence on variation in life history tactics. *Am. Nat.* 128:137-142.

Wagner, G. P. (1989). Multivariate mutation-selection balance with constrained pleiotropic effects. *Genetics* 122:223-234.

Webster, A. J. F. (1993). Energy partitioning, tissue growth and appetite control. *Proc. Nutr. Soc.* 52:69-76.

Williams, G. C. (1966). Natural selection, the costs of reproduction, and a refinement of Lack's principle. *Am. Nat.* 100:687-690.

Winkler, D. W. (1985). Factors determining a clutch size reduction in California gulls (*Larus californicus*): a multi-hypothesis approach. *Evolution* 39:667-677.

Wright, S. (1977). "Evolution and the Genetics of Populations," Vol. 3. University of Chicago Press, Chicago.

Zera, A. J., and Denno, R. F. (1997). Physiology and ecology of dispersal polymorphism in insects. *Annu. Rev. Entomol.* 42:207-231.

Zera, A. J., Mole S., and Rokke, K. (1994). Lipid, carbohydrate and nitrogen content of long-winged and short-winged *Gryllus firmus*: implications for the physiological cost of flight capability. *J. Insect Physiol.* 40:1037-1044.

Zera, A. J., Potts, J., and Kobus, K. (1998). The physiology of life-history trade-offs: experimental analysis of a hormonally induced life-history trade-off in *Gryllus assimilis. Am. Nat.* 152:7-23.

Zwaan, B., Bijlsma, R., and Hoekstra, R. F. (1995a). Artificial selection for developmental time *Drosophila melanogaster* in relation to the evolution of aging: direct and correlated responses. *Evolution* **49**:635-648.

Zwaan, B., Bijlsma, R., and Hoekstra, R. F. (1995b). Direct selection on life span in *Drosophila melanogaster*. *Evolution* **49**:649-659.

6

CHARACTER IDENTIFICATION: THE ROLE OF THE ORGANISM*

GÜNTER P. WAGNER[1] AND MANFRED D. LAUBICHLER[2]

[1]Department of Ecology and Evolutionary Biology, Yale University, New Haven, CT 06520
[2]Program in History of Science, Princeton University, Princeton, NJ 08544

INTRODUCTION

Some areas of organismic biology are among the mathematically most sophisticated parts of biology. Examples include population ecology, behavioral biology (game theory, ESS), life history theory, and population and quantitative genetics (e.g., Crow and Kimura, 1970; Bulmer, 1980; Berryman, 1981; Hofbauer and Sigmund, 1988; Stearns, 1992; Falconer and Mackay, 1996; Hanski and Gilpin, 1997; Hartl, 1997; Hofbauer and Sigmund, 1998; Maynard-Smith, 1998). Nevertheless, many of these theories do not have the degree of rigor that is typical of other mathematicized sciences such as physics and chemistry. Here we argue that this lack of rigor is due to a shared structural weakness of these biological theories, and that to overcome this deficiency requires in many instances an organismic perspective in character identification.

* Reprinted from *Theory in Biosciences*, Vol. 119 (2000) pp. 20-40

The structural weakness of many mathematical theories in biology is due to the lack of an analytical theory about their range of application, i.e., an analytical method that would allow us to identify and measure those objects in nature that correspond to the ones postulated by the theory. In the case of Newtonian mechanics it is quite clear that the mathematical models only apply to objects that can be characterized by their mass, location in space, and velocity. Physical theories are as powerful as they are, in part because they implicitly contain a definition of the conditions of their validity. When applying physical theories to material phenomena, the required ontological commitments on part of the researcher are rather limited. One only has to assume that the objects described by the theory in fact exist, or at least that these theoretical objects are close enough abstractions of some observed natural phenomenon. In addition, one has to establish an equivalence relation between the mathematical operations of the theory and the natural processes they describe (Rosen, 1962, 1978).

In contrast, the application of mathematical models in biology has been largely ad hoc. In our biological models we postulate the existence of certain theoretical objects, such as populations, genes, or life history characters, but we usually do not reflect too much on the question which properties a biological object must have in order to be a legitimate instance of the model. As a result, few, if any, biological models can be rigorously tested, since it is often not clear whether any empirical result is actually a relevant instance of the model to be tested. In testing biological theories empirically we are thus prone to commit both, Type I and Type II errors (Sokal and Rohlf, 1981). In other words, we might either falsely reject or accept a model.

A particularly striking example of this problem is illustrated by the work of William Murdoch on population regulation in red scale mites in Californian orchards (Murdoch et al., 1996). Two different kinds of models have been proposed to explain population regulation in such situations: the refuge model (Bailey et al., 1962) and a model based on metapopulation dynamics (Gilpin and Hanski, 1991). Murdoch and his collaborators set out to test population regulation empirically. When they started their experiments, it was initially not clear whether the population size variable refers to the number of individuals in an orchard, as suggested by the metapopulation model, or to the number of individuals on a tree, as would be predicted by the refuge model. The experimental results showed that neither is the case, but that the population size variable, i.e., the functional unit that exhibits the phenomenon of population regulation (the dynamical unit), is the population of red scales in the outer branches of a tree. Neither the removal of the refuge population nor the isolation of an individual tree in a cage (to test for metapopulation effects) yielded the effects predicted by either model. Obviously, the problem with both models is that they do not contain a prescription how to find the limits of the unit to which they actually apply, a problem that has so far been overlooked because both models have rarely been tested in the field. However, as Murdoch points out, the fact that they did not find evidence for refuge or metapopulation effects in their experimental setting does not rule out that these dynamics can exist at another scale (which would be a Type I error). We just do not have any straightforward criteria how to identify the appropriate dynamical units in a given experimental setting.

Theories about the adaptive evolution of organismic properties, such as life history theory, are another example for the ad hoc nature of many models in biology. These theories represent a sophisticated mathematical framework developed to explain the origin and maintenance of life cycle characters in plants and animals (humans included). An ongoing problem with many life history models, however, is that model predictions and experimental results frequently disagree. In most cases these discrepancies are caused by an inappropriate identification of the relevant organismal traits (Stearns, 1992). As a result, the existing models are frequently patched up with ad hoc assumptions about "interactions" among the characters and "constraints" on variation. However, all that "interactions" and "constraints" indicate is that the way the objects where measured in the experiment did not correspond to the units considered in the model. A rigorous theory of life histories would therefore require an operational definition of the biological (i.e., dynamical) properties that are expected of the characters in the model. In addition, it would also have to include methods how to identify these characters in the real world, or at least a procedure of testing whether or not the conditions for the validity of the model have been met experimentally.

The only partial exception to this structural weakness of mathematical models in biology is the theory of natural selection. In the extensive literature about the units of selection one finds explicit definitions of the dynamical properties a class of objects must have in order to be a legitimate instance of the theory of natural selection. Despite continuing controversy about the relative importance of different units of selection (gene, individual, group, etc.), this is one area of biology where mathematical and empirical research have contributed equally to our understanding of the subject (e.g., Fisher, 1930; Dawkins, 1982; Sober, 1984; Endler, 1986; Brandon, 1990; Nagylaki, 1992; Stearns, 1992; Mitton, 1997).

In general, all productive theories should contain at least implicit definitions of the set of objects to which they apply. In biology, examples of such objects are genes, cells, characters, organisms, and populations. Unfortunately, in our empirical studies we are often guided more by a colloquial and less so by a rigorous technical interpretation of these concepts. Consequently, the problems we described earlier are endemic in biology. In this paper we want to investigate a specific question: how do we define the "right kind of object" in the light of a specific process that we want to understand and given that we have the nucleus of a theory to explain this phenomenon.

In many cases, and in particular in the most problematic ones, the theories we are concerned with refer to suborganismal objects, such as genes or cellular and organismal characters (traits). The relationship of these objects to the individual organism can be of one of two kinds: (i) the organism can be thought of as a composite entity "made up" of its traits and characters or (ii) the traits can be thought of as (conceptual) abstractions of the organism. These two scenarios differ as to which object—part or whole—is ontologically prior. In the first case the characters or parts are ontologically prior to the higher level object or the organism. The relationship between atoms and molecules is of this kind. Molecules are composed of atoms and many of their molecular properties can be derived from the properties of their atomic parts. In the second case the higher level unit is ontologically prior. In this instance the suborganismal objects (characters) are defined as conceptual abstractions of a higher level integrated whole and thus ontologically secondary. Here we argue that most biological objects at the suborganismal scale are of the second kind. In other words, we assume the ontological primacy of organisms and derive the objects relevant to the theory, i.e., the biological characters, by means of a conceptual decomposition of the organism. We further argue that if we define biological characters as conceptual abstractions, we also need an appropriately defined organism concept within biological theory.

BIOLOGICAL EXAMPLES FOR THE ONTOLOGICAL PRIMACY OF HIGHER LEVEL OBJECTS

In the previous section we stated that the organism as a whole is ontologically prior to its composite parts. Here we will give additional examples of the ontological primacy of higher level entities from other areas of biology. We thus argue that the ontological primacy of higher level entities is a general principle that characterizes most (if not all) of biology. For a systematic discussion of these issues and the related question of emergence we refer to the book by Maynard-Smith and Szathmáry (1995).

Arguably the cell is the most fundamental biological unit. The cell is also the unit of ontological primacy in molecular biology. The subject matter of molecular biology is the investigation of the molecular mechanisms that underlie the fundamental processes of life, such as DNA replication, protein synthesis, regulation of gene expression, cross-membrane transport, metabolic pathways, and intracellular communication. All these processes take place within or between cells. Furthermore, they are both enabled and constrained by this cellular milieu. The cellular context not only guarantees the specific physicochemical and spatial conditions that are required by these highly specific chemical interactions, it is also the level at which the *functional* roles of these processes can be assigned. The specific nonmolecular, i.e., *functional*, characterizations—*messenger* RNA, *metabolic* enzymes, *transcription* factors, *transport* molecules, etc.—of all objects in molecular biology represent exactly the kind of abstractions that are derived by means of a theory-guided conceptual decomposition of a higher level unit, the cell, that is ontologically prior to its component parts. Molecular biology therefore differs from biochemistry in that it does not study the interactions between these molecules *qua* molecules, but investigates them within the functional context of a cell. In this sense the level of the cell can be said to be ontologically prior to its component parts.

Similarly, a population is a biological object that is ontologically prior to its functionally characterized traits. Any functional concept in population biology, such as refuge, migration balance, predator-prey relation, and demographic pattern, is a conceptual abstraction that presupposes the reality of a population and therefore its ontological primacy. The same is also true for the functional concepts in quantitative and population genetics. Notions, such as the average effect of an allele, used to identify theoretically relevant properties of lower level objects, such as alleles, are all population specific.

In our view, these examples, although generally known to biologists, deserve further attention in that they point to a common feature of biological objects, one that, in our opinion, should also be reflected by the structure of biological concepts and theories. The objects and functional characters that are at the center of every biological theory are not just *found objects* waiting to be collected in nature, rather they are conceptual abstractions determined by the parameters of a specific theoretical interest (Griffiths, 1997; Laubichler, 1997b; Wagner, 1997; Griesemer, 2000; Laubichler and Wagner, 2000a,b; Wimsatt, 2000). As such, they can only be identified within the context of the object that is ontologically prior to them. Therefore, a precise formulation of these objects of ontological primacy has to be part of every theory as well.

The organism concept plays a central role in many biological theories, even though it is rarely defined explicitly. The reason for the centrality of the organism concept in biology lies in the fact that a large number of biological phenomena only occur in an organismal context. Biological characters or traits simply do not exist independent of the organism in the way atoms can (and do) exist outside of molecules. It is therefore rather surprising that we currently do not have an analytical concept of the organism that would reflect its importance. The following section discusses the role of the organism with regard to the problem of character identification in evolutionary biology but first we will briefly outline the significance of the organism as a unit of integration in the context of biological theories.

THE ANALYTICAL ROLE OF THE ORGANISM CONCEPT WITHIN A THEORY OF BIOLOGICAL OBJECTS

The organism is a privileged biological object that deserves special attention. Living objects face two fundamental challenges: (i) to maintain their integrity in the face of a changing internal and external environment and (ii) to assure their propagation in the face of inevitable death. Self-maintenance and reproduction are thus the two most crucial biological processes. Among present-day biological objects, the cell is the minimum level of complexity that is capable of both reproduction (cell division) and self-maintenance (the maintenance of an internal cellular milieu.) Accordingly, the cell is a privileged biological object with respect to self-maintenance and reproduction. Here we restrict our treatment to present-day biological objects and do not investigate any objects that might have been important during the early stages of life's history on earth (Miller and Orgel, 1974; Gesteland *et al.,* 1999).

Organisms are also capable of self-maintenance and reproduction. Furthermore, they are composed of cells. We have seen before that the cell is the object of ontological primacy for many of the processes studied by molecular biology. The crucial next question is, when do we need to refer to the organism in order to explain a biological phenomenon because it cannot be explained on any other level of complexity?

A situation where this question becomes relevant is when one asks what is the difference between a multicellular animal, a so-called metazoan, and a colony of cells? What is added that makes metazoans so distinctly organismic compared to a mere aggregation of cells? Both the colony as well as the metazoan organism are multicellular and in both there is a certain degree of cooperation between the cells. The metazoan animal, however, is considered an organism, while a colony is put in a separate category. A possible answer to this question can be found when one asks what organizational features are necessary in a metazoan animal in order to realize, at the multicellular level, the functions that are carried out by the cell in single celled organisms. A necessary function of cellular organization is the maintenance of the chemical gradient between the inside and the outside of the cell. In the cell a necessary component of this function is the cell membrane, which separates the cytosole from the outside and regulates the traffic of molecules between inside and outside with the aid of receptor and channel molecules. We propose that the epithelial surface, which completely surrounds the body cavity of all metazoans and actively regulates the exchange of molecules between the body cavity and the outside, fulfills the same function in metazoan organisms. In contrast, cell colonies and mesozoan organisms like sponges are unable to regulate the chemical composition of their inside. In fact, there is no well-defined inside or body cavity in these animals. We thus propose to define metazoans as animals that have the potential to regulate the chemical composition of their body cavity through the activity of an epithelial body surface. Note that the gut and internal genital organs like the uterus are topologically part of the body surface and that the gut lumen and the uterine cavity are spaces external to the body cavity. Regulation of the chemistry of intercellular spaces through epithelial covers is an example of an emergent biological phenomenon that cannot be explained at any level of complexity other than the organism. In addition, this view helps identify the stages of evolution at which a new organismal level of organization has been established, with all its consequences for the further differentiation of this group of organisms.

In the case of multicellular organisms reproduction and self-maintenance are also more elaborate processes; reproduction frequently involves highly specialized organs and behavioral patterns, and self-maintenance is achieved through a variety of different cellular, physiological, and behavioral mechanisms that utilize highly complex structures, such as the nervous and endocrine systems. In addition, multicellularity requires a whole new kind of biological process—development.

Higher organisms have to be rebuilt in every generation. During development the potential conflict between the competing interests of cells and the organism as a whole becomes apparent. The functional and structural integration of the organism requires the specialization of cells, which in turn leads to the loss of their potential for independent cellular reproduction, a phenomenon already recognized by Weismann in his distinction between the germ plasm and the somatoplasm (Weismann, 1892). As a result of this complex integration, the organism itself becomes the privileged biological object with respect to the processes of reproduction and self-maintenance. Leo Buss described this process as the "evolution of individuality" and suggested that many intricate features of development can be understood as direct consequences of this conflict of interest between individual cells and integrated multicellular organisms (Buss, 1987). In cancer we see the effects of a reversal of the organism's control over the reproductive interests of its cells (see, e.g., Weinberg, 1998).

The principle of functional integration is not confined to higher organisms. It is by now standard wisdom in biology that the eukaryotic cell is the result of a symbiotic integration between two or more formerly independent prokaryotic organisms (Margulis, 1970, 1981). In addition, Maynard-Smith and Szathmáry (1995) suggest that life's history on earth is marked by a series of "major transitions in evolution" that have led to the emergence of newly integrated levels of complexity (see also Margulis and Fester, 1991). These phenomena raise important technical and conceptual problems, such as the question under what conditions does the integrated "whole" gain functional precedence over its composite parts, i.e., when can we say that the integrated "whole" acts as a biological individual in a specific biological process? Within the context of an integrated higher organism certain parts, such as organs or organ systems, acquire the role of functional biological characters. They are involved in specific biological processes, such as reproduction and self-maintenance. However, their very existence depends on the ontologically prior object, the organism. We identify these biological characters by means of a theory-guided conceptual abstraction. We also want to note that defining biological character through conceptual abstraction is formally equivalent to the mathematical problem of subsystem decomposition (Klir, 1985).

In theoretical biology the organism concept functions as a unit of integration. Units of integration, such as the cell and the organism, are ontologically prior objects in the context of specific biological processes. The qualifier "in the context of specific biological processes" is important because not every biological phenomenon can be understood at the same level of complexity. Finding the appropriate level of description is ultimately an empirical question. Privileging any particular level of description *a priori* is bad metaphysics and even worse science (Brandon, 1995). In the biological hierarchy the role of any particular object can shift relative to the specific process under investigation. The cell can be the liminal reference frame, i.e., the object of ontological primacy, for the process of DNA replication or gene transcription, but it can be a functionally characterized part (character) in the context of an organismal process, such as sexual reproduction. Similarly, the organism can act either as an integrated whole or it can be a part within a larger object, such as a population. Our success in developing consistent biological theories that explain central biological processes ultimately depends on our ability to establish the appropriate level of description and to define the range of objects that are legitimate instances of the theory.

The following section discusses two concrete examples—the questions of quantitative inheritance and adaptive evolution—of how functionally relevant characters can be identified in the context of a mathematical theory representing a specific biological process.

CONTEXT-DEPENDENT UNITS OF SELECTION AS AN EXAMPLE OF FUNCTIONAL CHARACTER DECOMPOSITION

So far we have argued for the need of an analytical method that would allow us to identify the range of objects a given biological theory applies to, but we have not given any concrete examples of such a method. In this section we discuss how recent work on the unit of selection question can help to illustrate the decomposition problem and the related question of the individualization of specific characters (Laubichler, 1997a; Wagner et al., 1998; Laubichler and Wagner, 2000b). The unit of selection question is particularly well suited for this purpose because the process of natural selection is among the best understood biological phenomena (Darwin, 1859; Fisher, 1930; Williams, 1966; Dawkins, 1976; Sober, 1984; Endler, 1986; Grant, 1989).

Traditionally two different units of selection have been recognized: the replicator and the interactor (Dawkins, 1976; Hull, 1980). Both concepts are defined to capture important aspects of the process of natural selection, accurate replication of an entity type and selective interaction with the environment, respectively. In addition, Brandon (1982) introduced the conceptually important distinction between units and levels of selection. Here we analyze the role of a different kind of unit of selection, one that can be defined as a *unit of inheritance* (Laubichler, 1997a; Wagner, Laubichler et al., 1998; Laubichler and Wagner, 2000b) for technical details, but see also Wimsatt (1980) and Lloyd (1988) for further discussions of these issues.

The concept of a unit of inheritance refers to objects that contribute to the additive genetic variance of a character in a population. Consequently, these objects play an important role in the process of natural selection. The additive genetic variance is defined as the fraction of the total phenotypic variance that determines the heritability of a phenotypic (quantitative) character and therefore its ability to respond to natural selection. It can be expressed as the variance of the breeding values in a population which in turn are determined by the average effects of all the alleles in question (Falconer and Mackay, 1996; Lynch and Walsh, 1998). The technical reason why the additive genetic variance of a population has to be expressed in the form of the average effects of alleles lies in the fact that, in the case of sexual reproduction, genotypes do not cross the generational boundary intact, only genes do. In each generation, the gametes combine to reassemble the genotypes. Therefore, the breeding value of any particular genotype depends on the values of the individual genes it carries, in other words their average effects.

This is the reason why genes are privileged units of inheritance, but are they the only ones that matter? This is ultimately an empirical question. In order to address this problem we have to develop a conceptual framework that would enable us to ask this question in the first place. Over the last decades quantitative genetics has been biased toward the assumption that most problems can be approximated by a theory based on additive effects of genes (Bulmer, 1980; Falconer and Mackay, 1996). One aspect of this view is the idea that the single locus paradigm for defining average effects and breeding values can easily be extended to a multilocus case. It is simply assumed that the average effects of all alleles on a gamete add up to the average effect of each gamete and the multilocus genotype, respectively. Such a view supports the idea that genes are the relevant units of selection (Dawkins, 1976, 1982). Here we report how, by reconceptualizing the problem of defining average effects, we were able to address the question of the unit of inheritance operationally.

In our analysis we did not *a priori* assume that the average effects of the gametes can be approximated by the sum of the average effects of their alleles, i.e., we did not privilege the gene as a unit of inheritance. We were thus required to identify the average effects of objects other than individual alleles as well. Accordingly, we defined average effects for whole gametes, for functionally relevant associations of alleles on these gametes, and for all the individual alleles that make up these gametes (Laubichler, 1997a; Wagner *et al.,* 1998). Within our framework of analysis the identification of individual contributions to the total additive genetic variance thus becomes a decomposition problem. What fraction of the total additive genetic variance of the genotypes can be assigned to each of those individual entities that cross the generational boundary, such as, gametes, chromosomes, or single alleles?

To frame the question this way represents a conceptual shift in the way we analyze the unit or selection problem. Rather than defining a specific set of potential units of selection *a priori*, we now begin our analysis by identifying the most inclusive functionally relevant object. We then have to decompose the total additive genetic variance into the irreducible contributions of all objects in question.

However, neither the gametes nor the single alleles are the objects of ontological primacy in this case. Heritability is defined at the level of the phenotype, i.e., the organism. It is a phenotypic measurement, a regression of the midparent to the midoffspring phenotype (Falconer and Mackay, 1996). Organisms represented by their phenotypes are thus the ontologically prior objects in this case. However, phenotypes do not cross the generational boundary and therefore we need to identify functionally defined objects, i.e., characters, that can account for the observed phenotypic correlation between parents and their offspring. In other words, we need to identify units of inheritance, which in turn explain the observed heritability.

We have a well-established theory of quantitative genetics that describes this process (Bulmer, 1980; Falconer and Mackay, 1996; Lynch and Walsh, 1998). The next crucial step was to develop an analytical method that allows us to identify the range of objects our theory applies to. We have developed such an analytical method for the case of a two locus two allele system with epistasis and linkage disequilibrium (Laubichler, 1997a; Wagner *et al.*, 1998; Laubichler and Wagner, 2000b). To determine the relative importance of all possible units of inheritance—in the case of a two locus two allele model gametes and alleles—we calculated the average effects of both and then defined (i) the additive gametic (Agam), (ii) the additive genic (Agen), and (iii) the irreducible additive gametic value (AEgam) of a genotype. The irreducible additive gametic value of a genotype represents the fraction of the additive gametic value that cannot be accounted for by an additive combination of the average effects of the alleles. The total additive genetic variance can then be decomposed into its components:

$$VA(total)=VAgen+VEAgam+Cov(Agen \times EAgam). \qquad (1)$$

The question which biological object (gamete, allele, or both) is a functionally relevant unit of inheritance can now be addressed operationally. It is now no longer an either/or problem, but rather a question of determining the relative importance of all elements of analysis. We found that the relative importance of different units of inheritance is population specific. Depending on the amounts of linkage disequilibrium and the values for the epistasis coefficient (a measure for the strength of nonadditive genetic effects), as well as the allele frequencies in a population, the magnitude of the irreducible gametic component varies. In general, we can refute the assumption that the additive genic variance is always a good approximation of the total additive genetic variance. In Fig. 1 we give an example that illustrates the change in the relative importance of different units of inheritance in the course of evolution. The model in question is a so-called corridor model, used to study the role of pleiotropic effects on the fitness of complex adaptations (Wagner, 1988, 1989; Baatz and Wagner, 1997). Figure 1 shows a simulated evolutionary trajectory of a population from a point far away from its equilibrium until it reaches its fixed point. During the course of evolution the relative importance of different components of the additive genetic variance in fitness varies dramatically. While initially most of the additive genetic variance can be attributed to the irreducible gametic component, by the time the population nears its equilibrium the genic component dominates.

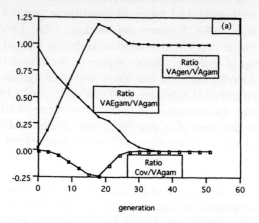

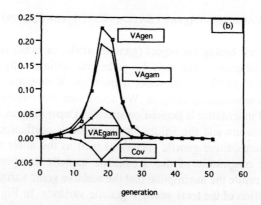

FIGURE 1 Additive genetic variance of the genotypic values for fitness in a corridor model. Shown are the fractions of the total additive genetic variance (VAgam) that can be attributed to the genic (VAgen), the irreducible gametic (VAEgam) and the covariance components (Cov) (a) and the absolute values for these variance components. The relative importance of different units of inheritance varies in the course of evolution. Initially, when the population is far away from its equilibrium point, the irreducible gametic component dominates, while closer to the equilibrium point the total additive genetic variance can be approximated by the additive contributions of individual alleles. Details can be found in Laubichler (1997a).

This type of analysis shows the kind of results we obtained when we turned the unit of selection question into a decomposition problem. We identified the functionally relevant characters (in our case units of inheritance) by their contributions to the total additive genetic variance. Units of inheritance are thus defined as abstractions of an ontologically primary object, the organism. In the next section we will generalize this theory of character decomposition. We will present the outline of an analytical theory that allows us to identify functionally relevant objects based on the notion of quasi-independence of characters.

DECOMPOSING ORGANISMS INTO A SET OF CHARACTERS

Adaptation is the result of spontaneous variation and natural selection. Since competitive success in natural selection depends on the overall lifetime performance of the organism, or even a group of organisms, adaptation is an organism level property. Nevertheless, it is often assumed that it is possible to study the adaptation of individual parts more or less independently of the rest of the organism. This assumption is at the heart of the character concept in evolutionary biology. The postulate of the independence of parts also underlies many research programs in biology (occasional lip service to the wholeness of the organism not withstanding). It raises the question under what conditions certain parts can be conceptually isolated from the rest of the organism and still fulfill the conditions of validity of the theory that we apply to them, for instance the theory of natural selection. Lewontin (1978) gave a heuristic answer to this question with the notion of *quasi-independence*. Quasi-independence refers to the assumption that it is possible to selectively adapt a specific character without simultaneously disturbing the remaining characteristics of the organism. Here we want to outline a mathematical approach that gives a precise meaning to this intuitive notion of quasi-independence. We also note that the problem of character identification in evolutionary biology is structurally the same as the so-called natural subsystem decomposition in general systems theory (Klir, 1985; Zwick, this volume).

In the context of the adaptationist research program in evolutionary biology the problem of character identification can be interpreted as a generalization of the unit of selection problem (earlier discussion). However, in contrast to much of the unit of selection debate, we do not assume *a priori* that certain objects are relevant characters, i.e., adaptations. Rather we start our investigations with the organism as the primary unit of analysis and construct characters as abstractions from the ontologically prior object. We use the following criterion for constructing characters: *given the equations that describe the dynamics of natural selection among individuals (see later), how can we lump the organism "types" (be it genotypes or phenotypes) into equivalence classes (i.e., character states), such that the dynamics of these abstract types is still predicted by the same equations without any loss of generality.* The answer to this question reveals that the symmetries of the selection equations determine the conditions under which abstraction (i.e., identification of a character) is theoretically justified.

Let $T = \{t_1, ...\}$ be a set of "types," i.e., the set of possible states of a complex adaptive system (CAS), such as genotypes or phenotypes. Associated with the set shall be a Malthusian fitness function $m:T \rightarrow R$ with $m_i = m(t_i)$, and a frequency distribution $x_i = x(t_i) \in [0,1]$. Ignoring the transmission function (i.e., the influence of mutation and recombination) the frequency distribution shall obey the Crow-Kimura differential equation

$$\dot{x}_i = x_i(m_i - \overline{m}) \tag{2}$$

with $\overline{m} = \sum_{t_i \in T} x_i m_i$ (Crow and Kimura, 1970). For the theory developed here it is important to note the invariance properties of this equation. It is easy to show that the equation is invariant to a translation of the fitness values, i.e., the dynamics of the type frequencies does not change, if each fitness value is changed by the same additive constant.

A partitioning of T will be denoted as $\Pi = \{\pi_1, ..., \pi_N\}$, where each π_j is an equivalence class of T. Each equivalence class π_j can be thought of as an allele in classical population genetics. The frequency of an equivalence class is $p_j = \sum_{t_i \in \pi_j} x_i$.

Following Rosen it is clear that any natural subsystem decomposition requires that the set T can be represented as a Cartesian product of at least two sets: $T = T_1 \times T_2$ (Rosen, 1978; see also Kim and Kim this volume). The same conclusion can also be reached from considerations about "independent variational characters" (Kim and Kim, this volume). In our context we want to represent T by the product of the focal partitioning and its "orthogonal complementary partitioning": $T = \Pi \times \overline{\Pi}$. Later we briefly discuss how one can obtain an *oc*-partitioning from a given one.

It is easy to see that an *oc*-partitioning can only exist if each equivalence class of the original partitioning has the same size M or if the number of elements of T is NM. An *oc*-partitioning can be constructed from a family of invertible 1-1 functions $F = \left\{ f_{kj} \middle| f_{kj} : \pi_j \to \pi_k \right\}$. Each function in this family maps each element of one equivalence class to its corresponding member in another equivalence class. A possible interpretation of such a function is that we seek two genotypes that are identical at a second locus. If F is also transitive, i.e., if $s = f_{kj}(t)$ and $t = f_{jl}(u)$ then $s = f_{kl}(u)$, then F defines a complementary partitioning $\overline{\Pi} = \left\{ \overline{\pi}_1, ..., \overline{\pi}_M \right\}$. Each class in this partitioning is $\overline{\pi}_k = \left\{ s \cong t \text{ if there is a } f_{kj} \in F \middle| s = f_{kj}(t) \right\}$. This definition leads to an equivalence class since the functions in F are invertible and transitive. It is also clear that for each t in T there is an $i \leq N$ and $j \leq M$ such that $\{t\} = \pi_i \cap \overline{\pi}_j$ and the type set T can be represented as $\Pi \times \overline{\Pi}$.

Using the concept of a *oc*-partitioning, we can define two important terms: Pi-additivity and Pi-LE (linkage equilibrium), i.e., additivity of fitness and linkage equilibrium with respect to a given *oc*-partioning.

Definition 1: a fitness function is additive with respect to a partitioning Π (Pi-additive) if there is an associated *oc*-partitioning and $m\left(f_{kj}(t) \right) = m(t) + c_{kj}$ for all $t \in \pi_j$.

Definition 2: a frequency distribution over T is in linkage equilibrium with respect to a *oc*-partitioning (Pi-LE) if for all $t \in \pi_j$ and all k, $x_k\left(f_{kj}(t) \right) = x_j(t)$ with $x_j(t) = \dfrac{x(t)}{p_j}$.

Note that Pi-LE implies that $x\left(t \in \pi_j \cap \overline{\pi}_k \right) = p_j \overline{p}_k$, the same criterion as in the classical definition of linkage equilibrium.

Let us now consider the conditions under which the dynamics of the frequency distribution over a partitioning is independent of the information lost because of the partitioning. In other words we want to understand under what conditions the equations governing the frequency distribution over a partitioning are dynamically sufficient. Each partitioning can be seen as an abstraction. We concentrate on the differences among the members of separate equivalence classes, but ignore the variation that exists among the members of the same equivalence class. It will be shown that the functional independence of a subsystem is closely tied to the invariance properties of the dynamical equations that govern the system. In our case the relevant equations are the Crow-Kimura differential equations, introduced earlier. These equations are invariant under a translation of the scale of Malthusian fitness values. From this property we obtain the following Lemma:

Lemma: the dynamics of the frequency distribution over a partitioning of T is independent of the rest of the system if each equivalence class has the same variance in fitness.

Proof: it is easy to show that, ignoring transmission, the selection equation for the frequency distribution over that partitioning is

$$\dot{p}_j = p_j\left(\overline{m}_j - \overline{m}\right) \tag{3}$$

with $\overline{m}_j = \dfrac{1}{p_j}\displaystyle\sum_{t \in \pi_j} m(t)x(t)$. These equations, however, do not account for the

selection that is going on between members of the same equivalence class. Rather, competition between members of the same class leads to a dynamics of the mean fitness of the equivalence classes. This dynamics, ignored by the partitioning, induces a transformation of the parameters of the system of equations (3). It has already been mentioned earlier that these types of equations are invariant under any transformation of the form $\overline{m}_j \to \overline{m}_j' = \overline{m}_j + c$.

Consequently the equations for the class frequencies remain unaffected by this dynamics, if the parameters all change with the same rate, i.e., the equations remain invariant if $\dot{\overline{m}}_j = \dot{\overline{m}}_k$ for all j and k.

It is well known that the rate of change in mean fitness is equal to the variance in fitness (one version of the fundamental theorem of natural selection)

$$\dot{\overline{m}}_j = \sum_{t \in \pi_j} m(t)^2 x_j(t) - \overline{m}_j^2 = V\left(m\middle|\pi_j\right) = V_j. \tag{4}$$

Consequently, the dynamics of the frequencies over the partitioning is dynamically sufficient if $V_j = V_k$.

Next we will consider what conditions need to be met by the frequency distribution over T and the fitness function in order to establish the partitioning as an independent subsystem. We will show that Pi-additivity and Pi-LE are necessary and sufficient in the sense defined next.

Proposition: Pi-additivity of the fitness function and Pi-LE are necessary and sufficient for the dynamical independence of the partitioning in the following sense:

1. Pi-LE is necessary and sufficient to guarantee $V_j = V_k$ for all Pi-additive fitness functions.
2. Pi-additivity is necessary and sufficient to make $V_j = V_k$ for all Pi-LE frequency distributions and to maintain Pi-LE under selection.

The proof of this proposition is somewhat technical and can be found in the Appendix. It is important to note that this result does not state that Pi-LE and Pi-additivity are the only conditions that fulfill the conditions of the lemma, since it is only stated that if one is given, than the other is necessary and sufficient. There are a number of special cases that formally fulfill the criterion of equal genetic variance, but are highly singular and dynamically unstable. A more general theorem has not yet been proven.

This result explains why additivity and linkage equilibrium play such a prominent role in theories of population and quantitative genetics. First, they are the conditions that assure that the dynamics of allele frequencies at a locus, or the frequency distribution of a quantitative character, are dynamically independent of the rest of the genotype (in the case of genes) or the phenotype (in the case of quantitative characters). In other words, additivity and linkage equilibrium are the conditions under which the theory of natural selection can be applied to sub-organismal characters. Second, these conditions are directly dictated by the invariance properties of the dynamical equations themselves, i.e., the Crow-Kimura equations for overlapping generations with weak selection (Crow and Kimura, 1970). Analogous results can be derived for Wright's selection equation for non-overlapping generations (Carter and Wagner, unpublished). These conditions also guarantee that those properties of the organism that have been ignored by the abstraction of characters have no impact on the dynamics of the system, since the equations that govern the dynamics of these characters are invariant to their influence.

It is interesting to note that the conditions for dynamical independence require a match between the frequency distribution of the types and the fitness function. The fitness function is determined by the physiology of the organism and its interactions with the environment, whereas the frequency distribution is determined largely by the transmission function, i.e., the probability distribution with which certain types are generated by a certain parental type (Altenberg and Feldmann, 1987). This means that, only if a character defined by the dynamics of selection is simultaneously also a unit of transmission, such as a gene under Mendelian patterns of inheritance, can it be a dynamically independent character. Hence, there is a strong relationship between a unit of inheritance and a unit of selection, not in the material sense, but as defined by their dynamical properties that may, but do not need to, correspond to a physical unit.

Another interpretation of this result is that the conditions of Pi-additivity and Pi-LE guarantee the independence of one character from the other and that this enables these characters to act as independent modules of phenotypic adaptation (Wagner and Altenberg, 1996). Our results thus define the conditions under which complex organisms can maintain their evolvability, despite their physical complexity. Furthermore these conditions imply what kind of evolutionary changes can lead to the origin of a new independent character. The origin of a new character is based on the convergence of a functional property (Pi-additivity) and a property of the genetic/developmental system that maintains Pi-LE. Only if the genetic representation (Wagner and Altenberg, 1996) that leads to Pi-LE matches the pattern of functional constraints (Pi-additivity) can one speak of a new independent entity. This statement has obvious consequences for the definition of evolutionary novelties.

CONCLUSION

We began this paper with the claim that the structural weakness of many mathematical theories in biology is caused by their lack of an analytical theory that identifies the objects the theory applies to. We argued that in order to overcome this deficit, in many cases one needs an organismal perspective. One of the roles that we ascribed to the organism in theoretical biology is to provide the functional context from which biological objects (characters) can be abstracted. We also argued that in these cases the organism is ontological prior to its composite parts and that the individualization (abstraction) of biological characters is in essence a subsystem decomposition problem.

We gave two analytical examples for such a decomposition theory, the unit of inheritance and a generalized theory of biological characters. In both cases, we defined the relevant objects through their role within an organismal process represented by dynamical equations. Classes of objects are defined by their invariance with respect to specific transformations, i.e., by their stability within the process. We argued that this process specificity is characteristic for all suborganismic characters and even for biological objects in general.

Finally, we argued that the cell and the organism are privileged biological objects with respect to the processes of reproduction and self-maintenance. Self-maintenance implies the regulation of an internal milieu. We therefore propose that, in the case of animals, the organismal level of complexity arose with the origin of epithelial surfaces that separate the interior, or the body cavity, from the external environment.

ACKNOWLEDGMENTS

We thank the following individuals for helpful discussions on the subject: Lee Altenberg, Robert Brandon, James Griesemer, Paul Griffiths, and Bill Wimsatt. GPW would thank Junhyong Kim for his intellectual companionship in developing many of the ideas presented in this paper. JK presented his own interpretation of these issues in another paper (Kim and Kim, this volume). He also acknowledges his collaboration with Ashley Carter on extending the results on character decomposition (results not yet published). This is contribution # 62 of the Yale Center for Computational Ecology.

APPENDIX

Remark: this formulation of the proposition recognizes that there are other constellations of fitness functions and frequency distributions which make $V_j = V_k$, but they are all highly singular and do not guarantee this condition under a wide variety of conditions.

PROOF OF PROPOSITION:
Part 1: since the fitness function is assumed to be Pi-additive the variance of the kth equivalence class is

$$V_k = \mathrm{Var}\left(m|\pi_k\right) = \mathrm{Var}\left(m\left(f_{jk}(s)\right) + c_{jk}\middle| s \in \pi_k\right)$$
$$\mathrm{Var}\left(m\left(f_{jk}(s)\right) + c_{jk}\middle| s \in \pi_k\right) = \mathrm{Var}\left(m\left(f_{jk}(s)\right)\middle| s \in \pi_k\right)$$

The difference $V_k - V_j$ can be written as

$$\sum_{t \in \pi_j} m(t)^2\left[x_k\left(f_{kj}(t)\right) - x_j(t)\right] - 2\overline{m}_j \sum_{t \in \pi_j} m(t)\left[x_k\left(f_{kj}(t)\right) - x_j(t)\right]$$

The fitness function can be arbitrarily scaled such that $\overline{m}_j = 0$, leading to

$$V_k - V_j = \sum_{t \in \pi_j} m(t)^2 \left[x_k \left(f_{kj}(t) \right) - x_j(t) \right]$$

Since $m(t)^2 \geq 0$ the only condition which makes $V_k = V_j$ for all Pi-additive fitness functions is $\left[x_k \left(f_{kj}(t) \right) - x_j(t) \right] = 0$ for all $t \in \pi_j$, i.e., Pi-LE.

Part 2: assume the frequency distribution over T is Pi-LE, then the variances can be written as

$$V_k = \sum_{t \in \pi_j} \left(m \left(f_{kj}(t) \right) - \overline{m}_k \right)^2 x_k \left(f_{kj}(t) \right)$$

$$V_j = \sum_{t \in \pi_j} \left(m(t) - \overline{m}_j \right)^2 x_j(t)$$

we now replace the fitness values of the types in the class π_k by

$$m \left(f_{kj}(t) \right) = m(t) + d_k \left(f_{kj}(t) \right)$$

then we have

$$V_k = \sum_{t \in \pi_j} \left(m(t) + d_k \left(f_{kj}(t) \right) \right)^2 x_j(t) - \overline{m}_k^2$$

$$\overline{m}_k^2 = \overline{m}_j^2 + 2\overline{m}_j \sum_{t \in \pi_j} d_k \left(f_{kj}(t) \right) x_j(t) + \left(\sum_{t \in \pi_j} d_k \left(f_{kj}(t) \right) x_j(t) \right)^2$$

which can be simplified to

$$V_k = V_j + 2 \mathrm{E}_j \left(m(t) d_k \left(f_{kj}(t) \right) \right) - 2 \overline{m}_j \, \mathrm{E}_j \left(d_k \left(f_{kj}(t) \right) \right) +$$
$$+ \mathrm{E}_j \left(d_k \left(f_{kj}(t) \right)^2 \right) - \mathrm{E}_j \left(d_k \left(f_{kj}(t) \right) \right)^2$$

$$V_k = V_j + 2 \mathrm{Cov}_j \left(m(t), d_k \left(f_{kj}(t) \right) \right) + \mathrm{Var}_j \left(d_k \left(f_{kj}(t) \right) \right)$$

Hence, $V_k = V_j$ if and only if

$$2 \mathrm{Cov} \left(m(t) d_k \left(f_{kj}(t) \right) \right) + \mathrm{Var}_j \left(d_k \left(f_{kj}(t) \right) \right) = 0.$$

This equation is trivially fulfilled if there is no variation in any of the equivalence classes because then the variances and covariances are also equal to zero. Another condition is that $d_k\big(f_{kj}(t)\big) = \text{const}$, which is the same as Pi-additivity. Note that zero variance also implies zero covariance. Finally, the condition is fulfilled if

$$\frac{\text{Cov}_j\big(m(t)d_k\big(f_{kj}(t)\big)\big)}{\text{Var}_j\big(d_k\big(f_{kj}(t)\big)\big)} = -\frac{1}{2}$$

which is the regression coefficient of d on m equal to -1/2, or if the linear regression equation reads

$$m(t) = -\frac{1}{2}d_k\big(f_{kj}(t)\big) + \text{const},$$

or if

$$m\big(f_{kj}(t)\big) = -m(t) + 2\,\text{const}$$

This is an interesting condition that could be called reverse additivity. It is the same as additivity, but the fitness gradient is exactly reverse. Even if this condition fulfills the requirement $V_k = V_j$, it is incompatible with Pi-LE, since it leads to the selection of noncorresponding elements in the classes π_j and π_k. Thus only Pi-additivity is both sufficient for $V_k = V_j$, as well as necessary for arbitrary Pi-LE distributions and compatible with the maintenance of Pi-LE. This proves the proposition.

LITERATURE CITED

Altenberg, L., and Feldmann, M. W. (1987). Selection, generalized transmission, and the evolution of modifier genes. I. The reduction principle. *Genetics* **117**:559-572.

Baatz, M. and Wagner, G. P. (1997). Adaptive inertia caused by hidden pleiotropic effects. *Theor. Popul. Biol.* **51**:49-66.

Bailey, V. A. *et. al.* (1962). Interaction between hosts and parasites when some host individuals are more difficult to find than others. *J. Theor. Biol.* **3**:1-18.

Berryman, A. A. (1981). "Population Systems: A General Introduction." Academic Press, New York.

Brandon, R. (1982). "The Levels of Selection." Biennial Meeting of the Philosophy of Science Association, Philosophy of Science Association.

Brandon, R. N. (1990). "Adaptation and Environment." Princeton University Press, Princeton.

Brandon, R. N. (1995). "Concepts and Methods in Evolutionary Biology." Cambridge University Press, Cambridge.

Bulmer, M. G. (1980). "The Mathematical Theory of Quantitative Genetics." Calderon Press, Oxford.

Buss, L. (1987). "The Evolution of Individuality." Princeton University Press, Princeton.

Crow, J. F., and Kimura, M. (1970). "An Introduction to Population Genetics Theory." Harper and Row, New York.

Darwin, C. (1859). "The Origin of Species." John Murray, London.

Dawkins, R. (1976). "The Selfish Gene." Oxford University Press, Oxford.

Dawkins, R. (1982). "The Extended Phenotype." Oxford University Press, Oxford.

Endler, J. A. (1986). "Natural Selection in the Wild." Princeton, Princeton University Press.

Falconer, D. S., and Mackay, T. F. C. (1996). "Introduction to Quantitative Genetics." Longman, Edinburgh.

Fisher, R. A. (1930). "The Genetical Theory of Natural Selection." Calderon Press, Oxford.

Gesteland, R. F., Cech, T. R. *et al.*, eds. (1999). "The RNA World: The Nature of Modern RNA Suggests a Prebiotic RNA." Cold Spring Harbor Laboratory Press, Cold Spring Harbor.

Gilpin, M., and Hanski, I., eds. (1991). "Metapopulation Dynamics: Empirical and Theoretical Investigations." Academic Press, London.

Grant, P. (1989). "Ecology and Evolution of Darwin's finches." Princeton University Press, Princeton.

Griffiths, P. (1997). "What Emotions Really Are? The Problem of Psychological Categories." University of Chicago Press, Chicago.

Griesemer, J. (2000). "Reproduction and the Reduction of Genetics in Development" (unpublished manuscript).

Hanski, I., and Gilpin, M., eds. (1997). "Metapopulation Biology: Ecology, Genetics, and evolution." Academic Press, San Diego.

Hartl, D. L. (1997). "Principles of Population Genetics." Sinauer, Sunderland, MA.

Hofbauer, J., and Sigmund, K. (1988). "The Theory of Evolution and Dynamical Systems: Mathematical Aspects of Selection." Cambridge University Press, Cambridge.

Hofbauer, J., and Sigmund, K. (1998). "Evolutionary Games and Population Dynamics." Cambridge University Press, Cambridge.

Hull, D. (1980). Individuality and selection. *Annu. Rev. Ecol. Syst.* 1:311-332.

Kim, J., and Kim, M. (2000). The mathematical structure of characters and modularity. *In* "The Character Concept in Evolutionary Biology" (G. P. Wagner, ed.). Academic Press.

Klir, G. (1985). "The Architecture of Systems Problem Solving." Plenum Press, New York.

Laubichler, M. D. (1997a). Identifying Units of selection: Conceptual and Methodological Issues. *Biology.* Yale University, New Haven:vii+197.

Laubichler, M. D. (1997b). The nature of biological concepts." *Eur. J.Semiotic Stud.* 9(2):251-276.

Laubichler, M. D., and Wagner, G. P. (2000a). Levels of selection in a two locus two allele system (in preparation).

Laubichler, M. D., and Wagner, G. P. (2000b). Organisms and character decomposition: steps towards an integrative theory of biology. *PSA* (in press).

Lewontin, R. (1978). Adaptation. *Sci. Am.* 239:156-169.

Lloyd, E. (1988). "The Structure and Confirmation of Evolutionary Theory." Greenwood Press, New York.

Lynch, M., and Walsh, B. (1998). "Genetics and Analysis of Quantitative Traits." Sinauer, Sunderland, MA.

Margulis, L. (1970). "The Origin of Eucaryotic Cells." Yale University Press, New Haven.

Margulis, L. (1981). "Symbiosis in Cell Evolution." W. H. Freeman, New York.

Margulis, L., and Fester, R., eds. (1991). "Symbiosis as a Source of Evolutionary Innovation." MIT Press. Cambridge, MA.

Maynard-Smith, J. (1998). "Evolutionary Genetics." Oxford University Press, Oxford.

Maynard-Smith, J., and Szathmáry, E. (1995). "The Major Transitions in Evolution." Oxford University Press, Oxford.

Miller, S. L., and Orgel, L. E. (1974). "The Origins of Life on Earth." Prentice-Hall, Englewood Cliffs.

Mitton, J. (1997). "Selection in Natural Populations." Oxford University Press, Oxford.

Murdoch, W. W. et al. (1996). Refuge dynamics and metapopulation Dynamics: an experimental test. Am. Nat. 147(3):424-444.

Nagylaki, T. (1992). "Introduction to Theoretical Population Genetics." Springer, Berlin.

Rosen, R. (1962). "Church's thesis and its relation to the concept of realizability in biology and physics. Bull. Math. Biophys. 24:375-393.

Rosen, R. (1978). "Fundamentals of Measurement and Representation of Natural Systems." North-Holland, New York.

Sober, E. (1984). "The Nature of Selection." MIT Press, Cambridge, MA.

Sokal, R. R., and Rohlf, J. F. (1981). "Biometry. The Principles and Practice of Statistics in Biological Research." W. H. Freeman, New York.

Stearns, S. C. (1992). "The Evolution of Life Histories." Oxford University Press, Oxford.

Wagner, G. P. (1988). The influence of variation and of developmental constraints on the rate of multivariate phenotypic evolution. J. Evol. Biol. 1:45-66.

Wagner, G. P. (1989). Multivariate mutation-selection balance with constrained pleiotropic effects. Genetics 122:223-234.

Wagner, G. P. (1997). The structure of biological concepts and its relation to the dynamics of biological organizations. Eur. J. for Semiotic Stud. 9:299-320.

Wagner, G. P., and Altenberg, L. (1996). Complex adaptations and the evolution of evolvability. Evolution 50:967-976.

Wagner, G. P., Laubichler, M. D. et al. (1998). Genetic measurement theory of epistatic effects. Genetica 102/103:569-580.

Weinberg, R. A. (1998). "One Renegade Cell: How Cancer Begins." Basic Books, New York.

Weismann, A. (1892). "Das Keimplasma: Eine Theorie der Vererbung." Gustav Fischer, Jena.

Williams, G. C. (1966). "Adaptation and Natural Selection." Princeton University Press, Princeton.

Wimsatt, W. (1980). "The Unit of Selection and the Structure of the Multi-level Genome." Biennial Meeting of the Philosophy of Science Association, Philosophy of Science Association.

Wimsatt, W. C. (2000). Emergence as non-aggregativity and the biases of reductionism. In "Natural Contradictions: Perspectives on Ecology and Change" (P. J. Taylor and J. Haila, eds.), (forthcoming).

Zwick, M. (2000). Wholes and parts in general systems methodology. In "The Character Concept in Evolutionary Biology" (G. P. Wagner, ed.), Academic Press.

7

FUNCTIONAL UNITS AND THEIR EVOLUTION

KURT SCHWENK

Department of Ecology and Evolutionary Biology, University of Connecticut, Storrs, CT 06269

INTRODUCTION

According to the neo-Darwinian ideal, organisms can be decomposed into atomistic traits or characters which, by definition, represent units of phenotypic evolution. Implicit in this view is the assumption that any given character can evolve independent of all other characters. As such, it is free to respond to

The Character Concept in Evolutionary Biology

selection according to its own, unique relationship with the environment, unfettered by association with its fellows. While such a characterization of the neo-Darwinian position could justifiably be regarded as overstated and few, if any, modern biologists would profess it absolutely, it is at the same time the unstated credo of practicing systematists who, as a matter of pragmatic routine, must atomize their organisms into characters with little or no regard as to how they might be associated genetically, developmentally, or functionally.

This paradox is at the heart of our ignorance of how phenotypes evolve and it manifests a growing philosophical schism in the field. At one extreme we have the traditional, atomistic approach as noted—organisms viewed as little more than "bags of characters," each character available for individual honing by environmental selection to create adapted phenotypes (Rieppel, 1986). Accordingly, the phenotype is held to be highly responsive to the exigencies of an ever-changing environment, virtually protean through evolutionary time, with diversity the inevitable outcome. Change, though inevitable, occurs incrementally, character by character. Phenotypic stasis is seen as exceptional and explicable only by reference to unusually persistent environmental conditions (Simpson, 1953). In short, the organism can be understood as the sum of its parts in relation to the environment.

Yet this view of the phenotype is transparently false. Characters are not diffused through space, but are grouped and bounded within organisms where they exist in ordered relations with one another to form tissues, organs, and systems (Whyte, 1965). These character associations are further related temporally through development and growth, and dynamically through functional interaction. As historical entities, organisms transmit not only their morphological features from one generation to the next, but their unique set of organizational properties, as well—the patterns of interaction among their characters. Thus, by virtue of being *organismal* attributes, characters *must* evince associations with other characters. Organisms thus manifest complexly nested webs of character interaction and integration, such that a phenotypic change in one character will almost certainly have an impact on others.

This inescapable conclusion, virtually a truism, has led to an extreme, structuralist conception of the phenotype in opposition to atomistic doctrine (e.g., Whyte, 1965; Allen, 1980; Wake *et al.,* 1983; Webster and Goodwin, 1996). In this view, rather than the character-environment relationship, it is character *integration* that is held to be paramount in phenotypic evolution. This integration is necessary to maintain adaptive phenotypes in the face of environmental changes that would tend to disrupt them. As such, organisms are viewed as self-maintaining, self-stabilized, "autopoietic" systems:

> The range of ontogenetic and phylogenetic change of one element is, therefore, determined by the structural and functional properties of all other elements. Each ontogenetic or phylogenetic change of the system must remain within the functional

limits of the process of circular production and maintenance of the elements, or the system itself will decompose (Wake *et al.*, 1983, p. 218).

Thus, successful phenotypes "gel," i.e., they become increasingly self-stabilized and resistant to phenotypic change due to the internal dynamics of their character associations. Long-term phenotypic stasis is expected and evident in *Baupläne* within which diversity is limited to "variations on a theme." Significant phenotypic change is seen as rare and the result of unusual processes or circumstances that perturb intrinsic stability. Implicit is that such perturbations will cause a rapid phenotypic reorganization followed by return to an integrated steady state (Arthur, 1997). Character integration is an emergent property of whole organisms, hence in this view the organism is more than the sum of its parts—it is, moreover, a set of organizational and coordinative conditions (Whyte, 1965).

In this dialectical view of the organism we are confronted, on the one hand, by evolutionarily malleable phenotypes that evolve in lock-step with the environment, and on the other, by evolutionarily static phenotypes that are intrinsically resistant to evolutionary change. In the former view an adaptive state is maintained dynamically by matching individual characters to the environment, while in the latter an adaptive state is maintained by matching suites of characters to the internal demands of coordination and integration, irrespective of the environment. In the first, phenotypic change is the null model of evolution, diversity its outcome, and stasis exceptional; in the second, stasis is the null model, disparity its outcome, and phenotypic change exceptional.

Although, as noted, the atomistic view cannot be wholly correct because organisms are integrated wholes, neither can the structuralist view be accepted entirely. It is clear that traits sometimes do behave atomistically, that organisms do maintain adaptive phenotype-environment matching and that *Baupläne* are remarkably diverse. Thus, while characters might not be truly atomistic, in many cases they are at least "quasi-independent" entities able to respond adaptively to environmental selection pressures (Lewontin, 1984; Wagner, 1996). Furthermore, there is no denying the efficacy and power of atomistic approaches, imperfect as they are, in evolutionary analysis; the success of cladistic methods in morphological systematics is proof enough of this. Clearly there exists a middle ground for phenotypic evolution.

To a large extent the philosophical differences just outlined reflect different emphases. The atomistic approach focuses on characters and adaptive evolution, whereas the structuralist approach focuses on *Baupläne* and nonadaptive forces in evolution, such as constraint. These foci follow logically from the fact that characters and body plans are most apparent as stable historical entities, i.e., phenotypic "units." However, there is a third level of organization that also manifests elements of stability, namely the character complex (Wagner and Schwenk, 2000). Evolutionary attributes of character complexes have received comparatively little attention. As Wagner and Schwenk (2000) pointed out, the emergent stability evident at each level (character, character complex, *Bauplan*) is believed to reflect different causal mechanisms, hence each requires a differentempirical and conceptual characterization. The object of this essay is to explore the middle ground of phenotypic evolution reflected in character complexes. Specifically, I consider the notion of "functional units," or character complexes identified by shared functional interactions. Examples are drawn from vertebrate zoology because this is where my expertise lies, but the concepts should be no less applicable to other animal systems and, with some modification, to plants as well.

WORKING CHARACTER CONCEPT

Although the character concept is by no means resolved (the subject of this book), neither is it of central concern here. It is enough to know that there are such things as characters and that these represent individuated units of phenotypic evolution (Wagner, 1999). As such, they are "quasi-independent" (Lewontin, 1984) and "quasi-autonomous" (Wagner, 1995, 1996, 1999) parts of the phenotype, meaning, in the first place, that they are capable of evolutionary change somewhat independent of change in other "characters," a property largely contingent upon the fact that they are, in the second place, developmentally autonomous units individuated from other such units. In short, my use of "character" in this paper follows the biological homology concept (e.g., Roth, 1984, 1986; Wagner, 1986, 1989a,b, 1999; see also Bock, 1963): characters are units of phenotypic evolution that are individuated by a unique set of developmental constraints. Homology among characters can be recognized at the phenotypic level at which these constraints are shared, i.e., homologues share the same set of constraints and variational properties (Wagner, 1999).

THE FUNCTIONAL UNIT CONCEPT

Gans (1969) attributed the concept of the "functional unit" to van der Klaauw (1945) and in a modern, formalistic sense this is true. However, that organisms comprise groups of interrelated parts was recognized as early as Aristotle (Russell, 1916). Etienne Geoffroy Saint-Hillaire saw the relationship among parts as the basis for "unity of plan" among diverse species: "Now it is evident that the sole general principle one can apply is given by the position, the relations, and the dependencies of the parts, that is to say, by what I name and include under the term of *connections*" (Geoffroy, 1818; in Russell, 1916, p. 53). However, Geoffroy's "principle of connections" was nonfunctional in the sense that the "relations" and "dependencies" to which he referred were purely morphological. These morphological connections imposed on organs a particular function, rather than the converse.

In contrast, Georges Cuvier elaborated the view that it is the functional interrelationships among parts that are the primary determinants of form. As such, form followed from function. Thus the notion of functional integration is most clearly identified with Cuvier and his "principle of the conditions of existence": "It is on this mutual dependence of the functions and the assistance which they lend one to another that are founded the laws that determine the relations of their organs; these laws are as inevitable as the laws of metaphysics and mathematics, for it is evident that a proper harmony between organs that act one upon another is a necessary condition of the existence of the being to which they belong" (Cuvier, 1800; in Russell, 1916, p. 34). This principle is the basis for Cuvier's better known   "principle of correlation" with which he asserted that one could (in most cases) reliably reconstruct the whole organism from one of its parts: "This must necessarily be so: for all the organs of an animal form a single system, the parts of which hang together, and act and re-act upon one another; and no modifications can appear in one part without bringing about corresponding modifications in all the rest" (Cuvier, 1826; in Russell, 1916, p. 35). Owen (1837) suggested that Cuvier's ideas were actually a refinement of much earlier 18th century ideas, but this is not wholly justified; in any case, the source of Cuvier's "functionalism" lies ultimately with Aristotle.

It is apparent from the quote that Cuvier's principal concern was the whole organism, which he viewed as a composite of interrelated parts. Most relevant here, however, is that Cuvier also applied his principle of correlation to lower hierarchical levels such as among parts within a single organ or organ system (Russell, 1916). Furthermore, although Cuvier believed that organisms were "adapted" to their environments, his "conditions of existence" related not to the Darwinian notion of the external environment ("the struggle for existence"), but to *intraorganismal* function. As summarized by Russell (1916, p. 34), "The very condition of existence of a living thing, and part of the essential definition of it, is that its parts work together for the good of the whole." Thus Cuvier perceived organisms as hierarchically organized into systems and "characters" whose phenotypes were ultimately determined by the internal dynamics of functional interaction. Although Cuvier was not an evolutionist, his work foreshadowed not only the functional unit concept, but systems-level approaches to the phenotype (e.g., Dullemeijer, 1974; Riedl, 1977; Wagner, 1986) and the structuralist notion of "internal selection" (Whyte, 1965; Wake *et al.,* 1983; Arthur, 1997; Wagner and Schwenk, 2000), i.e., the idea that the internal dynamics, or temporal and functional relationships among parts, create selection pressures on character phenotypes independent of the external environment (see later).

Other workers such as Owen (1837, 1866) embraced Cuvier's principles, but most 19th century morphologists continued to focus on anatomy and embryology while functional analysis was relegated to the separate tradition of physiology. Functional morphology as a discipline did not emerge until mid-20th century (e.g., Dullemeijer, 1974; Lauder, 1982; Schwenk, 2000b). Van der Klaauw's (1945) study was pioneering in this regard because he considered how subdivisions within an apparently unitary structure, such as the vertebrate skull, were to be recognized. Whereas most previous workers had resorted to topological or embryological arguments, van der Klaauw (1945, p. 31) concluded that function provided a rational basis for decomposing the skull into separate moieties: "however important the distinction between facial and cerebral skull may be, more important still is the consideration of the skull as a combination of a great number of *comparatively separate functional units*" (italics added). Indeed, van der Klaauw (1945) identified minimally 36 such units in the mammalian skull! Nonetheless, he saw the organization of the skull as hierarchical: functional units impose strong demands on the form of their component elements, but the form of each unit is comparatively less influenced by its relationships with other functional units. Overall skull form must, nevertheless, dictate some degree of integration among functional units, the strength of which van der Klaauw (1945) regarded as taxonomically variable dependingon skull function (see later).

Pieter Dullemeijer, van der Klaauw's student and successor at the University of Leiden, elaborated the concept of "functional components" (e.g., 1956, 1959, 1974, 1980) in which each morphological element, or character, is the center of a "field of decreasing...constructional consequences" (Dullemeijer, 1980, p. 226). Although functional components are characterized by functional (and other) relationships among elements, they do not constitute a functionally closed or "comparatively separate" system in the sense of a functional unit (*contra* Gans, 1969[**]). They are manifest by the identification of a central element from which extends its functional component without boundary. Nonetheless, Dullemeijer strongly emphasized the functional interrelations among parts, particularly within systems. In Dullemeijer's view, a functional system is composed of overlapping functional components. Thus in considering a functional unit, such as the feeding system of viperid snakes, he portrayed the complex relationships among nearly all parts constituting the system (Fig. 1). This conception of the functional unit stresses their global integration rather than their individuation or "separateness" from other such units. Nonetheless, Dullemeijer (1974, 1980) recognized that functional units were, to some extent, evolutionarily individuated, stable entities.

[*] Gans (1969) conflates the terms "functional unit" and "functional component" owing to the fact that van der Klaauw (1945) does not make a distinction. However, Dullemeijer (1956, 1959 and later works) develops functional component as a separate concept along the lines noted. I have preserved this important distinction here.

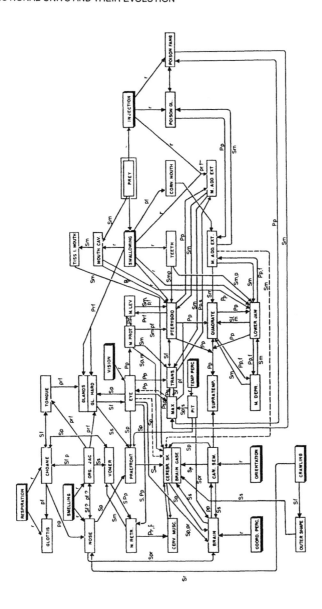

FIGURE 1 Dullemeijer's representation of the functional and morphological relationships within the cranial feeding system of a crotalid snake. Shaded boxes indicate activities which have a strong impact on head form and rectangles represent morphological elements. Arrows indicate the direction of influence, capital letters the property that is influenced, and small letters the influencing property. f, function; m, mechanical influence; p, position; r, relation between form and function, or the realization of a function in a form; s, shape and structure. From Dullemeijer (1974), *Concepts and Approaches in Animal Morphology*, Van Gorcum and Co. B.V., with permission.

Olson and Miller (1958) wrote an ambitious book that developed a quantitative, statistical approach to the concept of morphological integration within organisms. They posit at the outset what is, in effect, Cuvier's "principle of the conditions of existence," little modified: "Each component of the organism must be formed so that the part it plays in the existence and function of the whole is carried out properly with respect to all other parts. It would seem logical that the degree of interdependency of any two or more morphological components in development and function would bear a direct relationship to the extent of their particular morphological integration" (p. v). They went on to show that certain groups of morphological characters within an organism were more tightly correlated, in a statistical sense, than other such sets of characters and that tightly correlated (integrated) character sets usually corresponded to functionally interrelated suites of characters, or "F-groups." F-groups therefore represent functional units which they found to have a coherence not evident in functionally unrelated characters. Their system of correlation and covariance emphasized the hierarchical organization of such relationships within organisms with centers of integration organized around functional complexes and the complexes, themselves, integrated within the whole organism.

Bock (1963) acknowledged that characters in a system compose a set of functionally interrelated parts whose mutual interactions tend to influence patterns of phenotypic evolution in the system (and the organism). Bock (1964) and Bock and von Wahlert (1965, p. 272) formalized this idea in their own concept of the "functional complex or functional unit" which they suggested is composed of "a group of smaller features which act together to carry out a common biological role." The essential point is that all participating characters in a functional unit subscribe to the same "biological role," i.e., the part played by the character complex in the natural life of the animal which determines a set of selection pressures to which it is exposed (for example, the function of horns in mountain sheep might be to absorb the energy of impact incurred during head-butting behavior, but their biological role is to establish dominance in a polygynous breeding hierarchy, the "source" of selection for the evolution of horns). Individual characters might participate in more than one functional unit, but a functional unit is uniquely characterized by a set of characters that all function together to perform a particular biological role. The application of "biological role" to functional units is similar to Wagner and Schwenk's (2000) use of "proper function" to characterize "evolutionarily stable configurations" (see later).

Importantly, Bock and von Wahlert (1965, p. 272) explored the relationship between part and whole in their conception of the functional unit. They noted that:

> No real distinction exists between individual features and character complexes; forms, functions, and biological roles may be described for each. Usually, however, the biological roles of the individual features are the same as those of the character complex; this concurrence of biological roles results from the traditional division of the major functional units of the organism into smaller but still distinguishable morphological units.

They thus express the hierarchical organization of organismal phenotype into systems, or functional units, and these into individuated (morphological) characters. Moreover, by suggesting that all individual characters of the complex share the same set of biological roles as the functional unit, itself, they highlight the "comparatively separate" nature of functional units within the organism, i.e., not only the characters are individuated, but the functional units are, to some extent, as well.

Bock and von Wahlert (1965, p. 272) go on to acknowledge the difficulty inherent in accurately atomizing such systems into characters ("features"): "Should, for example, a muscle-bone unit be considered as a single feature or should this unit be considered as a character complex that can be divided into smaller units—the muscle and bones to which it attaches. Is the hind limb of a tetrapod, which everyone would agree is a character complex, a single feature or a series of smaller features?"

With the concept well-established and sophisticated functional analyses on the rise, functional units were increasingly invoked in the empirical literature (Gans, 1969). Gans (1969), however, complained that the concept was being misapplied by most workers. His objection seems partly to have stemmed from his erroneous conflation of "functional component" and "functional unit" concepts (see earliler discussion). In any case, he suggested that separate concepts were required to distinguish between essentially anatomical subdivisions and true, functional subdivisions. He proposed to call the former "mechanical units" which were to be recognized by relative degrees of movement: relatively little movement among component parts within a unit and relatively greater movement between and among units. "Functional unit" was to be reserved for cases in which a much greater knowledge of function in a biological context was available. Gans (1969) objected to the use of "functional unit" when an individual element might participate in different functions. However, overlap among functional units has been recognized and even emphasized by all workers noted earlier. Despite his concerns about overlap, Gans (1969, p. 366) recognized that "there is a basic coordination of parts that obviously circumscribes certain regions."

Finally, Wagner and Schwenk (2000) identified a type of functional unit they called "evolutionarily stable configurations" (ESCs). ESCs are character systems united in their performance of a particular "proper function" (Millikan,

1984), roughly, the purpose for which the complex evolved ["proper function"combines elements of both "function" and "biological role," *sensu* Bock and von Wahlert (1965)]. Wagner and Schwenk (2000) argued that the functional integration among characters constituting an ESC exerts intraorganismal selection pressures on individual characters. This selection derives from the need of characters to function in a coordinated way with the other characters of the system to produce the proper functional output. Too great a phenotypic deviation in one character would disrupt the functionality of the system, as a whole, and would therefore be selected against. This type of selection occurs with little regard to the particular external environment because it is imposed by an intrinsic, organismal attribute, hence it has been called "internal selection" to distinguish it from typically Darwinian, adaptive "external selection" (Whyte, 1965; Dullemeijer, 1980; Arthur, 1997; Wagner and Schwenk, 2000). ESCs are putatively self-stabilized phenotypic units that resist modification in a wide range of environments, thus promoting phenotypic stasis in the system through time and cladogenesis. This stability arises from an emergent property of the system which is evident as functional integration. Thus, ESCs are similar to Bock and von Wahlert's (1965) conception of a functional unit with the additional conditions of functional integration and self-stabilization. The ESC concept is developed further (see later).

FUNCTIONAL UNITS RECONSIDERED

While it is unlikely that there will be universal agreement on the nature of functional units, there appears to be little doubt about their reality. In this sense, functional units are no different from characters and other putative "unitary" entities which must be defined and characterized within a specific context and with reference to particular processes (e.g., Wagner, 1999, for characters; Wagner and Schwenk, 2000, for ESCs). Functional units are recognizable as such in two different contexts: within an individual organism in the context of function and within a population of organisms through evolutionary time, during which functional units manifest some level of phenotypic stability. Thus functional units are potentially definable with reference to particular functional and evolutionary processes. A characterization of these processes provides the grist for generating testable hypotheses about the nature of functional units, particularly in regard to the phenotypic and phylogenetic patterns we expect them to produce. In this portion of the essay I make a preliminary foray into this arena and identify three general "types" of functional unit (Table I). Each proposed type should be taken as an hypothesis, ideally to be dissected, tested, modified, refuted, or supported.

TABLE I   Types of Functional Unit

Type	Components	Attributes	Putative examples
1. STRUCTURAL UNIT	a. morphogenetic units	a. morphological unity	a. rodent mandible
		b. intrinsic functional integration	b. snake maxilla
		c. phylogenetic stability	c. some skull roof bones
		d. may exhibit phylogenetic fission	
2. MECHANICAL UNIT	a. characters	a. morphological composite	a. mammal skulls
	b. structural units	b. intrinsic functional integration	b. primitive fish skulls (e.g. *Amia*, *Acipenser*, fossil Dipnoi)
		c. phylogenetic stability	c. lizard skull/parietal foramen position
		d. solidity/little instinsic movement	d. turtle shell
		e. strong selection on total form, weak selection on components	e. limb girdles
		f. total form canalized, components variable	f. snake braincase
3. EVOLUTIONARILY STABLE CONFIGURATIONS (ESC)	a. characters	a. morphological composit	a. mammalian masticatory system
	b. structural units	b. intrinsic functional integration	b. lizard lingual feeding
	c. mechanical units	c. phylogenetic stability	c. kinetic skulls
	d. other ESC's	d. kinetic/high levels of instrinsic movement	d. cognitive systems (?)
		e. internal selection on components	
		f. phenotypic stability of components within threshold limit	
		g. phenotypic stability of total form	
		h. buffered from environmental change	

The conceptual and especially operational ambiguity of functional units reflects the complexity inherent in hierarchically organized systems that are not only integrated through function, but also temporally through development and historically through phylogenesis. Thus, to invoke function exclusively as a causal explanation for the coherence exhibited by certain sets of characters and the stability they evince through evolutionary time is to ignore, rather egregiously, several facts: (a) some characters are likely to be included within one or more additional functional units, i.e., the functional domains of such "units" overlap; (b) all characters identified as integrated through function are simultaneously nested within potentially noncongruent sets of developmentally correlated characters; (c) in the vast majority of cases we have no knowledge of the genetic architecture underlying the functional unit, nor of its heritability; nor do we know how the genetic variance-covariance structure of the characters might change in different environments (Schlichting and Pigliucci, 1998); (d) we rarely have information about the "variational properties" (*sensu* Wagner, 1999) of constituent characters since we lack developmental information for most characters, let alone suites of characters; in any case, we typically view the characters in a population after natural variation has been reduced by selection; (e) functional units comprise elements deriving from many tissue types and organ systems, e.g., bone, muscle, connective tissue, nerve, and blood vessel, but we rarely consider more than one or two of these at a time (Gans, 1969; Dullemeijer, 1974). In short, even a well-developed characterization of a functional unit is a one- or two-dimensional projection of a multidimensional entity.

Despite the cautionary pessimism just expressed, it is remarkable how much explanatory power functional units seem to have in evolutionary morphology, at least on a qualitative level. A clearer conceptualization of functional units and heightened recognition of their importance in evolutionary character analysis should facilitate the design of empirical studies that elaborate more fully their multidimensional nature. The three broad categories of functional unit identified here are based on the literature review and my own thoughts (Table I). These types are best considered modal patterns of functional unit organization that blend one into the other and are not intended to be mutually exclusive. It is likely that there are others. Although the three modal types have different attributes, they are united in that each represents a complex of actually or potentially separate (i.e., individuated) characters that are integrated by participation in a common function. Their reality as historical, evolutionary units is supported circumstantially by the stability they exhibit in comparative analysis (see later).

Type 1: Structural Units

The simplest form of functional unit is an apparently single morphological structure, unitary in its adult form, but arising through fusion of two or more developmental precursors. Such functional units are most clearly manifest in certain bones whose precursors take the form of mesenchymal condensations, sometimes evident as separate centers of ossification (Fig. 2). It could reasonably be argued that developmental fusion of multiple *Anlage* into single adult structures forms a separate class of developmentally integrated structures not related to functional units, *per se*. I suggest they are worth considering here for two reasons: first, because in many cases the fusions are causally related to functional demand, and second because structural units comprising more than one individuated, morphogenetic "unit" are likely to differ from truly unitary structures in their "variational properties" (*sensu* Wagner, 1999) and therefore in their evolutionary potential.

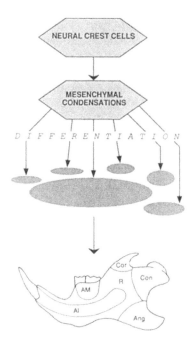

FIGURE 2   The mouse mandible as a structural unit (see text). Separate mesenchymal condensations coalesce into a single dentary bone, the solidity of which is presumed to reflect its functional role in the mammalian masticatory system. AI, alveolar-incisor; AM, alveolar-molar; Ang, angular process; con, condylar process; cor, coronoid process; R, ramus. After Atchley (1993).

The case of the mammalian (mouse) mandible is an especially well-developed example (summarized by Atchley and Hall, 1991; Atchley, 1993). The mammalian mandible consists of a single bone, the dentary. Indeed, a mandible comprising a single bone is a diagnostic character of the crown group Mammalia. The dentary bone, however, is a developmental composite of six, separate morphogenetic units that coalesce into a single adult structure (Fig. 2). These morphogenetic components arise as condensations of neural crest cells that contribute to the body (ramus) of the dentary, its two alveolar (tooth-bearing) portions, and its three processes (condylar, coronoid, and angular). Given the nature of the mammalian feeding system, particularly the function of mastication, there can be little doubt that the solidity of a single, fused dentary is functionally adaptive, hence one can argue that the composite dentary represents a functional unit. A consequence of maintaining separate morphogenetic units within a unitary structure, however, is enhanced evolutionary potential. Atchley and Hall (1991) proposed that interspecific differences in mandibular form among mammals result from modifications in individual morphogenetic units, and work summarized by Atchley (1993) supports this view.

The skull provides many cases of bones with multiple centers of ossification that correspond to ancestrally separate bones that have fused in a descendant taxon. In addition, there are examples of the reverse polarity, i.e., bones that are ancestrally single with multiple centers that in the derived condition are represented by separate bones. Whether such phylogenetic fusions and fissions have a functional basis is an open empirical question. However, their existence highlights the hierarchical complexity of character individuation. Of more relevance here is that they represent a cryptic source of evolutionary variation that can be exploited when novel functional demands arise. A good, if unusual, example is the case of the bolyeriid snake maxilla. Bolyeriids (Round Island boas) comprise only two known species, one recently extinct. They are unique among tetrapod vertebrates in having a maxilla (upper jaw bone) divided into two, articulated parts (Frazzetta, 1970; Cundall and Irish, 1989). This arrangement permits the anterior end of the upper jaws to be depressed beyond the rest position, an ability unique among snakes (Cundall and Irish, 1989). Dietary and other ecological data suggest that bolyeriids are skink specialists. Skinks are unusually hard, slippery lizards due to their smooth, chain mail-like integument. The unique bolyeriid jaw mechanism allows the upper jaws to bend around the lizard when it is held transversely in the mouth in order to retain it in the jaws' grasp (the lizard tending otherwise to slip out). Analogous adaptations are found in other snakes that specialize on skinks (Savitzky, 1983; Cundall and Irish, 1989). This feeding mechanism arguably constitutes a novel and specialized functional unit but how could it have evolved? A developmental study of an unrelated snake with a typical, unitary maxilla found that the bone formed from "two distinct and independent centers of ossification" (Haluska and Alberch, 1983, p. 54). The maxilla extends farther posterior in snakes than in lizards (representing the ancestral condition) and it was suggested that the derived, elongated snake maxilla may have arisen by addition of a new, posterior center of ossification (S. B. McDowell, in Haluska and Alberch, 1983). A speculative scenario is that strong selection on the feeding apparatus for dietary specialization in bolyeriids, whose food options were limited to terrestrial skinks (Cundall and Irish, 1989), drove a paedomorphic shift in maxillary development, leading to a jointed, two-part bone, with each segment corresponding to an ancestral ossification center. In fact, heterochronies in late stage ossifications are strongly implicated in the evolution of snake skulls generally (Irish, 1989).

In the case of the mammalian mandible it was argued that the fusion of separate morphogenetic units into a unitary dentary is maintained by functional (as well as developmental) integration, thus the dentary represents a functional (structural) unit, but in the bolyeriid maxilla the opposite is true. In this case, an ancestrally fused structural unit became divided into two separate elements, revealing its ancestrally composite nature. The cryptic variation represented by separate centers of ossification in the snake maxilla may have permitted a repatterning of the feeding system into a novel form in bolyeriid ancestors. Thus, the patterning of morphogenetic units within a structural unit is likely to influence both its own direction of phenotypic evolution as well as that of the larger functional unit within which it is embedded as a constituent element. In effect, structural elements increase the evolutionary degrees of freedom inherent to the functional unit (see later).

Type 2: Mechanical Units

Mechanical units are complexes of individual elements which form a more-or-less solid or unitary structure. As such, I use mechanical unit in the sense of Gans (1969) to indicate that this type of functional unit is characterized by limited intrinsic movement (i.e., among component elements) as compared to the relatively greater movement possible between units. The best examples of mechanical units are found in the vertebrate skull, but there are others (discussed later). In developing the example of the skull I will concentrate on the bony elements, as have previous authors, but as noted above, this must be taken as a pragmatic simplification of the true complexity of any functional unit. Skulls are obviously associated with a variety of soft tissues that are integrated along with the bony elements in the performance of particular functions (e.g., Fig. 1).

The essential quality of mechanical units is that they are relatively solid structures despite the fact that they visibly comprise multiple constituent parts. The mammalian skull, in particular, exemplifies this condition in that it is composed of many separate bones, but these are joined in such a way that the skull, as a whole, forms a single, akinetic structure. That the skull functions as a mechanical unit cannot be doubted, but the skull obviously subserves multiple functions and biological roles. It was this factor which led van der Klaauw (1945) to recognize 36 separate, if overlapping, functional units within the skull, each unit corresponding to a principal function.

In solid, akinetic skulls such as in mammals, subdivision along functional lines is problematic due to its very solidity (although there is no denying that certain functions are concentrated in particular regions of the skull). Van der Klaauw (1945, p. 35) suggested that some functional demands (such as the need to generate powerful crushing forces by the jaws) could lead functional units to combine in forming "a higher morphological whole" or "higher functional unit." He made the point that overall skull shape could be determined in two different ways: (a) relatively independent units have a particular shape imposed by their function and these combine to form a composite skull whose form is, in this sense, the sum of its parts ["it is not the circumference of the whole that is the determining factor, but when parts comparatively independent in the development of size and shape have developed separately and the circumference and shape are, as it were the secondary result" (p. 20]; (b) overall skull form is, itself, the overriding functional imperative and the morphology of individual component parts is less important ["the circumference of the whole is fixed and the component parts conform to this whole" (p. 19)]. In the latter case one expects the skull, in its entirety, to represent a higher order functional unit with the form of intrinsic, subordinate units subjugated to its higher function. Van der Klaauw (1945) expected this situation especially in aquatic and fossorial vertebrates in which head shape must conform to the demands of streamlining. In this case the functional demands on the skull (e.g., mechanical stresses) are integrated across the whole structure rather than being restricted to individual parts or limited regions of the skull.

I defer discussion of the first case for the subsequent section of this paper where it fits more appropriately, but the latter case bears scrutiny here. To restate it in evolutionary terms, the hypothesis is that when selection on overall skull form is strong, selection on individual, component elements is relaxed so long as functional integrity of the whole is maintained. This hypothesis can be tested in a rudimentary way by considering patterns of intraspecific variation in skull morphology. Relaxed selection on component bones of the skull should be evident as a relatively high frequency of among-individual variation. Unfortunately, studies of such variation are rare (Hanken and Hall, 1993); in particular, studies that draw on single populations in which it is reasonably inferred that selection regimes are equivalent are lacking (L. Roth, pers. comm.). However, suggestive results are provided by some available data. Roth and Thorington (unpubl., in Roth, 1991), for example, tentatively found that overall skull form was quite stable in some mammals whereas the size and shape of individual bones were variable (e.g., in the position of sutures). Roth (1991) pointed out that the individuated bones that we would consider homologous characters might not be separable in such structures. In mammals the overriding functional importance of skull shape is not always clear, but to a large extent it probably relates to braincase function. It is certainly well known that the human cranium is highly variable in the number and shape of bones (Berry and Berry, 1967), particularly with regard to the presence of so-called "sutural bones" especially common within the lambdoidal and sagittal sutures and said to "have no true morphological significance" (Warwick and Williams, 1973, p. 303). A better example is provided by some primitive bony fish which are subject to the streamlining demands suggested by van der Klaauw (1945) and which, like mammals, have solid cranial vaults. Two species for which there are data, *Amia* and *Acipenser*, show profound intraspecific variation in the morphology of individual roofing bones, including asymmetries, in conformity with the prediction of relaxed selection on these elements (Jain, 1985; Grande and Bemis, 1998; Hilton and Bemis, 1999) (Fig. 3). A final example is the position of the parietal foramen in the skull roof of lizards. This has been widely used as a "character" in phylogenetic analysis (e.g., Estes *et al.,* 1988; Lang, 1989; Lee, 1998). The foramen provides a passage for light to pass into a photoreceptive parietal "eye." Among lizards its position is variable: within the parietal bone, within the frontal bone, or within the fronto-parietal suture (Estes *et al.,* 1988). Usually the position is fixed for higher taxa, but within the closely related basiliscine lizards it is found in all three positions (Lang, 1989). In one species, *Laemanctus serratus*, its position is highly variable: in some specimens it is wholly within the frontal, in some it is near to the suture, and in others it is within the suture.

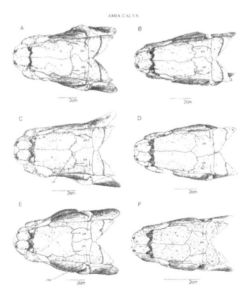

FIGURE 3 A putative mechanical unit. The skulls of six individuals of *Amia calva* show extreme variation in dermal roofing bones, especially the parietal (pa) and frontal (fr) bones. Such variation probably reflects relaxed selection on component parts of a mechanical unit (see text). From Grande and Bemis (1998), *A Comprehensive Phylogenetic Study of Amiid Fishes (Amiidae) Based on Comparative Skeletal Anatomy. An Empirical Search for Interconnected Patterns of Natural History.* Society of Vertebrate Paleontology, with permission.

What is not known is whether the samples derive from single populations, but nonetheless it is likely that such "irregularities" appear with such frequency because the form of individual bony elements is relatively unimportant to total skull form and function. This situation obtains because the skulls in such animals are mechanical units whose integrated functional characteristics override those of the developmentally individuated characters from which they are constructed.

Interspecific variation provides only indirect evidence of relaxed selection, but it may speak to the question of character individuation. For example, Romer (1936) found that it was virtually impossible to homologize the many, small cranial roofing bones of fossil lungfish, there being too many small, irregular bones to permit a one-to-one correspondence among related species. This situation is pronounced among many fossil fish groups and between fossil fish and their early tetrapod relatives (Thomson, 1993). These taxa tend to have many, rather than few, bones in the dermal skull and all are characterized by relatively solid, heavily ossified skulls presumably subject to streamlining selection. The implication, again, is that individual skull bones were neither well individuated nor canalized in such taxa because selection acted on overall skull form rather than the form of individual bones. These circumstances would mitigate against one's ability to identify homologous bones among taxa, as noted by Roth (1991).

Other mechanical units might be found in the axial skeleton. The limb girdles are possibilities, but the example of the turtle shell is intriguing because of its analogy to the skull. The shell is a composite of axial skeleton components (vertebrae and ribs), ossifications which extend from these, and overlying dermal (integumentary) "shields" whose sutures are noncongruent with those of the bones, providing additional strength to the shell. In addition, there are anterior and posterior shell components that develop independent of the axial skeleton, although they are integrated with the rest of the shell to form a solid, more-or-less unitary structure (Zangerl, 1969). Pattern formation of most of the carapace (the upper portion of the shell) seems to be dictated by the segmental axial skeleton. As predicted by the mechanical unit model, Zangerl (1969) found that the form of the superficial dermal shields was highly variable among individuals. Given their role in total shell function (stiffness and rigidity, among others) this is not a surprising result. In contrast, on first examination the bony elements of the shell do not seem to conform to the prediction. Most of the shell bones are, in fact, highly stable (Zangerl, 1969). However, the stable elements are those directly associated with the elements of the segmented axial skeleton (vertebrae and ribs). Stability might be expected in these elements because of independent developmental constraints obtaining for elements of the axial skeleton. In fact, those regions of the carapace free of the axial skeleton and presumably not subject to the same constraints were "notably variable" among individuals

(Zangerl, 1969, p. 322). Thus the bony elements of the carapace seem also to support the hypothesis of relaxed selection on individual elements, at least where other sets of constraints do not apply (see later).

Additional circumstantial evidence is provided by interspecific comparisons. In general, higher chelonian taxa tend to be relatively stable in the pattern of elements composing the carapace (with the exception noted) (Zangerl, 1969). However, Owen (1866, p. 61) observed significant interspecific variation in the number and position of the peripheral, marginal elements, which he attributed to the fact that "these marginal pieces are the least essential parts of the carapace," i.e., their individual form is not important to total carapace function.

Following Berry and Searle (1963), Hanken and Hall (1993) called the kinds of intraspecific variation described earlier "epigenetic polymorphisms" to emphasize their origin in the complex interaction between genome and environment. They suggested that such polymorphisms are of great importance in phenotypic divergence among taxa. As noted previously for structural units, such polymorphism represents a source of variation that selection can act upon in changing circumstances.

For heuristic purposes I have concentrated on large, obvious mechanical units such as the skull and the turtle shell, but as van der Klaauw (1945) pointed out, larger-scale functional complexes can be composed of several, smaller constituent mechanical units. As such, mechanical units might be movably articulated with one another. This would be the case, for example, in most vertebrate skulls which are characterized by varying degrees of intracranial mobility known as cranial kinesis. The introduction of movement among component parts of a functional unit, however, leads to greater complexity in its attributes and our expectations for such units are different in important respects. These types of functional units are considered next.

Type 3: Evolutionarily Stable Configurations

ESCs are complexes, often comprising diverse characters, that function together in a concerted manner to produce a functional output (Wagner and Schwenk, 2000) (Fig. 4). The key element is that individual parts, or characters, are functionally integrated, i.e., each part must coordinate with others in order to produce the desired output. The temporal scale of this coordination is short, in the sense of "behavior" and "function." In other words, functional integration is distinct from the kind of temporal ordering implied by developmental integration.

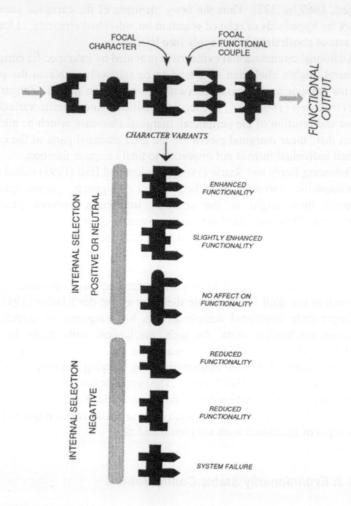

FIGURE 4 Conceptual diagram of an ESC showing the relationship between functional integration among component parts (characters) and the range of variation permitted in an individual part by internal selection. A chain of functional couples required by a system to produce an appropriate functional output is illustrated across the top of the figure. Vertically below are shown a series of possible (random) variants in the phenotype of one character. Only those variant forms that maintain or improve the functional couple will maintain functionality of the system as a whole, promote survival of the individual, and persist in the population. Variants that disrupt the functional couple will be selected against to the extent that reduced functionality of the ESC has an impact on fitness. This selection arises intrinsically as a manifestation of the functional interrelationships among characters. Thus, "internal selection" is a component of natural selection largely independent of the external environment and therefore relatively buffered from external environmental changes. As such, the ESC is internally self-stabilizing. Based on Wagner and Schwenk (2000).

As a simple example consider mammalian teeth. The teeth are part of the unique mammalian masticatory system, a complex of characters involved in the processing of food within the mouth. For mammalian mastication to occur, upper and lower teeth must occlude, i.e., their cusps and concavities must intermesh precisely so that the mechanical output of the system (e.g., grinding or shearing of food) is possible. Malformities that interfere with precise occlusion can adversely affect the health and survival of an individual. In rodents, for example, it is necessary for the ever-growing upper and lower incisors to slide past one another in order to maintain a sharp, shearing mechanism for the "chiseling" function of the teeth. Failure to occlude, which occurs through disease, injury, or developmental anomaly, not only prevents the teeth from proper function during feeding, but it also prevents the teeth from being worn down so that eventually they grow back into the skull, killing the animal (Owen, 1868; Nowak, 1991). Excessively worn molars in herbivores, such as elephants, can lead to premature death because the loss of occlusal grinding surfaces prevents proper comminution of the food. These anecdotes indicate simply that in order for the entire masticatory system to function, the individual parts, including upper and lower teeth, must be precisely integrated in their form and movement. Phenotypic deviations in these parts are likely to have serious fitness consequences. In other words, a mutation leading to a phenotypic variant in one character that interferes with its ability to function in concert with other characters will precipitate a cascade of disrupted functional integration, causing reduced system performance or total failure. As such, most variant phenotypes will have lower relative fitness as compared to individuals whose characters remain within the strictures dictated by functional integration.

Another way of expressing this important quality of an ESC is that system function imposes stabilizing selection pressures on the form of the elements participating in the system. This selection obtains from the coordination, or integration, *of the system itself* and is not directly related to selection pressures arising from a particular habitat, or environment. As such, the selection is intrinsic to the organism. It moves with the organism wherever it goes and is maintained consistently in different environments. Such selection is called "internal" to distinguish it from "external" selection deriving from the extrinsic environment (selection in the typical Darwinian sense) (Whyte, 1965; Dullemeijer, 1980; Arthur, 1997; Wagner and Schwenk, 2000). Internal selection is defined by Wagner and Schwenk (2000) as rank-invariance of fitness across a range of environments and rank-dependence of fitness on the particular combination of traits posessed (see also Arthur, 1997). In other words, given a group of individuals who vary in some character, the rank order of morphs according to fitness remains constant across environments (although absolute fitness values might vary in the different environments). The invariance of relative fitness among morphs in different environments reflects the fact that internal selection is an intrinsic, organismal attribute and not a function of the environment as it is for typical external selection. In addition, the fitness dependence of character variation on the other characters with which the character is associated in the ESC is a reflection of the functional integration uniquely manifested by a particular suite or complex of characters.

The significance of internal selection in characterizing ESCs as functional units is the implication that, in contrast to mechanical units, the phenotypes of individual component parts are subject to strong, stabilizing selection. Although the phenotype of the whole system is likely to become stabilized as well, this follows secondarily from stabilization of the individual characters. As such, ESCs exemplify van der Klaauw's (1945) first pattern of functional unit form (see earlier discussion) in which individual components that are comparatively fixed in form sum to create the form of the whole functional unit, as opposed to the subordination of component parts to the form of the whole characteristic of mechanical units (see earlier discussion). The key point is that, in contrast to mechanical units, we expect phenotypic stabilization of ESC component elements rather than intraspecific variation. Such stabilization derives from internal selection on the elements. Over the long term, such stabilizing selection could lead to canalization in development of a character (e.g., Wagner *et al.,* 1997) so that variant forms do not arise in the first place (see later); however, this is not a necessary feature of an ESC.

It is worth noting in this context a supporting anecdote. In the section on mechanical units, variation in the position of the parietal foramen was cited as an example of relaxed selection on component characters. However, another suggestive pattern emerges from this example (Schwenk, 1987; unpublished data). In the largest, most diverse clade of lizards (Scleroglossa) the foramen is usually in the parietal bone and is never found within the fronto-parietal suture (Estes *et al.,* 1988). In the sister clade (Iguania) its position is variable, as in the basiliscine example, but it is often found within the suture. Experimental evidence shows that many, if not most, scleroglossans have kinetic skulls with active flexure occurring at the fronto-parietal suture. In iguanians there is no evidence for active flexure at this joint and anatomical considerations suggest that they are incapable of such movement (Schwenk, 2000a). The implication is that when there is relaxed selection on parietal foramen position due to its subordinate position within a larger mechanical unit it is free to vary within certain limits (as in iguanians). However, in the kinetic Scleroglossa, the skull is divided into two or more mechanical units movably articulated at several joints. The internal selection that arises in this case stabilizes the position of the foramen so that it does not occur in the suture where function of the parietal eye would be compromised.

Stability of ESC characters does not imply immutability. Divergence in ESC characters is possible as long as the functional output of the system remains the same or is improved (Fig. 4). For each functional coupling in an integrated system there is a threshold of phenotypic variation beyond which the functional relationship is affected, causing decreased performance or a breakdown in the system. However, sub-threshold variations, or variations that do not affect the functional coupling, would be permitted by system function. Threshold values will be different for each coupling. In the rodent incisor example noted earlier, the functional coupling between upper and lower teeth is very tight, implying a low threshold; even slight deviations in tooth position break the functional couple. Incisor position and movement should be highly canalized in order to preserve the functional couple and therefore stable among all taxa exhibiting the ESC. Nonetheless, as long as the functionally relevant criteria are maintained, other aspects of incisor phenotype might evolve. For example, they may become proportionately larger or smaller without breaking the functional couple, hence we might expect incisor size variation among taxa. In other systems the functional couple may not be so tight. Many herbivorous lizards, for example, cut vegetation with laterally compressed, multicuspate teeth (e.g., Edmund, 1969) that lack precise occlusion in the mammalian sense. Thus position and form of upper and lower teeth can vary considerably before their cropping function is impaired. This implies a larger functional threshold and one might expect both intraspecific variation and evolutionary divergence in the form and position of these teeth among taxa that nonetheless retain the ESC. Hence, one quality of ESCs is that their component characters are phenotypically stabilized to values imposed by functional thresholds, but within these limits both intraspecific variation and evolution are possible (Wagner and Schwenk, 2000).

Recognition of an ESC as a functional unit depends on its intrinsic integration and the unique relationships of its elements in performing a particular functional role in the life of the organism called the "proper function" (Wagner and Schwenk, 2000). The proper function of a system refers to its historically based, adaptive "purpose," essentially the architectural *raison d'être* of the system (Millikan, 1984). By definition, the ESC is represented by the particular suite of characters whose coordinated function uniquely serves a given proper function. Examples of proper functions might include "the capture of prey with the tongue," "the mastication of fibrous plant matter," or "the generation of powerful digging forces by the forelimb." Hence proper functions are typically characteristic of organs and organ systems. In this sense, ESCs are the type of functional unit typically identified by functional morphologists who study "feeding systems" or "locomotory systems," etc. Such systems often involve complex kinematic chains and multiple parts that move relative to one another, but this characterization of ESCs may be too narrow (see later).

Wagner and Schwenk (2000) developed an ESC case study at length. They argued that the lingual feeding system characteristic of one monophyletic group of lizards (Iguania) forms a functionally integrated character complex that, through internal selection, is evolutionarily self-stabilized. Its proper function is the capture of prey with the tongue. Prey capture is mediated by extra-oral tongue protrusion and adhesion of the prey onto its anterodorsal surface. The behavior is extremely rapid and requires a precisely choreographed series of kinematic events involving movements of the head, jaws, tongue, and hyobranchial apparatus (tongue/throat skeleton). An essential feature of the system is the tongue's adhesive mechanism. Long papillae covered by a mucous epithelium on the forepart of the tongue mediate prey adhesion through a combination of frictional interlocking and glue-like mucus (Schwenk, 1988; 2000a). During prey capture the tongue is deformed around the mandible so that its adhesive surface is presented toward the prey item (Schwenk, 2000a; Schwenk and Throckmorton, 1989). Without an effective adhesive mechanism the system cannot perform its proper function. The lingual feeding system, overall, is remarkably uniform within the entire clade. Nonetheless, members evince nearly every diet, habitat preference, and ecological specialization possible for a lizard, hence there is no overarching or common set of ecological (i.e., external) conditions that could account for the system's long-term phenotypic stability. To the contrary, part of the argument in support of the lingual feeding system as an ESC is strong, circumstantial evidence for ongoing external selection in this lineage for modification of the foretongue to subserve a different function as part of the vomeronasal chemosensory system (Schwenk, 1993, 1995a, b). However, based on mechanical considerations, optimization of the tongue for chemoreception would quickly disrupt its adhesive mechanism, thereby causing system failure (Schwenk, 1995; Wagner and Schwenk, 2000). Hence, despite external selection pressure for modification, the lingual feeding system has remained stable within its threshold limits throughout the adaptive radiation of the iguanian clade. The functional constraint of the ESC was apparently broken only once in the ancestor of its sister clade (Scleroglossa) in which the lingual feeding system was putatively first duplicated, then replaced by a jaw-prehension system of prey capture (Wagner and Schwenk, 2000). This released the internal selection maintaining the stability of the ESC since the tongue was no longer necessary as a prehensile organ, precipitating disintegration of the ESC. Once freed of this functional constraint, component characters were available for adaptive remodeling, including radical modification of the foretongue for chemoreception and other purposes.

As noted by Wagner and Schwenk (2000), identification and delimitation of ESCs is problematic and minimally depends on a combination of comparative character analysis and functional analysis to document both the stability (and therefore the historical reality) of the functional unit and the nature of the functional interactions underlying the stabilizing internal selection. However, many such systems could probably be identified and once again the skull is a likely source. Advanced snakes, for example, are typified by extreme cranial kinesis in which both halves of the mandible are separable, the upper jaw joint bones (quadrates) are free to swing laterally and anteroposteriorly, the upper jaws can move independently, the tip of the snout can flex dorsoventrally, and the entire apparatus is movably suspended from a solid, bony braincase (e.g., Albright and Nelson, 1959; Frazzetta, 1966; Cundall and Shardo, 1995). The system permits these snakes both to engulf huge prey relative to their head and body size and, by means of asymmetrical movements of the upper jaws, to pull the snake over the subdued prey in order to swallow it. The obvious complexity of movement among these parts suggests that their dynamic relationships will impose internal selection on individual elements so that their phenotypes are compatible with system-level function. Again, for reasons of simplicity (and ignorance) I have ignored the soft tissue elements of this system (but see Dullemeijer, 1956, 1959, and Fig. 1).

Of interest in the snake skull example is that this putative ESC is composed of multiple bony elements including individually articulated bones, such as the quadrates, structural units, such as the maxillae, and several mechanical units, notably the braincase, which comprises several tightly sutured bones. The solidity of the braincase unit is a derived condition of snakes relative to lizards and is attributed to the need for complete protection for the brain both from prey being engulfed and from the stresses of burrowing presumed to characterize the ancestry of snakes (e.g., Bellairs and Underwood, 1951; Romer, 1956). As such, the elements of an ESC can, themselves, be hierarchically organized. Some component parts of the ESC are likely to be characters in the traditional sense, but others might be structural units or mechanical units. Complexes of mechanical units which are functionally integrated can constitute an ESC. Thus ESCs potentially represent higher-order functional units with other types of functional unit embedded within them as subsystems. It is conceivable that some ESCs even include other ESCs as elements within their integrative realm just as computer programs can include subroutines within their set of instructions. The central requirement in such cases is that the higher order ESC maintains a single proper function that uniquely relates all component parts, whether they are characters or functional units, themselves. Although an argument would have to be developed for a specific case, such higher-order ESCs are a virtual necessity given the complexly integrated hierarchy embodied in organisms and *Baupläne*.

ESCs need not be restricted to systems comprising hard parts and articulations. The concept might be expanded to include other functional realms, such as cognitive systems. It has been shown, for example, that apparently unitary cognitive processes, such as the ability to use language, are actually based on the interaction of separate brain centers located in different hemispheres (papers in Beeman and Chiarello, 1998). Disruption of integration between these centers (e.g., through resection of the corpus callosum) causes a breakdown of this system such that spoken and written language become dissociated (Baynes and Eliassen, 1998; Baynes *et al.,* 1998). Thus the functional output of the cognitive system derives from the interaction and integration of subordinate elements. It is difficult to know whether this functional integration imposes internal selection on the individuated, component brain centers, but I suggest that it is worth exploring the possibility that such neural systems, and other nonmusculoskeletal systems, represent ESCs. It is possible, however, that such nontraditional functional systems will require formulation of another type of functional unit.

EVOLUTIONARY ATTRIBUTES OF FUNCTIONAL UNITS

A universal feature of the functional unit concept historically and as developed here is that functional units are characterized by (a) some degree of autonomy or individuation relative to other such units and (b) some degree of evolutionary stability. As such, the starting point for recognizing functional units is not so much their functional integration, but rather their phylogenetic persistence. It is this persistence that betrays their coherence and justifies their identification as a "unit". Thus, in an evolutionary sense, functional units behave uncomfortably like *characters*. However, the patterns and causal bases of evolutionary stability in functional units are hardly uniform. In the case of structural units the stability is as much developmental as functional, although it is argued that functional interaction provides the selective basis for building or maintaining the coherence of developmentally fragmented subunits. However, to distinguish structural units as defined here from characters in the sense of biological homology is admittedly to split hairs. For mechanical units and ESCs, however, there is no denying their composite nature as character complexes. Phylogenetic stability arises in each case in different ways. For mechanical units there is putative selection, in most cases external, on total unit form with relaxed selection on component characters. As noted previously, the maintenance of stabilizing selection tends to lead to canalization, hence the overall form of mechanical units may be developmentally canalized while individual, constituent elements are not (Roth, 1991). Some evidence for this was adduced earlier. This is an important point because it suggests that the mechanical unit as a whole has different variational properties than the characters that compose it. As for ESCs, the converse is true. It is the component parts that are stabilized via internal selection with system-level phenotypic stability arising secondarily. In this case the phenotypes of the parts are likely to become canalized (though they need not be). However, the critical distinction for ESCs is that they are *evolutionarily self-stabilizing*. In other words, internal selection acts to cull deviant phenotypes as they arise. The stabilizing process is intrinsic and occurs consistently in a wide range of environments, *including those that impose selection pressures that would tend to change some component characters* (as in the lizard feeding example). Hence internally selected systems are, to some extent, buffered from environmental changes through time, and this is the basis for their persistence as historical entities.

In general, the more tightly integrated the characters within a functional unit are, the more evolutionarily stable the unit is likely to be and the harder it will be to disrupt. The integration is likely to reflect some combination of developmental constraint (canalization), functional constraint (internal selection), and external selection (e.g., the need for streamlining). It is the functional contribution to maintaining stability of a character complex that underlies the functional unit concept.

Although it is their stability and persistence that identifies functional units and makes them significant in the context of character evolution, they are nonetheless mutable. This must be so or organisms would long ago have congealed into a very few *Baupläne* which would be highly prone to extinction. Ultimately, the evolutionary potential for functional units lies with their composite nature. Since they comprise multiple, individuated characters the degrees of freedom inherent in their variational properties are greater than in a single character. This might ultimately enhance the evolvability of functional units as opposed to characters. Structural units provide a cryptic source of potentially individuated characters. The intrinsic variation of mechanical units and ESCs represents a phenotypic resource for future selection regimes outside those previously experienced by the organism. Even very tightly integrated ESCs, for example, can be disrupted by key changes in the environment (Wagner and Schwenk, 2000) and the ability to exploit a new environment might be tied to the composite nature of the functional unit. Wagner and Schwenk (2000) identified a set of additional conditions and circumstances that could lead to disintegration of an ESC so that its component characters are freed to behave as autonomous units in adaptive evolution. It is sufficient to note here that stability is a relative concept and that functional units, although stable and sometimes self-stabilizing, do not necessarily represent evolutionary *cul-de-sacs*.

Thus, functional units embody an evolutionary paradox. On the one hand, they are notable for their evolutionary stability and in most cases they are intrinsically self-stabilizing. On the other hand they maintain more degrees of freedom for phenotypic evolution than truly unitary characters and, as such, functional units differ from characters in their variational properties and ultimately in their evolvability. The apparent paradox of stability versus evolvability may be resolvable in the context of relative time frame. While it is essential that organisms maintain a certain level of adaptive fit between their phenotypes and the environments they occupy, there may at the same time be an evolutionary advantage in failing to respond immediately to capricious shifts in the environment. Functional units may provide a certain amount of phenotypic inertia that serves to integrate the effects of changing environments over generations. They might accomplish this by "tuning" the quality of phenotypic variation available in a population so that variants maintain high levels of integration and, therefore, functionality. Such variants reflect general grades of system-level adaptation rather than pointilistic, character-by-character responses to immediate environmental conditions, thus viability of organismal phenotype in the short term may be enhanced. In the long term, however, a major environmental shift outside the adaptive sphere of a functional unit might trigger its disintegration into component characters, thereby enabling a rapid phenotypic reorganization to a new adaptive state.

WHY AREN'T FUNCTIONAL UNITS CHARACTERS?

The premise for this essay was that functional units represent a level of organization intermediate between characters and *Baupläne*. By definition, they are character complexes. However, as noted earlier, they seem to behave suspiciously like characters in comparative evolutionary analysis: for a given in-group they are stable, homologous, phenotypic units that exist in different states whose distribution is concordant with phylogeny. A partial solution to this paradox lies in the biological homology concept. Characters are taken to represent irreducible units of phenotypic evolution that are "quasi-independent" and "quasi-autonomous" relative to other such units (see earlier discussion). Their identity and hierarchical delimitation are determined by their variational properties, or their degrees of freedom for natural variation, which are, in turn, a function of their developmental constraints. Do functional units satisfy the criteria of quasi-independence and quasi-autonomy in this sense? Three lines of reasoning suggest that they do not.

First, the phylogenetic stability inherent to functional units does not necessarily derive from developmental constraint. Indeed, in many cases stabilizing selection is sufficient to account for the phenotypic stability evinced in the parts and in the whole. For example, in a fish skull it is the persistence of a constant, strong external selective pressure (locomotion through viscous water) that promotes maintenance of a streamlined skull. In other cases, internal selection might maintain phenotypic stability. A comparative analysis of an ESC might indicate that a component part is virtually invariable within and among taxa. Nonetheless, it is perfectly possible that the part is developmentally unconstrained and, in fact, varies considerably during ontogeny so that a population of organisms potentially expresses multiple variants of the character. However, internal selection acting during early life stages will weed out most such variants. The tighter the functional couplings for that particular part, the stronger the internal selection and the lower the threshold for variation. Hence little variation will be apparent in adult populations despite the lack of canalization. Thus the variational properties of functional units are not necessarily determined by intrinsic, developmental constraints, but can result from some combination of external and internal stabilizing selection.

Second, the variational properties of a functional unit are not uniform. Component parts will almost certainly vary in the nature and degree of their developmental constraint. Some parts may be canalized while others are not. This suggests that parts vary in their evolutionary potential independent of their position in the functional unit. Furthermore, the variational properties of the whole may differ from those of the parts. In the case of some mechanical units, such as streamlined skulls, the whole may be more canalized than its component parts. In this case it is tempting to apply the character concept to the mechanical unit in its entirety, but the final argument would tend to refute this conclusion.

By broadening the scale of comparative analysis it will usually be seen that, for functional units, the criteria of independence and autonomy are applicable only at lower taxonomic levels and that as one descends to deeper nodes in phylogeny these criteria disappear. As such, the stability and autonomy of the functional unit, which imbue it with character-like attributes, are found to be clade specific, and consideration of a wider range of related taxa reveals the unit to be decomposable into smaller evolutionary entities, each of which more clearly manifests the properties of biological homology.

The case of the eye exemplifies this. An eye clearly represents a coherent phenotypic unit. Recent work has shown that eye development in *Drosophila* is triggered by a single gene whose signal can induce development of the entire complex even when expressed in inappropriate locations (Halder *et al.,* 1995). Thus the eye, as a whole, would appear to be an autonomous entity. Wagner (1989a, b) suggested that this organ-level autonomy is a result of a closed cycle of epigenetic interactions, particularly a series of inductions. Thus, the eye is such a phenotypically stable and developmentally autonomous unit that it would seem to satisfy our criteria for a character, and within a particular clade it might be justifiable to treat it as such. In frogs (anuran amphibians), for example, the eye retains its character-like stability. However, in a broader comparison including all amphibians, this stability is not maintained. In the limbless and fossorial caecilians (Gymnophiona), for example, eye structure is greatly reduced, as is typical of animals living in lightless or light-reduced environments. Importantly, however, the caecilian eye is reduced and modified mosaically, i.e., it does not simply shrink and disappear in the way expected of an individuated character. Rather, the pattern of eye reduction is highly variable among taxa (Wake, 1985). In fact, many of the components of the eye that we would have supposed to be parts of an irreducible whole are appropriated for use in another system, the chemosensory tentacular apparatus unique to caecilians (Wake, 1985, 1992; Billo and Wake, 1987). Again, the pattern of use of former eye parts by the tentacular apparatus is variable in the group. Thus, a structure which behaves as a character at a lower taxonomic level, or in one clade, is shown in broader phylogenetic context to comprise a complex of individuated parts that can be dissociated, redistributed, and evolutionarily modified in different functional circumstances (see also Bock and von Wahlert, 1965). We can infer in this case that the multiple parts of the eye (e.g., lens, iris, photoreceptors, sclera, etc., in vertebrate eyes) are integrated in a functional, as well as a developmental, sense and are therefore subject to internal selection. Such selection would help to maintain the epigenetic cascade noted earlier by weeding out mutations that do not conform to the functional demands of vision/light reception. When the functional context is changed, however, as in a key environmental shift (in this case to a lightless environment), the stabilizing internal selection is lost so that the unitary nature of the organ vanishes (Wagner and Schwenk, 2000). As such, the eye is revealed by comparative analysis to represent a functional unit (an ESC) rather than a character in the strict sense.

In sum, comparative analysis can reveal that the stability, independence, and autonomy inherent to functional units, as opposed to characters, are potentially transitory phenomena that, for the most part, are underlain by functional integration. Even developmentally autonomous complexes can be disrupted by loss or modification of the functional milieu. Dissociation, redistribution, and modification of parts revealed by broader comparison are strong evidence that functional units are character composites and not characters themselves. They nonetheless manifest many of the properties of characters and therein lies the hierarchical complexity of the character concept.

CONCLUSIONS

This essay has provided only a cursory exploration of the gray area between character and *Bauplan*. It is sufficient, however, to suggest that the functional association of characters within organisms contributes a level of emergent complexity that must be acknowledged if we are to disentangle the causal bases of phenotypic evolution. The atomistic methods and assumptions implicit in most types of character analysis are based on an unrealistic model of organismal evolution. Indeed, the entire notion of "character analysis" might be profitably recast as the study of character associations. Obviously there are several interrelated ways in which characters can be associated and I have considered only function here with little regard to genetic architecture and developmental integration. However, these are all facets of the same fundamental issue, which is that organisms are hierarchical entities whose boundaries circumscribe interwoven webs of character integration, the complexity of which confounds our attempts to decompose them into evolutionary units. One strength of functional analysis is that it can provide an *a priori* basis for identifying character correlations rather than the *post hoc* approach typical of phylogenetic character analysis. Functional units provide a rational framework for a first pass at organismal atomization which can be amplified through a combination of phylogenetic, developmental, and genetic approaches.

In conclusion the functional unit concept provides a tentative rapprochement of the atomist-structuralist dialectic posed at the outset of this essay. A synthetic view might recognize a hierarchy of "natural phenotypic units," e.g., characters, functional units, and *Baupläne* that represent causally homeostatic, stabilized entities in the context of phenotypic evolution (G. Wagner, in litt.). We can distinguish units, such as characters and ESCs, by the mechanistic bases of their homeostasis, i.e., developmental constraint and internal selection, respectively. Since groups of characters can become integrated and therefore subject to internal selection, and internal (stabilizing) selection within an ESC can lead to canalization (a developmental constraint), we must further recognize that there are transitional forms, leading us to the notion of a continuum between characters and pure ESCs. As such, phenotypic evolution can potentially travel in both directions along this continuum: characters can become functionally integrated with other characters to create an ESC (see Wagner and Schwenk, 1999), or a tightly integrated ESC can become canalized and thus transformed into a developmentally constrained, unitary character. On the other hand, an ESC can also become fragmented by disintegration into its component characters. In general, however, we might expect an evolutionary cycling of initially disintegrated characters into functional units and these, through stabilizing selection, into developmentally contrained units increasingly difficult to disrupt and therefore increasingly conservative. From this pattern of transformations would congeal the *Bauplan*, with those elements most persistently stable the most developmentally constrained.

Although recognition of functional integration is nearly as old as biology itself, we have only recently developed the conceptual and analytical tools necessary to make significant progress. The challenges are primarily methodological (e.g., Wagner and Schwenk, 2000). We must begin to formulate research programs designed explicitly to test predictions arising from our conceptual models (e.g., Wagner, 1999, for homology). Functional and morphological analysis should play a central role in this process (e.g., Wake and Larson, 1987) as it has historically in laying the conceptual foundation for character integration.

ACKNOWLEDGMENTS

I thank Günter Wagner for inviting me to contribute to this book and for stimulating discussions, timely contributions of single malt scotch, and his ability to see clearly when all around is fog. This paper could not have been attempted without his prior influence. I also gratefully acknowledge general discussions with Frietson Galis, Harry Greene, Cindi Jones, Carl Schlichting, David Wake, and Marvalee Wake whose thinking and ideas have insinuated themselves here and in all my work. Margaret Rubega and Günter Wagner kindly read through the entire manuscript and offered critical comments. Many years ago Fran Irish discussed with me the evolution of bolyeriid snakes and I have drawn upon those discussions here (although she would not necessarily agree with my conclusions). Preparation of the manuscript was supported by grants from the University of Connecticut Research Foundation and the National Science Foundation (IBN-9601173).

LITERATURE CITED

Albright, R. G., and Nelson, E. M. (1959). Cranial kinetics of the generalized colubrid snake *Elaphe obsoleta quadrivittata*. II. Functional morphology. *J. Morphol.* **105**:241-292.
Allen, G. E. (1980). Dialectical materialism in modern biology. *Sci. Nat.* **3**:43-57.
Arthur, W. (1997). "The Origin of Animal Body Plans. A Study in Evolutionary Developmental Biology." Cambridge University Press, Cambridge.

Atchley, W. R. (1993). Genetic and developmental aspects of variability in the mammalian mandible. *In* "The Skull" (J. Hanken and B. K. Hall, eds.), Vol.1, pp. 207-247. University of Chicago Press, Chicago.

Atchley, W. R., and Hall, B. K. (1991). A model for development and evolution of complex morphological structures. *Biol. Rev.* **66:**101-157.

Baynes, K., and Eliassen, J. C. (1998). The visual lexicon: its access and organization in commissurotomy patients. *In* "Right Hemisphere Language Comprehension: Perspectives from Cognitive Neuroscience" (M. Beeman and C. Chiarello, eds.), pp. 79-104. Lawrence Erlbaum Assoc., Mahwah, N. J.

Baynes, K,. Eliassen, J. C., Lutsep, H. L., and Gazzaniga, M. S. (1998). Modular organization of cognitive systems masked by interhemispheric integration. *Science* **280:**902-905.

Beeman, M., and Chiarello C. (1998). "Right Hemisphere Language Comprehension: Perspectives from Cognitive Neuroscience." Lawrence Erlbaum Assoc., Mahwah, N. J.

Bellairs, A. d'A., and Underwood, G. (1951). The origin of snakes. *Biol. Rev.* **26:**193-237.

Berry, A. C., and Berry, R. J. (1967). Epigenetic variation in the human cranium. *J. Anat.* **101:**361-379.

Berry, R. J., and Searle, A. G. (1963). Epigenetic polymorphism of the rodent skeleton. *Proc. Zool. Soc. Lond.* **140:**577-615.

Billo, R., and Wake, M. H. (1987). Tentacle development in *Dermophis mexicanus* (Amphibia, Gymnophiona) with an hypothesis of tentacle origin. *J. Morphol.* **192:**101-111.

Bock, W. J. (1963). Evolution and phylogeny in morphologically uniform groups. *Am. Nat.* **97:**265-285.

Bock, W. J. (1964). Kinetics of the avian skull. *J. Morphol.* **114:**1-41.

Bock, W. J., and von Wahlert, G. (1965). Adaptation and the form-function complex. *Evolution* **19:**269-299.

Cundall, D., and Irish, F. J. (1989). The function of the intramaxillary joint in the Round Island boa, *Casarea dussumieri. J. Zool. (London)* **217:**569-598.

Cundall, D., and Shardo J., (1995). Rhinokinetic snout of thamnophine snakes. *J. Morph.* **225:**31-50.

Dullemeijer, P. (1956). The functional morphology of the head of the common viper, *Vipera berus* (L.). *Arch. Neerl. Zool.* **11:**388-497.

Dullemeijer, P. (1959). A comparative functional-anatomical study of the heads of some Viperidae. *Morphol. Jb.* **99:**881-985.

Dullemeijer, P. (1974). "Concepts and Approaches in Animal Morphology." Van Gorcum, Assen, The Netherlands.

Dullemeijer, P. (1980). Functional morphology and evolutionary biology. *Acta Biotheor.* **29:**151-250.

Edmund, A. G. (1969). Dentition. *In* "Biology of the Reptilia" (C. Gans, A. d'A. Bellairs and T. S. Parsons, eds.), Vol. 1, pp. 117-200. Academic Press, New York.

Estes, R., K. de Queiroz, and Gauthier, J. (1988). Phylogenetic relationships within Squamata. *In* "Phylogenetic Relationships of the Lizard Families" (R. Estes and G. Pregill, eds.), pp. 119-281. Stanford University Press, Stanford, Calif.

Frazzetta, T. H. (1966). Morphology and function of the jaw apparatus in *Python sebae* and *Python molurus. J. Morphol.* **118:**217-296.

Frazzetta, T. H. (1970). From hopeful monsters to bolyerine snakes? *Am. Nat.* **104:**55-72.

Gans, C. (1969). Functional components versus mechanical units in descriptive morphology. *J. Morphol.* **128:**368.

Grande, L., and Bemis, W. E. (1998). "A Comprehensive Phylogenetic Study of Amiid Fishes (Amiidae) Based on Comparative Skeletal Anatomy. An Empirical Search for Interconnected Patterns of Natural History." Society of Vertebrate Paleontology Memoir No. 4, Soc. of Vertebrate Paleontology, Chicago. (*J. Vertebr. Paleontol.* **18** [Suppl.]:1-690).

Halder, G., Callerts, P., and Gehring, W. J. (1995). Induction of ectopic eyes by targeted expression of the eyeless gene in *Drosophila. Science* **267:**1788-1792.

Haluska, F., and Alberch, P. (1983). The cranial development of *Elaphe obsoleta* (Ophidia, Colubridae). *J. Morphol.* **178**:37-55.

Hanken, J., and Hall, B. K. (1993). Mechanisms of skull diversity and evolution. *In* "The Skull (J. Hanken and B. K. Hall, eds.), Vol. 3, pp. 1-36. University of Chicago Press, Chicago.

Hilton, E. J., and Bemis, W. E. (1999). Skeletal variation in shortnose sturgeon (*Acipenser brevirostrum*) from the Connecticut River: implications for comparative osteological studies of fossil and living fishes. *In* "Mesozoic Fishes: Systematics and the Fossil Record" (G. Arratia and H.-P. Schultze, eds.), Verlag Dr. Friedrich Pfeil, Munich.

Irish, F. J. (1989). The role of heterochrony in the origin of a novel *Bauplan*: evolution of the ophidian skull. *Geobios Mem. Spec. No.* **12**:227-233.

Jain, S. L. (1985). Variability of dermal bones and other parameters in the skull of *Amia calva*. *Zool. J. Linn. Soc.* **84**:385-395.

Lang, M. (1989). Phylogenetic and biogeographic patterns of basiliscine iguanians (Reptilia: Squamata: "Iguanidae"). *Bonn. Zool. Monogr. No.* **28**:1-172.

Lauder, G. V. (1982). Introduction. pp. xi-xlv. *In* "Form and Function" by E. S. Russell. Reprinted by the University of Chicago Press, Chicago.

Lee, M. S. Y. (1998). Convergent evolution and character correlation in burrowing reptiles: towards a resolution of squamate relationships. *Biol. J. Linn. Soc.* **65**:369-453.

Lewontin, R. (1984). Adaptation. *In* "Conceptual Issues In Evolutionary Biology" (E. Sober, ed.), pp. 234-251. MIT Press, Cambridge, Mass.

Millikan, R. G. (1984). "Language, Thought, and Other Biological Categories." MIT Press, Cambridge, Mass.

Nowak, R. M. (1991). "Walker's Mammals of the World," 5th ed., Vol. 1. Johns Hopkins Univ. Press, Baltimore.

Olson, E. C., and Miller, R. L. (1958). "Morphological Integration." University of Chicago Press, Chicago.

Owen, R. (1837). "The Hunterian Lectures in Comparative Anatomy, May and June 1837." Edited and commentary by P. R. Sloan, 1992. University of Chicago Press, Chicago.

Owen, R. (1866). "On the Anatomy of Vertebrates," Vol. 1. Longmans, Green, and Co., London.

Owen, R. (1868). "On the Anatomy of Vertebrates," Vol. III. Longmans, Green, and Co., London.

Riedl, R. (1977). A systems-analytical approach to macro evolutionary phenomena. *Q. Rev. Biol.* **52**:351-370.

Rieppel, O. (1986). Atomism, epigenesis, preformation and pre-existence: a clarification of terms and consequences. *Biol. J. Linn. Soc.* **28**:331-341.

Romer, A. S. (1936). The dipnoan cranial roof. *Am. J. Sci.* **32**:241-256.

Romer, A. S. (1956). "Osteology of the Reptiles." University of Chicago Press, Chicago.

Roth, V. L. (1984). On homology. *Biol. J. Linn. Soc.* **22**:13-29.

Roth, V. L. (1988). The biological basis of homology. *In* "Ontogeny and Systematics" (C. J. Humphries, ed.), pp. 1-26. Columbia University Press, New York.

Roth, V. L. (1991). Homology and hierarchies: problems solved and unresolved. *J. Evol. Biol.* **4**:167-194.

Russell, E. S. (1916). "Form and Function. A Contribution to the History of Animal Morphology." John Murray, London.

Savitzky, A. H. (1983). Coadapted character complexes among snakes: fossoriality, piscivory, and durophagy. *Am. Zool.* **23**:397-409.

Schlichting, C. D., and Pigliucci, M. (1998). "Phenotypic Evolution: A Reaction Norm Perspective." Sinauer Assoc., Sunderland, Mass.

Schwenk, K. (1987). Evolutionary determinants of cranial form and function in lizards. *Am. Zool.* **27**:105A.

Schwenk, K. (1988). Comparative morphology of the lepidosaur tongue and its relevance to squamate phylogeny. *In* "Phylogenetic Relationships of the Lizard Families" (R. Estes and G. Pregill, eds.), pp. 569-598. Stanford University Press, Stanford, Calif.

Schwenk, K. (1993). The evolution of chemoreception in squamate reptiles: a phylogenetic approach. *Brain Behav. Evol.* **41:** 124-137.

Schwenk, K. (1995a). A utilitarian approach to evolutionary constraint. *Zoology* **98:**251-262.

Schwenk, K. (1995b). Of tongues and noses: chemoreception in lizards and snakes. *Trends Ecol. Evol.* **10:**7-12.

Schwenk, K. (2000a). Feeding in lepidosaurs. *In* "Feeding: Form, Function, and Evolution in Tetrapod Vertebrates" (K. Schwenk, ed.), pp. 175-291. Academic Press, San Diego.

Schwenk, K. (2000b). Tetrapod feeding in the context of vertebrate morphology. *In* "Feeding: Form, Function, and Evolution in Tetrapod Vertebrates" (K. Schwenk, ed.), pp. 3-20. Academic Press, San Diego.

Schwenk, K., and Throckmorton, G. S. (1989). Functional and evolutionary morphology of lingual feeding in squamate reptiles: phylogenetics and kinematics. *J. Zool. (London)* **219:**153-175.

Simpson, G. G. (1953). "The Major Features of Evolution." Columbia University Press, New York.

Thomson, K. S. (1993). Segmentation, the adult skull, and the problem of homology. *In* "The Skull" (J. Hanken and B. K. Hall, eds.), Vol. 2, pp. 36-68. University of Chicago Press, Chicago.

van der Klaauw, C. J. (1945). Cerebral skull and facial skull: a contribution to the knowledge of skull-structure. *Arch. Neerl. Zool.* **7:**16-37.

Wagner, G. P. (1986). The systems approach: an interface between development and population genetic aspects of evolution. *In* "Patterns and Processes in the History of Life" (D. M. Raup and D. Jablonski, eds.), pp. 149-165. Springer-Verlag, Berlin.

Wagner, G. P. (1989a). The biological homology concept. *Annu. Rev. Ecol. Syst.* **20:**51-69.

Wagner, G. P. (1989b). The origin of morphological characters and the biological basis of homology. *Evolution* **43:**1157-1171.

Wagner, G. P. (1995). The biological role of homologues: a building block hypothesis. *N. Jb. Geol. Paläont. Abh.* **195:**279-288.

Wagner, G. P. (1996). Homologues, natural kinds and the evolution of modularity. *Am. Zool.* **36:**36-43.

Wagner, G. P. (1999). A research programme for testing the biological homology concept. *In* "Homology" (G. R. Bock and G. Cardew, eds.), pp. 125-140. Wiley, Chichester (Novartis Foundation Symposium No. 222).

Wagner, G. P., and Altenberg, L. (1996). Complex adaptations and the evolution of evolvability. *Evolution* **50:**967-976.

Wagner, G. P., and Schwenk, K. (2000). Evolutionarily Stable Configurations: functional integration and the evolution of phenotypic stability. *In* "Evolutionary Biology" (M. K. Hecht, R. J. MacIntyre, and M. T. Clegg, eds.), pp. 155-217. Vol. 31. Plenum Press, New York.

Wagner, G. P., Booth, G., and Bagheri-chaichian, H. (1997). A population genetic theory of canalization. *Evolution* **51:**329-347.

Wake, D. B., and Larson, A. (1987). Multidimensional analysis of an evolving lineage. *Science* **238:**42-48.

Wake, D. B., Roth, G., and Wake, M. H. (1983). On the problem of stasis in organismal evolution. *J. Theor. Biol.* **101:**211-224.

Wake, M. H. (1985). The comparative morphology and evolution of the eyes of caecilians (Amphibia, Gymnophiona). *Zoomorphology* **105:**277-295.

Wake, M. H. (1992). "Regressive" evolution of special sensory organs in caecilians (Amphibia: Gymnophiona): opportunity for morphological innovation. *Zool. Jb. Anat.* **122:**325-329.

Warwick, R., and Williams, P. L., (1973). "Gray's Anatomy," 35th ed. Longman, Edinburgh.

Webster, G., and Goodwin, B. (1996). "Form and Transformation. Generative and Relational Principles in Biology." Cambridge University Press, Cambridge.

Whyte, L. L. (1965). "Internal Factors in Evolution." George Braziller, New York.

Zangerl, R. (1969). The turtle shell. *In* "Biology of the Reptilia" (C. Gans, A. d'A. Bellairs, and T. S. Parsons, eds.), Vol. 1. pp. 311-339. Academic Press, New York.

THE CHARACTER CONCEPT: ADAPTATIONALISM TO MOLECULAR DEVELOPMENTS

ALEX ROSENBERG

Department of Philosophy, Duke University, Durham, NC 27708

WHY CLASSIFY?

"Element" is a technical term from chemistry, which secures its meaning in the periodic table of the elements. It is the fundamental classificatory term of chemistry, and the classifications it provides rest on a more fundamental set of classifications in atomic theory. When Mendelev fist propounded the periodic table of the elements he could not know that his classificatory system was the right one in the way that we know it today. Not only could he not foresee that the very blemishes in his system—the blank spaces in the table—would turn out to be among its strongest confirmations as they were filled, nor could he foresee that spaces in the period table beyond the heaviest elements of which he knew would also be filled in accordance with the families he had enunciated. Finally, at the time the period table was propounded no one knew of the theory of electrons which underwrites the common chemical properties of the families of elements and no one understood the nature of the nucleus which explains atomic weight and atomic number, as well as isotopic differences within elements. The periodic table has adorned the front of chemistry class rooms in schools and universities for nearly a century because Mendel got it right, that is, carved nature at the joints, where the laws of nature tell us its parts are individuated from one another.

Before Mendelev, the work of Faraday, Dalton, and others had suggested some fundamental laws of chemical proportionality: certain chemicals combine in fixed proportions, which enable us to attach masses to them in certain whole number accounting units which were chosen largely for calculational convenience, e.g., hydrogen was given the mass unit 1 and oxygen 16. Dalton famously claimed these to be atom weights and Dulong and Petite showed that the product of atomic weights and specific heats for the then known elements equals approximately 6.3. Faraday's experiments suggested the further law that in electrolysis the weight of an element is directly proportional to the among of electricity required to liberate it. However, nothing was known about the "atoms" which Dalton alleged elements where composed of, and the notion of an atom was little less speculative than that of Democratus. Mendelev began his work by uncovering a periodic law: that various properties of elements recur periodically through the sequence of elements graded by their atomic weights: Elements which did not combine with any other element in a chemical reaction are all gases. For each of these inert gases, the next heavier element is always a metal with a shiny luster that conducts both electricity and heat and reacts vigorously with water to form basic solutions and with hydrochloric acid to form salts. These elements are the alkali metals. For each of the inert gases, the immediately lighter element is a nonmetal, a poor conductor, and reacts with hydrogen to form acids. This group of elements is the halogens. The elements between each alkali metal and the halogen which follows it show a gradation of properties such as electrical conductance and acidity.

Each of these properties—in fact, characters of the elements—provided parts of their definitions. To establish whether a sample of some pure substance is sodium, one establishes degree of luster, conductance, results of reactions with hydrochloric acid in water, etc. However, the laws relating these characteristics to elements do not explain what it is for a substance to be sodium. They report what are more in the nature of symptoms or effects of being sodium than causes or explainers for why the substance is sodium and not neon or magnesium.

Mendelev parlayed his period law into the periodic table which grouped elements both by atomic weight and these periodically recurring characters. The vertical columns in the table are "groups" of which there are eight, along with a set of "transition elements" composed of subgroups. Elements in each of the columns—the groups and subgroups—have characters in common. The periodic table has seven horizontal "periods" which reflect the recurrence of the characters Mendelev used to sort the then known elements into the eight groups (and the transitional elements). These descriptive generalizations about the physical properties of elements and the ratios in which they combined made Mendelev's table a promising proposal for classifying the elements.

However, it was the discoveries of the electron structure, the proton and the neutron, subsequent to Mendelev's work which vindicated his classification as the right one. The electron structure of each element's atoms explains their chemical characteristics, that is, determines membership in the groups and subgroups, and the nuclear structure explains atomic weight, atomic number, and isotope differences, which fixes the order of elements within the seven periods. This theory about structure explains the empirical generalizations Mendelev started from.

As a result, Mendelev's table of the elements came to include information about electron shells, as well as proton and neutron numbers. For it is these underlying causes which now come to define the elements, to underwrite Mendelev's classification as the uniquely correct one, and to explain the characters, the effects which once defined the elements "symptomatically" in terms of more fundamental causes. The periodic table is the right classification of the elements because it is underwritten by a theory of underlying causes.

Why do we want the right classification? It's not just a desire for neatness, or even to organize the information we have in a compendious way. We want the right classification because of our explanatory and predictive interests. A classification is right if its classifications are explained by an underlying theory and if the classification or the theory which underlies it leads to new insights about the objects of classification and their relations to one another in the form of general laws that we can apply to explain and control phenomena hitherto unexplained or uncontrolled. Thus, the blank spaces in the periodic table are not defects, but predictions that elements not known at the time the table was established would be isolated; the underlying theory turned the "imperfection" of isotopes into a well-understood phenomenon, and the theory eventually led to the synthesis of transuranium elements which had never existed before the experiments undertaken to synthesize them.

The periodic table is an important classificatory system to keep in mind as we wrestle with classification in biology. It may variously provide us with a paradigm of what biology should aim for, or an example of what is neither possible nor desirable in biological classification. Let's begin by considering why the biologist wants a system of classification. I suggest that the reasons should be broadly similar to the aims of the chemist: to help frame and test underlying theories and to provide explanations and predictions of biological regularities hitherto unsuspected.

It seems unarguable to an outsider like me that one object of taxonomy is to provide a classificatory system for biodiversity that substantiates the Darwinian evolutionary tree of life, that shows how each species is descendant from a prior species working back in a network that becomes the ever thickening branches of a tree until we reach two or three trunks merging into one tap root at the beginning of organic evolution on this planet. That some systematists—presumably pheneticists—dissent from the construction and testing of phylogenies as the aim of taxonomy is deeply perplexing. I will briefly sketch why this dissent is untenable in any case. Besides reflecting the fact of evolution on this planet, a classification needs to do more if it is to the right one in anything like the way Mendelev's periodic table is the right classification for chemistry. It needs to have explanatory and predictive power, or give rise to generalizations, laws, and theories with such powers. In particular, classification should help evolutionary biologists uncover regularities in the pursuit of adaptive strategies, mechanisms of speciation, the identification of and quantification of evolutionary forces. Classification should have important predictive ramifications for ecology and environmental regulation, both through the identification of species so that we may assess threats to them and the testing of generalizations about the consequences of competition, cooperation, and other interactions among them. Without a prospective payoff in scientific explanation and prediction, classification would have no more scientific interest than a very large, very full stamp album.

It is immediately obvious, however, that biological classification faces an utterly different set of problems from chemical classification, though it is not clear that the chemists knew this when they set out to classify the elements. Samples of most of the elements are to be found distributed throughout the galaxy. In some cases there are no samples of some elements now extant, but there have been samples and will be again any where and any time in the history of the universe. Each atom is a token—a particular sample, a type, or an element—and shares in common with other tokens a common structure. For organisms and species you can negate almost everything I have just said about atoms and elements. Instances of all species can only be found on the Earth. None, we believe (and the theory of evolution attests), are found elsewhere. Even a perfect specimen big-eared elephant on a planet circling Alpha Centari would not count as a member of the species, *Elephas africanus*. Where no instances of a species are now extant, the species as a whole is not extant, but extinct (unexemplified elements are neither nonexistent nor extinct) and there will never be instances of that same species again some where and some when in the history of the universe, even if organisms with the same nucleic acid sequence genes emerge here or any where. Each organism is a part, not an instance, of a line of descent, a component of a larger individual item – the line of descent. Arguably the species is a line of descent, an individual item composed of its members: not a kind with instances but a thing with parts. Most important, an organism does not appear to share in common with other members of its species a common structure. Heritable variation is not only the observed rule, it is also the theoretically required fact about the members of a species.

These facts about organisms and species, by contrast with atoms and elements, make the prospects for a classification of species that resembles the periodic table of the elements in theoretical significance, explanatory or predictive power slim, or so I shall argue.

"ALL TRUE CLASSIFICATION IS GENEALOGICAL" (Darwin, 1859, p. 410)

That the right classification should ultimately reflect Darwinian considerations—descent with modification, gradualism, and perhaps most of all, adaptation—is true but controversial for several reasons. Leaving aside those who question the theory of natural selection much as we would ignore people who argue that the earth is flat, there remain serious biologists who hold that such a classification is question begging. To begin with, numerical taxonomists will hold that our classification system should be theory free. For if it simply reflects a theory like Darwin's it cannot be employed to test the theory. These proponents of theory-free taxonomy will point to Mendelev and note that the period table strongly confirmed atomic theory, even when we had only the slimmest indirect evidence for it because its independently established classification of the elements jibed with the claims of atomic theory about electrons, protons, and neutrons. There is something to this point, but not much. Classification independent of theory is an unattainable ideal, and neither Mendelevian nor biological taxonomy can approach it very closely. It is true that Mendelev's classifications were based on "observational generalizations" about chemicals which had been well established, but Mendelev's achievement was to pick and choose among a welter of empirical generalizations about chemical substances a subset to which the attached importance. Why this subset of chemical generalization, and not others? Because Mendelev had a theory or at least some hypotheses about which chemical properties were more fundamental and this theory was supported by the atomism of 19th century chemistry. Mendelev wasn't right by accident, he was right because he was simultaneously advancing a classification and uncovering the theory that subserves that classification. Explanation and classification interanimate; separating them is artificial and obscures the process through which both develop together. That the language of theoretical explanation and the terms of observational description march together has long been recognized in the philosophy of science. It is the reason why philosophers of science have despaired of providing a logic of confirmation – a recipe for testing theories by comparing their predictions for observation with observations described in a theory neutral language. We cannot describe data in ways that don't presuppose theory, and the way in which data confirm theory is mediated by a great deal of further theory. So, the biologists who seek a theory-neutral classification are asking the impossible.

What of those biologists who seek a classificatory system free from one particular theory which they hope thereby to test more crucially? Darwin's theory. There are those who seek a theory neutral about matters such as adaptation or gradualism. A classificatory system imbued with these presuppositions cannot accommodate data which undermine "the modern synthesis." If we obliged S.J. Gould and his true believers to adopt a slavish consistency, then they would have to object to a Darwinian inspired taxonomy because they reject adaptationalism and gradualism. That is, Gould, along with Richard Lewontin (1979), famously has accused Darwinian biology of Panglossianism – the sin of seeing every trait of organisms as adaptations as the solution to a problem set by the environment required to be solved for survival of every lineage that faces it. By contrast, they have insisted that the traits of organisms are as much the result of constraints, and in particular developmental constraints within which adaptation may push and pull the shape of these traits. A classificatory system should be neutral, it would be held, between treating as adaptation or constraints on them the components into which we anatomize organisms for taxonomic purposes. Similarly, the Darwinian thesis of gradualism is one Gould and others have challenged. Presumably, a taxonomy that reflected Darwinian gradualism would, like Mendelev's classification, have blank spaces for missing intermediate forms, at equal intervals between fossils discovered in the record. However, if, as Gould has held, what the fossil record shows is that evolution occurred elsewhere, that evolution is characterized by bursts of rapid change punctuating long periods of changelessness, then the blank spaces of a Darwinian system of classification will not be filled. Notice that this is a classificatory outcome that would tend to undermine Darwinian theory. It shows clearly how classification and explanation go hand in glove. Too many empty cells in a classification and it becomes clear that there may be no explanatory theory underwriting it.

Can we really have a theory—free taxonomy—free not just from Darwinian inspired interests in genealogy, but free from all biological theory? About the only attempt to provide one, proposed by the numerical pheneticists, came to grief for the same reason that all other purportedly theory neutral, purely behavioral or observational attempts to collect data have failed. In the absence of a theory, we simply don't know what to observe, what measurements to take. Simply classifying in accordance with the categories of common sense or purportedly immediate observation, even when limited to cardinal scales, is not theory-free classification. Indeed it is classification in accordance with a theory we can't even articulate, one which is both complicated and difficult even to recognize. This theory is not only folk biology – the one whose kind terms have evolved in our species as a result of the long term adaptation of its perceptual apparatus, it is even more the creature of our more invisible cognitive structure. The classificatory terms of color, shape, smell, sound, taste, and touch are plainly "subjective" in the philosopher's terms—"secondary qualities" reflecting our perceptual apparatus as much as the traits of the things with which they come into contact. Of course there is no getting away from this theory—the one embedded in our perceptual and cognitive apparatus. Every classificatory system, from that of grand unified theory in physics to those of the classical philologists, embeds this theory. For these classificatory systems are ours, they are ones we have framed, and we understand, by the lights of our own intelligence. If this apparatus some how prevents us from carving nature at the joints, then wherever it does so, the prospects of a classification with the objectivity and rightness of Mendelev's become nil.

Securing a taxonomy prior to and independent of any theory is a will-o-the wisp. So, on which theory should a taxonomy rest? The trouble here is that the Darwinian theory of natural selection is the only game in town. A classification free from this theory would be a classification free from any biological theory. Darwinian theory is the only biological theory that there is. Its reach is so great and so firm that almost any way of individuating traits, properties, and "characters" to test biological theory will already presuppose Darwinian theory. To see this we need to understand how our biological concepts have emerged.

Consider any of the characters, traits, and properties of organisms which biologists or, for that matter, common sense employ to classify them: their distinctive appendages, organs, behaviors, etc. These terms, like almost all terms in ordinary and biological language, are "functional" ones. That is, they individuate characters by their effects, and not just any effects, but those effects which are salient to us, salient either because they are evidently adaptations of the organism or they are traits of the organism it is relatively easy for us to discriminate. Thus, color, size, shape, head, neck, wing, fin, leg, heart, lung, blood, nerve, neural tube, cell, cilia, contractile vacuole, cell-surface receptor, gene, enzyme, and behavioral character labels such as clutch size and mating display, every trait down to the level of the biologically active macromolecular is in whole or in part named by a functional term, one that individuates its referents by one or more effects selected for in evolution, or a trait that happens to be detectable by creatures like us. [Lewontin (1980) recognized this, but he seems to think that functional classification in biology, and outside it for that matter, is optional.]

Biological classification will have to reflect considerations from the theory of natural selection for several reasons: first, a good classification must have a theoretical base, and if the theory is the most fundamental or perhaps the only general theory in biology, then only it can provide the base a classification requires. Second, the characters in which classification is to be affected will be functional traits, and so identifying them requires either an appeal to the theory of natural selection or the argument from design to the existence of God. Even where we remain agnostic about the adaptational history (if any) of a character, we are already committed to its having one by the only biological means we have to individuate it.

Now the first thing to recognize about characters functionally individuated is that there are no interesting, manageable, strong empirical generalizations about such traits. This is easy to show, but its consequences for taxonomy and systematics are very far reaching. Recall Mendelev's achievement: the period table gets started because there was already a body of well-confirmed and fairly precise empirical generalizations about the characters of the then known elements. Mendelev's table is accepted as the right classificatory system in chemistry because beyond the empirical generalizations there is a deeper theory of underlying causes which explains its rightness and which underwrites its predictive powers as well. If there are neither empirical generalizations in taxonomy nor a deeper theory of underlying mechanisms, we cannot anticipate a single correct classificatory system in biology with the scientific usefulness we may have hoped for.

The reason we can expect no very strong empirical generalizations or laws about any character or trait that we might functionally individuate is a fairly obvious one: natural selection makes such laws highly improbable.

Natural selection "chooses" variants by *some of their effects*, those which fortuitously enhance survival and reproduction. When natural selection encourages variants to become packaged together into larger units, the adaptations become functions. Selection for adaptation and function kicks in at a relatively low level in the organization of matter. As soon as molecules develop the disposition to encourage the production of more tokens of their own kind, say by template matching or self-catalysis, selection begins to operate. However, among such duplicating molecules, at apparently every level above the polynucleotide, multiple *physically distinct* structures are frequently found, which foster copies of themselves and of large ensembles that include them at (nearly) identical rates. The particular ways these molecules foster copying of assemblages that contain them becomes their functions—their selected effects— and if slightly differing molecules can have the same rate of copying, then there will be physically distinct structures with the same functions.

In fact it is a serious mystery that the function of information storage and transmission seems to be carried out only by one sort of structure, the polynucleotide. Either there is only one way in our environment for hereditary information to be carried or long ago the polynucleotide won the selective race for carrying genetic information, and all competing information carrying molecules became extinct.

Above the polynucleotide, from the lowest level of organization onward, there are always *ties* between structurally different molecules for first place in the race to be selected. Sometimes these molecules are only slightly different, as in the case of allozymes, sometimes they are very different in parts of their structures that "don't matter" for their functions, as in the case of the various hemoglobins, and sometimes utterly different molecules perform the same functions. This is because, as with many contests, in case of ties, duplicate prizes are awarded. For the prizes are increased representation of the selected types in the next "reproductive generation," which will be true up the chain of chemical, all the way to the organelle, cell, organ, organism, kin group, etc.

It is the nature of any mechanism that selects for effects that *it cannot discriminate between differing structures with identical effects*. Functional equivalence combined with structural difference will always increase as physical combinations become larger and more physically differentiated from one another. Moreover, perfect functional *equivalence* isn't necessary. Mere functional similarity will do. If two different assemblages are composed of quite different substructures, molecular or otherwise, but these substructures have only slightly different payoffs for replication of the assemblages in a given environment, selection will not discriminate between them for a very long time depending on the size of the populations and the stability of the environment remains. It is crucial to see that in a Darwinian biocosm functional equivalence-cum-structural diversity will be universal at every level of organization above the molecular. By contrast, physical and chemical processes will not produce structural differences with equivalent effects, for there is no scope for nature to select by effects until molecules begin to help make copies of themselves.

Most objects of physics, chemistry, biology, and ordinary affairs are picked out, referred to, "individuated" in the philosopher's jargon, by their effects: an acid is a proton donor, and a magnet is something which attracts iron filings. However, the objects of biology and ordinary life are individuated by an important subset of their effects, those which have been selected for and those which are not just effects, but functions. That cognitive agents of our perceptual powers individuate by effects should be no surprise. Cognitive agents seek laws relating natural kinds. Observations by those with perceptual apparatus like ours reveal few immediately obvious regularities. If explanations require generalizations, we have to theorize. We need labels for the objects of our theorizing even when they cannot be detected because they are too big or too small, or mental. We cannot individuate electrons, genes, ids, expectations about inflation, or social classes structurally because we cannot detect their physical features. We can, however, identify their presumptive effects. This makes most theoretical vocabulary "causal role" descriptions. Biology, common sense, and the social sciences generally differ in their individuation from the physical sciences in that they appeal to functional traits, those which have been selected for either consciously, such as fork or chair, or naturally, such as wing, cilia, or hormone.

Now it is easy to show that there will be no strict exceptionless generalizations incorporating functional kinds. Suppose we seek a generalization about all $F's$, where F is a functional term, such as gene, wing, or cell-surface receptor. We seek a generalization of the form, all $F's$ are $G's$. In other words, everything in the extension of Fs is also in the extension of some other property G; everything which is an F is also a G. Such generalizations are obviously crucial to taxonomy: for example, "all mammals have fir." Now, G will have to be either a functional term, one which picks things out by their selected effects, or a nonfunctional term, which picks things out by their structure, their causes, or some nonselected effect.

G cannot be a nonfunctional term. If F is a functional kind, then due to the fact that selection for effects is blind to structure, there will be no single or even a small number of structural kinds shared by all members of F. G cannot be a structural kind. If F's do not share the same physical structure they cannot share the same causes. Recall the principle of same cause, same effect: different structures are different effects, they can't have the same causes. Nor will the members of F share some other unselected effect in common, or at least no scientifically interesting one. For on the same principle of same cause, same effect, for them to share another unselected effect in common, they would have to share a structural feature in common. Were we to find some structural feature shared by most or even all of the members of F, it will be a property shared with many other things, such as mass or electrical resistance, properties which have little or no explanatory role with respect to the behavior of members of the extension of Fx. For example, weighing less than 100,000,000,000 grams is a structural property all mammals share in common, but the only other unselected effects they share in common can't be joined with this property to make a scientifically interesting law about all mammals.

Could G be a functional term? That is, could there be a distinct functional property different from F shared by all items in the extension of the functional predicate F? The answer must be that the existence of such a functional property distinct from F is highly improbable. If F is a functional kind, then the members of the extension of F are physically diverse due to the blindness of selection to structure. Since they are physically different, any two F's have nonoverlapping sets of effects. If there is no item common to all these nonoverlapping sets of effects, selection for effects has nothing to work with. It cannot uniformly select all members of F for some further adaptation. Without such a common adaptation, there is no further function all $F's$ share in common.

Whether functional or structural, there will be no predicate G that is linked in a strict law to F. We may conclude that any science in which kinds are individuated by causal role will have few if any exceptionless laws. In particular, there will be few laws of the sort on which Mendelev capitalized in framing the periodic table of the elements, and which atomic theory subsequently provided to underwrite his classification. Without such laws, however, classification is itself neither open to explanation nor does it have explanatory power.

The problem is obviously the role of functional individuation. Perhaps a science seeking to provide itself with a classificatory scheme that mirrors nature should surrender functional individuation? This, however, is not a suggestion we can take seriously. To begin with, most nouns in most languages are functional, as are many verbs describing the behavior of organisms. The preponderance of functional vocabulary in ordinary language reflects a very heavy dose of anthropomorphism, or at least human interests, but it is an anthropomorphism vindicated and now naturalized by Darwin's substitute for design. In addition, it is not just selected effects which our vocabulary reflects but selected effects important to us either because we can detect them unaided and/or because we can make use of them to aid our survival. We cannot forego functional language and still do much biology about phenomena above the level of the polynucleotide and amino acid. Surrendering functional individuation is surrendering biology, including molecular biology in favor of organic chemistry.

ALL CLASSIFICATION SHOULD BE DEVELOPMENTAL

The absence of laws commits classification in biology to a fate altogether different from classification in chemistry. To begin with, classification in biology is historical; classification in chemistry is ahistorical. That is, while the items elaborated in the periodic table are types, kinds, and abstract entities whose instances are particular atoms which may exist everywhere and no where throughout the universe and all time, the items elaborated in any biological classification are spatiotemporally restricted to this planet over some billions of years. There are no more likely to be laws about species or higher taxa than there are to be laws about Napoleon or all Frenchmen.

Almost whatever character we may hit upon to classify organisms, or their components, right down to the level of specialized cells, will turn out to be functionally individuated, and therefore to figure in no strict laws. This means that biological classification will never attain the sort of neatness, agreement, and theoretical support characteristic of chemical classification. Even if and when we secure the classification that matches the phylogeny of this planet, the classification will not have the sort of explanatory and predictive payoff that Mendelev's classification has come to have for chemistry.

When Darwin wrote that all classification is genealogical he was more right than he knew. It's not just that the purposes of classification are ultimately to reveal the tree of descent which relates all living things to one another and to a single ultimate ancestor. Beyond its ends, taxonomy's mean is genealogical.

Not only are they genealogical, but they also reflect the parochial interests, and more significantly, the cognitive, perceptual, and computational limits of one of the species which taxonomy aims to taxonomize: *Homo sapiens*. The characters employed to classify are functional: hereditary effects selected by the environment. However, they are also selected by their salience for us classifiers. In Dawkin's terms, characters are "memes" which have colonized us because of their usefulness to us in getting about the biocosm we inhabit. Or rather the characters that organisms bear are selected for, and the concepts which label them in our classificatory schemes are selected for as well. This dual and distinct selection of character and label makes for much of the difficulty, dispute, and inconclusiveness of taxonomy.

Again, compare the characters that the periodic table relies upon. Acidic, conductor, gas, soluble: these too are memes which before the advent of scientific chemistry survived and spread among us because of their practical use. Our need for more reliable knowledge, and therefore better classification, subjected these characters to the test of whether they would enable us to form a predictively and explanatorily powerful classificatory scheme, or at least one that lent itself to a theory with such powers.

Give that in biology we now recognize both the hold on us which character names that biology and ordinary thought have had and the limitations to which these characters and the terms that name them face, what should we do? We need not do anything if we are uninterested in securing generalizations or anything closer to them then those which already grace biology. If we are satisfied with descriptions of historical sequences in terrestrial evolution, and mathematical models of limited applicability and less long-term unification, then we need not surrender functional individuation of the characters on which classification will be based. However, if our aim is classification with an illuminating underlying causal mechanism, with a greater payoff in prediction and explanation, then we need to rethink our classificatory character terms. If we think there are generalizations about evolutionary mechanisms, and implications of applied ecology, we need to rethink our character commitments.

Can we go physical in our character-based classifications? That is, can we eschew all functional vocabulary in favor of a purely structural characterization of the flora and fona of this biocosm? The suggestion has its attractions, and we even have a good idea of how to proceed. All we need do is establish the primary sequence of all the organisms that now exist and have existed in the past on this planet. Then we sort them into classes based on sequence homologies. Once we have done this, we establish the simplest pattern of descent and relationship consistent with this primary sequence data. Of course the "we" in the previous sentences refers to our computers. The output they will provide is unlikely to be a unique set of hierarchial taxonomies of descent, but this is no worse than the taxonomies given by nonprimary sequence-based character classifications. Presumably the differences among alternative nucleic acid based taxonomies will be small. Moreover, the taxonomies will at the same time be our best guides to phylogeny, thus further vindicating Darwin's conception of its intellectual tasks.

Doubtless the molecular biologically sophisticated reader has been suppressing a fealing of mirth while reading the previous paragraph. For the counsel there offered to taxonomy is far from reasonable. To begin with, sequencing specimens of all organisms now extant would require the time and resources of the gross domestic product of many nations over a millenium. Just sequencing the genome of a specimen of *Homo sapiens* is currently priced at $3,000,000,000. Classification is important but not that important. Second, within species there are large differences in nucleic acid sequence, so that sequencing specimes is not sufficient to establish the physcial characters of a species. Third, it is a regrettable fact of paleonotology that only the hard parts are preserved in the fossil record. We have no way of establishing the primary sequence of any species no longer extant. Accordingly, even if we had the primary sequence data that enabled us to uniquely taxonomize all extant species, the most we can do without importing nonmolecular information is to hypothesize the primary sequence data of the now extinct species to which any two extant species might trace their lineage. Here too the number of alternative possiblities is huge, due at least to the redundancy of the genetic code and the large body of nonfunctional junk DNA in almost every species. It would be to underestimate the genius of modern science to conclude that these obstacles to a purely physical taxonomy with phylogentic implications are insurmountable, but they do suggest that we seek alternate routes to classification.

If our aim is a taxonomy available in the foreseeable future one piece of advice we cannot follow is that of eschewing all functional vocabulary all together. It is true that the failure to provide a taxonomy which could uncover general mechanisms of evolution and other biological laws is largely the result of functional-role vocabulary of biology hitherto. However, consider the suggestion that we attempt to minimize the amount of functional individuation involved in classificatory characters. Suppose we limit ourselves to classification in terms of those properties which minimize the heterogeneity of structures to which selection is blind. Since heterogeneity of structural packages similar in function obviously increases as the size of the packages increases, we should employ descriptive terms that pick out characters of the smallest functional packages selected for. This self-imposed limitation will make taxonomy another compartment of molecular biology. For molecular biology is just organic chemistry plus natural selection. That is, the descriptive vocabulary of molecular biology differs from that of organic chemsitry just insofar as it identifies molecules in terms of their functions, i.e., their selected effects: "messenger RNA," "cell-surface receptor," "DNA polyerase," etc.

As noted previously at the level of the macromolecule, nature has less scope than elsewhere to select for two or more different structures with the same or similar effects. This is just because the number of different types of effects a small molecule can have on its reproduction or that of the larger assemblages into which it is packaged must be smaller than the number of different effects of a larger one or a package of them. Thus, for example, histone 3, one of the proteins around which the DNA is wrapped, is structurally the same in its primary sequence of amino acids across almost the entire animal kingdom. This makes it a very poor character for purposes of taxonomy, but an excellent one in terms of grounding generalizations about the structure of all organisms. Similarly, the mysterious fact that the only way genetic information is packaged in our bioshpere is in nucleic acids suggests strongly that nucleic acids are the only molecules which have high-fidelity information storage as an effect that can be selected. Again, this universality makes "having a nucleic acid-based genetics" a poor character for classification; it won't distinguish *E. coli* from *Cygnus olor*.

Alternatively, suppose we seek characters of a much more complicated but still molecular biological sort, for example, we might classify by the character "employs the TCA cycle," where this can be understood as a purely biochemical classification fixed by the specific chemical reactions and catalysts we now know compse it. Again, this character will not sort organisms very finely. Also, what we know about the structural similarity of biochemical processes in cells from the simplest to the most complex organs suggests that none is likely to provide the kinds of discrimination that will either shed light on our actual genealogy or else enable us to frame laws. Once we get to the level of particular species we have already identified characters too specific to give us a topology with either of these features. For example, suppose we employ hemoglobin sequences as classificatory characters. The result I suggest will be either to fail to distinguish between humans and other primates or to distingush within the class of *Homo sapiens* between those who suffer from various blood disorders as members of different species.

However, this conclusion suggests an insight that should be made to work well if any classificatory schema employing molecular biological characters will. What distinguishes organisms from one another across the phylogenetic spectrum must not be their structural genes but their regulatory ones. If we and the chimpanzees share 98 % of our primary sequences in common then the substantial differnces between us must result in large measure from differences in regulatory genes that control the expression of those structural genes we share in common. *Mutatis mutandis* for the rest of creation.

Though it will follow from this that the most revealing and most well-founded taxonomy to which the diversity of life on this planet submits of will reflect sequence homologies and differences among structural and regulatory genes, it does not follow that we should now immediately adopt this standard for the characters taxonomy should employ. The reason of course is that we don't know enough about the genome to distinguish regulatory from structural sequences in even a small number of organisms, let alone enough to establish a taxonomy that meets our geneological needs.

However, this conclusion does underwrite the centrality of developmental biology in the establishment of the best taxonomy we can expect to produce in the immediate future, and one which will most easily be adapted to a taxonomy of structural gene sequences when one does begin to emerge in the future.

The centrality of developmental considerations to taxonomy is not news. Developmental biologists have been arguing for it implicilty and explicitly at least since von Baer. Darwin held that "the community of embryonic structure reveals community of descent", and that the presence of embryonic eyes in moles and teeth in whale embryos reflected descent from organisms which had eyes and teeth, thus securing a classificatory relation to animals that lacked these organs. Moreover, Darwin noted that characters with an adaptive role in relatively specialized environments typically emerged late in embryological development. These early and late embryological traits reflect homologies and adaptations that provide a basis for phylogenetical classification. By the late 19th century embryologists were arguing that when the same organs emerge from the same group of cells in two different species, the result is a homology on the basis of which shared descent may be hypothesized. Embryological cleavages that result in different structures from apparently similar embryos reflect species differentiations. However, the sample pattern of homology and adaptation that embryologists hoped to find disappeared in the welter of developmental data coming in through the remainder of the 20th century.

Developmental data as a source for taxonomy went into exclipse, stimatized as the vestage of von Baer and speculative *Naturphilosophie* more redolent of Goethe than Ghiselin. Now perhaps, with the advent of developmental molecular biology, we can recast the taxonomic strategy in terms not so different from those of the 19th century embryologists. Instead of distinguishing organs into those which reflect homologies and those which reflect adapations, we can substitute the distinction between structural genes, which more or less code for the constituents of structures homologous across species, and regulartory genes, which shape these homologous structures within species into more locally suitable adaptations. In fact the structural/regulatory gene distinction is a much finer grained one than the homology/adaptational organ distinction of 19th century developmental evolutionary biology. It will operate within species as well as across them. In doing so, however, it will more finely grain our phylogenies and reveal commonalities in evolutionary mechanisms of the sort we expect from the right classification–mechanisms that operate at or at least via the physiology of the gene.

This sort of a shift interestedly mirrors the vicissitudes of the one gene-one phenotype distinction in genetics. Although completely repudiated at the level of functional biology, the one gene-one character model turns out to be an accurate reflection of the relationship between molecular genes and their immediate protein products, out of which macromolecular and sub- and cellular phenotypes are built. Similarly, the homology/adapation distinction superceded at the level of the larger functional unit in taxonomy becomes a viable basis for classification at the level of the gene sequence lineage.

Of course acquiring a complete catalog of the structural and regulatory genes for any one species is a vast undertaking. For *Homo sapiens* the task is nothing short of the Human Genome Project. However, we don't need all the structural and regulatory genes to begin to fill in a molecular developmental taxonomy. We do need structural and regulatory gene data from several species. Indeed the way in which we distinguish structural from regulatory genes is by molecular homologies. However, a homology at the level of the molecular gene is a base pair for base pair structural isomorphism, not a functionally characterized identity! This is what gives molecular developmental characters the advantage over higher level characters. With some structural and some regulatory gene data in hand we seek structural isomorphism and structural differences in sequences, and employing methods already well understood in designing phylogenies on the basis of differences in these sequences, establish a taxonomy which reflects our closest approach to a window on actual speciations throughout the whole past history of the biosphere, and on mechanisms we can expect to operate in the future of evolution on this planet. Whether these eventuate in interesting generalizations of use in ecology or elsewhere remains to be seen, but the result is a taxonomy as near to what Mendelev did for chemistry as biological classification can ever approach.

LITERATURE CITED

Darwin, C. (1859). "The Origin of Species." Murray, London.
Gould, S.J., and Lewontin, R. (1979). The spandrels of San marco and the Panglossian Paradigm: a critique of the adaptational program. *Proc. Roy. Soc. London B* **205**:581-598.
Lewontin, R. (1980). "Adaptation." *The Encyclopedia Einaudi*, Milan.

9

THE MATHEMATICAL STRUCTURE
OF CHARACTERS AND
MODULARITY

JUNHYONG KIM[1] AND MINHYONG KIM[2]

[1]*Department of Ecology and Evolutionary Biology, Yale University, New Haven, CT 06520*
[2]*Department of Mathematics, University of Arizona, Tucson, AZ 85721*

INTRODUCTION

In a review of biological usage of the world "character," Colless (1985) classifies three different types of usage. The first kind, which he calls an *attribute*, denotes "a distinctive attribute, quality, or property of structure, form, material, or function." The second kind, which he calls *feature*, denotes "a part, often of a more or less physical nature, but generally ... more abstract." The final kind, which he calls a *character variable*, denotes "a set of mutually exclusive attributes constituting a logical or mathematical variable." This rather innocent word "character" seems to be sufficiently slippery in our minds as to require some critical examination (and to generate this volume). In the somewhat enchiridion "Keywords in Evolutionary Biology" (E. F. Keller and E. A. Lloyd, eds.), L. Darden (1992) notes that a "unit-character" concept, a view of organisms as a composition of variable parts, was a new view appearing in works of Darwin and Mendel, contrasting earlier views of the organism as "exhibiting the whole essence of the species." Indeed, Brooks (1883) writes that an essential component of Darwin's theory of pangenesis was the idea that the development of an organism proceeds not from a "perfect miniature" of the adult, but from a mixture of distinct germs from each distinct cell or structure. In fact, the notion of a "gene" seems to parallel the notion of a character albeit with an emphasis on transmission for the former and on measurement for the latter.

In all of the discussions, a meaning close to "set of mutually exclusive attributes" predominates scientific usage of the term character, especially in systematics with its association to the notion of homology (Sneath and Sokal, 1973; Patterson, 1982; Wagner, 1995). Within the concept of genes as well, with the establishment of Mendelism, the idea of a gene seems to have become closely tied with the idea of nondivisible and independent factors (Punnett, 1911; Carlson, 1966). The theme of nondivisible independent entity is common, for example, the idea of a module appearing in studies of morphology and development. "Modules are discrete subunits of the whole..." (p. 326, Raff, 1996) but perhaps with some coarser sense of granularity than that of characters. That is, the word module seems to imply a decomposition of the whole but also an amalgamation of atomic parts. A reason for this central role of the "hypothesis of factor decomposition by indivisible entities" might be that it gives us a reasonable handle on biological diversity and complexity. Combinatorial complexity is a quick way to generate diversity (as noted earlier by Punnett, 1911) and engineering by modular parts seems an efficient scheme for generating complexity ("building blocks"; Wagner, 1995).

Biological characters (or modules) have the additional twist that those things we call characters are usually not immortal (contra say the mass of an element) but, at least at first glance, seem to be "born" and "destroyed" through the action of evolution. The idea of the evolution of a character (by which we mean this birth-death process of character-entities rather than the common meaning of changes in the states of a particular character) asks us to answer the rather practical question: how do we know when a new character is born or when it has died? The trivial answer is that we might recognize a new character when we are able to make new, previously unknown, measurements either through the development of a new tool or through the observation of a new organism. However, this does not automatically bring about the properties of "indivisible entity" and "independence." In a similar vein, Quine (1987, p. 22) asks "but what if everything that has this property has that one as well and vice versa? Should we then say that they are the same property?" This rhetorical question is obviously meant to point out that what we recognize as discrete, independent units are dependent on the variational context of our observations. If two separate measurements, e.g., eye color and hair color, perfectly covary within our observations we would not recognize them as separate entities. Fristrup (1992) notes "...characters are always defined with reference to a sample of biological entities." If we are then given a set of organisms how do we proceed to catalog these characters?

In a paper discussing the homology concept, G. P. Wagner (1995) made a key observation that allows an answer to this question. He recognized that characters are collections of "mutually exclusive" attributes. That is, blue eyes and brown eyes are mutually exclusive in that they cannot be observed on the same individual, thus they are part of a single character. He called this the property of *demographic replaceability*. In terms of the language of genes, he notes that "...two genes may be different but compete for representation in the population, because they tend to replace each other. Genes of this type are called alleles ... . On the other hand there are genes that are different but do not compete with each other. These are genes at different loci...".

Motivated by Wagner's observation, in the following, we wish to set up the language in a way that makes the notion of a character more precise with the idea that it may make it easier to incorporate such notions as modularity of organisms and especially allow the possibility of stating the problem *evolution of modularity* with some degree of precision. We will describe a procedure whereby one arrives at a notion of characters, starting from *any* representation scheme for organisms, coarse or fine, phenotypical or genotypical. The idea we pursue is that we are first given a set of *descriptors* for organisms which are similar in meaning to Colless' *attributes*. That is, a mechanism for describing the organisms in a casual manner ("getting a handle" on the organisms) without particular formal properties. We are then given a set of organisms – which we call a population (abusing the terminology from population genetics), to which we can apply our descriptors. We then define conditions under which a combination of descriptors or decomposition of descriptors can be derived such that "mutual exclusivity" and "demographic replaceability" are satisfied. We reserve the word *characters* or the phrase *character systems* for these "well-behaved" combination/decomposition of descriptors. Given the dependence on the population, we conjecture the existence of processes in which descriptors that were not characters for a population during some non-equilibrium phase of evolution become characters as the population "settles down." Other forms of evolution through equilibrium states may also cause descriptors to evolve into characters. Similar construction can be applied to the idea of modules that we view as aggregations of characters.

DESCRIPTORS AND REPRESENTATION

We will call any map $g : O \rightarrow S$ from the set O of organisms to some other set, S a *primitive* descriptor. The set S is then called the state space of the descriptor, so that we say that an organism c has the state s, if c maps to s and the reference to g is understood. That is, we view g as the act of (possibly indirect) measurement, therefore, we assume that the map is realizable in some measurement sense. The set O denotes the set of realized or possible organisms, where organisms are used in the sense of "whole reproductive units" in a commonsensical manner, such as "Junhyong Kim," a cat, an individual fruitfly, etc. The set O may have additional structures (say evolutionary relationships) but we will ignore them for this paper. Instead we will just assume that the set is for practical purposes finite and has the counting measure. The nature of S can be quite diverse. It could, for example, be a set of numbers, e.g., the number of bristles on *Drosophila,* a set of colors (eye colors), a set of labels for the alleles of a genetic locus, or even a set of "shapes," say the chin shape for humans. In most situations, the state space will have an obvious correspondence to directly measurable quantities (which we might call *natural descriptors)* while in other situations the state space may be abstract (e.g., "additive effects" in quantitative genetics). In any case, we assume S will usually be equipped with a measure from O in the following manner: the measure $\mu(s) = \sum \chi\left(g^{-1}(s)\right)$ where χ is the counting measure for O. In this paper, for the ease of exposition, we will deal with the case where S is a finite set. We will also treat S only as a set, although in many cases S may naturally have additional structures such as ordering, topology, etc. In practical terms, many *a priori* complicated value sets for primitive descriptors can be reduced to the finite set by an appropriate encoding. For example, if the set of values consists of various contours (for chin shapes), one could expand the curve into Fourier series and keep the set of coefficients corresponding to some fixed cut-off of frequencies, the coefficients themselves being discretely divided following the degree of precision desired.

By saying that a primitive descriptor is any map from the state set to a measurable set of organisms, we assume that a primitive descriptor is, in a sense, primary, without necessarily having additional desirable properties such as independence, homology, invariance, etc. Our objective is to first start with a "jumbled" collection of primitive descriptors and extract subsets or agglomerations of the descriptors that in fact have those desirable properties. One of the first properties of interest is that the elements of the state space, S, have the demographically replaceable property described earlier. To make this more precise, define the relative measure of an element $s \in S$ as $v = \mu(s) / \mu(S)$. That is, the relative measure is the value of the element divided by the measure of the whole set. Now we will say that the elements of the set S are mutually demographically replaceable if $v(s) = 1$ implies $v(t)=0$ for all $t \in S, t \neq s$. That is, if the relative measure of s is 1, the relative measure of all other elements is zero. This fits in with the idea that if we "select" for fixation of the relative frequency of the state $s,$ the relative frequency of all other states should go to zero. We can immediately see that this condition will be satisfied if and only if the map g from O to S is a proper function (i.e., it is not one to many). When the map g is a proper function we will call the map a *proper descriptor*. In the following we will only deal with proper descriptors and simply call them descriptors.

We find it important to fix the notion of a collection of descriptors as a surrogate for the collection of organisms and, therefore, we define a *representation* as the total set of descriptors that is deemed sufficient to "represent" the set of organisms of interest. (Thereby, we avoid the need to discuss what exactly we mean by the set of organisms, O, except in the "counting units" sense.) In particular, we assume that the representation is sufficient to contain all organisms within an evolutionary history of interest. That is, it is able to represent organisms "before and after" evolution. The representation, of course, is not complete in the sense that not every detail of the organism is included. If we pursue an analogy with physics, the S's that arise in practice will be "coarse grained" in a sense similar to macroscopic state spaces in thermodynamics, and each point in S will be occupied by many different organisms described microscopically (for example, in terms of their full molecular description). Similar examples can be found in classic population genetics where we typically concentrate on a few genetic loci, and thereby regard the state space for those loci as the phase space, ignoring differences that would arise by enlarging the collection of loci under study. (The state spaces are thus referred to sometimes as "partial" genotypes of phenotypes; Lewontin, 1992.) In any case, we take the viewpoint that fixing a representation is prerequisite to any further discussion of organisms.

Associated to any descriptor f is its frequency space V_f, which we will usually just denote by V when the reference to f is understood, or by V_s, when we want to emphasize the state space. Although a bit abstract, it will be convenient to define V as the real vector space with basis S, so that an arbitrary vector $\mathbf{v} \in V$ can be written $\mathbf{v} = \sum v_s[s]$ for real coefficients v_s. (Our convention will be to write a vector in boldface, while its components will be denoted by the corresponding roman letter with subscripts from the set S. Also, $[s]$ denotes the basis element given by s, but which we surround by square brackets for psychological reasons.) Mathematically, we can think of V as being the space of measures on S, while the physical analogy would regard V as "phase space for populations." In fact, any actual population p gives rise to a vector $\mathbf{v} = \mathbf{v}_p = \sum v_s[s] \in V$, where v_s is the relative frequency of organisms of type s.

(Conversely, we might define a population as some $\mathbf{v} \in V$.) From the definition, we get that $\sum v_s = 1$, so that v actually lies on the unit simplex $\Delta_S \subset V_S$. A far-fetched analogy with physics would identify a descriptor with an observable, S with it spectrum, V with the Hilbert space on which the observable acts and Δ with the projective space of lines in the Hilbert space, describing the set of pure states for a quantum mechanical system.

PRODUCT DESCRIPTORS AND FACTOR DESCRIPTORS

Essential to our use of the term "descriptor" is that we should be able to talk about things like how an organism might be described by an "aggregation" of descriptors or how a descriptor might "decompose" into several descriptors, etc. (Wagner and Altenberg, 1996, call the dynamic version of these two phenomena "integration" and "parcellation".) Therefore, very important for our purposes is the notion of product descriptors. That is, given a set of descriptors $f_1, \ldots, f_n$ with corresponding state spaces $S1, \ldots, S_n$, the product descriptor

$$f = f_1 \times f_2 \times \ldots \times f_n$$

assigns to an organism c the element

$$\left(f_1(c), f_2(c), \ldots f_n(c)\right)$$

in the product state space

$$S = S_1 \times S_2 \times \ldots \times S_n$$

Thus, even when we speak of a single descriptor, it might be constructed "as a package" out of many conventional descriptors. Often, the notion of a product descriptor rises when the set of descriptors $f_1, \ldots, f_n$ are natural descriptors, i.e., descriptors with states directly measurable from the organism. Then, it becomes natural to talk about a descriptor consisting of "brown eyes and brown hair and median height," which is the product descriptor. When we have formed the product, we can recover the original descriptors by composing f with the n projections, $p_i : S \to S_i$.

Given a product descriptor $f = f_1 \times f_2 \times \ldots \times f_n$ with state space $S = S_1 \times S_2 \times \ldots \times S_n$, there is a natural isomorphism

$$V_f \cong V_{f_1} \otimes \ldots \otimes V_{f_n}$$

obtained by sending a basis element $\left(s_1^{i_1}, s_2^{i_2} \ldots, s_n^{i_n}\right)$ of V_f to the basis element $s_1^{i_1} \otimes \ldots \otimes s_n^{i_n}$ of the tensor product. That is, the first element is a basis of the frequency space of the product descriptor formed by n-tuple combination of states from each state space S_i. The latter basis element (the corresponding basis element by the isomorphism) is a basis of the tensor product space formed as the "product" of the basis of each descriptor frequency space (i.e., the states of the descriptors).

We will casually identify the two vector spaces and their natural bases using this isomorphism. If we are now given a population p, then we will get a frequency vector $\mathbf{v} \in V = V_f$ as well as frequency vectors $\mathbf{v}_i \in V_i = V_{f_i}$. Assuming a normalized frequency (i.e., the vectors are on a simplex), each $\mathbf{v}_i$ can be recovered from $\mathbf{v}$ using the operator $t_i : V \to V_i$ which is defined by using the "sum functionals" $s_j : V_j \to R$ that send $\sum v_s s$ to $\sum v_s$, and applying them to each factor in the tensor product but the i-th one.

This discussion on product descriptors naturally leads to the idea of a decomposition of descriptors. We say a descriptor f is *decomposable* if it can be written as a product descriptor of two (or more) nontrivial descriptors, say $f = f_1 \times f_2$, such that $V_f \cong V_{f_1} \otimes V_{f_2}$. (Here, "nontrivial" refers to the condition that each descriptor should have at least two states.) Now, different ways of writing f as a product of two descriptors corresponds to ways of writing the state space S for f as a product of two sets $S = S_1 \times S_2$. Since there are many ways of doing this, there are potentially many ways in which f might be decomposable. For convenience, we will introduce the terminology of *bi-product structure* to refer to a particular isomorphism $S \cong S_1 \times S_2$. We will also call an n-product structure (on S or f) an isomorphism $S \cong S_1 \times ... \times S_n$. When we are starting thus with an f and considering its possible product structures, we will also refer to the set of descriptors f_1, ..., f_n obtained via a product structure on S as its *factor descriptors* (or just *factors* when the meaning is clear) of f.

In the product descriptors, we said that the descriptors on the right hand side of $f = f_1 \times ... \times f_n$ are natural descriptors. The motivation for a decomposition of a descriptor is that here the descriptor on the left hand side of $f = f_1 \times ... \times f_n$ is a natural descriptor and we are looking to see whether its state space has some other natural or abstract decomposition. For example, we might have $f : S = \{0,1,2,3\}$ for a quantitative trait. Observing this we think $0 = 0+0$ and $1 = 0+1$, etc., therefore $S = S_1 \times S_2$ with $f_1 : S_1 = \{0,1\}$, $f_2 : S_2 = \{0,1\}$ is a reasonable abstract decomposition. As another example, we may often have a label descriptor on the left hand side $f : S = \{$Ingrid, Larissa, Peter, ...$\}$ which we know to have the natural decomposition $S = S_1 \times S_2 \times S_3$ with $f_1 : S_1 = \{$blue eyes, brown eyes$\}$, $f_2 : S_2 = \{$blond hair, brown hair$\}$, and $f_3 : S_3 = \{$tall, medium, short$\}$.

The possible product structure for S depends on its cardinality, $|S|$. For example, suppose f has the product structure $f = f_1 \times f_2$ with the respective state spaces S_1 and S_2. Then it has to be the case that $|S| = |S_1| \times |S_2|$. That is, the cardinality of S is the product of the cardinality of S_1 and S_2. Therefore, if $|S| = N$, then the cardinality of the factors of the possible product structures can be determined from the prime number decomposition of N. For example, if $N = 12$, then $N = 3 \times 2^2$ and we can have the following possible product structures:

$$f = f_1 \times f_2 \times f_3; \quad |S_1| = 2, |S_2| = 2, |S_3| = 3$$

$$f = f_1' \times f_2'; \quad |S_1'| = 3, |S_2'| = 4$$

$$f = f_1'' \times f_2''; \quad |S_1''| = 6, |S_2''| = 2$$

An interesting observation can be made from this cardinality decomposition. Suppose f is a label descriptor for an organism, for example, all of the human population. Then S might consist of {"*Junhyong Kim*", "*Minhyong Kim*",...} until we account for everybody and the cardinality $|S|$ = a few billion. We immediately see that f can be decomposed into at most $log_2 |S|$ number of nontrivial factor descriptors. Say $|S|$ = 10 billion, then $log_2 10^{10} < 40$. Or, at most, we have 40 or so possibly independent descriptors for the human population. On one hand, this is making the trivial statement that combinatoric complexity increases exponentially with the number of elements. On the other hand, it gives some pause to think that for most populations, the size of the population greatly limits the independently selectable set of characters.

DECOMPOSITION OF FREQUENCY VECTORS

As discussed earlier, we associated with each descriptor, f, its frequency space, V_f, and a population was represented as a vector, v_p, in this space. We also stated that if we have a product structure on the descriptors, $f = f_1 \times ... \times f_n$, then we have the isomorphism $V_f \cong V_{f_1} \otimes ... \otimes V_{f_n}$. However, it is important to note that $v \neq v_1 \otimes ... \otimes v_n$, $\left(v \in V_f, v_i \in V_{f_i} \right)$, in general. That is, a vector in V_f cannot be written as a tensor product of vectors in $V_{f_1},...,V_{f_n}$. This will hold if and only if v is itself a decomposable vector for the tensor product decomposition of V which will be true if and only if each descriptor f_i is statistically independent of all the f_j. Because of the "functional" correspondence between the state spaces, the frequency spaces, and their products, we will place primary emphasis on the decomposability properties of frequency vectors and deemphasize direct usage of statistical notions.

An example will clarify things a bit. Suppose we have descriptors f_1 and f_2 with the corresponding state spaces $S_1 = \left\{ s_1^1, s_1^2 \right\}$ and $S_2 = \left\{ s_2^1, s_2^2 \right\}$, respectively. A product descriptor $f = f_1 \times f_2$ has the frequency space $V_f \cong V_{f_1} \otimes V_{f_2}$. Now suppose we have two vectors $v_1 \in V_{f_1}$ and $v_2 \in V_{f_2}$ with the coefficients (a, b) and (c, d), respectively. Then the tensor product $v = v_1 \otimes v_2 \in V_{f_1} \otimes V_{f_2}$ has the coefficients (ac, ad, bc, bd), but we see that this implies $ac \times bd - ad \times bc = 0$. That is, if $v = v_1 \otimes v_2$ then the coefficients of v

satisfy a quadratic equation. An arbitrary vector $\mathbf{v} \in V_{f_1} \otimes V_{f_2}$ with arbitrary coefficients that does not satisfy this quadratic equation cannot be written as a "pure" tensor product (but of course can be written as a linear sum of tensor products). More generally, $\mathbf{v} = \mathbf{v}_1 \otimes ... \otimes \mathbf{v}_n$ if and only if the coefficients of $\mathbf{v}$ satisfy a system of quadratic equations of the form $v_{ij}v_{kl} - v_{ik}v_{jl}$ (additional indices are involved when there are more than two factors). We see that the measure of linkage disequilibrium in population genetics (cf. Lewontin, 1988, 1995) is a special case of these quadratic forms. (The subspace defined by these equations in the two locus case has been dubbed "the Wright manifold" in the literature; Sigmund, 1988.) Suppose that the dimensions of the frequency space $V_{f_1},...,V_{f_n}$ are $m_1,...,m_n$, respectively. Then the dimensions of the product frequency space $V_f = V_{f_1} \otimes ... \otimes V_{f_n}$ are $m_1 \times ... \times m_n$. The decomposable vectors lie on a $m_1 + ... + m_n - n - 1$ dimensional subspace of V_f. We see that as the number of factors increases the decomposable vectors lie in an increasingly smaller dimensional proportion of the frequency space.

An important component of a descriptor decomposition is whether the frequency vector $\mathbf{v}_p \in V_f$ of population p is also decomposable with respect to the product structure of the factor descriptors. As noted earlier, the population frequency vector $\mathbf{v}_p \in V_f$ is decomposable with respect to the product structure $f = f_1 \times f_2 \times ... \times f_n$ if it can be written as a tensor product of the vectors $\mathbf{v}_p^{f_i} \in V_{f_i}$. We will say $f = f_1 \times f_2 \times ... \times f_n$ is a *complete decomposition* with respect to the population if $\mathbf{v}_p = \mathbf{v}_p^{f_1} \otimes ... \otimes \mathbf{v}_p^{f_n}$. In principle, it is a finite algorithm to find a complete decomposition if it exists: We have fixed vector in the frequency space V for f. Different product structures on S will induce different sets of decomposable vectors in V. We need only go through the finitely many possibilities for product structures on S and see if v becomes decomposable for any of these structures by asking whether it satisfies the quadratic equations (which can be derived in a straightforward, albeit, tedious manner).

A complete decomposition with respect to a population is critical for our development. As mentioned earlier, a descriptor f with the state space S can be arbitrarily decomposed into factor descriptors and corresponding factor state spaces as long as we respect the prime number decomposition of the cardinality of S. More likely than not, however, the factor descriptors for any particular decomposition will be abstract and imaginary. For example, we might have the descriptor $f: S = \{Tall, Medium, Small, Tiny\}$ which can have the nontrivial decomposition into two factors: $f_1 : S_1 = \{0, 1\}$ and $f_2 : S_2 = \{A, B\}$ with the correspondence $\{0,A\} \leftrightarrow \{Tall\}$, $\{0,B\} \leftrightarrow \{Medium\}$, etc. However, f_1 and f_2 are unlikely to be natural descriptors for which we can find direct measurements. On the other hand, if this decomposition is a complete decomposition with respect to a population p with v_p at an evolutionary quasi-equilibrium (i.e., v_p hasn't been determined purely by chance), then we might have some hope that the factor descriptors are actually natural descriptors even if we haven't quite made the measurement yet. (Consider, for example, how Mendel's factors for green and yellow peas were determined.)

CHARACTERS

We are now ready to define characters. Given a population p, a collection $\chi_1, ..., \chi_n$ of descriptors is called a *character system* for p if it is a complete decomposition of some descriptor $\chi = \chi_1 \times ... \times \chi_n$ such that the frequency vector

$$\mathbf{v}_p \in V_\chi \cong V_{\chi_1} \otimes ... \otimes V_{\chi_n}.$$

is decomposable. In the simplest case, the product descriptor χ can be the *labeling* descriptor that assigns the label 1 to the 1st individual, 2 to the 2nd individual, etc. The complete decomposition implies that distinct χ_i are statistically independent on p as desired. By definition, a single descriptor by itself is always a character system. However, we will reserve the term character to refer to a descriptor which belongs to a character system with at least two nontrivial elements. A character system can be thought of as a partial coordinate system, if we think of the organisms as varying in a geometric space.

Going through an example will clarify these notions.

So let $S := \{a, b, c, d\}$ which therefore is also the basis for the frequency space V. Ordering the letters naturally, we get an isomorphism $V \cong \mathbf{R}^4$. Suppose the frequency vector is given by $\mathbf{v} = (1, 2, 2, 1)$. (We do not normalize the frequency here for clarity.) The only nontrivial product structure $S = S_1 \times S_2$ is when both S_1 and S_2 have two elements. Since the labels clearly are irrelevant for our considerations, we set $S_1 = \{1, 2\}$ and $S_2 = \{1, 2\}$. So different product structures correspond to different ways of matching up $\{a, b, c, d\}$ with $\{(1, 1), (1, 2), (2, 1), (2, 2)\}$. Now, inside the tensor product space, a vector

$$\mathbf{v} = x_{11}[1] \otimes [1] + x_{12}[1] \otimes [2] + x_{21}[2] \otimes [1] + x_{22}[2] \otimes [2]$$

is decomposable iff $x_{11}x_{22} - x_{12}x_{21} = 0$. So a decomposition of f will be given by a correspondence between S and $S_1 \times S_2$ which makes this equation hold for the vector $(1, 2, 2, 1)$. So, for example, if we make the correspondence

$$a \leftrightarrow (1,1)$$

$$b \leftrightarrow (1,2)$$

$$c \leftrightarrow (2,1)$$

$$d \leftrightarrow (2,2)$$

Then we get $1 \cdot 1 - 2 \cdot 2 \neq 0$, while the correspondence

$$a \leftrightarrow (1,1)$$
$$b \leftrightarrow (1,2)$$
$$c \leftrightarrow (2,2)$$
$$d \leftrightarrow (2,1)$$

will give us the equation $1 \cdot 2 - 2 \cdot 1 = 0$, so that v is decomposable for this product structure. Denoting by f_1 and f_2 the two elements of the character system resulting from this product structure, this labeling tells us how to assign values to organism. For example, $f_1(z)=1$ if $f(z)=a$ or $f(z) = b$, and $f_2(z) = 2$ if $f(z) = b$ or $f(z) = c$.

We get therefore a way of "pulling out" character systems from an descriptor, insofar as a complete decomposition exists. As we noted in the previous section, in the case of an arbitrary descriptor with state space S = {1, 2, ...n} there will again be a system of quadratic equations corresponding to each possible product structure on S. Thus the question of whether or not a character system exists reduces to checking whether or not our frequency vector satisfies at least one such system of quadratic equations. (We remark at this point that in practice, one would usually not require strict decomposability in our sense in the definition of a character system, but merely some kind of "smallness" for the quadratic system evaluated on the frequency vector. In this paper, however, we will continue to work with ideal situations and definitions for the sake of conceptual simplicity.)

In order to discuss another important practical issue, we need one more definition: Given a descriptor f with state space S, a *refinement* of f is a descriptor g with state space T and a surjective map $\pi : T \rightarrow S$ with the property that for any organism $x, f(x) = \pi \circ g$. Thus, g assigns many possible states to any given state with respect to f. We refer to f in this case as a coarse graining of g. A natural example would be the situation where we refine a numerical measurement, leading to a finer range of value intervals, say in millimeters rather than centimeters. Another good example of a refinement might be the case of the color character of Mendel's peas. The observed frequency distribution in this case might suggest that we should refine our two state descriptor of {*yellow peas, green peas*} to a four state descriptor of {yellow$_1$ peas, yellow$_2$ peas, yellow$_3$ peas, green peas}. This example especially illustrates that refinements may often be necessary before the character decomposition shows up. In Mendel's time, the two factor descriptors, the diploid chromosome pair with its allelic states, would not have been directly measurable. Thus, we can also *infer consequences* of a character decomposition even when it isn't directly measurable and use it to theorize the existence of characters. We will not deal with this issue in this paper, but merely outline the formalism for constructing character decompositions for "sufficiently fine" descriptors. It should be noted that the existence of a character system can be a *theory* for explaining observed phenomena, even when it is not directly measurable.

A special case of deriving character systems from a set of descriptors is the familiar situation where one groups together descriptors to form descriptor complexes. That is, we may start out by describing organisms using a collection $f_1, \ldots f_n$ of descriptors. We package this into one descriptor $f = f_1 \times \ldots f_n$ with state space $S = S_1 \times \ldots \times S_n$. Now if we have a frequency vector for a population v inside the frequency space V for f, a product structure on S which makes v decomposable might be simply a grouping

$$ S = \left(S_1 \times \ldots \times S_{k_1} \right) \times \left(S_{k_1 + 1} \times \ldots \times S_{k_2} \right) \times \ldots \times \left(S_{k_{m-1} + 1} \times \ldots \times S_{k_m} \right) $$

(Possibly after permutation.) That is, we may have

$$ \mathbf{v}_1 \in V_{f_1 \ldots f_{k_1}}, \ldots, \mathbf{v}_m \in V_{f_{k_{m-1}+1} \ldots f_{k_m}} $$

such that $\mathbf{v} = \mathbf{v}_1 \otimes \ldots \otimes \mathbf{v}_m$. In that case

$$ \left\{ f_1 \times \ldots \times f_{k_1}, f_{k_1+1} \times \ldots \times f_{k_2}, \ldots, f_{k_{m-1}+1} \times \ldots \times f_{k_m} \right\} $$

forms a character system in our sense. Each of the collections of descriptors occurring in this grouping, regarded exactly as a collection rather than put together into a product descriptor, is what one might otherwise call a character complex of descriptors.

In general, even when we are starting out with a descriptor having a product decomposition as described earlier, we may end up with character systems whose members are not simple "factors" of the original descriptor. In the numerical example shown earlier we saw that the state space $S = \{a, b, c, d\}$ may have arisen in the first place just as the product state space of the two descriptors f'_1, f'_2 given by the first assignment, $a \leftrightarrow (1,1)$, $b \leftrightarrow (1,2)$, $c \leftrightarrow (2,1)$, $d \leftrightarrow (2,2)$. That is, the first product structure may have been initially "given" by measurements on the individuals which led to the descriptors. We see then that the characters f_1 and f_2 derived from the assignment , $a \leftrightarrow (1,1)$, $b \leftrightarrow (1,2)$, $c \leftrightarrow (2,2)$, $d \leftrightarrow (2,1)$ are related in a rather complicated way to the f_i's.

MODULARITY

We first wish to extend the notion of decomposability to character systems. A set of descriptors $\{\chi_1,...,\chi_n\}$ is a *decomposition* of the character system $\{f_1,...,f_m\}$ (with respect to a population, p) if there exists a partition of $\{1,...,n\}$ into m subsets such that for each $i\in\{1,...,m\}$, there exists an element $\{j_1,...,j_k\}$ of the partition such that $f_i = \chi_{j_1} \times ... \times \chi_{j_k}$. That is, we are requiring that the χ_i's arise from the f_i's via product structures. If, in addition, $\{\chi_1,...,\chi_n\}$ is also a character system with respect to a population, p, then we will call it a *character decomposition* of the character system $\{f_1,...,f_m\}$.

From the viewpoint of the frequency vector $\mathbf{v}$ in the frequency space V of $f = f_1 \times ... \times f_m$, we know that $\mathbf{v}$ is already decomposable for the tensor product structure on V induced by the given product structure of f. Decompositions of the character system arise when we find that $\mathbf{v}$ is in fact decomposable for a further refinement of $\{f_1,...,f_m\}$. We say a character system is *decomposable* if a nontrivial decomposition exists. The problem of *evolution of modularity* is that of explaining how character systems for certain organisms become decomposable over time as the population evolves.

Sometimes we would like to refer to certain organisms as being modular. We believe it best to formulate this as a relative concept as follows. Consider a collection of descriptors $\{f_1,...,f_n\}$ and the frequency space V corresponding to the product $f = f_1 \times ... \times f_n$. Now there is a subset Z of V of vectors that are decomposable with respect to the given product structure. Another way to describe this set is to say that these vectors correspond to populations for which the given descriptors already form a character system. Thus, we are fixing a reference character system and concentrating on the populations for which this term applies. Now suppose we are given a further decomposition of the system $\{f_i\}$ into factor descriptors. We shall refer to such a population as modular with respect to the decomposition if it is in fact a character decomposition. We are thus requiring that its frequency vector belongs to some subset, Z' ($Z'\subset Z$), of vectors which are decomposable for the finer tensor product structure induced by the character decomposition. That is, the modular populations are those which are "more decomposable" than a reference group. This relative notion seems to make the discussion of evolution of modularity more clear than trying to formulate some absolute sense in which certain organisms are modular. An important special case is when the initial character system is just a single descriptor. Then the modular populations are those whose frequency vectors are decomposable for some nontrivial product structure of our single descriptor.

It has been suggested by Maynard Smith (1994) that many organisms do not really undergo significant evolution into greater complexity, and therefore, the fact that *some* organisms become complex is not surprising. However, our formulation indicates that as far as modularity is concerned, if it evolves for any substantial collection of organisms, this is a highly unlikely phenomenon requiring explanation. This is because inside the frequency space for any descriptor, if we consider the set of vectors that are decomposable for some character decomposition of the descriptor, the ones among them which are decomposable for some nontrivial refinement is a measure-zero set, being given by the zeros of quadratic equations. Therefore, if there is any population at all which is driven into such a thin set by evolution, this highly unlikely phenomenon demands explanation. The intuitive ease of making this point is the primary motivation for encasing all the discussion of characters and modularity in the "geometric" language that we have chosen.

INVARIANT CHARACTERS AND MODULAR OPERATORS

In the previous discussion on characters and modules, we emphasized the dependence of the notions on the variational context, that is, the frequency vector of the population. The evolution of a population will be described by the motion of a frequency vector in the frequency space with respect to a descriptor. Therefore, a character at a given population vector may not be a character at a different population vector after evolution. (We use evolution in a narrow sense to mean this change in the population vector.) This is in fact the property that we wished to emphasize. However, those measurements that we casually call characters, say the alcohol dehydrogenase gene, would not have such an ephemeral characteristic. This is because the evolution of the population vector *respects* the decomposition of the vector that generated the character. We now define this by saying that if the evolution of the population vector proceeds along some subspace of the frequency vector space and the subspace respects some character decomposition, then we call such characters *invariant* characters.

We next discuss briefly how the evolution of the population vector might respect character decomposition. First we assume that the evolution of the population vector can be described as an action of a linear operator (i.e., a transformation from V to V) that we call the *evolution operator*. For example, the standard notions of selection and transmission can be seen as a linear operator on the frequency space. It is sometimes reasonable to model the action of transmission and selection separately as operators on the frequency space. For example, one standard model for transmission takes the following form: Fix a unit time scale representing the intervals at which we make observations (it could be a "generation," for example). Let f be a descriptor with state space S. We can associate with each pair $s, t \in S$ a number m_{ts} with the following interpretation: $\sum_t m_{ts}[t]$ is the frequency vector after transmission for a unit length of time of a population that starts out entirely in state s. Since the basis S for the frequency space V is understood, we will denote the vectors merely by their components: $(v_s)_{s \in S} = \sum_{s \in S} v_s[s]$. Then the action of transmission on such a vector is described by multiplying the column vector (v_s) by the matrix (m_{ts}).

In general, if $S = S_1 \times ... \times S_n$ is a product structure on S which induces the tensor product structure

$$V \cong V_1 \otimes ... \otimes V_n$$

on V, then we say a linear operator M is modular with respect to this product structure if M is of the form $M = M_1 \otimes ... \otimes M_n$ for operators M_i on V_i. We say M is modular if it is modular for some nontrivial product structure on S. One simple paradigm for the evolution of modularity is the case where our space of organisms is the space of genotypes for diploids, the only transmission we consider is recombination, and the effects of selection are ignored. We will take S to be all possible states for some particular segment of a chromosome (which we assume that we can consistently locate on any of the organisms of interest). Then the space of genotypes for a diploid is $S \times S$. When considering possible product structures in this case, a preferred type is one of the form

$$(S_1 \times S_1) \times ... \times (S_n \times S_n) \qquad (*)$$

where $S = S_1 \times ... \times S_n$ is a product structure of S. This induces on the frequency space a tensor product decomposition

$$V \cong (V_1 \otimes V_1) \otimes ... \otimes (V_n \otimes V_n).$$

Among these decompositions are the ones arising from partitioning the chromosomal segment into disjoint pieces. That is, if the segment c is written as a union $c = c_1 \cup ... \cup c_n$, and the possible states for c_i is encoded by S_i, then the state space for c itself decomposes as above. Now suppose that the elements of the given partition actually correspond to the units of recombination. One way of remodeling this is by postulating a recombination operator $R : V \to V$ of the form

$$R = ((1 - r_1)I + r_1\sigma_1) \otimes ... \otimes ((1 - r_n)I + r_n\sigma_n),$$

where $\sigma_i : V_i \otimes V_i$ is the "switching operator" that interchanges the two factors of the tensor product and the r_i's are numbers between 0 and 1. To relate this to conventional models, one notes that the recombination rate for two fixed loci, say i and j, is given by $r_i + r_j - 2r_ir_j$. In any case, R is clearly modular for the product decomposition given previously (with the brackets). Now, assuming that our population consists of diploids amounts to saying that the frequency vector v is decomposable for the *a priori* product structure on $S \times S$.

The approach to linkage equilibrium, well known from population genetics (Sigmund, 1988), is expressed in the following Lemma:

Lemma 1 *Suppose we apply R repeatedly to a vector $v \in V$ representing a diploid population. Then $\lim R^n v$ is a decomposable vector for the product structure (*). In fact, it is modular for the full product structure*

$$S = S_1 \times S_1 \times ... \times S_n \times S_n$$

(without parantheses)

(The proof in the reference is given just for two loci decompositions, but it is easy to see that if linkage equilibrium holds for any two loci, then the vector must be decomposable for the n-product structure.)

One way of interpreting this lemma is to say that when natural variations occur in a modular way, then a population will converge toward greater modularity (at the level of the genotypes). This is of course ignoring the effects of selection and mutation. However, it gives a sense, albeit a bit trivial, of the kind of mechanisms that can induce an "increase of modularity" in populations that is also invariant.

There is strong relationship between the evolution operator and the set of charactors (or modules). The characters and modules are formed by some product structure of descriptors and a tensor product decomposition of the frequency vector that respects this product structure. The evolution operator determines which of the many possible product structures will be continuously respected by the tensor product decomposition. Therefore, in some sense, the evolution operator *is* the character system of the invariant characters. Again, as a remote analogy, in quantum physics an observable, i.e., a character, is attached to a linear operator with a particular structure of being self-adjoint.

GENOTYPE-PHENOTYPE MAP REVISITED

Previous treatments of complex evolutionary phenomena such as the evolution of quantitative characters, evolution of modularity, and evolution of evolvability involve the notion of a phenotype-genotype map (e.g., Lande, 1975; Lewontin, 1992; Wagner and Altenberg, 1996; Asselmeyer *et al.*, 1996). It is obvious that what we call the phenotype and what we call the genotype are both descriptors of the organism. The idea of the phenotype-genotype map distinguishes the two types of descriptors by two different properties. The first, and obvious, property is that the transmission operator is assumed to operate only on the genotypic descriptor. Conversely, the natural-selection operator (which we did not explicitly discuss in the previous sections) is assumed to operate only on the phenotypic descriptors. The transmission operator acts on the phenotypic descriptors through a composition with the phenotype-genotype map (and vice-versa for the selection operator). The second, less obvious property is that we usually assume that the genotypic descriptors are already modular and the transmission operator is also modular. That is, we assume that the genotype consists of factor descriptors called loci whose product structure is respected by the transmission operator. In our terms, the genotype is a character system. However, no such assumptions are made for the phenotypic descriptors and the idea is that the nature of the phenotype-genotype map may or may not induce modularity of the phenotypic descriptors (see Wagner and Altenberg, 1996).

To continue this discussion further in our framework, fix a class of organisms. Let g be a genotypic descriptor with state space G and f a phenotypic descriptor with state space S. A *genotype-phenotype* map (a "G-P map") is a map $\phi: G \rightarrow S$ with the property that for any organism x, $\phi \circ g(x) = f(x)$. This rather trivial definition formalizes the process of "assigning a phenotype to a genotype" in a way that emphasizes that we need to explicitly fix descriptors corresponding to genotypes and phenotypes for the map to make sense. This is in line with our view that the state space of phenotypes is well defined only when we have chosen some descriptors. It also emphasizes the need to model a G-P map explicitly (or a class of reasonable maps). As mentioned, we could argue that the space of genotypes is well defined as the product $G = G_1 \times ... \times G_n$ of the set of states G_i for a well defined collection of "genes." (Or at least, this is the well-accepted model paradigm for genes.) Thus, the "universal" genotypic character assigns to each organism a point in this product space. However, it seems misleading to use language that suggests that a corresponding model for a "space of phenotypes" is known. It is indeed a difficult and important problem to construct a reasonable space P which could be used as THE model for all possible phenotypes. If it were, the only thing that would need additional modeling would be the genotype-phenotype map from G to P. However, given that no natural candidate for P seems to be agreed upon, we give the definition mentioned earlier, which is descriptor dependent.

It was mentioned earlier that the evolution of modularity could be understood in terms of possible character decompositions of f. One possible framework for discussing this problem would be to construct for each gene G_i (here we are identifying the gene with its state space) a phenotypic descriptor f_i with state space S_i. (Of course, for most cases these S_i would be abstract.) Then our space of phenotypes would be $S = S_1 \times ... \times S_n$ and we could further attempt to define a genotype-phenotype map "product wise" from maps $\phi_i : G_i \rightarrow S_i$. Recall the special character systems on S given by the "grouping" product structure on S,

$$S = \left(S_1 \times ... \times S_{k_1} \right) \times ... \times \left(S_{k_{m-1}+1} \times ... \times S_{k_m} \right)$$

If such a character system were to exist, we would interpret the modularity as arising from "pleiotropy" between the genes G_1 through G_{k_*} etc. For such an interpretation to make precise sense and be experimentally demonstrable, however, the setup described previously for a genotype-phenotype map, which assigns a descriptor to each gene, would need to be explicitly modeled.

This brings up an important point. We sometimes speak of "evolution of the genotype-phenotype map." There seems to be at least two senses in which this phrase is used, one of which seems reasonable and the other less so. One situation is when we are compelled to literally change the descriptor f (or the descriptors f_i) over time for a natural reason. For example, one may need to replace f by a refinement f', which accounts for the evolution of observed and *measured* phenomena. This would indeed be a case of the G-P map "evolving," and needs to be thought of as a "deep-level" dynamics, which might be difficult to model. However, there is another sense in which the G-P map might be said to evolve: this is when our descriptors are fixed, but the character system of the phenotype side evolves. For example, evolution of modularity would be such a situation where a nontrivial character system emerges out of a trivial one. In this case, the perception that the G-P map itself evolves seems to have arisen from trying to identify the "space of phenotypes" with the set of values of the character system, which can indeed evolve as the product structure that gives rise to a character system evolves. That is, suppose we start out with a fixed G-P map in the sense defined earlier. Then, a character system arises from a product structure which is an isomorphism of P and $\prod_i S_i$. At this point it is natural to think of this particular product of S_i's as being the phenotypical state space for the reference population. Now as the population evolves the character system can evolve in the sense that we described in the previous sections, i.e., the product structure that is compatible with the population frequency vector decomposition can change, in which case we might be tempted to model it as a change in the G-P map (since the structure of the image of the map is changing). However, this is a matter of viewpoint, and it seems more clear to view the map as fixed and the character system as evolving. To summarize this argument, if we have fixed genotypic descriptors for which we have natural measurements and fixed phenotypic descriptors for which we also have natural measurements and we observe changes in the population vector decomposition structure, it may be proper and useful to model this as a change of the G-P map. On the other hand, if we are trying to understand and classify models of evolution of modularity without reference to fixed measurement of phenotypes, then it seems more clear to model it as a fixed G-P map with an evolution of modularity by an evolution of the product structure of the phenotypic descriptors.

Having spent the entire paper merely setting up the formalism in which the problem of evolving modularity can be given a precise statement, we should give at least a vague indication of how one would attack the problem itself. In brief, we need to find a dynamical system on the simplex of populations Δ_s inside the frequency space V_s of a descriptor which models the effects of evolution and has basins of attraction clustered around the union of the subspaces corresponding to vectors that are decomposable for the various product structure on S. More generally, we need to describe dynamics on some set of decomposable vectors which tends to drive them toward basins of greater modularity (decomposability). We discussed one such example in the previous section which arises from the recombination dynamics of diploids. In the general case we would like to investigate the effect of dynamical systems on V which have intimate relations to discrete dynamics on S itself. The case of the recombination operator itself was one such example where it can be seen as a permutation dynamics on S coupled to some "probability weighting" on the frequency side. That is, we are looking at convex linear combinations of certain permutation transformations.

Two weaknesses of this particular example are:

1. The effects of selection and mutation are completely ignored.
2. On the formal side, the dynamics on S is arbitrarily constructed to "anticipate" the modularity desired.

If we wish to describe the evolution of modularity in a biologically compelling manner, we would need to identify a class of dynamical systems which do not have modularity "built-in" but whose generic member tends to drive the generic population vector toward modularity. If we accept the standard view that genotypes are *a priori* modular, then the main challenge will be finding such dynamics on the phenotypic frequency space, where the various ingredients of evolutionary theory interact in a complicated manner.

We propose that such dynamical systems should "emerge" from model "theories of evolution" in the following sense.

A *theory of evolution* consists of:

- a set G, called the space of genotypes;
- a set P, called the space of phenotypes;
- a map $\phi: G \rightarrow P$ called the genotype-phenotype map;
- a dynamical system T on $\Delta_G \subset V_G$, called the transmission dynamics;
- and a "fitness structure" F on P,

from which should emerge dynamical systems, called "evolution," E_G on $\Delta_G \subset V_G$ that are "compatible" with T, F, and ϕ.

Of course, not all of these terms are defined and should be modeled flexibly. For example, a "fitness structure" might be just a fitness function or it might be a fitness "landscape" (Kauffman, 1993). The sense in which the evolution operators are compatible with the other structures can also be modeled in a variety of ways, some of which are well studied in the literature. (For example, Sigmund (1988), where one just writes down a specific dynamics.) However, the studies seem to usually confine themselves to either G or P alone or, equivalently, $G = P$ in some sense, rather than discussing the possible complex interactions arising via a nontrivial (fixed) G-P map. The G-P map makes the compatibility difficult and interesting to formulate, regardless of whether we view it as fixed or evolving. In any case, rather than concentrating on specific models of the sort one encounters in population genetics, a more desirable direction would be to capture general properties of models of evolution in terms of the processes we would like to understand, for example, the evolution of modularity or, more generally, of complexity. That is, it would be more desirable to "qualitatively" classify the models rather than pursue detailed dynamics. This would require a parameterization of the model space, perhaps the G-P map, such that we obtain invariants that capture interesting phenomena such as the evolution of modularity. This is very much in line with other branches of complexity theory, where one meets many model systems that exhibit different classes of behavior depending upon various parameters occurring in their constructions, and which one tries to classify in terms of these parameters. [For example, the study of random Boolean networks in Kauffman (1993). The vague comparison with Kauffman's theory is especially tantalizing since our situation also calls for a dynamics that drive the states toward a very confined region of phase space in a very precise sense.]

It is hard to resist concluding with yet another purely speculative analogy: A rather popular view in current complexity theory is to seek the origin of complexity in "phase transitions." As far as the evolution of modularity is concerned, the formulation we propose is very much in philosophical (if not even deeper) harmony with this view. This is because phase transitions in physics are often associated with the phenomenon of *symmetry breaking*. That is, when the states of a system are described by a Hilbert space H which is acted on by a symmetry group G, then a breaking of symmetry to a subgroup K of G often accompanies a phase transition, which in turn depends on factors like the dropping of temperature (or other parameters). While the original H may have been an irreducible representation of G, when viewed as a representation of K, it will generally break up into a direct sum of many spaces $H = \sum_i H_i$, each H_i being an irreducible representation of K. Now suppose further that K is a subgroup of the form $K_1 \times K_2$. Then each of the H_i's will further decompose into tensor products $H_i = (H_{i1}) \otimes (H_{i2})$ given by irreducible representations of the two subgroups K_i. Thus, one finds tensor product decompositions of state space, as well as decomposability of some natural vectors like the vacuum state, occurring naturally in conjunction with phase transitions. Although at present nothing beyond an amusing analogy, it is tempting to imagine a "biological symmetry breaking" that might occur as one changes a "temperature parameter" associated to different theories of evolution as an underlying mechanism for the evolution of modularity. We hope to be able to report on more concrete manifestations of this analogy in future work.

ACKNOWLEDGMENTS

Much of the motivation and ideas for this paper came from extended discussions with G. P. Wagner to whom we are grateful. This work has been supported in part by a Sloan Foundation Young Investigator Award to JK and NSF grant DMS 9701489 to MK.

LITERATURE CITED

Asselmeyer, T., Ebeling, W., and Rose, H. (1996). Smoothing representation of fitness landscapes: the genotype-phenotype map of evolution. *Biosystems* **39**:63-76.

Brooks, W. (1883). The law of heredity: a study of the cause of variation, and the origin of living organisms. J. Murphy, Blatimore, MD.

Carlson, E. (1966). "The Gene: A Critical History." W. B. Saunders Company, Philadelphia.

Colless, D. H. (1985). On "character" and related terms. *Syst. Zool.* **34**:229-233.

Darden, L. (1992). Character: historical perspectives. *In* "Keywords in Evolutionary Biology" (E. Keller and E. A. Lloyd, eds.,). Harvard University Press, Cambridge, MA.

Kauffman, S. (1993). "The Origins of Order: Self-Organization and Selection in Evolution." Oxford University Press, Oxford.

Lande, R. (1975). The genetic covariance between characters maintained by pleiotropic mutation. *Genetics* **94**:203-215.

Lewontin, R. C. (1992). Genotype and phenotype. *In* "Keywords in Evolutionary Biology." Harvard University Press, Cambridge, MA.

Lewontin, R. C. (1995). The detection of linkage disequilibrium in molecular sequence data. *Genetics* **140**:377-388.

Maynard Smith, J. (1994). Discussion of "the major transitions in evolution." *In* "Complexity: Metaphors, Models, and Reality" (G. Cowan, D. Pines, and D. Meltzer, eds.), volume XIX of *Santa Fe Institute Studies in the Sciences of Complexity,* Santa Fe, NM.

Patterson, C. (1982). Morphological characters and homology. *In* "Problems of Phylogenetic Reconstruction" (K. A. Joysey and A. E. Friday, eds.).

Punnett, R. (1911). "Mendelism." MacMillan Company, New York.

Quine, W. (1987). "Quiddities: An Intermittently Philosophical Dictionary." The Belnap Press, Cambridge, MA.

Raff, R. A. (1996). "The Shape of Life: Genes, development, and the evolution of Animal Form." University of Chicago Press, Chicago, IL

Sigmund, J. H. K. (1988) "The Theory of Evolution and Dynamical Systems." Cambridge University Press, Cambridge.

Sneath, P., and Sokal, R. (1973). "Numerical Taxonomy." W. H. Freeman, San Francisco, CA.

Wagner, G. (1995) The biological role of homologues: a building block hypothesis. *Neues Jahrb. Geol. Palaontol. Abh.* **195**:279-288.

Wagner, G., and Altenberg, L. (1996). Complex adaptations and the evolution of evolvability. *Evolution* **50**:967-976.

10

WHOLES AND PARTS IN GENERAL SYSTEMS METHODOLOGY

MARTIN ZWICK

Systems Science Ph.D. Program, Portland State University, Portland, OR 97207

RECONSTRUCTABILITY ANALYSIS

In general systems methodology, the decomposition of wholes into parts and the composition of parts into wholes is called *reconstructability analysis* (Klir, 1985). RA derives from the early work of Ashby (1964), and was developed by Broekstra, Cavallo, Conant, Jones, Klir, Krippendorff, and other researchers (see the citation in the reference list of an RA bibliography in the *International Journal of General Systems*). Here, a whole is a *relation* which is a *constraint* among a set of variables. The parts, which define the *structure* of the whole, are relations among subsets of the variables, subsets which may be either disjoint or overlapping. For a fully decomposable whole, the parts are simply the variables. For example, for variables, A, B, and C, the whole is the triadic relation ABC. The parts might be the dyadic relations, AB and BC, in which case the system has structure AB:BC. (A colon is used to separate the parts. The whole, when maximally decomposed, yields A:B:C.) Data defines a whole and RA reveals its parts. The parts taken together are specified by fewer numbers (*parameters*) than the whole, i.e., decomposition accomplishes "compression." An RA-generated structure is a *model* of the data.

The variables in RA are nominal (categorical, qualitative). Thus RA might be used, for example, in studying the interactions between biological characters or genes, and their relations to environmental conditions, since characters, genes, and conditions may be represented as nominal variables. RA has considerable value also for analyzing quantitative variables linked by unknown nonlinear relations. (For linear relations, standard methods are superior.) Quantitative variables are accommodated by "binning" their values into discrete states which are unordered, and binning can be done within the framework of fuzzy set theory (Zadeh, 1965) or by clustering techniques. RA can analyze not only static relations, but also dynamic relations: in ABC, variables might be state(t-2), state(t-1), and state(t). Relations can be deterministic or stochastic.

There are two main formalisms used in RA to define a relation (Conant, 1981; Klir, 1985; Krippendorff, 1986). These are here called the set-theoretic and the information-theoretic. Set-theoretically, a relation is a subset of a cartesian product: the combinations of possible variable values which are actually observed. Information-theoretically, a relation is a multivariate probability distribution. More precise labels for these formalisms are "crisp possibilistic" and "probabilistic," but the set- and information-theoretic labels are used here because they are more familiar and because this nomenclature correctly suggests parallelisms in the formalisms. The set- and information-theoretic perspectives are central to the general systems literature and can now also be viewed as components of a generalized information theory currently being developed (Klir and Wierman, 1998).

The information-theoretic aspect of RA overlaps with log-linear modeling (Bishop *et al.*, 1978; Knoke and Burke, 1980) which is widely used for analyzing nominal data in the social sciences. Log-linear modeling was developed in the same period as RA, but the general systems methodology literature is broader. Log-linear modeling is statistical, while reconstructability analysis also includes well-developed nonstatistical aspects, e.g., in its set-theoretic formalism, lattice explorations, advanced computational algorithms, and analysis of fuzzy distributions. It also differs from log-linear modeling in its extensive use of uncertainty measures, in its innovation of "state-based" modeling, and in its acceptance of the challenge of modeling data on many variables. On the other hand, the log-linear literature is very advanced statistically, while statistical considerations are sometimes absent in the systems literature where their presence would be desirable. Latent variable techniques, which are the nominal data analog of factor analysis, are well developed on the log-linear side, while on the systems side they are available mainly set-theoretically. Where information-theoretic and log-linear methods overlap, they are equivalent; compare, e.g., Knoke and Burke (1980) and Krippendorff (1986).

In Section II, RA is illustrated with simple examples of both set- and information-theoretic analyses, so the reader can easily grasp what is fed into RA and what it yields in return. To explain RA strictly in these input-output terms requires only a minimal mathematical description. The theory is then presented in Sections III and IV, which explain what relations and structures are. The practice of RA is described in Section V, i.e., the "black box" is opened and the analytical methods used to obtain the results of Section II are described. Section VI closes the paper with remarks on the current state of RA methodology.

EXAMPLES

Table I illustrates with two examples: (a) set-theoretic and (b) information-theoretic. In both examples, variables A, B, and C are dichotomous (binary). If variables had three states (values), e.g., $\{A_0, A_1, A_2\}$, the order of the states would be arbitrary because the variables are nominal.

The input to RA is shown on the left of the table. In (a) the data are the combinations (tuples, binary strings) observed for the variables—five tuples of the possible eight—without regard to how frequently they are observed. In (b) the A, B, and C values are labeled with 0 and 1 subscripts (instead of being taken as 0 or 1); all combinations are observed, but with the frequencies given in the contingency table.

TABLE I   System as data and model [two examples: (a) & (b)]

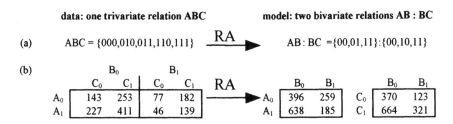

data: one trivariate relation ABC   model: two bivariate relations AB : BC

(a)   ABC = {000,010,011,110,111}   _RA_ →   AB : BC ={00,01,11}:{00,10,11}

(b)

		B_0		B_1	RA		B_0	B_1			B_0	B_1
		C_0	C_1	C_0	C_1							
A_0		143	253	77	182	→	A_0	396	259	C_0	370	123
A_1		227	411	46	139		A_1	638	185	C_1	664	321

The task of RA is to decompose the ABC relation into parts in such a way that the model, simpler than the data, still adequately agrees with the data. The information-theoretic analysis is statistical; the set-theoretic analysis is nonstatistical. Partial RA results are shown on the right side of Table I. Both data (a) and (b) are decomposable into structure AB:BC. The set-theoretic model consists of a set of AB tuples and a set of BC tuples. Taken together, these are *equivalent* to the ABC data (the 5 tuples) shown on the left of part (a) of the table. Model and data agree exactly, i.e., with no error.

The information-theoretic model analogously consists of the two 2-variable contingency tables, AB and BC, as shown on the right of Table I(b). Taken together, these are equivalent to a 3-variable table which *approximates* the data on the left. Data and model, expressed as probabilities and not frequencies, are shown in Table II. The data (the p distribution) and the model (the q distribution) do not agree exactly, but the error is statistically acceptable. Error is loss of constraint or, equivalently, loss of information.

TABLE II ABC (data) and ABC$_{AB:BC}$ (model) [data from Table I (b)]

	B_0		B_1				B_0		B_1	
	C_0	C_1	C_0	C_1			C_0	C_1	C_0	C_1
A_0	.097	.171	.052	.123		A_0	.096	.172	.049	.127
A_1	.154	.278	.031	.094		A_1	.154	.277	.035	.091

ABC data: p(A,B,C)   AB:BC model: $q_{AB:BC}(A,B,C)$

Detailed explanations of these set- and information-theoretic analyses are given in Sections IV and V. A complete RA does not merely yield a single model which adequately fits the data. It can determine how well *all possible* models fit the data or, if the number of possible models is too large to evaluate exhaustively, RA searches through a promising subset of them. Table III summarizes the analysis of all models for the set-theoretic example of Table I(a). There are 9 possible models, descending from the most complex (the triadic relation, ABC, i.e., the data, also called the "saturated" model) at the top to the simplest (three "monadic" relations, A:B:C, also called the "independence" model) at the bottom. After each model in parentheses is indicated the number of ABC tuples which the model specifies. The top model, ABC, is the data itself which contains 5 ABC tuples. Models AB:AC:BC and AB:BC also specify the same 5 tuples. These two models fit the data without error, but since AB:BC is less complex, it is preferred. Other models specify either 6 or 8 tuples, that is, they predict additional combinations which are not observed in the data and are thus in error. There is less constraint represented in these models than is in the data. Models which predict 8 tuples predict all possible combinations of A, B, and C and exhibit no constraint at all.

TABLE III   Reconstruction of the set-theoretic data of Table I(a). In parentheses, number of tuples in model

	ABC (5)	
	AB:AC:BC (5)	
AB:AC (6)	**AB:BC (5)**	BC:AC (6)
AB:C (6)	AC:B (8)	BC:A (8)
	A:B:C (8)	

Table IV summarizes the analysis of all models for the information-theoretic example of Table I(b). After each model three numbers are given in parentheses:

- Information^{norm}, the information (constraint represented) in the model, which ranges from 0 (no information) to 1 (complete information),

- α , the probability of making an error (called a Type I error) if one rejects the identity of the model with the data, and
- df, the degrees of freedom (complexity) of the model.

At the top of the list, the "saturated" model ABC has complete information. The probability of making an error in rejecting its identity with the data is 1, since it *is* the data. Its degrees of freedom, i.e., the number of probabilities needed to specify the 3-variable table, is 7. At the bottom of the list, the "independence" model A:B:C has no information. The probability of making an error in saying it is different from the data is 0. It needs 3 probability values for its specification. In between top and bottom models are all less-than-total decompositions. A good model is one which has high information content and small df (complexity). It is also a model which has high α. (The desirability of a high α is atypical of most statistical analyses, but if we compared a model not to the saturated model at the top but to the independence model at the bottom, the normal preference for low α values would be obtained.) The best model is AB:BC, but the analysis tells us more than this. It tells us how well all possible structures model the data.

TABLE IV   Reconstruction of the information-theoretic data of Table I(b). In parentheses, Information^{norm} , α, df

	ABC (1., 1., 7)	
	AB:AC:BC (.987, .382, 6)	
AB:AC (.827, .005, 5)	**AB:BC (.978, .518, 5)**	BC:AC (.153, .000, 5)
AB:C (.826, .014, 4)	AC:B (.000, .000, 4)	BC:A (.152, .000, 4)
	A:B:C (0., .000, 3)	

Relations

A. Lattice of Relations

For a system of two variables, there are only two possible structures: AB and A:B. For three variables, the relations which could exist are arrayed in the "lattice of relations" shown in Fig. 1 (Krippendorff, 1986). The number of variables in a relation is its *ordinality*. In this framework, relations can have arbitrary ordinality. By contrast, the conventional graph-theoretic representation which depicts systems in terms of nodes and links connecting nodes normally restricts ordinality to two. A relation obtained from a higher-ordinality relation by ignoring one or more variables is "embedded in" and "a projection of" the higher-ordinality relation, e.g., AC is embedded in ABC. Although a relation viewed as a *constraint* usually presumes the existence of at least two variables, it is useful to include "monadic relations," A, B, and C, and, for completeness, the "null relation," Φ.

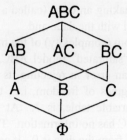

FIGURE 1 Lattice of relations for a 3-variable system.

Relations are *directed* if a variable acts on another, or an event causes another, or a discrimination is made between "generating" or "generated" variables. "Generating" vs. "generated" is used instead of the more familiar "independent" (IV) vs. "dependent" (DV) (variables) to reserve the word "independent" for its connotations of "mutually independent" or "independent of." Relations are *neutral* if directionality cannot be or is not indicated. Directed relation AB is distinguished from directed relation BA. Neutral relations are static, but directed relations can be static (e.g., nontemporal input-output pairs) or explicitly dynamic. Directed relations may be deterministic or stochastic.

B. Definition of Relation

Set-theoretically (Conant, 1981; Klir, 1985), $AB \subseteq A \otimes B$: a relation AB is a subset of the Cartesian product of sets, $A \otimes B$, where A and B signify also the sets of states the variables take on. The Cartesian product—call it H (for "heap")—is the set of all possible pairs of values of the two variables. A relation is a constraint which reduces the possible to the actual, i.e., AB reduces H to a smaller set of pairs $\{(A_i, B_j)\}$ which in fact occur. (If the pairs are allowed *partial* membership in set AB, the relation is fuzzy.) H can be written also as the independence model A:B, where the colon means mutually independent. The number of pairs (in general, "tuples") in AB is called its cardinality and is written $|AB|$.

Referring again to Fig. 1, projection into lower-ordinality relations ignores the projected variables. For example, AB is obtained from ABC by ignoring all C_k in $\{(A_i, B_j, C_k)\}$; the monadic relation (variable) A is obtained by ignoring both B_j and C_k. For the ABC relations shown on the left of Table I(a), the AB and BC projections are shown on the right.

Information-theoretically (Klir, 1985; Krippendorff, 1986), a relation is a multivariate probability distribution, i.e., AB is the set $\{p(A_i, B_j)\}$, for convenience sometimes written simply as $p(A,B)$. Technically "relation" means set-theoretic relation, but the word is here used more broadly. The number of probability values needed to specify the relation is its "degrees of freedom," $df(AB) = |AB|-1$, where one subtracts 1 from the number of (A_i, B_j) pairs since probabilities must add to 1. For example, if A and B are dichotomous, 3 probability values define the contingency table for relation AB although the table itself has 4 entries. Constraint is the deviation of the distribution from a reference distribution, usually A:B, but occasionally Φ, here the uniform distribution. The A:B distribution is $q_{A:B}(A,B) = p(A)\ p(B)$, where $q_{A:B}$ plays the same role as the Cartesian product H. "q" denotes a *calculated* distribution and "p" denotes data, i.e., an *observed* distribution and its projections. Projection into lower-ordinality relations sums over projected variables, e.g., $p(A,B) = \Sigma_k\ p(A,B,C_k)$. Again, in Table I(b), the ABC distribution shown on the left has AB and BC projections shown on the right.

Relations are characterized by *uncertainty*, a measure of variety or dispersion. Uncertainty is the nominal variable analogue of variance. Set-theoretic uncertainty is the Hartley entropy, $U = \log_2 |AB|$, i.e., the log of the number of pair values. Information-theoretic uncertainty is the Shannon entropy, $U = -\Sigma\ p(A,B)\ \log_2\ p(A,B)$. ("Uncertainty" is preferable to "entropy" as "entropy" evokes an association with the Second Law of Thermodynamics and is best reserved for physical systems.) Uncertainty measures can be defined for generalizations of the set- and information-theoretic formalisms. There is not just one concept of uncertainty and a complex relationship exists between its uses in different formalisms (Klir and Wierman, 1998).

A relation is a whole whose parts are projections. The uncertainty of a whole, U(AB), is less than or equal to the uncertainty of its parts, $U(A:B) = U(A) + U(B)$. The strength of constraint of AB in both formalisms is the uncertainty reduction, U(A:B)-U(AB); information-theoretically, this is also called "transmission" ("mutual information"), T(A:B). This is the *gain* of constraint in AB relative to A:B or equivalently the *loss* of constraint in A:B relative to AB. The uniform distribution, Φ, might serve as an alternative to A:B as a reference condition.

Structures

A. Lattice of Structures

A two-variable system can have only one (undirected) relation, but with three or more variables, systems can have multiple relations, i.e., *structure*. A structure is an unordered set of relations none of which is a projection of another. In representations of systems where relations are strictly dyadic (involve only two variables), variables are often shown as nodes (or circles) and relations as lines or arrows connecting nodes. In the present framework, relations may involve an arbitrary number of variables, and to facilitate focusing on relations rather than variables, structures are represented with relations as boxes and variables as lines, as shown in Fig. 2 (Klir, 1985; Krippendorff, 1986). One can add directedness by changing lines into arrows.

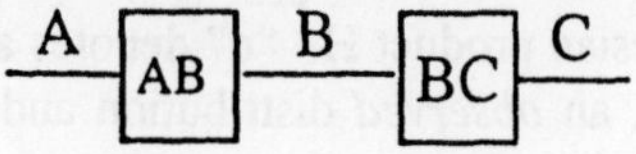

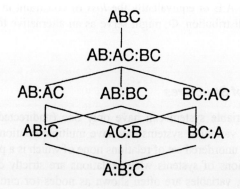

FIGURE 2   Specific structure AB:BC.

Relations may overlap in their variables, as in AB:BC, or may be disjoint, as in AB:C, or completely disjoint, as in the "heap," A:B:C.

The lattice of possible (undirected) structures with three variables is shown in FIGURE (Krippendorff, 1986). Tables III and IV simply list the possible structures; this figure shows how they are related to one another by descent.

ABC
|
AB:AC:BC

AB:AC   AB:BC   BC:AC

AB:C   AC:B   BC:A

A:B:C

FIGURE 3   Lattice of specific structures for a 3-variable system (undirected relations).

For three variables there are nine specific structures as show in Fig. 3 but only five *general* structures: (1) XYZ, (2) XY:XZ:YZ, (3) XY:YZ, (4) XY:Z, and (5) X:Y:Z. [NB: the "general" vs. "specific" nomenclature here differs from Klir's (1985).] Specific structures are obtained from general structures by permuting the assignments of specific variables, A, B, or C to the generic X, Y, or Z. For example, general structure (4) XY:Z subsumes AB:C, AC:B, and BC:A. For three variables, the lattice of general structures is small (5 general structures), but for four variables (add the variable W), there are 20 general structures (Fig. 4) (Klir, 1985; Krippendorff, 1986). If variables are all dichotomous (binary), the degrees of freedom of the structures range from 15 at the top to 4 at the bottom and decrease by 1 at every level. The figure also shows the acyclic structures, which are indicated with boxes in bold (10 of the 20). For four variables, there are 114 specific structures.

For directed systems the lattice is often simpler. Say W is generated from X, Y, and Z. It is assumed for directed systems that we are uninterested in relations among the generating variables, so structures always include an XYZ component which subsumes all such relations, and decompositions of this XYZ component are never considered. (If one wants to know about relations among the generating variables, the system is treated as neutral.) All the other relations in a structure necessarily involve the generated variable(s). For three generating variables and one generated variable, only 9 of the 20 structures of Fig. 4 apply: the top 6 and the next leftmost 3 (indicated in the figure with the generated variable—a line or connected lines uninterrupted by a box—in bold). Four of these 9 directed structures are also acyclic: (1) XYZ:W, (2) XYZ:XW, (3) XYZ: XYW, and (4) XYZW. These are the simplest decompositions of W's dependence on X, Y, and Z; they specify W as either independent of or dependent on 1, 2, or 3 of the generating variables. Note that XYZ:W (indicated on Fig. 4 by a *), and *not* X:Y:Z:W, is the bottom of this lattice of directed structures. Also, a structure like XYZ:XW:YW:ZW is *different* from XYZW, even though in both cases W depends upon X, Y, and Z (Zwick, 1996).

Combinatorial possibilities rapidly expand for five or more variables (Table V). Although exhaustive consideration of all structures becomes prohibitive at around 6 variables, intelligent heuristics can accommodate many more variables (Klir, 1985; Krippendorff, 1986; Conant, 1988). The lattice can be pruned as a search procedure descends or ascends so that consideration is restricted only to promising candidates or the search can be done first roughly between groups of structures and then finely within these groups (Klir, 1985).

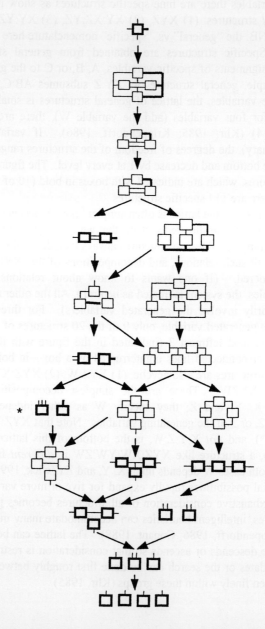

FIGURE 4 Lattice of general structures (4-variable system). A box is a relation; a line, with branches, uninterrupted by a box, is a variable. Arrows indicate decomposition. The top structure is XYZW; the fourth down is XW:XYZ:WYZ; the bottom is X:Y:Z:W. Generated variable W is shown in bold for the 9 structures of directed 4-variable systems. The 10 acrylic structures have all relations shown in bold.

TABLE V Numbers of structures. Only the bottom line is for directed structures.

# variables	3	4	5	6
# general structures	5	20	180	16,143
# specific structures	9	114	6,894	7,785,062
with 1 generated var	5	19	167	7,580

B. Cycles, Paths, Latent Variables, State Models

The possibility of (nontrivial) *cyclicity* emerges with 3 or more variables, as in AB:BC:CA. If the relations are directed, the structure's cyclicity is directed, showing *feedback*. Methodologically cyclicity is a source of complications (Krippendorff, 1986). *Mediation* can occur by overlapping relations, and if relations are dyadic and directed (a digraph), the structure has *paths*. In AB:BC, for example, B mediates between A and C; if AB:BC is directed, B transmits the *indirect* effect of A on C via the path A → B → C. (This is different from the effect of B on AC within an undecomposed triadic relation ABC.) In directed structure AB:BC:AC, there is also an A → C path which transmits the *direct* effect of A on C. This is the nominal analog of path analysis (Davis, 1985).

Consider directed structure BA:BC. B might be a *higher-level* latent variable (or "construct") which "chunks" together (Simon, 1981) A and C. In factor analysis B would be called a "common factor." Latent class analysis does a similar analysis for nominal variables. For example, a relation AC might be explained by a latent variable B and a posited relation ABC, which subsumes AC and is decomposable into BA:BC. The latent class procedure does not actually distinguish between the factor analytic and path analytic situations. An AC relation with latent B and inferred ABC describes also directed structures where B is prior to A and C, i.e., A ← B → C, or where B is intermediate between them, i.e., A → B → C.

The information-theoretic framework and its log-linear equivalent (Hagenaars, 1993) thus generalize to nominal data the more restricted methods of path analysis, factor analysis, and covariance structure modeling (Long, 1983) which apply only to linear relations. Latent variable methods apply also to set-theoretic relations (Grygiel *et al.*, 1999).

Normally, a structure requires the *complete* specification of its component relations. For example, AB:BC is defined information-theoretically by two distributions consisting of probabilities or frequencies for *all* (A_i, B_j) and (B_j, C_k), i.e., the full two tables shown in Table I (right). It is possible, alternatively, to define a model in terms of *any set of states* and their probabilities drawn from the top relation ABC and all of its projections (Jones, 1985). A small set of states might have probabilities unexpectedly high or low. These states represent salient "events" or "facts," and they, rather than complete projections, might be considered the "parts" of relation ABC. This approach is explained later in Section ANALYSIS. This *state-based* approach is more powerful than the conventional *variable-based* framework, which it encompasses as a special case. The cost, however, of the state-based approach is a sizable increase in the size of the lattice of possible structures. As presented by Jones, the reference distribution for state-based modeling is the uniform distribution, but other reference distributions can also be chosen.

C. Complexity; Constraint

Structures differ in *complexity*, where this word can be given different meanings. In this article, the complexity of a relation or a distribution is the number of tuples or probability values. Decomposition reduces this complexity, i.e., is *compression*. This notion of complexity has the sense of randomness and is different from Wolfram's (1986) or Langton's (1992) "edge of chaos" complexity.

The complexity of a structure, information-theoretically, is its degrees of freedom which is the sum of the degrees of freedom of its relations, corrected for overlap (Krippendorff, 1986). For example, df(AB:BC) = df(AB) + df(BC) - df(B). (Defining a set-theoretic analog, however, is not straightforward.) df depends only on the variable cardinalities and not on the actual relations. For $|A| = |B| = |C| = 2$, df(ABC) = 7 because a 2x2x2 table needs only 7 values to be specified (since probabilities sum to 1, and frequencies to the sample size). df(A:B:C) = 3 since only one probability value needs to be specified for each variable. df decreases by 1 at every level in Fig. 3. Normalizing the df measure to a 0-1 scale gives

$$\text{Complexity}^{norm} = [\text{ df(AB:BC) - df(A:B:C) }] / [\text{ df(ABC) - df(A:B:C) }].$$

Complexity reduction is achievable not only by decomposition of relations into simpler structures, i.e., by descending the lattice of structures, but also by the use of latent variables, which will be explained now, or by state-based modeling, which will be explained in the next section.

Complexity reduction can be accomplished by adding additional variables. A latent variable model typically *simplifies* the relation it explains. One cannot simplify in this way the relations of Table I, so this approach will be illustrated with a different example. Consider an AC relation, latent variable B, and posited relation ABC having the structure BA:BC. If $|A|=|C|= 4$ and $|B|=2$, then df(AC) = 15 while df(BA:BC) = 13, so BA:BC is less complex than AC.

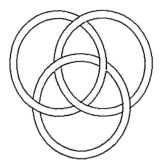

FIGURE 5 Borromean rings (removing one ring allows the other two to separate).

When considering some actual system, the lattice of structures indicates the possible decompositions of the top relation. Decomposition can be exact, i.e., with no constraint loss relative to ABC, or approximate, i.e., involving only small constraint loss. Where systems are highly decomposed, one can easily reduce the system to simple parts. There are systems, however, where the slightest decomposition results in *total* loss of constraint. This is illustrated by the set-theoretic relation ABC = {000, 011, 101, 110}, whose three dyadic projections, AB, AC, and BC are all heaps, i.e., {00, 01, 10, 11} which have no constraint at all. This can be visually represented by Fig. 5.

Constraint loss increases monotonically as one descends the lattice (Fig. 3). As indicated in Fig. 6, the constraint *lost* in AB:BC is T(AB:BC) = U(AB:BC) - U(ABC) and the constraint *retained* is T(A:B:C) - T(AB:BC) (Fig. 6) (Krippendorff, 1986). Constraint might alternatively be measured from a distribution other than A:B:C. For example, for directed systems where C is generated from A and B, the independence model is not A:B:C but AB:C.

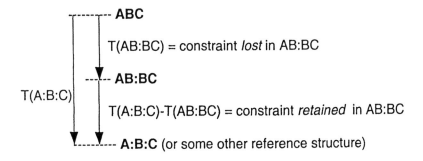

FIGURE 6 Constraint lost and retained in structures.

Retained constraint is information *in* the structure. Normalizing to a 0-1 scale gives

$$\text{Information}^{\text{norm}} = [T(A{:}B{:}C)\text{-}T(AB{:}BC)] / T(A{:}B{:}C) = 1\text{-}T(AB{:}BC)/T(A{:}B{:}C).$$

The transmission measure allows a precise formulation of Simon's (1962) observation that most systems are "partially decomposable." Consider a system ABCD having two parts, AB and CD, and consider also the identity,

$$T(A{:}B{:}C{:}D) = T(AB{:}CD) + [\ T(A{:}B) + T(C{:}D)\ ].$$

Partial decomposability means that T(AB:CD), the *between*-parts constraint, is small compared to T(A:B) + T(C:D), the *within*-parts constraint. This is analogous to a between-within decomposition in the analysis of variance.

ANALYSIS

A. Decomposition Losses

Structures encompass multiple relations, but every structure has an equivalent single (decomposable) relation linking all of the variables. This will be written as $\text{ABC}_{\text{model}}$ while the data itself will be written simply as ABC. For example, $\text{ABC}_{\text{AB:BC}}$ is the calculated ABC distribution for model AB:BC. The equivalent relation for any model can then be compared to the data. Difference between the two is the error in the model (the constraint loss or information loss).

The single relation equivalent to a structure is essentially the same in both set- and information-theoretic formalisms: it is the relation with maximum uncertainty, given the constraints imposed by the model. For example, $\text{ABC}_{\text{AB:BC}}$ is the relation (distribution) which maximizes the uncertainty, U, subject to the constraints of AB and BC. These constraints appear on the right of Table I as the two lists of tuples or the two 2-variable frequency tables.

Set-theoretically, the equivalent relation is the intersection of all relations in the structure, each relation being first "expanded" by Cartesian products with variables missing in it; here $\text{ABC}_{\text{AB:BC}} = (AB \otimes C) \cap (BC \otimes A)$. For example, the equivalent triadic relation for the AB:BC model in the set-theoretic example of Table I(a) is

$$
\begin{aligned}
\text{ABC}_{\text{AB:BC}} &= [\{00., \quad 01., \quad 11.\} \otimes \{..0,..1\}] \cap [\{\ .00, \quad .10, \quad .11\} \otimes \{0..,1..\}] \\
&= \{000,001, 010,011, 110,111\} \quad \cap \quad \{000,100, 010,110, 011,111\} \\
&= \{000, \quad 010,011, 110,111\}
\end{aligned}
$$

Here, as just noted, $ABC_{AB:BC}$ is the same as the original ABC, so model AB:BC has no error or constraint loss. Neither AB nor BC can be dropped because this would yield an equivalent ABC relation having 6 tuples, while $|ABC| = 5$. (Note also that AC is a heap and adds no constraint.) The set-theoretic analysis for all models is summarized above in Table III.

Information-theoretically, the equivalent single relation is the distribution $q_{AB:BC}(A,B,C)$ satisfying the conditions which maximizes

$$U(AB:BC) = -\sum q_{AB:BC}(A,B,C) \log q_{AB:BC}(A,B,C)$$

subject to the linear constraints: $q_{AB:BC}(A,B) = p(A,B)$ and $q_{AB:BC}(B,C) = p(B,C)$. Recall that p and q mean observed and calculated. Because model AB:BC is acyclic, the solution for this constrained maximization can be written directly

$$q_{AB:BC}(A,B,C) = p(A,B)\, p(B,C) / p(B).$$

For the information theoretic example of Table I(b), Table II gives the calculated distribution, $q_{AB:BC}(A,B,C)$ as well as the original observed probabilities, $p(A,B,C)$.

For any model the constraint of ABC_{model} is less than or equal to the constraint of ABC. Constraint loss is the "transmission" of the model, also called cross-entropy or mutual information (there is an analogous set-theoretic expression). For the set-theoretic example described earlier, the loss is 0, that is, the equivalent relation, $ABC_{AB:BC}$, is identical to ABC. For the information-theoretic example, the equivalent relation is not identical to the ABC, as can be seen by comparing the p and q distributions of Table II. The loss is

$$T(AB:BC) = -\sum p(A,B,C) \log [\, p(A,B,C)/q_{AB:BC}(A,B,C)\,].$$

T can be computed directly from p (without first obtaining q) by $T(AB:BC) = U(AB:BC) - U(ABC)$, where $U(AB:BC) = U(AB)+U(BC)-U(B)$ and where the U's are computed from projections of the data. However, *cyclic* structures do not have algebraic expressions for U and require the iterative generation of q and the calculation of T by the p log p/q expression. From T, normalized information is calculated. As above in Table IV, Information^{norm}(AB:BC)= 0.98. Very little constraint is lost in decomposing ABC to AB:BC. This, coupled with the greater simplicity of AB:BC compared to AB:BC:AC, is the basis for saying that AB:BC is the best model for the data of Table I(a).

State-based information-theoretic models are treated similarly. Calculated distributions, $q(A,B,C)$, have maximum U, constrained not by entire projections of $p(A,B,C)$ but by a set of selected individual probability values. The essence of how state-based decomposition can produce lower constraint loss is illustrated in Table VI.

TABLE VI   State-based decomposition

	B_0	B_1						
A_0	.1	.1	.2	.04	.16	.2	.1	.1
A_1	.1	.7	.8	.16	.64	.8	.1	.7
	.2	.8		.2		.8		
structure	AB			A:B			$[A_1,B_1]$	
df	3			2			1	
constraint loss, T	-			.087			0	

A:B, having probabilities $p(A)*p(B)$, is not identical to AB and thus exhibits constraint loss. The AB model has df=3, and to show this 3 cells (arbitrarily chosen) are shaded. The A:B model has df=2, i.e., it needs only two specified probability values, one (arbitrarily chosen and shown shaded) from each margin. A state-based model specifying the *single* probability value, $p(A_1,B_1)$=.7 (*not* arbitrary, shown shaded) forces the remaining (A_0,B_0), (A_0,B_1), and (A_1,B_0) probabilities, by the maximum uncertainty principle, to be .1. These are in fact correct and this state-based model thus has zero constraint loss even though it is simpler (has smaller df) than A:B. These data were of course "cooked" to produce this result, i.e., to show how a one-parameter state-based model could be superior to a two-parameter variable-based model. In the present example the state-based model has only one probability value, but in general such models can specify any (linearly independent) set of probabilities from an original table, its margins, its margins of margins, and so on. In the present example, $[p(A_0,B_0)$=.1, $p(B_1)$=.8] would be a legitimate df=2 state-based model.

State-based analysis (not shown here) of the earlier distribution of Table I(b) reveals that a four-parameter state-based model, consisting of $p(A_1,B_0)$, $p(A_0,B_1)$, $p(B_0,C_1)$, and $p(B_1,C_0)$, would capture virtually the same amount of information as the five-parameter AB:BC model. (Note that the four states come from the AB and BC relations.) So, state-based analysis would improve upon the RA results summarized in Table IV.

B. Reconstruction and Identification

RA includes *reconstruction* and *identification*. The two examples of Table I illustrate reconstruction. In reconstruction, one starts from a whole, and decomposes it into provisional parts (relations) and then recomposes these parts to see if they account for the whole. In identification the parts are given, and one does only composition; this is done by the calculation of the equivalent single relation for the model, as discussed in the previous section.

In reconstruction the objective is to find simple but low-error (high information) structures to model the data. There are different ways to balance the dual objectives of minimizing error and complexity. While conceptually one descends the lattice of structures and assesses the different decompositions, operationally one might actually either descend or ascend the lattice. If accuracy is the primary concern, or if systems are neutral, a modeling procedure might descend the lattice of structures until the error becomes too great. If simplicity is the primary concern, or if systems are directed, the procedure might ascend the lattice until further complexification is not forced or justified by the data. In information-theoretic/log linear modeling, which is Chi-square-based, the *statistical significance* of error, i.e., the probabilities of Type I and Type II errors, integrates error magnitude and complexity, but this does not resolve the issue: there are tradeoffs between these two kinds of errors (Knoke and Burke, 1980; Krippendorff, 1986).

Disallowing statistically significant error may prevent any simplification, so one might choose the simplest model whose error is acceptable. AB:AC:BC may be the simplest structure whose error is statistically *not* significant while the simpler AB:BC may have an error which *is* statistically significant but only slightly bigger. Statistical significance is not pragmatic significance, and AB:BC might be preferred for its simplicity.

As already noted, picking a best *particular* structure does not constitute a complete analysis. A system is fully described by the constraint losses (or conversely, the information captured) for *all possible* decompositions. Table IV shows the complete reconstruction analysis of the data given in Table I(b). Another illustration of a complete reconstruction analysis is provided by FIGURE 7 which plots Information^{norm} against Complexity^{norm} for data on medical and sociological characteristics of a sampled population (Zwick and Pope, 2000). The figure displays the entire lattice of possible decompositions for this data, treated as a neutral system. Models at the upper envelope of the cluster are plausible candidates for acceptance. That is, for any complexity, C, one wants a model with the maximum information, I.

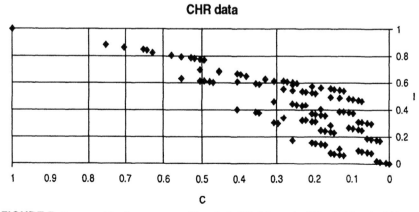

CHR data

c

FIGURE 7 Decomposition loss spectrum: (Complexity^{norm}, Information^{norm}); 114 four-variable structures. Data from Kaiser Permanente Center for Health Research.

A few words on identification. Structures can arise not from decomposition but from the composition of separate relations which may not be the projections of a single higher-ordinality relation. When wholes are thus composed of preexisting parts, there is no issue of constraint loss. If the relations overlap, e.g., AB and BC, they may be either consistent or inconsistent in their overlapping subrelation distributions, i.e., B. For example, suppose one were given the AB distribution shown on the right side of Table I(b), whose B margin is [1034, 444], but also an *altered* BC table whose B margin was, say, [1024, 454]. Such inconsistency could arise from sampling (or other) errors. When overlapping subrelations are identical, composition is straightforward, but if they are even slightly different—and inconsistency is to be expected—resolution of the inconsistency is first required (Klir, 1985; Anderson, 1996). Composition can also first utilize and then exclude extraneous variables in what might be called "reverse" latent class analysis. Suppose one is interested in the relation between A and C, but has inconsistent data only on AB and BC. After the inconsistency is resolved, the dyadic relations can be composed into a triadic ABC and then projected onto the desired AC relation (Anderson, 1996).

REMARKS

This article presents the essentials of reconstructability analysis, a particular set of procedures within general systems methodology (Klir, 1985). The framework presented here offers a very general approach to the multivariate modeling of nominal and quantitative data. This approach could be of significant use for analyzing the independencies among biological characters and genes and relevant attributes of the environment.

For further discussion on reconstructability analysis, see the special 1996 issue of the *International Journal of General Systems* on GSPS (General Systems Problem Solver). This framework is undergoing continued research and development, but tools which integrate what is already known are unfortunately not yet available. Ideally one wants a software implementation which can:

- do both set- and information-theoretic analyses and their fuzzy extensions
- use both variable- and state-based approaches
- in both confirmatory and exploratory (data mining) modes
- with efficient lattice search techniques for many variables
- on both nominal, ordinal, and quantitative data (and thus include effective binning)
- for static and dynamic (e.g., time series) applications
- with or without latent (supplementary) variables
- for both reconstruction and identification (including inconsistency resolution).

I am unaware of any software implementation which approximates these specifications, but the separate components exist. A software package aimed at this goal is being developed at PSU based on earlier PSU efforts and with external collaboration.

ACKNOWLEDGMENTS

I thank George Klir for illuminating discussions on reconstructability analysis and systems theory, Anthony Blake for stimulating conversations on wholes and parts (and for the Borromean rings), and both them and George Lendaris, Bjorn Chambless, Jeff Fletcher, Stanislaw Grygiel, Michael Johnson, and Tad Shannon for helpful comments on the manuscript. I'm grateful to Günter Wagner for his patience and his gentle insistence that I try harder to make this article accessible. Given all this help, I am solely responsible for whatever obscurity still plagues this presentation.

LITERATURE CITED

Anderson, D. R. (1996). "The Identification problem of Reconstructability Analysis: A General Method for Estimation and Optimal Resolution of Local Inconsistency." Ph.D Dissertation, Portland State University. Portland, OR.

Ashby, W. R. (1964). Constraint analysis of many-dimensional relations. *General Systems Yearbook* **9**:99-105.

Bishop, Y. M., Feinberg, S. E., and Holland, P. W. (1978). "Discrete Multivariate Analysis." MIT Press, Cambridge.

Conant, R. C. (1981). Set-theoretic structure modeling. *Int. J. Gen. Syst.* **7**:93-107.

Conant, R. C. (1988). Extended dependency analysis of large systems. *Int. J. Gen. Syst.*. **14**:97-123.

Davis, J. A. (1985). "The Logic of Causal Order" (Quantitative Applications in the Social Sciences #55). Sage, Beverly Hills.

Grygiel, S., Zwick, M., and Perkowski, M. (2000). Multi-level Decomposition of Relations." (In preparation).

Hagenaars, J. A. (1993). "Loglinear Models With Latent Variables" (Quantitative Applications in the Social Sciences #94). Sage, Beverly Hills.

International Journal of General Systems (IJGS). Special Issue on GSPS. (1996). **24**:1-2.

Jones, B. (1985). Determination of unbiased reconstructions. *Int. J. Gen. Syst.* **10**:169-176.

Klir, G. (1985). "The Architecture of Systems Problem Solving." Plenum Press, New York.

Klir, G., and Wierman, M. J. (1998). "Uncertainty-Based Information: Variables of Generalized Information Theory." Physica-Verlag, New York.

Knoke, D., and Burke, P. J. (1980). "Log-Linear Models" (Quantitative Applications in the Social Sciences Monograph # 20). Sage, Beverly Hills.

Knoke, D., and Kuklinski, J. H. (1982). "Network Analysis" (Quantitative Applications in the Social Sciences Monograph # 28). Sage, Beverly Hills.

Krippendorff, K. (1986). "Information Theory: Structural Models for Qualitative Data" (Quantitative Applications in the Social Sciences #62). Sage, Beverly Hills.

Langton, C. (1992). Life at the edge of chaos. *In* "Artificial Life II" (C. G. Langton, C. Taylor, J. D. Farmer, and S. Rasmussen, eds.), pp. 41-91. Addison Wesley, Reading.

Long, J. S. (1983). "Covariance Structure Models: An Introduction to LISREL" (Quantitative Applications in the Social Sciences #34). Sage, Beverly Hills.

McCutcheon, A. L. (1987). "Latent Class Analysis" (Quantitative Applications in the Social Sciences #64). Sage, Beverly Hills.

Simon, H. A. (1962). The architecture of complexity. *Proc. Am. Philos. Soc.* **106**:467-482. Reprinted in "The Sciences of the Artificial." (H. A. Simon)

Simon, H. A. (1981). "The Sciences of the Artificial." M. I. T. Press, Cambridge.

Wolfram, S (1986). "Theory and Application of Cellular Automata." World Scientific, Singapore.

Zadeh, L. A. (1965). Fuzzy sets. *Information and Control* **8** (3):338-353.

Zwick, M., and Shu, H. (1996). Set-theoretic reconstructability of elementary cellular automata. *Adv. Syst. Sci. Applic.* Special Issue **1**:31-36.

Zwick, M. (1996). Control uniqueness in reconstructability analysis. *Int. J. Gen.Syst.* **24**:1-2, 151-162.

Zwick, M., and Shu, H. (2000). Reconstructability and Dynamics of Elementary Cellular Automata. (Manuscript in preparation.)

Zwick, M., and Pope, C. (2000). Reconstructability Analysis on Medical Utilization [OPUS] Data. (Manuscript in preparation.)

OPERATIONALIZING THE DETECTION OF CHARACTERS

Whatever the abstract beauty of a concept, its first test of validity is whether it can be applied to real data. Whether operationalization follows abstract conceptual considerations or concepts are verbalizations of essentially operational notions is an open question and can only be decided by the success of the one or the other approach. In any case, the difficulty in operationalizing ideas and concepts is in itself an important source of empirical knowledge. Some lessons from them are discussed in this section.

Daniel McShea and **Edward Venit** develop a protocol for recognizing and counting morphological parts or characters. In the course of this project they clarify the relationship between parts and other concepts. The motivation for this attempt is to test hypotheses which predict a correlation between environmental and organismal complexity, defined as the number of characters.

Peter Wainwright and **John Friel** raise the question whether the notion of characters can be applied to motor patterns in a similar fashion as they are to the more tangible characters of morphology and molecular biology. They conclude that motor patterns are highly variable and are under behavioral control but nevertheless conserved between species. Comparability of motor patterns between species depends critically on the experimental conditions imposed on the animals.

Ward Wheeler discusses one of the apparently least problematic types of characters: DNA sequence data. He shows that the notion of homology of nucleotide sites across species is highly problematic because of insertions and deletions. Limits on our ability to recognize sequence homology are not only due to methodological limitation but are fundamental. Wheeler concludes that sequence homologies are at most taxonomically local.

Harold Bryant analyzes the various methods and their underlying assumptions of the polarizing character states. He concludes that methods for polarizing characters are equivalent to methods for rooting networks, and that individual methods for polarizing characters can be complementary.

11

WHAT IS A PART?

Daniel W. McShea and Edward P. Venit

Department of Zoology, Duke University, Durham, NC 27708

INTRODUCTION

Anyone familiar with bicycles would have little trouble listing the major parts: handlebars, frame, seat, front wheel, back wheel, pedals, drive chain, and so on. The hierarchical structure of parts in bicycles is also fairly obvious: for example, the links in the drive chain are parts of the chain and therefore what might be called "subparts" of the bicycle. In machines, the parts are usually quite distinct, and the hierarchical level at which a part occurs—part versus subpart—is fairly unambiguous. This is less true for organisms. For example, in a primate, a hand might seem to be a part of the whole animal, but it also seems to be a part of the arm, and therefore a subpart of the animal. In an individual in a bryozoan colony—a zooid—it is not clear whether the caecum (stomach) is merely a continuation of the pharynx (anteriorly) and of the pylorus (posteriorly) or whether it is a part distinct from both of them. (We will return to this example later.) What is a part, in principle? And how can we identify parts in practice and determine the hierarchical relationships among them in real organisms?

The Character Concept in Evolutionary Biology

For many studies of organismal structure, parts can be taken as unproblematic. Phylogenetic systematics is mainly interested in the homology of parts (and of their attributes), functional anatomy is interested in how parts interact to perform tasks, and studies of morphological and functional complexity are concerned with numbers of different types of parts in organisms. In these areas, we typically identify parts using precognitive perceptual mechanisms, or gestalts. The success of our analyses ordinarily leaves little reason to doubt that these gestalts are highly reliable, that the parts we have identified are biologically significant features of organisms. Also, in these areas, we can often choose parts opportunistically, incorporating into our analyses only structures which would be recognized as parts by any observer, the uncontroversial parts. For example, in studies of complexity, limb-pair types have been used as parts of the limb series in arthropods (Cisne, 1974), and vertebrae as parts of vertebral columns (McShea, 1993). At the level of the whole metazoan, cells have been used as the parts (Bonner, 1988; Valentine *et al.*, 1993; Bell and Mooers, 1997).

For certain purposes, however, reliance on gestalts and uncontroversial parts is quite limiting. In studies of complexity, it would be helpful to be able to count numbers of part types in whole organisms, not just in serial structures like vertebral columns, and to make these counts at any chosen hierarchical level, not just the cell level. In particular, our motivation in examining the parts issue here is mainly to develop methods for testing a certain hypothesis about the evolutionary relationship between numbers of part types and the emergence of new hierarchical levels.

The hypothesis is the following: In the history of life, as organisms combined to form higher-level functional entities, functional demands on these organisms would have been reduced. For example, as metazoans formed from clones of free-living eukaryotic cells, or as integrated marine invertebrate colonies formed from the budding of free-living polyps or zooids, functional demands on each cell, polyp, or zooid would have been reduced. Then, assuming that number of functions is correlated with number of part types (McShea, in review), the reduction in functional demands on the organisms should have been accompanied by a reduction in the number of part types they contain. That is, selection should have favored a loss of part types in the interest of economy. Thus, cells in metazoans and land plants should have fewer part types than free-living eukaryotic cells, and zooids or polyps in highly integrated colonies should have fewer part types than those in less integrated colonies. (For details and alternative formulations of the hypothesis, see McShea [in press].)

The hypothesis will not be tested here. For present purposes, the point is only that in order to test it, we need methods for identifying and counting part types, say, within bryozoan zooids, in a consistent fashion. Importantly, a proper test requires that we identify the parts at a specific hierarchical level. The hypothesis predicts that the loss of part types will be most manifest at the level closest to and just below the zooid (McShea, in press), roughly at what might be called the organ level (although see later). Thus, while it might be tempting to use cells as the parts in zooids, or even genes, because counting types is convenient, both lie a number of hierarchical levels too low.

Methods for identifying parts could be useful in other areas as well: for example, in phylogenetic studies, giving equal weight to parts at different hierarchical levels can bias the analysis, but it may be difficult to establish the level at which a part occurs. Thus, methods for identifying parts at a single level in a consistent way would help to weight parts appropriately.

Here, we propose a technical definition for the term "part," one which corresponds fairly well with its colloquial meaning and with our gestalts, and then briefly discuss the relationship between parts and other concepts, including the subject of this volume, characters. Next, we propose a way to (partly) operationalize the definition, a series of protocols for identifying certain kinds of parts, what might be called object parts or structural parts, at the hierarchical level just below the organism. We argue that, for certain comparative purposes, counts of object parts can be used as a proxy for counts of true parts (i.e., all parts fitting the technical definition). Then, we formalize the protocols with a "parts key," which shows how to use the protocols to identify object parts in apparently simple organisms. Finally, we use the key to produce a tentative list of object-part types for certain bryozoan zooids.

The protocols, key, and parts list were the major goals. However, this paper was also an exercise, an experiment, to investigate the following: given an *a priori* definition of parts, can protocols consistent with that definition be devised and then consistently applied to produce an intuitively reasonable list of part types? By this standard, the exercise was fairly successful, although—as will be seen—practical difficulties remain. However, a more telling test of the usefulness of the exercise lies ahead: if the parts identified in this way are biologically significant, then robust associations should be found between these parts and other variables. One example would be the correlation discussed earlier, between number of part types and the emergence of higher levels of functionality. Others can be imagined.

The intent here is not to introduce a new concept. In much of biology, including anatomy, systematics, and the study of adaptation, we routinely assume that organisms have identifiable parts. Rather, the point is to try to formalize and to (partly) operationalize our understanding of a notion that is already commonplace, indeed, that is foundational.

Our discussion is based on analytical treatments of individuality by Campbell (1958), Hull (1980), Mishler and Brandon (1987), and Ghiselin (1997), of integration and isolation by Olson and Miller (1958) and Bonner (1988), of modularity by Mittenthal *et al.* (1992), Wagner and Altenberg (1996), and Raff (1996), and of hierarchy by Wimsatt (1974, 1994) and Salthe (1985, 1993).

PARTS: IN PRINCIPLE

A. A Technical Definition

In the present discussion, a part is a system that is both integrated internally and isolated from its surround. Concretely, a crystal, such as a diamond, is a part. A free diamond is integrated internally in that its component carbon atoms are tightly bonded to each other, and it is isolated from its surround in that they are only weakly bonded, or not bonded at all, to the atoms in the surrounding air. By the same standard, most bicycle frames and individual organisms are parts.

Integration refers not just to bonds between components but to any interactions that produce correlations in the behaviors of the components (Campbell, 1958). Systems are integrated to the degree that interactions among components are many or strong, or both. Thus, interactions can also take the form of signals among components, in which case a spatially distributed system can also be a part. In organisms, possible examples include a hormone-mediated control system, or at a higher hierarchical level, a local population of crickets chirping in synchrony.

It has been pointed out that correlations may occur as a common response by components to an external cause, without any interactions among them (J. Padgett, personal communication). For present purposes, interaction-based parts seem sufficient; however, a more inclusive definition may be necessary at some point, as the investigation is extended.

Isolation is a reduction in, or termination of, integration. In some parts, isolation is a consequence of an intervening boundary, as in a gas-filled balloon or a cell. In others, the isolation is a consequence simply of the termination of integration, such as occurs at the surface of a crystal. In spatially distributed parts, isolation may be a consequence of the specificity of the signaling with which integration is achieved. The chirping cricket population may be a part, even if individuals of other species occupy the same space, perhaps the same patch of lawn, provided the chirp signals responsible for the integration are species specific, and therefore effectively isolate the population.

Integration and isolation are both required for parts. A group of people sitting in a train car may be isolated from a similar group in the next car, but neither group constitutes a part if its members are not interacting among themselves. Conversely, an arbitrary subset of guests at a cocktail party may be interacting strongly with each other, but they do not constitute a part if they are also interacting strongly with other guests outside the subset. Both integration and isolation may vary continuously, and therefore the extent to which a system is a part—its degree of partness—is likewise a continuous variable.

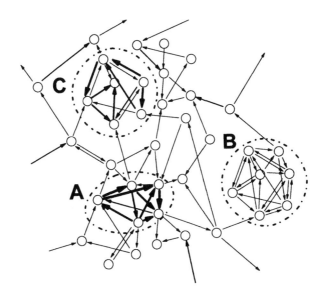

FIGURE 1 Parts. Small circles are components. Arrows show interactions among them; the thickness of the arrows corresponds to the strength of interactions. Parts are enclosed by dashed lines.

In the discussion and examples just given, it is understood that the patterns of integration and isolation which constitute parts are stable. Of course, stability is relative to some chosen time scale. Over the adult life of a multicellular organism, the structures we call organs tend to be fairly stable. Certain patterns of neural activation, such as might be responsible for a behavior, are also stable—and therefore parts—but on much shorter time scales.

Figure 1 shows parts abstractly. In the figure, the small circles are components and the arrows show interactions among them. The thickness of the arrows corresponds to the strength or intensity of the interactions. Parts are enclosed by dashed lines. What characterizes each part is the large number of interactions and/or their greater intensity relative to the number and/or intensity of interactions with the surround. Part A has few internal interactions, but they are strong relative to its weaker interactions with external components. Part B has weaker internal interactions, but they are many relative to interactions externally. Part C is intermediate.

B. Relationship to Other Terms

1. Characters

Colless (1985) distinguishes character-parts from character-variables and character-attributes. Variables and attributes refer to descriptions or measurements of aspects of organisms (Fristrup, 1992), while parts refers mainly to the physical structures that constitute organisms. Importantly, Colless notes that character-parts can be construed to include processes, or patterns of interaction, such as behaviors and metabolic cycles, as well as objects. Thus, parts as used here correspond well with Colless's character-parts.

2. Individuals

Parts are close to "individuals." Mishler and Brandon (1987; see also Hull, 1980; Ghiselin, 1997) list four criteria for individuality—spatial localization, temporal localization, integration, and cohesion. Parts meet some of these criteria. Like individuals, they must be integrated. Some parts will also be cohesive; more generally, the requirement is that they must be stable, as discussed. Also, parts are located in space, and therefore have boundaries, or at least limits of some kind; however, parts may also be spatially distributed and therefore not highly localized in the usual sense. Mishler and Brandon point out that biological entities meeting some but not all of the criteria may nevertheless behave as individuals in certain respects, and in certain contexts. Parts seem to fit this description fairly well, and therefore might be called individuals.

In many contexts, however, the term individual is applied exclusively to organisms. For example, it has been used to distinguish a well-integrated multicellular organism from a mere colony or aggregate of independent cells (e.g., Buss, 1987); only the organism would be called an individual. However, by the present definition, both could equally be parts.

3. Modules, ρ Groups, and F Groups

Our understanding of a part is also close to what Wagner and Altenberg (1996; see also Wagner, 1996), Raff (1996), and others have called a "module." The difference is that modules are internally integrated and externally isolated units in the development of an organism, in its generation or manufacture. On the other hand, parts are units in what might be called the "operation" of an organism, which is limited to processes occurring within the organism once it has been generated, such as adult physiology. Mittenthal et al.'s (1992) "dynamic modules" include both developmental and operational entities. In organisms, the distinction is not perfectly clean, but it is sufficiently so to make the use of different terms appropriate.

There is also some connection between parts and the entities that Olson and Miller (1958) call ρ groups and F groups. P groups are developmental entities, like modules. F groups are functional entities; parts may be functional but need not be, and therefore part is a more inclusive term.

4. Difficulties

Our choice of the term "part" is somewhat problematic. For one thing, as the term is used colloquially, it would not include most spatially distributed systems, such as a chirping cricket population or a hormonal regulatory system. For another, we usually think of parts as entities within larger wholes, entities that are "part of" something, but our technical definition includes many entities that we normally think of as standing alone, such as crystals and whole organisms. (Although arguably, all entities—except perhaps the universe itself—actually *are* parts of some larger whole, whether or not we choose to think of them that way.) On the other hand, almost any entity that would qualify as a part in colloquial usage would also qualify under the technical definition. In any case, other terms that might have been chosen instead of part, such as entity, thing, component, or system, would also have been problematic. In the present discussion, we use these other terms in their colloquial senses, in cases where a technical term is unnecessary or where ambiguity is helpful.

C. Part Hierarchies

Here, hierarchy refers to the physical nesting of parts within parts, or what Valentine and May (1996) called a cumulative constitutive hierarchy, Wimsatt (1994) described as "compositional levels of organization," and what Salthe (1985, 1993) called a scalar hierarchy (see also McShea, 1996a,b).

The parts occupying a given level in a hierarchy consist of internally integrated and externally isolated sets of components (often, subparts) from the level below. These may also be integrated with other parts at the same level to produce parts at the next level up. Notice that the level in a hierarchy at which a part occurs is purely a function of the topology of the system, not of the part's absolute size. Figure 2 shows a hierarchy in which C and D are parts of the whole, and A and B are parts of C. Thus, A and B occupy the hierarchical level below D, even though A, B, and D are all about the same absolute size. Indeed, if E is a part, it occupies the same level as C and D, although it is smaller in absolute terms than either. Further, notice that the hierarchical level at which a part is placed will be very sensitive to changes in our decisions at higher levels. For example, suppose that we discovered upon closer examination of the system that, contrary to Fig. 2, A and B are not highly connected, and therefore that part C does not exist. In that case, parts A and B would be promoted to the next higher level and would become parts on the same level as D.

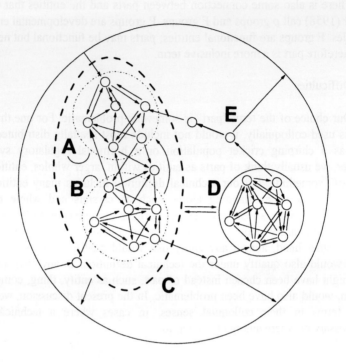

FIGURE 2 Hierarchical relationships among parts. Circles and arrows are as in Fig. 1 caption, except that strengths of interactions are not shown. D and E are parts; see text for discussion of the status of A, B, and C.

In Fig. 2, the occupation of hierarchical levels by parts is discrete, in that each part occurs exclusively at one level. In organisms, however, occupation of levels is often continuous in that parts may cross levels, existing simultaneously partly at one level and partly at another. For example, suppose that in Fig. 2, A and B were only weakly integrated, so that C would be only weakly a part. In that case, A and B would occupy two levels at once, the higher level at which D occurs and the lower level just below D. More precisely, they would occupy each level only to some degree. Alternatively, suppose that A and B were well integrated with each other but only at certain times, or in certain functional contexts. In that case, A and B might shift back and forth between levels. This crossing of hierarchical levels is undoubtedly quite common in organisms, and one of the main reasons for their apparent complexity (Wimsatt, 1974), for the difficulty we have in parsing them into their component parts at each hierarchical level.

Because level occupation may be continuous, we have no *a priori* reason to think that discrete levels will be identifiable at all in organisms. However, some levels do seem to be occupied more or less discretely, in particular those levels which have experienced especially intense selection (Maynard Smith, 1988). In multicellular organisms, for example, cells would seem to be parts at a single hierarchical level, presumably as a result of selection in the past, and to occupy that level fairly discretely, for the most part. Similarly, in most metazoans, the entire multicellular individual is a part, and individuals seem to occupy that level discretely. However, parts at levels between cells and multicellular individuals—parts at roughly the tissue, organ, and organ-system levels—have presumably not been selected at their own level, but instead seem to have been interpolated by selection at the individual level. Thus, they are more likely to cross levels and therefore to occupy any given level more ambiguously.

In the discussion of bryozoan parts in the next section, we assume that in bryozoans, selection occurs or has occurred in the past at the level of the zooid, and we take advantage of that fact to anchor the analysis hierarchically. In other words, we assume that all zooids in bryozoans occupy the same hierarchical level, and therefore that all parts one level down from the zooid occupy the same level. However, because parts are not expected to occupy the hierarchical level just below the zooid especially discretely, counting part types at that level in a consistent way is expected to be a rather uncertain undertaking.

Notice that the hierarchical structure of organisms based on this conceptual scheme need not accord with the classic object hierarchy in biology, also called "levels of organization" [not to be confused with Wimsatt's (1994) more precise "compositional levels of organization"]. Conventionally, these include: ... organelle, cell, tissue, organ, organ system, multicellular individual, ... Despite the intuitive appeal of this list, the identity of the structures that occupy each level is an empirical issue, one on which—for most levels—we have little solid evidence. For example, it is not obvious that all of the structures we call organs occupy the same hierarchical level in any organism.

D. Parts and Function

In the present scheme, functionality and partness are independent notions. Parts can be functional, as a bicycle tire is, but a nail that punctures and becomes buried in the tire is also a part. Likewise in organisms, an essential organ might be a part, but nonfunctional structures, or even deleterious ones, can also be parts. Indeed, a system of components with a randomly configured pattern of interactions—a random "wiring diagram," so to speak—would likely be entirely functionless, but would probably contain a number of internally integrated and externally isolated subsets of components, i.e., parts (Kauffman, 1993).

However, this is a conceptual separation only. Arguably, in organisms, functions are expected to be very closely associated with parts. Briefly, the reason is that in order for a system to function, it requires a certain amount of internal coordination, and therefore integration to achieve that coordination, and also some degree of isolation, to limit interference from other systems. Thus, for example, to the extent that an organism must be able to move and feed at the same time, and to the extent that these functions require different activities, components involved in locomotion are expected to be isolated from those involved in feeding. More generally, selection for function is expected to have isolated functions to some degree in parts. However, the relationship between number of different functions and number of part types is not expected to be simple, not one to one, for example. See McShea (in press) for a longer treatment of this issue.

This argument is a modification of a similar one by Wagner and Altenberg (1996). They suggest that selection for evolvability is expected to produce developmental modularity; the present argument is that selection directly for function itself is expected to produce a kind of operational modularity, i.e., parts, regardless of developmental organization. (Of course, this too is only an in-principle separation; in organisms, developmental modules may correspond closely with parts.)

The hierarchical level occupied by a part is, like partness itself, independent of function in principle. However, also like partness, there is reason to suspect that function is relevant in fact, in particular, that the hierarchical organization of parts in organisms is partly a consequence of selection. The argument is based on the notion of "screening off," a concept which has proved useful in analyzing levels of selection. [See Brandon (1996) for a more formal treatment; see also Roth (1991).] For present purposes, what is significant is that in organisms, higher-level functions often screen off lower-level functions. For example, in vertebrates, the malfunction of one or a small number of nephrons (possible "subparts") in a kidney (possible part) is screened off by, and would not interfere with, the function of the kidney as a whole. Likewise, the malfunction (e.g., occlusion) of a single small blood vessel ("subpart") is partly screened off by the function of the circulatory system as a whole (part), which has some capability to revascularize tissues deprived of sufficient blood supply. To the extent that upper-level functions are insulated from lower-level variation in this way, a hierarchical arrangement of functions is presumably advantageous. Thus, to the extent that parts organization follows functional organization in organisms, a hierarchical arrangement of parts is also expected.

E. Two Clarifications

1. Temporal Scale and Range

Just as stability is time-scale relative, as discussed, degree of integration and isolation—and thus partness itself and the hierarchical level of parts—are also time-scale relative. For example, a circulatory system might be a part in some very short time-scale process like respiration. However, in a (facultatively) longer time-scale process, such as a fight-or-flight response, the circulatory system might instead be a subpart in a larger part that includes the adrenal medula, the hypothalamus, and other systems. Parts are also expected to vary in absolute time; the parts in a sea urchin larva change dramatically during metamorphosis. Thus, any attempt to decompose an organism into parts must first specify a temporal scale and a range of absolute time over which parts will be identified.

2. Unique Decomposition

In the present scheme, the parts organization of a system is a function solely of the number, intensity, and configuration of interactions within it. It follows that—at a given temporal scale and range—every system has a single, unique decomposition into parts, subparts, and so on. (Granted, that decomposition may be difficult to discern in practice.) A possible objection is that, as Wimsatt (1974) has pointed out, the decomposition of organisms into parts varies with theoretical perspective. For example, viewed as a device for locomotion, a tetrapod limb is decomposable into one set of components, but viewed as a device for thermoregulation, it has a different set, and the two sets have noncoincident boundaries (Wimsatt, 1974). Actually, this insight raises no problem for present purposes. Different theoretical perspectives amount to different functional decompositions, and the fact that a system has many functional decompositions does not deny that it has a single wiring diagram like that in Fig. 2, or in other words, a single pattern of parts, subparts, and so on at each timescale.

Interestingly, it is not obvious that a tetrapod limb would, by itself, count as a part in the present scheme. A limb might amount to little more than a group of components which various parts share, perhaps with some overlap. For example, the bones, muscles, and other components associated with locomotion might constitute one part, while the circulatory system, sweat glands (if present), and so on constitute another. Another way to say this is that the relative reductions in integration at the boundaries of the locomotory and temperature-regulatory parts might be more significant than the reduction that occurs at the mechanical joints near the proximal end of the humerus or femur. Whether or not this is true is an empirical matter; in any case, the partness of a limb is not problematic in principle.

PARTS: IN PRACTICE

We propose a general strategy and a series of protocols for constructing a partial list of part types in an organism (i.e., a sample of all part types), in particular, the part types occupying the hierarchical level just below that of the whole organism. The temporal scale and range of interest is that portion of an organism's (presumably, adult) life over which its internal structure is relatively stable.

The protocols were devised for organisms that seem to be structurally fairly simple (at the level just below the organism), that is, for organisms containing only a small number of loosely packed objects, readily distinguishable from each other, in a fluid matrix. In particular, they were designed for identifying part types just below the zooid level in bryozoans. Applying them to more complex organisms and to other hierarchical levels might be possible, even straightforward, but it is also possible that difficulties not addressed here would arise.

Notice that our goal here is to produce a list of different part *types*, rather than of all parts (including duplicates of the same type). However, for simplicity, we will often refer simply to parts, omitting the word "types."

A. A Strategy: Object Parts

If we knew the complete pattern of interaction within an organism, that is, its complete wiring diagram (like Fig. 2), all parts would be equally visible and their hierarchical levels apparent. In the absence of such diagrams, we can still identify many parts using morphological boundaries, or physical demarcations, in other words, using objects. In practice, the objects used will be those that are typically visible in dissections, photographs, and anatomical drawings at a magnification that is low relative to the size of the organism. The assumption is that the appearance of an object is usually a consequence of the relatively tight integration among a set of components, and that the limits or boundaries of objects correspond to reductions in integration. Parts identified in this way might be called "object parts" or "structural parts."

This strategy has certain limitations. First, it cannot identify all parts, because many parts (i.e., the spatially distributed ones) will not be objects. Second, it cannot identify all objects: small objects will not be visible at low magnification, and many methods of specimen preparation will leave some of even the larger objects invisible. Third, on account of these two limitations, the hierarchical level of some parts will be misidentified; for example, some subparts will be identified as parts. Finally, it will produce some errors: some object parts will not be true parts at all, because integration will sometimes cross object boundaries. (Thus, object parts are technically not a proper subset of all true parts.)

However, the method should produce a representative sample of part types, one that should be useful for comparative purposes. At very least, lists of object parts should be useful for comparing part counts in closely related organisms, where the omission of small or invisible parts, and the inclusion of a small number of objects that are not parts, is unlikely to introduce a bias.

The use of object parts as a proxy for true parts is consistent with standard practice in various kinds of morphological analysis, including many phylogenetic analyses. One of its principal virtues is that it allows parts to be identified by direct observation, without elaborate experiments or equipment. It also allows procedures for identifying parts to be made fairly explicit, which minimizes subjectivity.

In the remainder of this paper (as well as in figures and the appendix), we refer to object parts simply as parts, on the assumption that they do represent a reasonable proxy.

1. Three Levels

Salthe (1985) argues that three levels are relevant in the analysis of hierarchical systems generally: the level of interest, or what he calls the focal level, plus one level above and one below. Here, the focal level is that of parts, the higher level is occupied by the organism as a whole, and the lower level by parts of parts, or subparts. Attention to level is crucial here in order to count parts in a consistent way. For example, a parts list that includes the circulatory system, the heart, and the blood vessels as distinct parts in effect counts the circulatory system twice.

Actually, for present purposes, it will not be necessary to establish the precise level at which entities below the focal level occur—that is, to establish whether they are subparts, sub-subparts, sub-sub-subparts, etc.—and it will be convenient to use the term "subparts" more broadly to describe the components at any level below parts. We place the word in quotes to emphasize this deliberate ambiguity.

2. Approximations

The strategy requires certain approximations. Ideally, a list of parts would take into account the fact that the degree to which each structure is a part and the degree to which it occupies the focal level are continuous variables. Perhaps quantitative measures of partness and of level occupation could be devised to assign a "weight" to each part. Lacking measures of this sort, we instead provide standards for making binary decisions, for deciding whether a given structure shall be counted as a part at the focal level or not. In effect, partness and level occupation are discretized, with the consequence that counts of parts are really approximations.

B. Protocols

1. Spatial Relationships and Compositional Differences

The first step is to identify the objects that might qualify as parts, or the "candidate parts," and for this purpose a liberal and somewhat vague standard is appropriate: a candidate part is an object that is distinct in some way from its surroundings and occupies a localized, definable region within the organism. Then, the partness and hierarchical level of a candidate is evaluated based on two types of spatial relationship: (1) enclosure and (2) contiguity with a difference in composition.

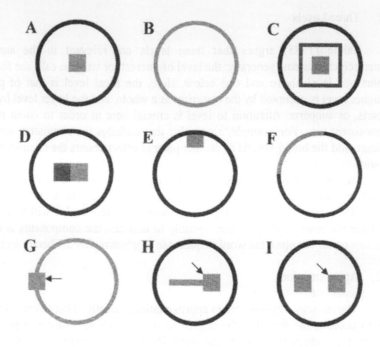

FIGURE 3 Enclosure and contiguity with a compositional difference. Outer structure of organism = dark circle; candidate part (CP) = light gray square (in A, C, D, E, G, H, & I) or light gray circle or arc (in B & F); dark gray = another part, differing in composition. A: CP = part (enclosed by outer structure); B: CP = part (outer structure itself); C: CP = "subpart" (enclosed by another part); D: CP = part (contiguous with another part but different in composition); E: CP = part (contiguous with outer structure but different in composition); F: CP = part (contiguous with the rest of the outer structure but different in composition); G: CP = nonpart (shape differences not considered sufficient for part status; however, entire outer structure, including gray square, is a part); H: CP = nonpart (shape differences not considered sufficient for part status; however, entire gray structure, including gray square, is a part); I: CP = part or duplicate part. See text for further discussion.

Figure 3 will be used to illustrate these relationships. The assumption is that the organism has a mainly fluid interior and that it is enclosed by an outer structure. In some organisms, this is a single structure, a shell, membrane, or wall (shown as a large black circle in the figure), which encloses the organism entirely. In others, however, the shell, membrane, or wall is incomplete, leaving a number of other structures in contact with the surrounding environment (e.g., the lophophore in bryozoans; see later). Importantly, here the term "outer structure" refers to the set of all structures that are in contact with the surrounding environment.

In the figure, a focal candidate part is drawn as a light gray square (and marked with an arrow in G, H, and I), except in B and F where the candidate is the outer structure itself or a segment of it (also shaded light gray). The darker gray shapes (C and D) are structures that differ in composition from the candidate part.

Before turning to the spatial relationships, two preliminary issues need to be addressed. First, we assume that the fluid-filled interior and other fluid-filled cavities are not objects and therefore do not by themselves count as parts. The rationale is that neither the integration among the fluid's components nor the isolation that begins at the edge of a fluid mass is likely to be sufficient to achieve part status. (Of course, if a fluid-filled cavity is bounded by a membrane, then fluid and membrane together might be a part.)

Second, candidate parts are understood to be "contained" within the organism, meaning that they occupy space inside an imaginary line marking the organism's outer perimeter, i.e., marking the limit of the outer structure. Containment is relevant, because a candidate part that is inside this perimeter is likely to interact most directly and strongly with parts in the same organism, and therefore is unlikely to be "outside" the organism in terms of its interactions. To put it another way, a part within the outer perimeter is unlikely to constitute an independent entity on the same hierarchical level as the organism or on a higher level. Therefore, it most likely occupies some lower level, either as a part or "subpart." All of the candidates in Fig. 3 are contained in this sense, including the outer structure (Fig. 3B) which technically lies within the organism's outer perimeter and therefore is itself a legitimate candidate for a part.

Now consider the first spatial relationship, "enclosure." As understood here, enclosure is different from containment; a contained candidate part merely lies inside an imaginary line, the perimeter of the organism, while an enclosed part lies inside a physical structure (e.g., the outer structure), which physically isolates it. The enclosing structure monopolizes the candidate's boundaries, so that all interactions between it and entities outside must go through the enclosing structure; such interactions are likely to be mediated by, and much attenuated by, the enclosing structure. Thus, the candidate part in Fig. 3A is not only isolated by virtue of being contained within the organism, as discussed earlier, but also by virtue of its enclosure within the organism's outer structure. Notice that the outer structure itself (Fig. 3B) does not enjoy this double isolation, but qualifies as a part anyway on account of being contained. The fact that it has a special role in enclosing and isolating other parts does not detract from its partness. Finally, notice that in Fig. 3C, the candidate is further enclosed by another structure, and therefore is a "subpart" rather than a part.

Now consider "contiguity with a compositional difference." Compositional differences between contiguous structures are relevant on the assumption that a change in the type of components that constitute a structure (i.e., a change in composition) is likely to be accompanied by a reduction in integration among them. In Fig. 3D, the candidate is contiguous with another structure, but differs from it in composition and therefore qualifies as a part. The same is true in Figs. 3E and 3F, except that in E the contiguous structure happens to be the organism's outer structure, and in F the candidate is a segment of the outer structure. In both, the candidate is a part by virtue of its distinctive composition. In cheilostome bryozoans, the operculum and the occulsor muscle that closes it are contiguous but differ in composition (see later).

Of course, composition may vary continuously. In principle, quantitative metrics might be used to discover compositional discontinuities, which in turn might correspond to part boundaries. Here compositional variation is assessed subjectively. Notice that in principle, small-scale compositional differences along structures that are compositionally homogeneous at a large scale would constitute "subparts." However, here we are interested only in the parts level and therefore disregard small-scale variation.

2. Shape Differences

Differences in shape might also be construed to indicate reductions in integration corresponding to part boundaries. A shape difference might consist of a local deformation of an object, one that is not compositionally distinctive (at least at a large scale). On this basis, an amoeba's pseudopod might count as a part distinct from the rest of the membrane. In bryozoans, the stomach is, on account of its shape, distinct from the rest of the digestive tract (see later). The use of shape to identify parts seems reasonable; on the other hand, shape differences also seem to be less substantial indicators of change in integration than compositional differences, and here we have chosen to disregard them. As a result, our protocols will undoubtedly miss some parts, but in any case, for comparative purposes, the main requirement is consistency not completeness.

In Figs. 3G and 3H, the gray square marked with an arrow is not a part, because it differs only in shape from the contiguous structures. There is a part in Fig. 3G, however: it is the entire outer structure *including* the gray square. In Fig. 3H, the entire gray shape, again including the gray square, is a part.

3. Duplicate Parts

As discussed, the goal is a list of distinct part *types*. Thus, if an arm of an octopus is a part, it has only one part of that type, not eight. In Fig. 3I, only one of the gray squares counts as a distinct part type. In principle, quantitative metrics can be used to search for discontinuities which might correspond to differences among part types; here, we make the evaluations subjectively.

The application of the protocols to bryozoans is formalized in Fig. 4 as a "parts key." Fig. 5 uses an abstract example to show how the key would classify various structures.

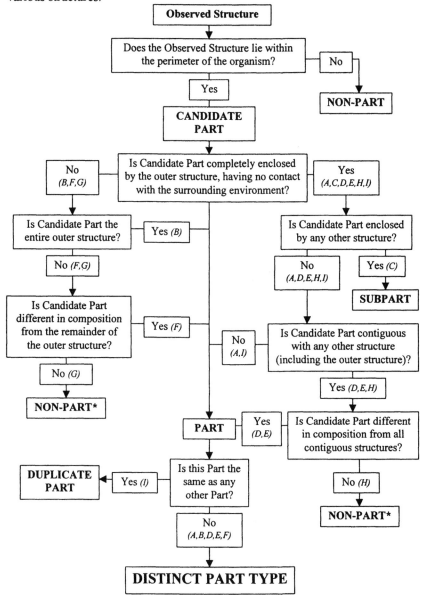

FIGURE 4 A key for identifying parts. A "nonpart" marked with an asterisk has no status as a part (or subpart) but may be a subset, or a region, of some larger entity which is a part (or subpart). Any such larger entities that can be identified are candidates for parts and should be tested using the key. For examples, see "nonparts*" and their likely larger parts (in parentheses) in column 2 of the Appendix.

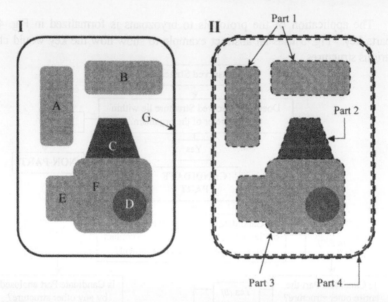

FIGURE 5 Abstract example to illustrate how the key works. Left drawing (I) shows the candidate parts, A-G. Right drawing (II) shows the distinct part types that would be identified by the key (Fig. 4). A and B are parts but are of the same type and therefore count as one distinct type (Part 1). C is a part (Part 2), because it is contiguous with and different in composition from the entire D-F group (Part 3). D is a "subpart" of the D-F group, because it is enclosed by it. E is distinct in shape from F, but the present protocols do not acknowledge shape differences, so E and F are not parts. G is the enclosing structure, which here qualifies as a distinct part (Part 4).

C. Bryozoans

1. Anatomy

Bryozoa is a phylum of usually sedentary colonial marine invertebrates. Colony morphologies are diverse, typically encrusting or branching, many of them calcified. In all species, the majority or totality of the colony is composed of (typically) box- or cylinder-shaped "autozooids," which feed, providing nourishment for the colony. In these autozooids, a tentacled lophophore extends through an orifice into the surrounding water, and cilia on the lophophore channel plankton and other suspended food into a mouth located at the tentacles' intersection. The mouth opens into a U-shaped digestive tract, which loops through the body cavity and terminates in an anus below the mouth and outside the tentacle crown. Lophophore extension is accomplished in various ways, but in almost all species, retraction is effected by a muscle that inserts at the base of the zooid (Fig. 6). Zooid walls are usually perforated in some way to allow interzooidal communication. For further descriptions, see Ryland (1970) and Boardman et al. (1987).

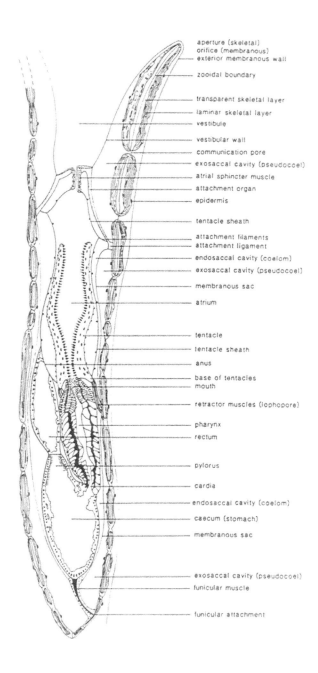

aperture (skeletal)
orifice (membranous)
exterior membranous wall

zooidal boundary

transparent skeletal layer

laminar skeletal layer

vestibule

vestibular wall

communication pore

exosaccal cavity (pseudocoel)

atrial sphincter muscle

attachment organ

epidermis

tentacle sheath

attachment filaments
attachment ligament

endosaccal cavity (coelom)

exosaccal cavity (pseudocoel)

membranous sac

atrium

tentacle

tentacle sheath

anus

base of tentacles
mouth

retractor muscles (lophopore)

pharynx

rectum

pylorus

cardia

endosaccal cavity (coelom)

caecum (stomach)

membranous sac

exosaccal cavity (pseudocoel)

funicular muscle

funicular attachment

FIGURE 6 Longitudinal section of a *C. elegans* autozooid. (From Boardman *et al.*, 1992; reproduced with permission.)

In addition to autozooids, many species also have one or more other differentiated zooid forms, collectively called heterozooids. In many species of cheilostome bryozoans, the most common type of heterozooids are avicularia, which are armed with claw-like mandibles, perhaps defensive in function.

2. Parts in Two Bryozoan Species

Eminooecia carsonae is a calcified, branching bryozoan in the class Gymnolaemata (order: Cheilostomata). *Cinctipora elegans* is also calcified and branching but in the class Stenolaemata (order: Cyclostomata). Part-type lists for two of the zooid types in *E. carsonae* (autozooids and avicularia) and for the one zooid type present in *C. elegans* (autozooids) appear in the Appendix. The lists were derived by applying the key (Fig. 4) to anatomical descriptions and drawings of *Hippadenella carsonae* (reclassified as *E. carsonae* by Hayward and Thorpe, 1988), mainly from Rogick (1957), and of *C. elegans*, from Boardman *et al.* (1992). Figure 6 shows a *C. elegans* autozooid with many of its structures labeled; many, but not all, of the labeled structures qualify as parts (see Appendix).

The following discussion illustrates how the protocols were used to make decisions about part status in certain cases, some of them problematic.

Eminooecia carsonae. As is common in the cheilostomes, *E. carsonae* autozooids have a trapdoor-like device called an operculum, which closes over the orifice when the lophophore is retracted. The operculum attaches to, and is thus contiguous with, a set of occulsor muscle fibers which provide the closing force. The two structures differ in composition: the operculum is composed of cuticle, while the muscle fibers are made of muscle cells. Thus, each is classified as a distinct part.

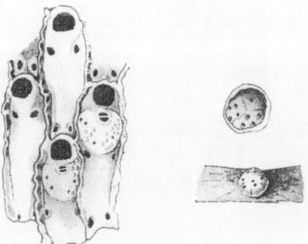

FIGURE 7 Left: arrangement of zooids along a branch in *E. carsonae*. Right: two views of the rosette plate. (From Rogick, 1957; reproduced with permission.)

Zooids in *E. carsonae* are arranged in longitudinal series along the colony branches. Adjacent longitudinal columns are offset by half of a zooid length such that the proximal half of any given zooid is flanked on the right and left by the distal halves of its neighbors (Fig. 7, left). Communication between zooids occurs through four openings in the walls of each zooid, two on each side. The two distal openings are sieve-like structures called rosette plates (Fig. 7, right). The two proximal openings are simple holes ringed by a raised annulus. On account of the offset between zooids and their neighbors, the rosette plate from one zooid aligns with the proximal opening of an adjacent zooid, forming a connection between zooids.

The status of the annulus surrounding the proximal opening and of the rosette plate might seem to be somewhat problematic. Both structures present prominent and distinctive gestalts, strongly suggestive of parts. However, the annulus seems to be composed of the same calcium carbonate as the zooid wall, and therefore is not a distinct part. Likewise, the rosette plate apparently does not differ in composition from the zooid wall, and therefore is not a part. Finally, the hole in the annulus might seem to qualify as a part, in that its composition differs from that of the annulus itself. However, the hole is not an object, and therefore cannot be a part.

Cinctipora elegans. Figure 6 shows the *C. elegans* digestive tract, consisting of a series of candidate parts: mouth, pharynx, cardia, caecum (stomach), pylorus, rectum, and anus. The cells of the pharynx are inflated and those of the upper pharynx are ciliated (Boardman *et al.*, 1992). The pylorus is also ciliated, while the remaining digestive-tract candidate parts are not.

The mouth and anus are openings, not objects, and therefore are not parts. Cell inflation was deemed a sufficient compositional difference for the pharynx to qualify as a distinct part. If ciliation were also deemed sufficient, then the digestive tract would actually consist of five distinct part types: (1) ciliated, inflated-celled, upper pharynx; (2) unciliated, inflated-celled, lower pharynx; (3) unciliated cardia and caecum; (4) ciliated pylorus; and (5) unciliated rectum. However, ciliation was deemed a minor compositional feature, and thus in our scheme, the digestive tract consists of only two parts: (1) inflated-celled pharynx and (2) the remainder of the tract (cardia, caecum, pylorus, and rectum).

SUMMARY

We have attempted to develop a scheme which formalizes a common intuition about the structural organization of organisms, namely, that they are divided into parts. We show how the scheme might be rendered partly operational, in particular, how it can be used to identify certain kinds of parts—object parts—in a consistent way at a hierarchical level just below that of the organism. The assumption is that, for comparative purposes, counts of object parts can be used as a proxy for counts of true parts.

The goal is to devise methods to enable us to answer evolutionary questions in which identification of parts, and counts of part types, would be useful, such as questions concerning the structural complexity of organisms. The scheme is expected to be especially useful for identifying parts in cases in which our precognitive mechanisms are not helpful or are even misleading, cases in which the question dictates the organism and the hierarchical level, and therefore we are not entitled to choose organisms or levels in which the parts are obvious.

Our approach has certain limitations. Although it represents an advance over purely subjective identification of parts, considerable subjectivity remains. Also, our protocols were designed to produce a parts list only for organisms that are apparently structurally quite simple, like bryozoans, and they may need to be modified for other organisms.

Despite these limitations, the approach seems worthwhile, at very least because of the explicitness of the protocols it proposes, and of the assumptions on which they are based. On account of this explicitness, areas where more conceptual work is necessary, or where revision is required, become easier to identify.

Identifying parts in organisms will only rarely be straightforward: part boundaries are often vague, and parts often occupy more than one hierarchical level at once. On the whole, organisms are not organized in cognitively congenial ways, as machines often seem to be, and thus it is not unreasonable to doubt that identification and listing of parts in organisms can be meaningfully done. Still, we are hopeful; we draw our optimism in part from an imaginative study by Schopf *et al.* (1975), who used number of anatomical terms—estimated mainly from glossaries in anatomy texts—as a proxy for morphological complexity, essentially taking a gestalt-based approach to counting parts about as far as possible. Despite the obvious subjectivity of their metric, they documented a correlation between morphological complexity and taxonomic turnover rate. Likewise, for the present scheme, the test of the meaningfulness of the parts list it produces will lie in whether robust correlations can be found with other biologically significant variables.

ACKNOWLEDGMENTS

For discussions, we thank A.M. Carroll, A.J. Dajer, M.A. Gutierrez, M.R. Larochelle, H.F. Nijhout, J. Padgett, E.J. Perkins, D.M. Raup, V.L. Roth, G.P. Wagner, and the participants in the "Agents and Aggregates Working Group" at the Santa Fe Institute's Integrative Themes Workshop (July 1998). We are also indebted to F.K. McKinney, H.F. Nijhout, V.L. Roth, and G.P. Wagner for thoughtful critiques of the manuscript.

APPENDIX

Classification of Various Observed Structures

Cinctipora elegans

candidate part name	status (larger part)	reasons for classification
annular muscle	part	enclosed, contiguous, different comp.
anus	nonpart	empty space
atrial sphincter muscle	part	enclosed, contiguous, different comp.
attachment ligaments	nonpart* (attachment organ)	enclosed, contiguous, not different comp.
attachment organ	part	enclosed, contiguous, different comp.
cardia	nonpart* (digestive tract)	enclosed, contiguous, not different comp.
caecum	nonpart* (digestive tract)	enclosed, contiguous, not different comp.
communication pore	nonpart	empty space
digestive tract (excluding pharynx)	part	enclosed, contiguous, different comp.
epidermis	part	enclosed, contiguous, different comp.
funiculus	part	enclosed, contiguous, different comp.
laminar skeletal layer	nonpart*(skeleton)	outer structure, contiguous, not different comp.
lophophore	part	outer structure, contiguous, different comp.
loph. retractor muscle	part	enclosed, contiguous, different comp.
membranous sac	part	enclosed, contiguous, different comp.
mouth	nonpart	empty space
orifice	nonpart	empty space
peristome	nonpart* (skeleton)	outer structure, contiguous, not different comp.
pharynx	part	enclosed, contiguous, different comp.
pylorus	nonpart* (digestive tract)	enclosed, contiguous, not different comp.
rectum	nonpart* (digestive tract)	enclosed, contiguous, not different comp.
skeleton	part	outer structure, contiguous, different comp.
tentacle	nonpart* (lophophore)	outer structure, contiguous, not different comp.
tentacle sheath	nonpart* (lophophore)	outer structure, contiguous, not different comp.
transparent skeletal layer	nonpart* (skeleton)	outer structure, contiguous, not different comp.
vestibular wall	nonpart* (lophophore)	outer structure, contiguous, different comp.

Autozooid total: 11 parts

Eminooecia carsonae

Autozooids:

part name	class (possible larger part)	reasons for classification
annulus	nonpart* (skeleton)	outer structure, contiguous, not different comp.
anus	nonpart	empty space
cardia	nonpart* (digestive tract)	enclosed, contiguous, not different comp.
caecum	nonpart* (digestive tract)	enclosed, contiguous, not different comp.
cardelle	nonpart* (skeleton)	outer structure, contiguous, not different comp.
digestive tract	part	enclosed, contiguous, different comp.
epidermis	part	enclosed, contiguous, different comp.
esophagus	nonpart* (digestive tract)	enclosed, contiguous, not different comp.
funiculus	part	enclosed, contiguous, different comp.
lophophore	part	enclosed, contiguous, different comp.

mouth	nonpart	empty space
occulsor muscles	part	enclosed, contiguous, different comp.
operculum	part	outer structure, contiguous, different comp.
opercular membrane	nonpart* (operculum)	outer structure, contiguous, not different comp.
oral gland	part	enclosed, contiguous, different comp.
orifice	nonpart	empty space
ovary	part	enclosed, contiguous, different comp.
ovicell	nonpart* (skeleton)	outer structure, contiguous, not different comp.
parietal muscle	part	enclosed, contiguous, different comp.
parieto-diaphagmatic	part	enclosed, contiguous, different comp.
parieto-vaginal band	part	enclosed, contiguous, different comp.
pylorus	nonpart* (digestive tract)	enclosed, contiguous, not different comp.
proximal hole	nonpart	empty space
rectum	nonpart* (digestive tract)	enclosed, contiguous, not different comp.
rosette plate	nonpart* (skeleton)	outer structure, contiguous, not different comp.
sphincter	part	enclosed, contiguous, different comp.
skeleton	part	outer structure, contiguous, different comp.
tentacle	nonpart* (lophophore)	outer structure, contiguous, not different comp.
tentacle sheath	nonpart* (lophophore)	outer structure, contiguous, not different comp.

Autozooid total: 13 parts

Avicularia:

part name	class (possible larger part)	reasons for classification
abductor muscle	part	enclosed, contiguous, different comp.
adductor muscle	part	enclosed, contiguous, different comp.
avicularial gland	part	enclosed, contiguous, different comp.
epidermis	part	enclosed, contiguous, different comp.
mandible	part	outer structure, contiguous, different comp.
mandibular membrane	nonpart* (mandible)	outer structure, contiguous, not different comp.
membrane	part	enclosed, contiguous, different comp.
pivot bar	nonpart* (skeleton)	outer structure, contiguous, not different comp.
rudimentary polypide	part	enclosed, contiguous, different comp.
skeleton	part	outer structure, contiguous, different comp.

Avicularia total: 8 parts

LITERATURE CITED

Bell, G., and Mooers A.O. (1997). Size and complexity among multicellular organisms. *Biol. J. Linn. Soc.* **60**:345-363.

Boardman, R.S., Cheetham, A.H., and Rowell, A.J., eds. (1987). "Fossil Invertebrates." Blackwell, Palo Alto.

Boardman, R.S., McKinney, F.K., and Taylor, P.D. (1992). Morphology, anatomy, and systematics of the Cinctiporidae, New Family (Bryozoa: Stenolaemata). *Smithson. Contrib. Paleobiol.* Number 70.

Bonner, J.T. (1988). "The Evolution of Complexity." Princeton Univ. Press, Princeton.

Brandon, R.N. (1996). "Concepts and Methods in Evolutionary Biology." Cambridge University Press, Cambridge.

Buss, L.W. (1987). "The Evolution of Individuality." Princeton Univ. Press, Princeton.

Campbell, D.T. (1958). Common fate, similarity, and other indices of the status of aggregates of persons as social entities. *Behav. Sci.* 3:14-25.

Cisne, J.L. (1974). Evolution of the world fauna of aquatic free-living arthropods. *Evolution* 28:337-366.

Colless, D.H. (1985). On "character" and related terms. *Syst. Zool.* 34:229-233.

Fristrup, K. (1992). Character: current usages. In "Keywords in Evolutionary Biology" (E.F. Keller and E.A. Lloyd, eds.), pp. 45-51. Harvard University Press, Cambridge.

Ghiselin, M.T. (1997). "Metaphysics and the Origin of Species." State University of New York Press, Albany, New York.

Hayward, P.J., and Thorpe, J.P. (1988). A new family of cheilostome Bryozoa endemic to Antarctica. *Zool. J. Linn. Soc.* 93:1-18.

Hull, D.L. (1980). Individuality and selection. *Annu. Rev. of Ecol. Syst.* 11:311-332.

Kauffman, S.A. (1993). "The Origins of Order." Oxford University Press, New York.

Maynard Smith, J. (1988). Evolutionary progress and levels of selection. In "Evolutionary Progress" (M.H. Nitecki, ed.), pp. 219-230. University of Chicago Press, Chicago.

Maynard Smith, J., and Szathmáry, E. (1995). "The Major Transitions in Evolution." Freeman, Oxford.

McShea, D.W. (1993). Evolutionary change in the morphological complexity of the mammalian vertebral column. *Evolution* 47:730-740.

McShea, D.W. (1996a). Complexity and homoplasy. In "Homoplasy: The Recurrence of Similarity in Evolution" (M.J. Sanderson and L. Hufford, eds.), pp. 207-225. Academic Press, San Diego.

McShea, D.W. (1996b). Metazoan complexity and evolution: Is there a trend? *Evolution* 50:477-492.

McShea, D.W. (In press). Parts and integration: consequences of hierarchy. In "Process from Pattern in the Fossil Record" (F.K. McKinney, S. Lidgard, and J.B.C. Jackson, eds.). University of Chicago Press, Chicago.

McShea, D.W. (In press). Functional complexity in organisms: parts as proxies. *Biol. Philos.*

Mishler, B.D., and Brandon, R.N. (1987). Individuality, pluralism, and the phylogenetic species concept. *Biol. Philos.* 2:397-414.

Mittenthal, J.E., Baskin, A.B., and Reinke, R.E. (1992). Patterns of structure and their evolution in the organization of organisms: modules, matching, and compaction. In "Principles of Organization in Organisms, Santa Fe Institute Studies in the Sciences of Complexity, Proceedings" (J. Mittenthal and A. Baskin, eds.), Volume XIII, pp. 321-332. Addison-Wesley, Reading, MA.

Olson, E., and Miller, R. (1958). "Morphological Integration." University of Chicago Press, Chicago.

Raff, R.A. (1996). "The Shape of Life." University of Chicago Press, Chicago.

Rogick, M.D. (1957). Studies on marine Bryozoa. X. Hippadenella carsonae, n. sp. *Biol. Bull.* 112:120-131.

Roth, V.L. (1991). Homology and hierarchies: problems solved and unresolved. *J. Evol. Biol.* 4:167-194.

Ryland, J.S. (1970). "Bryozoans." Hutchinson & Co., London.

Salthe, S.N. (1985). "Evolving Hierarchical Systems." Columbia University Press, New York.

Salthe, S.N. (1993). "Development and Evolution." MIT Press, Cambridge.

Schopf, T.J.M., Raup, D.M., Gould, S.J., and Simberloff, D.S. (1975). Genomic versus morphologic rates of evolution: influence of morphologic complexity. *Paleobiology* 1:63-70.

Valentine, J.W., Collins, A.G., and Meyer, C.P. (1994). Morphological complexity increase in metazoans. *Paleobiology* 20:131-142.

Valentine, J.W., and May, C.L. (1996). Hierarchies in biology and paleontology. *Paleobiology* 22:23-33.

Wagner, G.P. (1996). Homologues, natural kinds and evolution of modularity. *Am. Zool.* 36:36-43.

Wagner, G.P., and Altenberg, L. (1996). Complex adaptations and evolution of evolvability. *Evolution* **50**:967-976.

Wimsatt, W.C. (1974). Complexity and organization. *In* "Philosophy of Science Association 1972" (K.F. Schaffner and R.S. Cohen, eds.), pp. 67-86. D. Reidel Publishing Co., Dordrecht, Holland.

Wimsatt, W.C. (1994). The ontology of complex systems: Levels of organization, perspectives, and causal thickets. *Can. J. Philos*, Supplementary Volume **20**:207-274.

12

BEHAVIORAL CHARACTERS AND HISTORICAL PROPERTIES OF MOTOR PATTERNS

PETER C. WAINWRIGHT[1] AND JOHN P. FRIEL[2]

[1]*Department of Evolution and Ecology, University of California, Davis, CA 95616*
[2]*Cornell University Museum of Vertebrates, Department of Ecology and Evolutionary Biology, Ithaca, NY 14853*

INTRODUCTION

Organismal behavior has long been a controversial source of systematic characters (Brown, 1975; Masterton *et al.*, 1976; Brooks and McLennan, 1991; de Queiroz and Wimberger, 1993; Houch and Drickamer, 1996) that has never enjoyed the popularity of anatomy or DNA. This lack of popularity has both practical and philosophical causes. The chief practical obstacle is the difficulty in obtaining adequate observations on large numbers of taxa (Wenzel, 1992; Greene, 1994). Some authors have argued that behavioral characters are fundamentally poorly suited to use in systematics because they are plastic and often variable (Klopfer, 1969, 1975; Atz, 1970; Hailman, 1976; Aronson 1981), but in this chapter we take the point of view that neither attribute eliminates behavior as a source of useful characters (see also Lauder, 1986, 1994; Wenzel, 1992; de Queiroz and Wimberger, 1993; Greene, 1994). Although the effort involved in obtaining sufficient observations to assess behavioral characters in large numbers of species is likely to limit their widespread utility in systematics, we anticipate that studies of the evolution of focal behavioral traits will enjoy a dynamic future.

In this chapter we explore the properties of one type of behavioral data in some detail, motor patterns, when viewed in a phylogenetic context. In effect we ask if motor pattern characters distilled from behavioral observations can be shown empirically to contain phylogenetic information. We cover two major issues. First, we ask whether motor patterns (patterns of muscle activation that underlie movement) meet the criteria for potentially useful systematic characters. Second, two decades of comparative research on the evolution of the lower vertebrate skull and the behaviors associated with the skull permit a number of useful generalizations about the properties of these motor pattern variables and how to deal with them when searching for historical patterns. Our thesis is that behavioral variables are frequently phylogenetically informative but they require special attention to their description and variability. We illustrate the major conclusions that have come from research in this area with examples from our own research on behavioral evolution in tetraodontiform fishes.

The key property of a systematic character that establishes its utility in phylogenetic analysis is that similarity among taxa is due to homology. Thus a fundamental issue regarding the potential use of behavioral traits in systematics revolves around the problem of establishing homology of behaviors. This topic has been discussed recently by others (Lauder, 1986, 1994; Wenzel, 1992; Greene, 1994). Here we concur with Wenzel (1992) and take the view that "determining homology among behaviors is no different than determining homology among morphological structures". As with any character type, the utility of behavioral characters is best tested in the context of the total suite of characters that can be brought to bear on a phylogenetic problem (Patterson, 1982). Useful or informative characters will be those that tend to unite natural groups, and show relatively little convergence. Characters that tend to show unique states in each taxon of a clade, like any autapomorphy, will contribute little to resolving phylogenetic relationships.

The effort involved in obtaining thorough behavioral observations on large numbers of taxa will greatly limit the utility of behavior as a source of characters for use in systematic analyses. For this reason it is not our intention to argue for the extensive use of behavior in systematics, but instead we emphasize that phylogenies can be used to study the history of those behavioral traits that have the properties of good systematic characters. It is in the use of the phylogenetic perspective to study the evolution of animal behavior and it's component elements that we expect to see the greatest returns from viewing behavioral traits as systematic characters (Greene and Burghardt, 1978; Paul, 1981a, 1981b, 1991; Lauder, 1986, 1983; McLennan et al., 1988; Gittleman, 1989; Schweck and Throckmorton, 1989; Losos, 1990; McLennan, 1991; Brooks and McLennan, 1991; Schultz, 1992; Alves-Gomez and Hopkins, 1997; Wainwright and Turingan, 1997).

There are many ways that behaviors can be measured in such a fashion that the resulting metrics can be treated as characters with discrete states (Wenzel, 1992; Greene, 1994). As behaviors are particularly complex with several underlying components, one approach is to consider the various levels of activity that create animal movement (Paul, 1991; Stridter and Northcutt, 1991). Reilly and Lauder (1992) drew a useful distinction between kinematic patterns (physical movement), motor patterns (activity of muscles that cause movement), and morphology as three levels of animal behavior. Although theoretical considerations permit predictions about how modifications of morphology in specific systems are likely to affect kinematic patterns given a constant motor pattern (Barel, 1983; Westneat, 1990; Richard and Wainwright, 1995), it remains a relatively unstudied question the extent to which these components of behavior are independent (Reilly and Lauder, 1992; Smith, 1994). In this chapter we focus our examples and comments on muscle activity patterns because a substantial body of literature exists that examines historical changes in muscle activity patterns, but we expect our comments to apply more broadly to the evolution of kinematics and other components of behavior.

We use the term "motor pattern" to refer to patterns of muscle contraction that drive the behavior under study. Muscular contraction can be documented by several methods, ranging from direct measurements of muscle shortening to extracellular electrical recordings of muscle cell depolarizations, or electromyography. Electromyograms (Fig. 1) include information about the time course of nervous stimulation of muscles and the intensity of the stimulation (Loeb and Gans, 1986). By simultaneously determining the activity of several muscles, the relative activity period for each muscle can be determined. Often it is this sequence of muscle contraction that is of primary interest in studies of motor pattern. The data that we discuss here will all be of this latter type.

Triggerfish Feeding on Fiddler Crab

Levator Operculi

100 μv

100 ms

Adductor Mandibulae Section: A1α

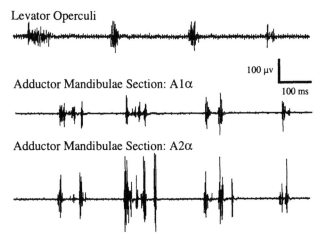

Adductor Mandibulae Section: A2α

Figure 1 Sample electromyogram of three muscles from a gray triggerfish, *Balistes capriscus,* feeding on a fiddler crab (*Uca* sp.). Each channel shows a record of voltage against time from a different muscle during several cycles of buccal manipulation behavior.

We distinguish motor patterns from "kinematic patterns," or the movements associated with animal behavior. Motor patterns represent the nervous stimulation of muscles that create kinematics. Note that some authors have used the term "motor pattern" to refer to what we define as kinematic pattern (Marler and Hamilton, 1966; Kardong, 1997), but our intention is to distinguish between these two elements of animal behavior. Candidate kinematic parameters that may evolve and be valuable characters include any aspect of motion that can be defined in context and measured. Thus, phylogenetically meaningful patterns could potentially be found in a variety of traits such as the angular excursion of a limb element during running, the extent of opening of the mouth during prey capture in a fish, or the time course and sequence of events in a complex movement such as tentacle projection by cephalopod molluscs during prey capture. Kinematic data can be collected by recording film or video sequences of animals executing behaviors. Image-by-image analysis of motion can then be conducted on the recordings to produce data on the extent, direction, and time course of motion. Typical kinematic variables distilled from recordings of prey capture in lower vertebrates might be the angular rotation of the jaws, the extent of mouth opening, the angle of dorsal rotation of the head, and the extent of forward body movement during the prey capture sequence. As noted earlier, motor patterns interact with the mechanical properties of the stiff elements of anatomy (e.g., bones or muscular hydrostats) to produce kinematic patterns.

Regardless of what type of behavioral trait is being studied there are a number of advantages to quantifying the qualities of interest and subjecting them to analyses that capitalize on the wealth of tools embodied within parameteric statistics (Zar, 1984; Sokal and Rohlf, 1985). Traits that are measured are more easily defined, and as behaviors are frequently influenced by numerous external stimuli it is crucial to be able to partition sources of variance when comparing behaviors across taxa.

EVOLUTION OF MOTOR PATTERNS IN THE SKULL OF TETRAODONTIFORM FISHES

We present two examples from our studies of behavioral evolution to illustrate the methods for analyzing motor patterns that have been developed in the past two decades and to illustrate some general results that have emerged concerning the evolution of muscle activation patterns. The first example involves a comparison of the activity pattern of a jaw adductor muscle in four species of tetraodontiform fishes. This case study focuses on the activity pattern of a single muscle and is used to illustrate steps in evaluating motor pattern evolution: methods of quantifying the activation pattern of the muscle, methods for comparing trait values across species, and the biological insights gained from such a comparison.

In the second example we review a study of the evolution of inflation behavior in pufferfishes and the role of motor pattern changes in the acquisition of this extraordinary behavior. This case study places motor pattern evolution into the broader context of behavioral evolution and illustrates several general results that have emerged from studies of motor pattern evolution: motor patterns are conservative during evolution and yet distinct modifications can be identified that had a key role in the origin of this novel behavior.

Activity Patterns of the Adductor Mandibulae Muscle in Tetraodontiform Fishes

The teleost order Tetraodontiformes is a cosmopolitan group of marine fishes made up of nine families, including such recognizable and divergent groups as the triggerfishes, (Balistidae), filefishes, (Monacanthidae), boxfishes, (Ostraciidae), pufferfishes, (Tetraodontidae and Diodontidae), and ocean sunfishes (Molidae). Tetraodontiform feeding is characterized by the use of a small mouth equipped with stout teeth and jaws that are controlled by well-developed musculature to grasp and reduce prey. In using direct oral jaw biting extensively during feeding these fishes differ from generalized teleosts which typically rely heavily on suction feeding for prey capture (Turingan and Wainwright, 1993).

Figure 1 shows a typical electromyographic recordings of the adductor mandibulae section 2α (hereafter A2α) and two other jaw muscles during four cycles of manipulation of a crab prey by a gray triggerfish, *Balistes capriscus*. Feeding triggerfish, like other tetraodontiforms, repeatedly bite their prey as they attempt to remove small pieces and generally reduce the prey to segments that can easily be swallowed. This behavior, termed buccal manipulation, involves repeated cycles of activity in the jaw adductor muscles. Each cycle of biting is associated with a discrete burst of activity in the adductor muscle. The activity pattern of the adductor mandibulae muscle and its antagonists drive this behavior and interact with the skeletal system to produce the biting movements.

We focus here on activity of section A2α during buccal manipulation. We quantified each burst of A2α activity by measuring the duration of the burst, the average amplitude of the electrical spikes (intensity), and the timing of the burst relative to a reference muscle, the levator operculi, which is active during jaw opening.

For this analysis we recorded and analyzed electromyograms during feeding by individuals of four tetraodontiform species: the planehead filefish *(Monacanthus hispidus)*, the gray triggerfish *(Balistes capriscus)*, the Southern puffer *(Sphoeroides nephalus)*, and the striped burrfish *(Chilomycterus schoepfi)* (see also Friel and Wainwright, 1998). Recordings were made from four individuals per species (for burrfish N = 3) feeding on three prey types. The prey, live panaeid shrimp, live fiddler crabs, and pieces of squid mantle, were chosen to present different escape abilities and toughness during biting. An average of 80 cycles of buccal manipulation behavior were recorded from each individual fish feeding on each prey type for a total of 6,837 cycles of activity analyzed in the entire data set. We compared the motor pattern variables across four species using an experimental design that partitioned variance from several sources: variation among species, among individuals, among prey type, and the interactions between the main effects. The overall experimental design was a two-way ANOVA (Species crossed with Prey Type) with Individuals nested within Species. This design also includes two interaction terms, Species X Prey Type and Individuals X Prey Type.

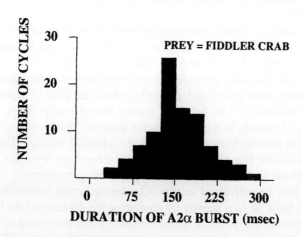

Figure 2 Frequency histogram of the burst duration of the A2a muscle. Data represent 93 cycles of muscle activity from four individual gray triggerfishes, *Balistes capriscus*, feeding on fiddler crabs (*Uca* sp.). Note that the value of this motor pattern trait is highly variable, as is typical of EMG variables taken from fish feeding behaviors. The variable has a mean of 152 milliseconds and a standard deviation of 52, for a coefficient of variation of 0.34.

The frequency distribution of a typical variable (e.g., burst duration) from this analysis is shown in Fig. 2. Note the broad range of values and high variance found for the burst duration of the A2α muscle. Table I reports the ANOVA results for the Species and Prey Type main effects run on the three variables that describe the activity pattern of the A2α muscle. There are significant species effects on the onset time and duration of the burst of activity, but not on the intensity of the burst. Post-hoc pairwise comparisons revealed that for both onset time and burst duration the burrfish differs from each of the other three species, but these other three species do not differ from each other.

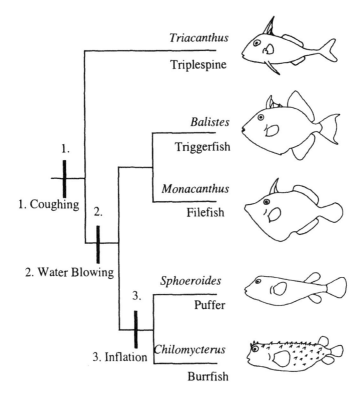

FIGURE 3 Phylogenetic relationships of the five tetraodontiform species studied by Wainwright and Turingan (1997). The distribution of the three oral compression behaviors is indicated by the numbered bars. 1. Coughing behavior is found in all tetraodontiform species as well as other teleost fishes. 2. Water blowing behavior is found in triggerfishes, filefishes, cowfishes (not shown), and members of the puffer lineage. 3. Inflation behavior is exhibited by all members of the two pufferfish families, the Tetraodontidae and the Diodontidae. As tetraodontiforms appear to retain a more plesiomorphic ability to cough, there are no anatomical specializations for this behavior in the order. There are also no anatomical novelties associated with water blowing behavior in the group above *Triacanthus*. In sharp contrast, inflation behavior is associated with a sweet of major morphological modifications, including hypertrophied muscles and bones, and kinetic pectoral girdle and a novel mechanism for depression of the floor of the mouth.

If we consider A2α onset time and burst duration to be two characters, each can be resolved into two character states from this analysis. On average, burrfish exhibit a significantly longer burst of activity in the A2α that has a later onset than seen in the other three species. If we "map" these two characters on a phylogeny of the four species (see Fig. 3 for a phylogeny of these taxa) it becomes apparent that activity of the A2α muscle is conserved through most of the tetraodontiform clade, while both traits show state changes and are autapomorphic for burrfish. At present neither character would contribute to resolving questions about relationships between these four taxa because both show a single state change that is unique to the burrfish. In other words, these characters do not suggest natural groupings within the four taxa. However, the characters do suggest the hypothesis that burrfishes and other diodontids exhibit a derived pattern of activity in the A2α muscle. This hypothesis could be explored in additional species within the burrfish clade.

These data illustrate three general features of motor patterns of the lower vertebrate skull. (1) Motor patterns are highly variable, (2) individual fish exhibit considerable control over them, and yet despite these two points (3) motor patterns show a strong tendency for conservation across species.

Figure 2 is a histogram of 93 cycles of buccal manipulation activity from four gray triggerfish feeding on fiddler crabs. Burst duration of the A2α muscle ranged from 54 to 298 milliseconds, with a mean of 154 ms and a standard deviation of 52. The coefficient of variation of this variable is 0.34. Despite this variability the triggerfish mean of 154 ms was found to be significantly different from the burrfish mean of 218 ms, though not different from the filefish or pufferfish means (119 and 132 ms, respectively).

The significant effects of prey type on burst onset and burst duration illustrate a result that is typical to studies of the effects of prey on motor patterns. Virtually every teleost fish species that has been studied quantitatively has shown the ability to modify muscle activation patterns in response to prey type (Liem, 1979; Sibbing, 1982; Lauder, 1983; Sibbing et al., 1986; Wainwright and Lauder, 1986; Sanderson, 1988; Wainwright and Turingan, 1993; Ralston and Wainwright, 1997; Wainwright, 1989; Friel and Wainwright, 1998). As indicated earlier, an important implication of this result is that interspecific comparisons of muscle activity must either account for prey type affects or hold this variable constant. In some cases species also differ in how they respond to prey type (Friel and Wainwright, 1998), a more subtle way that motor patterns may differ between taxa.

The result that activity of the A2α muscle is unchanged in three of the taxa also highlights a general finding. Previous interspecific studies of muscle activity patterns in feeding lower vertebrates have shown that, within fairly closely related groups (e.g., confamilial species), feeding motor patterns tend to be strongly conserved in that tests of the species effect with ANOVAs typically find significance in fewer than 15% of EMG variables (see also Wainwright, 1989; Lauder and Shaffer, 1993; Smith, 1994).

Evolution of inflation behavior in pufferfishes

Our second case study concerns the motor basis of pufferfish inflation behavior and its evolutionary origins (Wainwright and Turingan, 1997). All species in the two tetraodontiform families Tetraodontidae (smooth-skinned puffers) and Diodontidae (spiny puffers) have the remarkable ability to inflate their body with water, a behavior these slow swimming fish use to deter predators. Inflation is accomplished by the fish pumping several mouthfuls of water into its stomach. The stomach of puffers and the ventral skin of the body are modified to permit the tremendous volume change and expansion that occur as the animal inflates itself (Brainerd, 1994; Wainwright et al., 1995). However, here we focus on the mechanism of the oral pump that is used by puffers to repeatedly draw water into the mouth and pump it through the esophagus and into the stomach.

The sequence of events that occurs in a single pumping cycle begins with expansion of the oral cavity and opening the mouth to permit water to flow in. The mechanism of oral expansion is a novelty of pufferfishes and involves a suite of substantial anatomical modifications that permit the pectoral girdle to swing posteriorly, simultaneously with depression of the ventral region of the mouth by flexion of the hyoid apparatus. Once the oral cavity is full of water these actions are reversed: the mouth is closed and the ventral mouth region and pectoral girdle are elevated and protracted respectively. A stout oral valve is located just behind the jaws and helps to prevent water from escaping back out the mouth during oral compression. Instead the esophagus is relaxed and water is pumped through it into the stomach. This is followed immediately by another cycle of the same actions.

Among tetraodontiform fishes there are at least two additional behaviors that have similarities to the inflation pump and are phylogenetically more broadly distributed. All tetraodontiform fishes, indeed virtually all fishes, possess the ability to "cough." This is a behavior in which fishes expel unwanted material from the mouth by rapidly compressing the oral cavity and using the water flow to carry the debris away. Many tetraodontiform fishes also actively "blow" water jets out of their mouth and use the pulses of water to manipulate the environment. Perhaps the most spectacular use of water blowing is seen in many triggerfishes who will use water jets to expose prey that are buried in sand, and to manipulate large prey such as sea urchins that must be rolled over to expose the more vulnerable oral surface before the fish can successfully attack and consume them (Fricke, 1971, 1975; Frazer, 1991). Like inflation behavior, coughing and water blowing involve a cyclical pattern of oral expansion, when water is drawn into the mouth, and oral compression, when water is forced out of the mouth. The key difference is that during coughing and water blowing the mouth is held open during the compression phase, to permit water to rapidly exit the mouth, while the mouth is closed during the compressive phase of inflation to prevent water loss out the mouth.

The distribution of these three behaviors on a simplified phylogeny of tetraodontiform fishes (Fig. 3) suggests that the three behaviors may represent a transformation series. Coughing is seen in all tetraodontiform taxa, and more broadly in other teleosts as well. Water blowing is not known to occur in the Triacanthoidea, the sister group to all other tetraodontiforms. However, it does occur in all other tetraodontiform families. Finally, inflation has the most restricted distribution, being found only in puffers. A hypothesis that follows from the phylogenetic distribution of these behaviors is that water blowing and inflation behavior are increasingly specialized behaviors that evolved from coughing. This hypothesis has been explored by mapping the anatomical and motor pattern traits associated with these behaviors onto an independently derived cladogram of tetraodontiform fishes (Wainwright and Turingan, 1997).

There are five major results from the quantitative comparisons of muscle activity patterns and anatomical specializations associated with these three bahaviors in the five tetraodontiform taxa shown in Fig 3. (1) The motor patterns of the three behaviors were similar, with only minor differences distinguishing them. (2) The motor pattern for coughing did not differ among the five tetraodontiform taxa. (3) The motor pattern for water blowing did not differ between the two species studied (gray triggerfish and southern puffer) but differed from the coughing motor pattern by a change in a single motor pattern character, the burst duration of one of the muscles that compresses to mouth cavity (the AP; Fig. 4). (4) The motor pattern for inflation differed from the water blowing pattern in showing a briefer burst of activity in the mouth opening muscle (the LOP; Fig. 4). (5) There appear to be no anatomical specializations associated with coughing or water blowing but there are numerous major anatomical modifications associated with inflation behavior.

In this case the motor pattern characters clearly support the nested phylogeny of this group of five taxa (Fig. 3), although we note that variation within each component clade (the different families) has not been explored. The broad similarity in motor pattern across the three behaviors supports the hypothesis that the three behaviors share a common functional basis with slight modifications to create water blowing and inflation from the primitive and ubiquitous coughing pattern. However, the three behaviors can be distinguished from each other by a single modification in each case (Fig. 4).

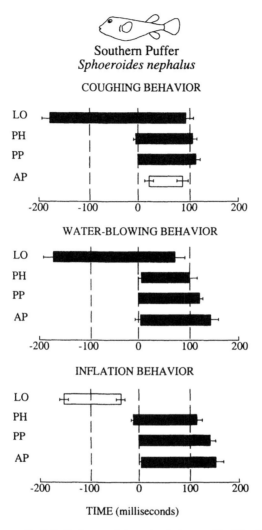

FIGURE 4 Patterns of activity for four cranial muscles during coughing (N = 17), water blowing (N = 19), and inflation behavior (N = 48) in four southern puffers, *Sphoeroides nephalus*. The length of bars indicates the mean burst duration of the muscle. Standard errors of burst duration are indicated to the right, while standard errors of relative onset time (onset times measured relative to the PP) are indicated to the left. Analyses of variance revealed that the motor pattern for these three behaviors is quite similar, but a single variable distinguishes coughing for the other two behaviors (AP burst duration) and a second variable distinguishes inflation from coughing and water blowing (LO burst duration). The LO muscle is the primary mouth opening muscle in these fishes and the reduced activity of it during inflation is associated with a major kinematic difference between the behaviors: during coughing and water blowing the LO muscle continues to contract into the compressive phase (indicated by activity of the PH, PP, and AP muscles), holding the mouth open so that water can be forced anteriorly. During inflation the mouth is closed during the compressive phase to prevent water from escaping out the mouth. This is indicated by the cessation of LO activity prior to the compressive phase.

HISTORICAL PROPERTIES OF MOTOR PATTERNS

We have illustrated with examples from our tetraodontiform research the observation that motor pattern parameters of the skull in aquatic lower vertebrates are highly variable within individuals and species. Further, this variability is under considerable control by the individual fish. Fish demonstrate an ability to respond to a wide range of environmental factors by modifying the motor patterns used during feeding behaviors. External factors such as prey type, prey position, and satiation level have all been shown to have significant effects on muscle activity patterns. Thus, the motor patterns involved in feeding are highly plastic and under considerable sensory control by the individual.

In the face of this tendency for high variability in motor patterns, quantitative analyses of feeding and related behaviors have, paradoxically, repeatedly found a strong tendency for patterns of muscle contraction to be conservative within fairly closely related groups of species, such as within families (Shaffer and Lauder, 1985; Wainwright and Lauder, 1986; Sanderson, 1988; Wainwright, 1989; Turingan and Wainwright, 1993; Friel and Wainwright, 1998). Although motor pattern characters are quite variable within individuals and within species, mean values tend to be similar across species. For example, our analysis of the activity of the A2α muscle in four species of tetraodontiform fishes revealed that burst duration and burst onset time were significantly greater in the burrfish than in the other three species, but that there was no significant difference among the other three species in any of the three variables (Table I). The simplest interpretation of these results given the phylogenetic relationships among these taxa is that the average activity pattern of the A2α muscle has been conserved across much of the tetraodontiform radiation.

Burst duration of the A2α had means of 154, 119, and 132 ms, respectively, in the three species but variance was high in each species with coefficients of variation around 0.3 for every variable. Thus, high variance makes it difficult to distinguish the mean trait values in these three species. In fact, with sufficiently high sample sizes it may be possible to demonstrate that the mean burst durations differ. Calculations of power in these ANOVAs suggest that samples of over 30 individuals per species would be needed to distinguish these means, given the observed variability of the motor pattern. Nevertheless, reasonable sample sizes rarely show more than 10–15% of EMG variables differing significantly among closely related species while broader phylogenetic comparisons inevitably show a greater number of differences in motor pattern (e.g., Sanderson, 1988; Wainwright et al., 1989; Wainwright and Turingan, 1997). Those differences in motor pattern that do occur, as in our analysis of muscle activity in oral compression behaviors by tetraodontiform fishes, are usually consistent with independent phylogenetic hypotheses. Empirically then, motor pattern traits associated with the skull tend to be conserved, though changes that are observed frequently support independently derived phylogenies. No cases have yet to be reported of the convergent evolution of a distinct, derived motor pattern.

Table I Summary of Prey Type and Species effects on EMG variables of the A2a section of the adductor mandibulae muscle in triggerfishes *(Balistes)*, filefishes *(Monacanthus)*, pufferfishes *(Sphoeroides)*, and burrfishes *(Chilomycterus)*. Effects are based on univariate ANOVAs of EMG data recorded during prey processing of live fiddler crabs, pieces of squid tentacle, and live shrimps. Onsets for muscles are relative to the activity of the levator operculi. See also Friel and Wainwright, 1998.

EMG Variable	Prey Effect			Species Effect		
	F ratio	df	P	F ratio	df	P
Relative Onset	7.39	2,20	<0.01	4.24	3,10	0.04*
Duration	5.02	2,20	0.02*	7.96	3,10	<0.01
Relative Intensity	0.95	2,20	0.40	2.91	3,10	0.09

* = significant effects at £ 0.05 level

Can a behavioral trait that is highly plastic be meaningfully argued to be evolutionarily conserved? We suggest that as long as (1) the conditions under which the trait is measured are controlled and defined and (2) the conclusion of conservation is based on a quantitative analysis that accounts for the measurable sources of variation, then findings of no significant interspecific difference are accurately interpreted as a conserved character (see also Wenzel, 1992; Greene, 1994; Lauder, 1994). The high intraspecific variance of motor pattern variables does suggest that increased sampling of individuals within species is likely to enhance the ability to discriminate species means that differ by up to 50%. This variability suggests that relative conservativeness or historical lability of behavioral traits should be based on observations from comparable numbers of individuals per species. However, the key point is that, even if intraspecific variation is high, adequate experimental design can result in a robust test of the hypothesis that the motor pattern trait is the same across the study taxa.

Why are average motor pattern trait values relatively conservative across species, given their high variability and the control that animals have over them? Particularly when the anatomy of the feeding mechanism is as different among species as in the case of the tetraodontiforms reported here (Fig. 3) one might expect to see modifications of the motor pattern that suit species-specific anatomy. Nevertheless, there are numerous striking examples of conserved motor patterns in clades of fishes that have undergone extensive anatomical evolution of the feeding apparatus and patterns of prey use (Sanderson, 1988;

Westneat and Wainwright, 1989; Wainwright and Lauder, 1992). Only in a few cases have workers been able to attribute a historical origin of a novel feeding ability or significant shift in feeding performance to changes in motor pattern (Liem, 1980; Lauder, 1983).

The best documented example of changes in motor pattern playing a central role in the origin of a novel feeding behavior concerns the origin of molluscivory in the freshwater sunfishes, Family Centrarchidae (Lauder, 1983). Two species of sunfish consume mostly molluscs in the wild, redear and pumpkinseed sunfish. These species are believed to be sister taxa (Mabee, 1993). In addition to possessing hypertrophied muscles and jaws used in mollusc crushing these species also exhibit a modified motor pattern that is used during the cracking of mollusc shells (Lauder, 1983).

Other work with centrarchids has found that the motor pattern used in prey capture has undergone only minor modification across a sample of four species that encompass the phylogenetic, morphological, and ecological range within the family (Wainwright and Lauder, 1986). For example, the motor pattern used during the capture of fish prey does not differ between the largemouth bass and bluegill sunfish, despite the fact that the morphology, feeding abilities, and patterns of prey use are different in these two species. Adult largemouth bass are specialized piscivores and bluegill are generalized invertebrate predators. Numerous phenotypic differences between the species have been identified that underlie the differences in feeding ability (summarized in Wainwright and Lauder, 1992). The body form and swimming abilities differ and the jaws of the largemouth bass are approximately 50% larger than the bluegill, have a different shape, and the mechanical properties of the lever system in the mandible differ between the two species (Wainwright and Richard, 1995). However, the two species use the same pattern of muscle activity when feeding on the same prey, despite a well-documented ability to modulate the motor pattern. This suggests the possibility that, across the rather broad range of anatomical designs of the feeding apparatus in sunfishes, the most effective motor pattern for prey capture is the same. Similarly, although the planehead filefish, gray triggerfish, and Southern puffer differ considerably in feeding morphology, they utilize the same activity pattern of the A2α muscle.

Analyses of motor patterns often find differences among individuals within species to be a major source of variation, as in the studies reported here. However, it should be pointed out that to our knowledge there has never been an empirical attempt to measure the heritability of a motor pattern or kinematic trait. This would be a worthy research direction as a key assumption concerning characters that are used in systematics is that the variation one documents among taxa is heritable. It would also be valuable to know the extent to which motor pattern and kinematic traits are heritable as this will determine their potential for responding to natural selection.

We have presented one approach to quantifying and comparing behavior. An important goal of future research will be to compare the properties of motor patterns to other types of behavioral traits, such as kinematic variables. Existing studies of feeding behaviors in aquatic feeding lower vertebrates suggest that kinematics show levels of variation comparable to, or slightly less than those seen in motor pattern traits (Shaffer and Lauder, 1985; Richard and Wainwright, 1995). For example, the time taken to completely open and close the jaws during prey capture in a representative 78-mm spotted sunfish, *Lepomis punctatus* (N = 108 feeding sequences), varied from 18 to 72 ms with a mean of 46 ms and a coefficient of variation of 0.24 (Wainwright, unpublished observations). Reilly and Lauder (1992) found comparable levels of variation in kinematic and motor pattern variables in several families of salamanders. It is tempting to believe that the profile of a single feeding sequence captures the key elements of the behavior, but the studies cited here of motor patterns and kinematics show the importance of adequately assessing variance in behavioral variables.

LITERATURE CITED

Alves-Gomes, J., and Hopkins, C. D. (1997). Molecular insights into the phylogeny of mormyriform fishes and the evolution of their electric organs. *Brain Behav. Evol.* **49**:324-351.

Aronson, L. R. (1981). Evolution of Telencephalic function in lower vertebrates. *In* "Brain Mechanisms of Behavior in Lower Vertebrates" (P.R. Laming, ed.), pp. 33-58. Cambridge University Press, Cambridge.

Atz, J. W. (1970). The application of the idea of homology to animal behavior. *In* "Development and Evolution of Behavior: Essays in Honor of T. C. Schneirla" (L. R. Aronson, E. Tobach, D. S. Lehrman, and J. S. Rosenblatt eds.), pp. 53-74. W. H. Freeman, San Franscisco.

Barel, C. D. N. (1983). Towards a constructional morphology of the cichlid fishes (Teleostei, Perciformes). *Neth. J. Zool.* **33**:357-424.

Brainerd, E. L. (1994). Pufferfish inflation: functional morphology of postcranial structures in *Diodon holocanthus* (Tetraodontiformes). *J. Morphol.* **220**:243-262.

Brooks D. R., and McLennan, D. A. (1991). "Phylogeny, Ecology and Behavior." University of Chicago Press, Chicago.

Brown, J. (1975). "The Evolution of Behavior." W. W. Norton, New York.

de Queiroz, A, and Wimberger, P. H. (1993). The usefulness of behavior for phylogeny estimation: levels of homoplasy in behavioral and morphological characters. *Evolution* **47**:46-60.

Frazer, T. K., Lindberg, W. J., and Stanton, G. R. (1991). Predation on sand dollars by gray triggerfish, *Balistes capriscus*, in the northeastern Gulf of Mexico. *Bull. Mar. Sci.* **48**:159-164.

Fricke, H. W. (1971). Fische als feinde tropischer seeigel. *Mar. Biol.* **9**:328-338.

Fricke, H. W. (1975). Losen einfacher probleme bei einem fisch (freiswasser an *Balistes fuscus*). *Zool. Tierpsychol.* **38**:18-33.

Friel, J. P., and Wainwright, P. C. (1998). Evolution of motor pattern in Tetraodontiform fishes: does muscle duplication lead to functional diversification? *Brain Behav. Evol.* **53**:159-170.

Gittleman, J. L. (1989). The comparative approach in ethology: aims and limitations. *In* "Perspectives in Ehtology" (P. P. G. Bateson, and P. H. Klopfer, eds.), pp. 55-83. Plenum Press, New York.

Greene, H. W. (1994). Homology and behavioral repertoires. *In* "Homology: The Hierarchical Basis of Biology" (B.K. Hall, ed.), pp. 369-381. Academic Press, San Diego.

Greene, H. W., and Burghardt, G. M. (1978). Behavior and phylogeny: constriction in ancient and modern snakes. *Science* **200**:74-77.

Hailman, J. P. (1976). Homology: logic, informaiton and efficiency. *In* "Evolution, Brain and Behavior: Persistent Problems" (R.B. Masterton, W. Hodos, and H. Jerison, eds.), pp. 181-198. Lawrence Erlbaum, Hillsdale, New Jersey.

Houck, L. D., and Drickamer L. C. (1996). "Foundations of Animal Behavior: Classic Papers with Commentaries." University of Chicago Press, Chicago

Kardong, K. V. (1997). Evolution of a motor pattern within squamates: the tell tale tongue. *Am. Zool.* **37(5):81A.**

Klopfer, P. (1969). Review of R. F. Ewer, Ethology of Mammals. *Science* **165:887.**

Klopfer, P. (1975). Review of J. Alcock, Animal Behavior: An Evolutionary approach. *Am. Sci.* **63:578-579.**

Lauder, G. V. (1983). Functional and morphological bases of trophic specialization in sunfishes (Teleostei: Centrarchidae). *J. Morphol.* **178:1-21.**

Lauder, G. V. (1986). Homology, analogy and the evolution of behavior. *In* "Evolution and Animal Behavior: Paleontological and Field Approaches" (M. H. Nitecki and J. A. Kitchell, eds.), pp. 9-40. Oxford University Press, Oxford.

Lauder, G. V. (1994). Homology, form and function. *In* "Homologies: The Hierarchical Basis of Comparative Biology" (B. K. Hall, ed.), pp. 151-196. Academic Press, San Diego.

Lauder, G.V., and Shaffer, H.B. (1993). Design of feeding systems in aquatic vertebrates: major patterns an their evolutionary interpretations. *In* "The Skull" (J. Hanken and B. K. Hall, eds.), Vol. 3. Univ. of Chicago Press, Chicago, IL.

Liem, K. F. (1979). Modulatory multiplicity in the feeding mechanism in cichlid fishes, as exemplified by the invertebrate pickers of Lake Tanganyika. *J. Zool. Lond.* **189:93-125.**

Loeb, G.E., and Gans, C. (1986). "Electromyography for Experimentalists." University of Chicago Press.

Losos, J. B. (1990). Concordant evolution of locomotor behavior, display rate and morphology in *Anolis* lizards. *Anim. Behav.* **39:879-890.**

Mabee, P. M. (1993). Phylogenetic interpretation of ontogenetic change: sorting out the actual and artefactual in an empirical case study of centrarchid fishes. *Zool. J. Linn. Soc.* **107:175-291.**

Marler, P., and Hamilton, W. J. (1966). "Mechanisms of Animal Behavior." John Wiley, New York.

Masterton, R. B., Hodos W., and Jerison, J. (1976). "Evolution, Brain, and Behavior: Persistent Problems." Lawrence Erlbaum Associates, Hillsdale, N.J.

McLennan, D. A. (1991). Integrating phylogeny and experimental ehtology: from pattern to process. *Evolution* **45:1173-1789.**

McLennan, D. A., Brooks, D. R., and McPhail, J. D. (1988). The benefits of communication between comparative ehtology and phylogenetic systematics: a case study using gasterosteid fishes. *Can. J. Zool.* **66:2177-2190.**

Patterson, C. (1982). Morphological characters and homology. *In* "Problems of Phylogenetic Reconstruction" (K. A. Joysey and A. E. Friday, eds.), pp. 21-74. Academic Press, New York.

Paul, D. H. (1981). Homologies between body movements and muscular contractions in the locomotion of two decapods of different families. *J. Exper. Biol.* **94:159-168.**

Paul, D. H. (1981). Homologies between neuromuscular systems serving different functions in two decapod of different families. *J. Exper. Biol.* **94:169-197.**

Paul, D. H. (1991). Pedigrees of neurobehavioral circuits: tracing the evolution of novel behaviors by comparing motor patterns, muscles and neurons in members of related taxa. *Brain Behav. Evol.* **38:226-239.**

Ralston, K. R., and Wainwright, P. C. (1997). Functional consequences of trophic specialization in pufferfishes. *Funct. Ecol.* **11:43-52.**

Reilly, S. M., and Lauder, G. V. (1992). Morphology, behavior, and evolution: comparative kinematics of aquatic feeding in salamanders. *Brain Behav. Evol.* **40:182-196.**

Richard, B. A., and Wainwright., P. C. (1995). Scaling the feeding mechanism of largemouth bass *(Micropterus salmoides):* kinematics of prey capture. *J. Exp. Biol.* **198:419-433.**

Sanderson, S. L. (1988). Variation in neuromuscular activity during prey capture by trophic specialists and generalists (Pisces: Labridae). *Brain Behav. Evol.* **32**:257-268.

Schultz, J. W. (1992). Muscle firing patterns in two arachnids using different methods of propulsive leg extension. *J Exper. Biol.* **162**:313-329.

Schwenk, K., and Throckmorton, G. S. (1989). Functional and evolutionary morphology of lingual feeding in squamate reptiles: phylogenetics and kinematics. *J. Zool. London* **219**:153-175.

Shaffer, H. B., and Lauder, G. V. (1985). Aquatic prey capture in ambystomatid salamanders: patterns of variation in muscle activity. *J. Morphol.* **183**:273-284.

Sibbing, F. A. (1982). Pharyngeal mastication and food transport in the carp *(Cyprinus carpio):* a cineradiographic and electromyographic study. *J. Morphol.* **172**:223-258.

Sibbing, F. A., Osse, J. W. M., and Terlow, A. (1986). Food handling in the carp *(Cyprinus carpio):* its movement patterns, mechanisms and limitations. *J. Zool. Lond.* **210**:161-203.

Smith, K. K. (1994). Are neuromuscular systems conserved in evolution? *Brain Behav. Evol.* **43**: 293-305.

Sokal, R. R., and Rohlf, F.J. (1985). "Biometry." Freeman and Co., New York.

Striedter, G. F., and Northcutt, R. G. (1991). Biological heirachies and the concept of homology. *Brain Behav. Evol,* **38**:177-189.

Turingan, R. G., and Wainwright, P. C. (1993). Morphological and functional bases of durophagy in the queen triggerfish, *Balistes vetula* (Pisces, Tetraodontiformes*). J. Morphol.* **215**:101-118.

Wainwright, P. C. (1989). Prey processing in haemulid fishes: patterns of variation in pharyngeal jaw muscle activity. *J. Exp. Biol.* **141**:359-376.

Wainwright, P. C., and Lauder, G. V. (1986). Feeding biology of sunfishes: patterns of variation in the feeding mechanism. *Zool. J. Linn. Soc.* **88**:217-228.

Wainwright, P.C., Sanford, C. J.. Reilly, S. M., and Lauder, G. V. (1989). Evolution of motor patterns: aquatic feeding in salamanders and ray-finned fishes. *Brain Behav. Evol.* **34**:329-341.

Wainwright, P. C., and Turingan, R. G. (1993). Coupled versus uncoupled functional systems: motor plasticity in the queen triggerfish *Balistes vetula. J. Exp. Biol.* **180**:209-227.

Wainwright, P. C., and Turingan, R. G. (1997). Evolution of pufferfish inflation behavior. *Evolution* **51**:506-518.

Wainwright, P. C., Turingan, R. G., and Brainerd, E. L. (1995). Functional morphology of pufferfish inflation: mechanism of the buccal pump. *Copeia* **1995**:614-625.

Wenzel, J. W. (1992). Behavioral homology and phylogeny. *Annu. Rev. Ecol. Syst.* **23**:361-381.

Westneat, M. W. (1990). Feeding mechanics of teleost fishes (Labridae: Perciformes): a test of four-bar linkage models. *J. Morphol.* **205**:269-295.

Zar, J. H. (1984). "Biostatistical Analysis." Prentice-Hall, Englewood Cliffs.

13

HOMOLOGY AND DNA SEQUENCE DATA

WARD WHEELER

Department of Invertebrate Zoology, American Museum of Natural History, New York, NY 10024

INTRODUCTION

The phylogenetic analysis of DNA sequences, like that of all other comparative data, is based on schemes of putative homology which are then tested via congruence to determine synapomorphy schemes and cladistic relationships. Unlike some other data types, however, the matrix of putative homologies or "characters" is not directly observable. When sequences are unequal in length, the correspondences among sequence positions are not preestablished and some sort of procedure is required to determine which positions are "homologous." This is the traditional province of multiple sequence alignment (= alignment here). Alignment generates a collection of column vectors through the insertion of gaps, which form the character set. Whether accomplished manually, or via some computational algorithm, these characters are then submitted to phylogenetic analysis in the same manner as other forms of data. This scheme of correspondences or putative homologies has two salient features. First, alignment precedes the phylogenetic analysis (i.e., cladogram search) and is never revised in light of systematic hypotheses. Second, alignment-based homology schemes rest on a notion of base-to-base homology where individual nucleotide bases transform among five states (A, C, G, T/U, and gap) within a single character. Two methods have recently been proposed ("Optimization-Alignment," Wheeler, 1996 and "Fixed-State Optimization," Wheeler, 1999) which avoid multiple alignment altogether and question these two tenets of sequence analysis. Although these approaches are parsimony methods, and rely on testing homology through synapomorphy, they differ in the entities they propose for testing and this has implications for the interpretation of DNA sequence homology.

In discussing these concepts, a shorthand will be used. To describe those correspondences among states frequently referred to as putative homologies, the lowercase "homology" will be used. To describe those correspondences that have been tested through congruence on a cladogram (i.e., synapomorphy), the uppercase "Homology" will be used. The discussion here is mainly concerned with methods of deriving homology statements, but all of these would then be tested with other data to determine which homologies are Homologies.

STATIC VERSUS DYNAMIC HOMOLOGY

The standard precursor to the phylogenetic analysis of DNA sequences is alignment. This procedure takes the unequal length strings of nucleotide bases and inserts place-holding gaps ("-") to make the corresponding (homologous) bases line up into intelligible columns. These columns (characters) comprise the data used to reconstruct cladograms. However this alignment is created, once phylogenetic analysis has begun, it will not be revised. That is, the homologies explicitly defined in the alignment will not be reexamined during the cladogram-search process.

Consider four sequences: I GGGG, II GGG, III GAAG, and IV GAA. An alignment can be generated to be supplied to standard phylogenetic analysis. In this case, insertion-deletion events are given a cost of two and base substitutions one. The most parsimonious (minimum cost) cladogram relating these four taxa would be that which holds I and III to be sister taxa with an overall length of six (1 indel and 4 base changes—Fig. 1.) Given this alignment, the two other phylogenetic scenarios are less favored (7 and 8 steps). There is another alignment, however, which generates the same minimum length for topology ((((I III) II) IV) yet yields the same length (6 steps) for one of the other two possible topologies (Fig. 2). Using this alignment, two of the topologies are equally parsimonious.

		Topology		
Alignment		((I II) III) IV)	(((I III) II) IV)	(((I IV) II) III)
I	GGGG	7	6	8
II	--GGG			
III	GAAG			
IV	--GAA			

Insertion-Deletion events cost 2
Base changes cost 1

FIGURE 1 Possible alignment for four simple sequences and the cladogram cost (length) for the possible topologies for these taxa.

The point of this example is that the alignment process yields static homology schemes which is not optimized for any particular topology. Once the alignment is determined, all testing of the alignment itself stops. Although homologies are tested on each cladogram, there may be no single homology matrix which optimizes Homology (yields the most parsimonious result) for each cladogram. In order to give each topology its shortest length, homologies need to be generated which are optimal for that particular topology. It is this need which motivates the method of "Optimization-alignment" (Wheeler, 1996).

		Topology		
Alignment		((I II) III) IV)	(((I III) II) IV)	(((I IV) II) III)
I	GGGG	7	6	8
II	--GGG			
III	GAAG			
IV	--GAA			
I	GGGG	6	6	8
II	GGG--			
III	GAAG			
IV	GAA--			

Insertion-Deletion events cost 2
Base changes cost 1

FIGURE 2 Comparison of the implications of two different alignments on the cladogram costs for the sequences in Fig. 1.

In the method of optimization-alignment, the diagnosis of each cladogram attempts to find the lowest cost hypothetical ancestral sequences possible. This is accomplished by examining all possible homologies between the nucleotide bases of the two descendent nodes. Dynamic programming is used (in a step akin to pairwise sequence alignment) to optimize each HTU sequence for the minimum weighted number of insertion-deletion events and base substitutions (Fig. 3). At each node down the cladogram, all possible hypothetical ancestral sequences are implicitly constructed and their costs determined. The minimum cost sequence is retained and used to optimize the next node down the cladogram. This algorithm is greedy (only using descendent sequence information) and can overestimate the real cladogram length (as described by Wheeler, 1996, 2000). The case of the four example sequences shows the local optimization of homology (Fig. 4). Since no *a priori* homology statements are made, each cladogram can create those homologies most advantageous to its topology. Hence, the two best cladograms at length 6 arise without fuss.

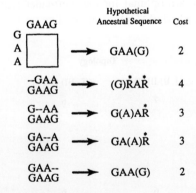

FIGURE 3 An example of HTU sequence optimization at an internal node via the optimization-alignment (Wheeler, 1996) procedure.

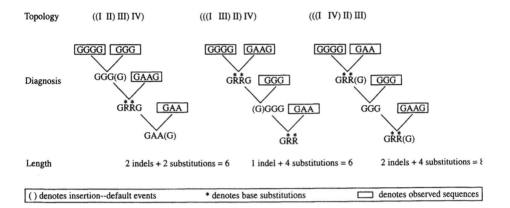

FIGURE 4 Optimization results for the three topologies and sequences of Fig. 1. Note the possible ambiguities in HTU sequence determination.

Searching for Homology can directly test this dynamic optimization scheme. Since the method attempts to more efficiently derive homologies, and Homologies, it can be tested by parsimony. Cladograms should be more parsimonious and homoplasy less prominent with dynamic homology. Although this will have to be tested by multiple data sets, the cases presented by Wheeler and Hayashi (1998) on chelicerates and the examples in Wheeler (2000) and here show this pattern of more parsimonious solutions for dynamic homology than for static alignments.

BASE-TO-BASE VERSUS FRAGMENT HOMOLOGY

Both the static and dynamic homologies mentioned earlier rely on a notion of homology which derives from nucleotide base correspondences. This need not be the case, however. Homology could be viewed as a phenomenon existing at the level of the sequences themselves as opposed to base-to-base statements. This view sees entire strings of DNA nucleotides as characters. Such entities as the small subunit rDNA locus could be a single character. The locus would then be homologous among all the taxa, and the actual observed sequences themselves would constitute the character states. In the context discussed earlier, this would constitute a "static" homology scheme since the character vectors would be preordained. The complexity of the sequences would allow for homology statements more like those of other forms of character analysis, where position and complexity aid in character delimitation.

	Seq. 1	Seq. 2	Seq. 3
Seq. 1	0	C_{21}	C_{31}
Seq. 2	C_{12}	0	C_{32}
Seq. 3	C_{13}	C_{23}	0

$$C_{ij} = C_{ji}$$

FIGURE 5   Matrix of minimum transformation cost between sequence pairs.

When employing blocks of contiguous sequence as characters, with observed sequences as states, dynamic programming methods must be used to optimize cladograms and determine their length. The procedure is identical to the optimization of Sankoff-style characters ("step-matrix" characters), just modified for large numbers of states ($n_{states} \leq n_{taxa}$). This approach relies on the postulate that only observed sequences may be optimized to hypothetical ancestors. This restricts the possible world of reconstructed sequences, but also requires that these sequences exist. Each sequence becomes a state in an extremely complex character. The first step in the optimization procedure is the determination of the transformation cost matrix among all the states. This is defined as the minimum transformation cost (including all forms of base substitutions and insertion-deletion costs) between each pair of states (Fig. 5). Once these transformation costs are known, standard dynamic programming implicitly examines the assignment of each of these states to each internal node and determines the optimal set of states and cladogram length (Fig. 6).

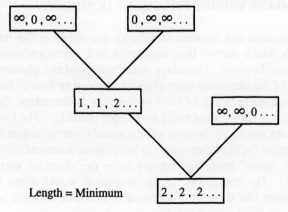

FIGURE 6   An example of down-pass cladogram optimization via the fixed-state approach.

Given that the cladogram length is based on nucleotide sequence, it might seem strange to say that the bases themselves are not homologous. This effect is derived from the pairwise nature of the character transformation matrix. Consider three sequences I AAATTT, II TTT, and III AAA. When transformation costs are determined, the first "T" in sequence II (position 1) corresponds with the first "T" of sequence I (position 4). This same "T" in sequence I also corresponds with the first "A" of sequence III (position 1). If our logic were transitive, this would imply that position 4 of sequence I would correspond to position 1 of sequence III. It does not. Position 1 of sequence III ("A") corresponds to position 4 of sequence I (also "A"). No circle of correspondence can be drawn among these nucleotides describing state transformations. They are not homologous (Fig. 7).

FIGURE 7 Scheme of base correspondences implied by the fixed-state approach.

Two other aspects of this approach affect ideas of homology. One of the salient features of base-to-base methods, whether built upon static or dynamic homologies, is difficulty in tracing complex homologies through the cladogram (or alignment)—in other words, messy data. When there is extensive sequence length variation coupled with base changes, tremendous uncertainty in homology can occur in both multiple-alignment and optimization alignment. The requirement that such variation be accommodated over the entire cladogram can make local uncertainties propagate throughout the analysis. Since the sequence level homology approach transforms the complex states with their variations in length and nucleotide base composition into simple numbered states with pairwise costs, this problem does not occur. Such seemingly confusing variation patterns will certainly lead to longer cladograms, but the homologies (at the fragment level) will remain clear.

A second feature of fragment level homology is the requirement that the character homologies be defined *a priori*. Whether entire loci, structurally or functionally defined regions are employed as homologies, they are determined by the investigator. This is akin to the delimitation of variation in complex morphological features. Are complex structures such as complete development in the endopterygote insects single or multiple characters? As with all such seemingly arbitrary decisions, what matters most is the effect of changing these character delimitations on phylogenetic results.

The notion of synapomorphy as a shared derived feature might also seem to be altered by the homology concept implicit in sequence fragment comparisons. Since each taxon may well express a unique character state, it might appear that synapomorphy (as a shared state) would be impossible. This criticism would only apply if the characters were completely unordered. State transformation costs are not equal among states, hence are more akin to synapomorphy in the context of ordered characters. Two taxa might present states 1 and 2 of an ordered series $0 \rightarrow 1 \rightarrow 2$. These taxa are united by the transformation implied by the ordering with 1 and 2 sharing special derived similarity not found in 0 (Platnick, 1979). The concept of synapomorphy (or Homology) is unaffected by the fixed-state approach.

COMPARISONS

For these distinctions (static versus dynamic; base-to-base versus fragment) to be anything more than nomenclature, some means of comparing these methods and judging superiority must be offered. Congruence could be that measure. When analyzing single data sets, via whatever method, the best solution is that which minimized discord among data (i.e., characters). This may be measured by simplicity (parsimony) or with respect to complex statistical models (likelihood). The three methods of viewing homology here define characters in somewhat different ways, hence simple counting of change for single data sets (i.e., cladogram length) cannot be used. The things that are counted are just not the same. This notion of character congruence, however, can be extended to the broader concept of congruence among data sets. Character congruence has been used to discriminate among analysis parameters (Wheeler, 1995; Whiting *et al.*, 1997; Wheeler and Hayashi, 1998) and could reasonably be used to compare the behavior of methods (although numerous other means could also be employed).

Two types of congruence measures can be used: character based and topological. The relative merits and demerits of these approaches have been explored in the literature (Mickevich and Farris, 1981; Wheeler, 1995) and character congruence will be used here due to its link with parsimony and combined data analysis. Phylogenetic methods are judged to be superior if they accommodate variation in multiple data sets efficiently as measured by the Mickevich-Farris incongruence length metric (Mickevich and Farris, 1981).

EXAMPLE—ARTHROPODS

In order to compare these three homology-determination methods, the arthropod data of Wheeler *et al.* (1993) are used. These data consist of 100 morphological characters, ~650 18S rDNA nucleotides, and 228 Ubiquitin nucleotides. To these data ~350 28S rDNA nucleotides were added. The 18S and Ubiquitin data were determined for 25 extant taxa and the morphological data scored for these taxa and "Trilobita," an extinct clade. The 28S rDNA data were determined for 15 of the extant taxa (Table I).

TABLE I Taxon List

Mollusca		
	Cephalopoda	*Loligo pealei*
	Polyplacophora	*Lepidochiton cavernae*
Annelida		
	Polycheata	*Glycera sp.*
	Oligocheata	*Lumbricus terrestris*
	Hirudinea	*Haemopis marmorata*
Onychophora		
	Peripatoidae	*Peripatus trinitatis*
	Peripatopsidae	*Peripatoides novozealandia*
Trilobita		groundplan of Ramsköld and Edgecombe, 1991.
		(morphological analysis only)
Chelicerata		
	Pycnogonida	*Anoplodactylus portus*
	Xiphosura	*Limulus polyphemus*
	Scorpiones	*Centruroides hentzii*
	Uropygi	*Mastogoproctus giganteus*
	Araneae	*Nephila clavipes*
	Araneae	*Peucetia viridans*
Crustacea		

	Cirrepedia	*Balanus sp.*
	Malacostraca	*Callinectes sp.*
Myriapoda		
	Chilopoda	*Scutigera coleoptrata*
	Diplopoda	*Spirobolus sp.*
Hexapoda		
	Zygentoma	*Thermobius sp.*
	Ephemerida	*Heptagenia sp.*
	Odonata	*Libellula pulchella*
	Odonata	*Dorocordulia lepida*
	Dictyoptera	*Mantis religiosa*
	Auchenorrhyncha	*Tibicen sp.*
	Lepidoptera	*Papilio sp.*
	Diptera	*Drosophila melanogaster*

Three analyses were performed. In each case, the insertion-deletion cost was set at two and all base substitutions set at one. When morphological characters were used, character transformations were set at two. In the first analysis, the data were aligned (via MALIGN; Wheeler and Gladstein, 1994) and phylogenetic analysis was performed using PHAST (Goloboff, 1996). The second analysis employed optimization-alignment as implemented in POY (Gladstein and Wheeler, 1996). The third used the fixed-state optimization technique also as implemented in POY. Gaps/indels were included and given the same weight (2) in all length calculations. All searches employed TBR branch swapping and 10 random addition sequences. The results of the individual data partitions, combined results, and congruence calculations are summarized in Table II and Figs. 8-10.

TABLE II Comparison of Methodologies

Data	Alignment	Method Optimization-alignment	Fixed-state
18S rDNA	503	501	584
Ubiquitin	387[1]	392	484
28S rDNA	919	848	943
Morphology	252[2]	252	252
Combined	2123	2007	2271
Incongruence[3]	0.0292	0.00698	0.00352

[1] This length of 387 steps is shorter than that of the optimization-alignment purely due to the treatment of ambiguities. When all ambiguities are treated as missing data, both alignment (MALIGN-PHAST) and optimization-alignment (POY) yield the same length of 387 steps.

[2] This length is 2 times the length of 126 steps.

[3] Calculated as (Combined − 18S rDNA − Ubiquitin − 28S rDNA − Morphology)/Combined.

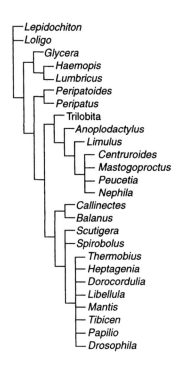

FIGURE 8　Morphologically based cladogram of arthropod relationships (Wheeler *et al.*, 1993).

The dynamic homology approach of optimization-alignment resulted in more parsimonious cladograms in all the cases where sequences were unequal in length. This is due, no doubt, to the simultaneous optimization of synapomorphy and homology uniquely for each topology. The cladograms derived from the fixed-state approach were the longest. The restriction on the possible range of internal node (HTU) sequences is responsible for this. Since internal node sequences are chosen from the range of observed terminal sequences, longer cladograms frequently arise (Wheeler, 1999). Overall character incongruence was lowest (0.00352 vs. 0.00698 and 0.0292) for the fixed-state analysis.

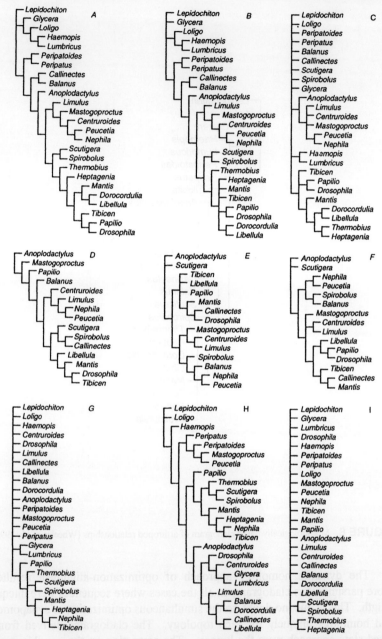

FIGURE 9 Cladograms of individual data partitions when subjected to different analytical techniques. A. 18S rDNA and multiple sequence alignment. B. 18S rDNA and optimization-alignment. C. 18S rDNA and fixed-state optimization. D. 28S rDNA and multiple sequence alignment. E. 28S rDNA and optimization-alignment. F. 28S rDNA and fixed-state optimization. G. Ubiquitint and multiple sequence alignment. H. Ubiquitin and optimization-alignment. I. Ubiquitin and fixed-state optimization.

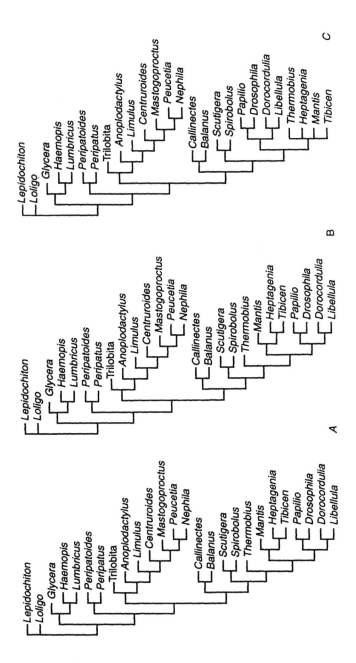

FIGURE 10 Cladograms of combined data (18S rDNA, Ubiquitin, 28S rDNA, morphology) for arthropod taxa when subjected to different analytical techniques. A. Multiple sequence alignment. B. Optimization-alignment. C. Fixed-state optimization.

DISCUSSION

Clearly, the way we view sequence homology has tremendous implications for the elucidation of phylogenetic pattern. The three modes discussed here (static, dynamic, and fragment level) imply different patterns of relationship in the small test case used here. Furthermore, since the reconstructions of hypothetical ancestral sequences vary with the method, the types of evolutionary events reconstructed on these patterns differ as well.

An additional feature of the base-to-base methods, the nonindependence of the nucleotide characters, remains largely unexplored. In both alignment and optimization-alignment, the homology scheme for each nucleotide is determined in concert with all the other bases that surround it. The relative position and number of indels and nucleotide substitutions in adjacent sequence positions fundamentally affect positional homology. Clearly, such character statements are not independent. However, when cladograms are constructed, the cost of changes and indels are summed linearly over the data—an assumption of rigid character independence. Since the individual bases play no role in homology with the fixed-state approach, this character dependence problem vanishes. The indels and base substitutions only determine the cost of transformation between states, there is no requirement that these changes be independent. This inconsistency of base-to-base homology is avoided.

With character incongruence levels at half or lower levels than the other techniques and character nonindependence removed, the fixed-state approach to sequence homology is clearly worth consideration.

ACKNOWLEDGMENTS

I would like to acknowledge the contributions of Daniel Janies, Gonzalo Giribet, Norman Platnick, Lorenzo Prendini, Randall Schuh, Susanne Schulmeister, and Mark Williams to this work through discussion and abuse. I would also like to thank Portia Rollins for expert art work.

LITERATURE CITED

Gladstein, D. S., and Wheeler W. C. (1997). "POY: The Optimization of Alignment Characters." Program and Documentation. New York, NY. Available at "ftp.amnh.org" /pub/molecular.

Goloboff, P. (1996). PHAST. Program and Documentation. Version 1.5.

Mickevich, M. F., and Farris, S. J. (1981). The implications of congruence in *Menidia*. *Syst. Zool.* **30:**351-370.

Platnick, N. I. (1979). Philosophy and the transformation of cladistics. *Syst. Zool.* **28:**537-546.

Ramsköld, L., and Edgecombe, G. D. (1991). Trilobite monophyly revisited. *Hist. Biol.* **4:**267-283.

Wheeler, W. C. (2000). Heuristic reconstruction of hypothetical-ancestral DNA sequeces: sequence alignment versus direct optimization. *In* "Homology and Systematics: Coding Characters for Pylogenetic Analysis" (R. W. Scotland, ed.), pp. 106-113. Taylor and Francis, London.

Wheeler, W. C. (1999). Fixed character states and the optimization of molecular sequence data. *Cladistics* **15**:379-385.

Wheeler, W. C. (1996) Optimization alignment: the end of multiple sequence alignment in phylogentics? *Cladistics* **12**:1-10.

Wheeler, W. C. (1995). Sequence alignment, parameter sensitivity, and the phylogentic analysis of molecular data. *Syst. Biol.* **44**:321-332.

Wheeler, W. C., and Gladstein, D. S. (1994). MALIGN: A multiple sequence alignment program. *J. Hered.* **85**:417.

Wheeler, W. C., and Gladstein, D. M. (1992-1996). Malign: A Multiple Sequence Alignment Program. Program and Documentation. New York, NY. available ftp.amnh.org /pub/molecular/malign

Wheeler, W. C., and Hayashi, C. Y. (1998). The phylogeny of the chelicerate orders. *Cladistics* **24**:173-192.

Wheeler, W. C., Cartwright, P., and Hayashi, C. (1993). Arthropod phylogenetics: a total evidence approach. *Cladistics* **9**:1-39.

Whiting, M. F., Carpenter, J. C., Wheeler, Q. D., and Wheeler, W. C. (1997). The Strepsiptera problem: phylogeny of the holometabolous insect orders inferred from 18S and 28S ribosomal DNA sequences and morphology. *Syst. Biol.* **46**:1-68.

14

CHARACTER POLARITY AND THE ROOTING OF CLADOGRAMS

HAROLD N. BRYANT

Royal Saskatchewan Museum, Regina, Saskatchewan, Canada S4P 3V7

INTRODUCTION

Character polarity refers to the cladistic relationship among character states within characters, the identification of the member of a pair of alternative character states that is relatively plesiomorphic, and the member that is relatively apomorphic. Character polarity is fundamental to cladistic analysis because it identifies the synapomorphies that diagnose clades or monophyletic groups on cladograms. Within an evolutionary context, polarity refers to the direction of character state transformations, which in turn identifies the character state phylogeny within characters.

The Character Concept in Evolutionary Biology

In this chapter I provide some historical perspective to the issue of character polarity, review the rationale and assumptions of the most widely used methods for polarizing characters, and consider the relationship between polarity and the rooting of trees. I also discuss issues that have been central to debates and controversy regarding the validity and applicability of these methods within a cladistic framework. Much of this debate has been concerned with the use of ontogenetic evidence to polarize characters and, secondarily, the role of stratigraphic (paleontological) evidence.

HISTORICAL BACKGROUND

The methods or criteria considered valid or appropriate for polarizing characters depend on the axiomatic assumptions ("unproblematic background knowledge" of Lakatos, 1970) of the particular systematic approach. In the precladistic era (e.g., Maslin, 1952; Simpson, 1961; Mayr, 1969, 1974; Crisci and Stuessy, 1980) a wide variety of methods were used to polarize characters, including not only various versions of outgroup comparison and the ontogenetic method, the most widely accepted methods today, but also methods that are now generally considered either invalid or highly unreliable. There was much reliance on the fossil record (e.g., Simpson, 1961; Gingerich, 1976; Szalay, 1977), and functional or adaptational criteria that, based on assumptions of particular evolutionary processes, were used to infer the most probable direction of character transformations (e.g., Simpson, 1961; Bock, 1977; de Jong, 1980; Gutmann, 1981; Bishop, 1982).

Since the advent of phylogenetic systematics (Hennig, 1966) there has been an increasing tendency to avoid assumptions that are based on particular evolutionary processes, and hence the polarization methods that depend upon them, because phylogenetic analysis is now seen as primarily an investigation of pattern, rather than process (e.g., Hennig, 1966; Eldredge and Cracraft, 1980; Wiley, 1981; Fink, 1982; Ax, 1987). If process is avoided in the inference of pattern, circularity is avoided in the use of that pattern to test hypotheses of process (e.g., Eldredge and Cracraft, 1980). Within a theoretical framework and methodology that is based on the axiom that biodiversity is the product of evolution (see also Wiley, 1975; Bonde, 1977; Gaffney, 1979), but avoids assumptions associated with specific evolutionary processes, only outgroup comparison, the ontogenetic method, and, to a lesser extent, the paleontological method have been widely accepted as appropriate methods of polarity determination (e.g., Eldredge and Cracraft, 1980; Wiley, 1981; Eldredge and Novacek, 1985). These three methods are valid within this axiomatic framework (de Queiroz, 1985; Bryant, 1992).

A   character state distribution

ontogenetic stage	taxa 1	2	3	4	5
X	·A	A	A	B	B
Y	A	A	A	A	A

B   Outgroup Comparison

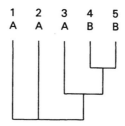

C   Generality Criterion (e.g., Ontogenetic Method)
     Paleontological Method

FIGURE 1   Polarity determination in an ingroup (taxa 3, 4, and 5) using outgroup comparison, the generality criterion (e.g., the ontogenetic method) and the paleontological method. Taxa 1 and 2 are outgroups. **A.** Character state distribution: At ontogenetic stage X, taxa 1, 2, and 3 display character state A and taxa 4 and 5 display character state B; at ontogenetic stage Y all taxa display state A. A is more generally distributed than B within the ingroup because A is found in all taxa, whereas B occurs in only a subset of those taxa. Outgroup comparison and the paleontological method involve comparisons at ontogenetic stage X. **B.** Outgroup comparison: The occurrence of only A in the outgroups results in an unequivocal inference that this state is plesiomorphic within the ingroup; B is considered a synapomorphy for (4, 5). **C.** Generality criterion (e.g., the ontogenetic method; see text): B is less generally distributed within the ingroup and therefore using Weston's rule (see text) B is inferred to be a synapomorphy for (4, 5); using Nelson's rule (see text) polarity can be inferred only if ontogenetic stage Y precedes stage X. Paleontological method: If the earliest known fossils in the ingroup have character state A then this state is considered plesiomorphic within the group and state B is considered a synapomorphy for (4, 5). See Table I for the auxiliary assumptions of these methods. This figure is modified from Fig. 1 in Bryant (1992).

Pattern cladists (Platnick, 1979; Nelson and Platnick, 1981; Patterson, 1982; Rieppel, 1988; Williams *et al.*, 1990) reduce the reliance on assumptions still further. All assumptions associated with evolution, including the basic assumption that evolution has occurred, are seen as unnecessary for the inference of hierarchical pattern, and systematics becomes completely independent of evolutionary theory (Brady, 1985). As a result, the use of paleontology to polarize characters is considered invalid because of its explicit evolutionary assumption that plesiomorphy precedes apomorphy in the fossil record. Outgroup comparison is considered an "indirect" method because of its reliance on assumptions of hierarchical pattern at higher systematic levels (Nelson, 1973, 1978). Ontogeny is considered the primary evidence for polarity and rooting. The perceived parallelism between ontogeny and order in nature is seen as a "direct" method of pattern analysis (Rieppel, 1990). Although the debate between phylogenetic systematists and pattern cladists has not been resolved, most systematists do rely primarily on outgroup comparison, and to a lesser degree ontogeny, to root cladograms. Both methods can be justified on the assumption that organismic diversity forms a nested taxonomic hierarchy. The controversial issue is the degree to which this assumption can or should be independent of the assumption that evolution is the source of that hierarchy (e.g., de Queiroz and Donoghue, 1990).

RATIONALE OF METHODS FOR POLARIZING CHARACTERS

The rationales for outgroup comparison, the ontogenetic method, and the paleontological method are outlined here and illustrated in Fig. 1; their auxiliary assumptions are listed in Table I. The validity of the auxiliary assumptions determines the applicability or utility of each method in a particular instance (Eldredge and Novacek, 1985; Bryant, 1992); these assumptions include necessary conditions implicit in the rationale of the method or the adequacy of available information (see Bryant, 1992, for detailed discussion). Other methods, common equals primitive and midpoint rooting in particular, are discussed briefly.

A. Outgroup Comparison

In outgroup comparison the polarity among alternative character states within the group of interest (the ingroup) is inferred from the distribution of those character states in one or more additional taxa that are assumed to be cladistically outside the group of interest (the outgroups). In the simplest case, if only one of two character states found within the ingroup is found in the outgroups, the most parsimonious inference is that the state found in both the ingroup and the outgroups is plesiomorphic and that the state found only in the ingroup is apomorphic and diagnoses a clade within that group (Watrous and Wheeler, 1981; Maddison *et al.*, 1984; Fig. 1). Maddison *et al.* (1984) presented a set of "rules" for determining the plesiomorphic character state given particular character-state distributions among multiple outgroups (see also Kitching, 1992). Outgroup comparison is by far the most widely used method because the distributional information required is usually the most readily available, it can be applied to all types of characters, and systematists tend to have confidence in the validity of its auxiliary assumptions.

TABLE I   Auxiliary assumptions for selected methods of character polarization or rooting [modified from those of Eldredge and Novacek (1985) and Bryant (1992)]

Outgroup comparison:	1. Higher-level relationships: one or more taxa (outgroups) are assumed to be positioned cladistically outside the ingroup; if multiple outgroups are used, particular relationships among the outgroups may also be assumed
	2. Equivalent ontogenetic stages are being compared
	3. Distribution of character states among and within the outgroup(s) has been surveyed adequately
The ontogenetic method: (generality criterion)	1. The ontogenies of taxa (distribution of relevant character states) in the ingroup have been surveyed adequately
The paleontological method:	1. Fossils chosen are members of the ingroup
	2. Equivalent ontogenetic stages are being compared
	3. Relative ages assigned to fossils referred to the ingroup are correct
	4. The fossil record is complete enough that the character state displayed by the oldest known fossils is representative of the earliest members of the group

B. The Ontogenetic Method

The ontogenetic method uses the distribution of homologous or complementary character states through the ontogenies of members of the study group to infer polarity. This criterion has been the most controversial, largely because it has been interpreted in a variety of ways. Its use is restricted primarily to morphological characters that have observable ontogenies. Traditionally it was assumed that the plesiomorphic character state occurs earlier in ontogeny than the apomorphic character state (Hennig, 1966; Mayr, 1969). This implies that character state change occurs through terminal addition as in Haeckel's discredited biogenetic law of recapitulation. Nelson (1978, p. 327) reformulated the biogenetic law within a cladistic framework by emphasizing the relative generality of character states: "given an ontogenetic character transformation from a character observed to be more general to a character observed to be less general, the more general character is primitive and the less general advanced" (Nelson's rule; Wheeler, 1990). Nelson's (1973) original example involved two species, one (X) in which pharyngeal gill slits occur in both juvenile and adult stages and a second (Y) in which the gill slits occur in the juvenile but are closed in the adult. The occurrence of gill slits is the more general character state because it occurs in both species, whereas the closed gill slits are restricted to species Y. Assuming that the occurrence of gill slits is primitive implies a single phylogenetic transformation, the closure (loss) of the slits in the adult stage of species Y. Assuming instead that the absence of slits in adults is primitive requires two phylogenetic transformations, the appearance of slits in the juvenile stage and the failure of the slits to close in adults of species X. Thus, the inference that the occurrence of slits is primitive is more parsimonious.

Nelson's "generality" has been interpreted in a variety of ways (Kitching, 1992; Meier, 1997), leading to much confusion and nonproductive debate. Meier (1997) listed three interpretations of generality in Nelson's rule: (1) commonness across species; (2) commonness across ontogenetic stages; and (3) generality as defined by de Queiroz (1985). de Queiroz's (1985, p. 283) definition states that "character x is more general than character y if and only if all organisms possessing y (at some stage of ontogeny) also possess x and in addition some organisms possessing x do not possess y." The third interpretation of generality is the appropriate one because it follows logically from the character distributions inherent to a nested taxic hierarchy. Nelson's rule applies only to instances in which the plesiomorphic state precedes the apomorphic state in ontogeny. de Queiroz (1985), Neff (1986), and Weston (1988) argued that only relative generality, and not the sequence of ontogenetic change, is relevant to the inference of polarity. Thus, given an ontogenetic character transformation between two character states, if one state is more generally distributed among the ontogenies within the study group, while the alternative state is less generally distributed, the more general state is plesiomorphic (Fig. 1). This version of the ontogenetic method can be considered a special case of a more broadly applicable generality criterion (see Discussion).

The ontogenetic method has been interpreted and applied in additional ways. de Pinna (1994) proposed that most parsimonious networks be rooted based on the direction of the ontogenetic transformations observed in the terminal taxa, whereas Patterson (1996) rooted cladograms using an "ontogeny" that consisted of the absence of all characters in the data matrix. Both of these approaches appear to involve the conflation of ontogenetic and phylogenetic transformations (see Discussion).

B. The Paleontological or Stratigraphic Method

If the taxic hierarchy is interpreted from an evolutionary, and hence a temporal, perspective, the more plesiomorphic of two alternative character states is by definition the more ancient. Given that stratigraphic position and other dating techniques determine the relative age of fossils, this information is used to infer the plesiomorphic character state based on its occurrence in the presumably oldest fossils referrable to the ingroup. Putative plesiomorphic character states may occur in a single oldest fossil that acts as a surrogate ancestor or may be provided by a number of, often incomplete, early fossil members of the group. The method is applicable to only the subset of characters that is preserved in the fossil record. This method is usually avoided in cladistics because its appropriateness in a particular instance depends on auxiliary assumptions regarding the completeness of the fossil record that are considered difficult to test or justify (Schaeffer *et al.,* 1972; Patterson, 1981; Schoch, 1986; Forey, 1992).

C. Other Methods

The commonality principle (common equals primitive) states that the character state that occurs in the largest number of taxa within the ingroup is plesiomorphic. This "rule of thumb" has no theoretical support (Watrous and Wheeler, 1981; Bryant, 1992). It tends to produce symmetrical branching patterns (Stevens, 1980; Watrous and Wheeler, 1981) and cannot root a three-taxon network because this would require that the majority (two) of the three taxa share the apomorphic character state (Watrous and Wheeler, 1981). In a simulation study testing the accuracy of various methods for coding terminals that represent higher-level taxa, the common-equals-primitive method performed well at low and intermediate rates of character change (Weins, 1998). Although this study suggests that the criterion may often produce correct results, it provides no theoretical justification and therefore no basis for assuming in a particular instance that the more common of two character states is plesiomorphic.

Midpoint rooting (Farris, 1972) is sometimes used to root networks based on molecular data. This approach is based on the assumption of equal mutation rates across lineages or, in other words, a molecular clock model. Other approaches that would base the rooting of such networks on information within the molecular data itself are being explored (see Patterson, 1994). Additional methods rely on either specific evolutionary models or assumptions in addition to those associated with character distributions in taxic hierarchies (see Kitching,

1992). These methods have no justification in cladistic theory and are not considered further.

POLARITY AND ROOTING

Traditionally polarity determination has been viewed as a discrete step in cladistic analysis that follows character coding and precedes the construction of the cladogram (e.g., Hennig, 1966; Eldredge and Cracraft, 1980; Stevens, 1980; Maddison *et al.*, 1984; Neff, 1986; Bryant, 1989, 1992; Brooks and McLennan, 1991). Individual characters are polarized using appropriate methods, producing a hypothetical ancestor with the states inferred to be plesiomorphic; this ancestor is usually included in the parsimony analysis and used to root the tree. Although rooting using a hypothetical ancestor that summarizes the character-state distribution in multiple outgroups does require that polarity be determined prior to cladistic analysis, Nixon and Carpenter (1993) emphasized that polarity determination using specific outgroups can be seen instead as a direct consequence of the rooting of networks. Polarity has no effect on the length of trees or networks (de Pinna, 1994) and therefore in analyses that include outgroups there is no need to independently polarize characters. Through the rooting of a network using outgroups to form a particular cladogram (Fig. 2), character-state transformations become polarized; this procedure identifies both the synapomorphies and the clades they diagnose. This connection between polarity and rooting emphasizes that methods of polarizing character change are methods for rooting networks.

In outgroup comparison one or more outgroup taxa are included in the parsimony analysis and the resulting network is rooted at an internode in the outgroup portion of the network (Figs. 1 and 3). An *a priori* hypothesis of higher-level relationships is required to differentiate between the ingroup and one or more outgroups and, with multiple outgroups, to determine where rooting occurs within the outgroup region of the network. The use of multiple outgroups allows for the testing of ingroup monophyly. Information on character distributions in multiple outgroups is summarized at the outgroup node (Maddison *et al.*, 1984), the most recent common ancestor of the ingroup and the closest outgroup (Fig. 3). The outgroup node is separated from the most recent common ancestor of the ingroup (the ingroup node; Fig. 3) by one internode. If there is only one outgroup, its position in the network is equivalent to that of the outgroup node. Multiple outgroups are often replaced by a hypothetical ancestor that summarizes the outgroup information in one terminal taxon. This approach can have heuristic advantages (Bryant, 1997), but the added assumptions involved place constraints on the outcome of the analysis (Nixon and Carpenter, 1993). Replacement of outgroup taxa by a hypothetical ancestor should be avoided unless those added assumptions are justified (Bryant, 1997).

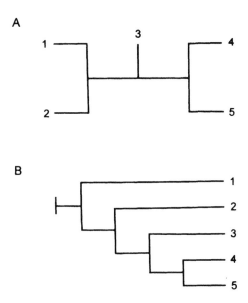

FIGURE 2   Distinction between a network and a cladogram. On the network (**A**) character distributions lack directionality; the topology describes only the adjacency and distances among taxa. Rooting is the conversion of a network into one of a number of possible cladograms by "pulling down" an internode or internal node to form the base of the tree. Rooting along the internode that connects taxon 1 to the rest of the network results in the cladogram (**B**); the position of the root determines the nested hierarchy of clades and the polarity of character change. A network is consistent with a set of cladograms that differ in the position of the root and the resulting clades and character polarities.

In contrast, rooting using the ontogenetic method, paleontological method, or the commonality principle considers only character distributions within the ingroup. In theory, any group of taxa can be considered an ingroup and, therefore, no *a priori* hypothesis of cladistic relationships is necessary for rooting using the ontogenetic and commonality methods. In practice, most ingroups are identified using previous cladistic analyses. In using the paleontological method, previous recognition of an ingroup is required to determine which fossils are relevant for inferring the root of the network. In simple cases (small data matrices, or when only a small number of characters can be polarized), the region of the network that contains the root inferred by these methods can be delimited by hand (e.g., figs. 8.7, 8.9, and 8.11 in Weston, 1994). In more complicated instances, inferences of plesiomorphy for individual characters are used to construct a hypothetical ancestor; unlike the use of a hypothetical ancestor in outgroup comparison, this ancestor is not included in the parsimony analysis. The network is rooted by finding the internode at which the hypothetical ancestor attaches most parsimoniously (Lundberg rooting; Lundberg, 1972; see Bryant, 1997, for additional examples).

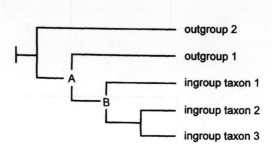

FIGURE 3   The locations of the outgroup (A) and ingroup (B) nodes.

The paleontological method can be employed in two ways.   The oldest known fossil taxon in the ingroup can be included in the analysis and used to root the ingroup network directly.   Alternatively, one can construct a hypothetical ancestor from the information provided by available early fossils referred to the ingroup and use it to root the ingroup network using Lundberg rooting.   The latter is more appropriate if the oldest fossil taxa are poorly known and no single taxon provides enough characters to adequately delimit the root of the tree.

The difference in the method of rooting between outgroup comparison and the other methods implies that inferences of polarity based on outgroup comparison cannot be combined with those based on the other methods to form a composite hypothetical ancestor (as in the empirical analyses of Wheeler, 1990, and Meier, 1997).   Such composite hypothetical constructs are mosaics of information pertinent to the outgroup node (outgroup comparison) and the ingroup node (the other methods).   Use of this chimera to root a network is invalid (Bryant, 1997).   Because the ontogenetic method, paleontological method, and the commonality principle all infer character states at the ingroup node, their inferences can be combined in a single hypothetical ancestor that would be used to root the ingroup network using Lundberg rooting.

DISCUSSION

A. The Generality Criterion

Generality is a hierarchical concept in which the taxic distribution of one character state is a subset of the distribution of another (Fig. 1). Thus, generality differs from commonality, which is a relative frequency concept concerning alternative character states (Bryant, 1992). Given the axiomatic assumption of a taxic hierarchy of groups within groups, the character state that characterizes an entire group is plesiomorphic relative to the alternative character state that characterizes a subset of that group. Relative generality can be observed directly only when the taxa that have the less widely distributed (apomorphic) state also have the more widely distributed (plesiomorphic) state. The study of ontogeny provides the opportunity to observe relative generality in instances where both the plesiomorphic and apomorphic states are observed in the development of individual taxa. This is the basis for the cladistic application of the ontogenetic criterion as first developed by Nelson (1973, 1978) and modified by de Queiroz (1985), Neff (1986), Weston (1988), and others.

Weston (1988, 1994) described other instances in which the relative generality of homologous character states can be observed. In *Acacia* some species develop only true compound leaves throughout the growth of the plant, whereas in others the compound leaves are replaced later in growth by serially homologous simple phyllodes (Weston, 1988). Compound leaves are more generally distributed than phyllodes because the compound leaves are produced by all species that produce phyllodes and also by others that do not. The more generally distributed compound leaves are inferred to be plesiomorphic. In this case one character state does not transform into the other; instead, both types of leaf develop from the same ontogenetic precursor. Some biosynthetic pathways and paralogous genes have similar character distributions. In the paralogous alpha and beta subunits of ATPase, various homologous residues differ in generality (Weston, 1994); in these instances one subunit has the same residue in all taxa, whereas in the second subunit a subset of the taxa have a different residue. As in the previous example, the residue that occurs in all taxa is more generally distributed and can be considered plesiomorphic; the residue with the more limited distribution is apomorphic.

Given this broader applicability of the direct observation of relative generality of character states, Weston (1988) proposed the following "methodological rule" that Wheeler (1990) called Weston's rule: "Given a distribution of two homologous characters in which one, x, is possessed by all of the species that possess its homolog, character y, and by at least one other species that does not, then y may be postulated to be apomorphous relative to x." (Weston, 1988, p. 45). Thus, the ontogenetic method becomes a particular instance of the generality criterion, which also encompasses other situations such as serial homologues and paralogous genes, in which relative generality can be observed directly in a set of taxa.

Weston (1994) argued that the core assumption of the generality criterion is that every homology represents a synapomorphy at some level in the taxic hierarchy, with the corollary that sets of ontogenetic and iterative homologies are the result of additive phylogenetic change. These assumptions are associated with the axiom that nature forms a nested taxic hierarchy and are therefore not equivalent to the auxiliary assumptions of individual methods listed by Bryant (1992; see Table I). Nonetheless, the generality criterion would seem to entail the auxiliary assumption that the distribution of relevant character states within the study group has been adequately surveyed. Whereas de Queiroz (1985) argued that the ontogenetic criterion assumes that plesiomorphic character states are retained in the ontogenies of descendants, Weston (1994) saw the retention of plesiomorphic features as a conclusion, rather than a premise, of the interpretation of ontogenetic data using the generality criterion.

Seeing the distribution of character states through ontogeny as one of several types of data that are applicable to the generality criterion emphasizes that the sequence of ontogenetic transformation is irrelevant to the inference of character polarity. Only the occurrence of character states throughout the ontogenies of taxa within the group is relevant. It is not ontogeny per se that provides the basis for inferring polarity based on relative generality, but rather the potential that ontogeny provides to observe multiple character states within characters in individual taxa in the ingroup.

B. Direct versus Indirect Methods

Nelson (1973) differentiated between outgroup comparison and the ontogenetic method by considering the former an indirect method for determining character polarity, whereas the latter was considered a direct method. Indirect methods involve taxa other than those of immediate concern (the ingroup), whereas direct methods involve only the ingroup. Nelson (1973) was responding to claims (e.g., Colless, 1969; Lundberg, 1972) that all cladistic methods for inferring polarity depend on *a priori* assumptions of relationships. If true, this claim would expose a fatal flaw because it would lead to infinite regress. Certainly the claim is true of outgroup comparison; by differentiating between the ingroup and one or more outgroups that are used to root the tree, outgroup comparison relies on assumptions of higher-level relationships. Nelson (1973) demonstrated that the ontogenetic criterion is a direct method for inferring character polarity; direct methods provide a means of rooting the "tree of life" at its highest levels, providing a cladistic framework for rooting subsets of the tree using indirect methods such as outgroup comparison.

Eldredge and Novacek (1985) argued that the ontogenetic method relies on the assumption that the ingroup is monophyletic; Eldredge and Novacek (1985) reasoned that because it is comparative the ontogenetic method assumes some hierarchical grouping of the organisms being studied. If true, this reliance on a previous hypothesis of relationships would make the method indirect. Similarly, Wheeler (1990) argued that Nelson's rule is not a direct method because it relies on comparisons that involve two taxa. Suggestions (Eldredge and Cracraft, 1980; Wiley, 1981; Watrous, 1982; Brooks and Wiley, 1985; Ax, 1987) that the ontogenetic method is dependent on outgroup comparison would have the same effect. However, ontogenetic comparisons relevant to rooting involve only the search for equivalent characters in a set of taxa and need not involve *a priori* assumptions of higher-level relationships.

de Pinna (1994) and Weston (1994) argued that only ontogeny or the generality criterion, respectively, actually root trees, whereas outgroup comparison, with its dependence on a previous cladistic hypothesis, can only extend the rooting implied by that previous hypothesis onto a new network. Certainly the generality criterion is necessary to provide a root for the most inclusive networks. However, indirect methods, by determining the internode that is pulled down to form a tree from a network, do provide a root. They simply need additional assumptions of hierarchy to do so and therefore cannot root the entire tree of life. The difference of opinion here appears to be related to different frames of reference, the entire tree of life on the one hand, and the less inclusive networks resulting from most cladistic analyses on the other.

Nelson (1973, p. 88) stated that "the study of ontogenetic character transformations" provides "the only valid direct technique of character phylogeny of which I am aware." Given that this use of ontogeny is a particular instance of a more broadly applicable generality criterion, is this criterion the only direct method for rooting cladistic networks? There was a long-standing tradition among paleontologists of reading ancestor-descendant relationships and relative apomorphy "directly" from the stratigraphic record (Schaeffer *et al.*, 1972; Gingerich, 1976; Patterson, 1981). However, implicit assumptions of phylogenetic relationships are involved in this approach (e.g., Patterson, 1981; Wiley, 1981; Eldredge and Novacek, 1985; Bryant, 1992). By basing those decisions on particular fossil taxa and not on others, one is relying in part on assumptions of higher-level relationships, indicating that the stratigraphic method is an indirect method. Closer analysis of the various approaches to the paleontological method (Bryant, 1997) suggests, however, that there is a one particular instance in which the method could be considered direct.

Lundberg rooting using a hypothetical ancestor inferred using the paleontological method entails the assumption that the fossils involved are members of the ingroup and approximate its ancestor. There is a clear decision to use the information provided by some fossils and reject that provided by others; this is an assumption of cladistic relationships, making the method indirect. Assumptions of higher-level relationship are usually also involved when a fossil taxon is included in a cladistic analysis and used to root the network; again, one is assuming that the fossil is a close approximation of the ancestor of the clade delimited by the other taxa. However, given that that oldest fossil will almost always be connected to the rest of the network by a branch of some length, it could be argued that by rooting the tree using that fossil, it becomes equivalent to the outgroup node and therefore that this method is a special case of outgroup comparison. In effect, the decision to root along the non-zero length branch that connects the fossil to the rest of the network entails the assumption that the fossil is in an outgroup position to the rest of the network. This assumption can be avoided, however, if the oldest fossil is connected to the tree by a zero branch length. In this instance the fossil is equivalent to a node on the network and there is no necessary differentiation between ingroup and outgroup. Rooting is based solely on the relative age of the taxon and might be considered direct. Although the choice of taxa in the analysis would normally be based on previous phylogenetic analyses, this is not necessary to the logic of the method. Any group of taxa could be analyzed; if the oldest fossil were connected to the resulting network by a zero branch length, the sole basis for rooting could be age without any necessary assumptions of relationship.

C. Ontogenetic versus Phylogenetic Polarity

Some of the confusion associated with the ontogenetic method concerns the failure to clearly differentiate between phylogenetic and ontogenetic transformations (Wheeler, 1990). Only the former are relevant to inferences of relative generality based on parsimony. Nonetheless, conflation between ontogenetic and phylogenetic transformations appears to underlie Wheeler's (1990) rejection of Weston's (1988) attempt to broaden Nelson's rule to include situations in which the less generally distributed character state occurs earlier in ontogeny. Wheeler argued that whereas Nelson's rule provides a definitive decision of polarity, Weston's rule is equivocal. Wheeler's (1990) conclusion was based on unstated assumptions regarding ontogenetic mechanisms and the counting of ontogenetic transformations as well as phylogenetic transformations. Consider taxa 3 and 4 in Fig. 1A and assume that stage X precedes stage Y in ontogeny; in taxon 4 there is a transformation from character state B to character state A, whereas in taxon 3 state A occurs throughout the same portion of ontogeny. Whereas Nelson's rule does not apply to this situation because the more generally distributed state (A) occurs later in ontogeny, Weston's rule considers state A as plesiomorphic because it disregards the direction of the ontogenetic transformation. Only one phylogenetic transformation is required, the appearance of B early in the ontogeny of taxon 4. If character state B is primitive, two phylogenetic transformations are implied, the appearance of A and the loss of B in taxon 3. Wheeler (1990) found this situation equivocal because he was concerned with the ontogenetic transformation from the apomorphic state B to the plesiomorphic state A in taxon 4. He wondered whether A in taxon 4 is equivalent to the plesiomorphic A observed in taxon 3 and argued that to account for the possibility that it is not, the ontogenetic transformation between B and A in taxon 4 should also be counted, making the two alternative polarities equally parsimonious at two steps each. Wheeler (1990) erred in counting both phylogenetic and ontogenetic transformations. His desire to include an ontogenetic transformation in the relative assessment of parsimony seems to have been based on the assumption that ontogeny proceeds epigenetically from the more general to the less general.

de Pinna (1994) was concerned that the determination of polarity or rooting based on the principle of generality (Weston, 1988) failed to adequately incorporate ontogenetic information because the direction of ontogenetic transformations was ignored. He proposed a new method of "ontogenetic rooting" in which most parsimonious networks are rooted by the direction of ontogenetic transformations observed in the terminals. de Pinna (1994, p. 170) posed the question "how much do relationships derived from ontogenetic rooting reflect historical relationships among organisms?" The answer would seem to depend on the extent to which the epigenetic mechanisms and terminal addition inherent in an isometry between the direction of ontogenetic and phylogenetic transformations reflect evolutionary history. This is best determined by avoiding the use of the direction of ontogenetic transformations to root trees, and using the resulting cladograms to compare the polarity of character evolution to that of ontogenetic transformations.

Patterson (1996) reassessed the empirical evidence from centrarchid fishes that Mabee (1989, 1993) had used to argue for the rejection of Nelson's rule. Patterson's (1996) interpretation of the ontogenetic method was based on the observation that absence is more generally distributed in ontogenies than presence, and as a result he rooted cladograms using an "ontogeny" consisting of absence for all characters in the matrix. However, in rooting a network of centrarchid fishes based on ontogeny, one is not interested in the absolutely most primitive character state (absence), but rather the primitive character state among centrarchids (see also Mabee, 1996). The universal absence of centrarchid features in the earliest ontogenetic stages is uninformative for the inference of relative generality within the group. Patterson (1996) seems to have also confused ontogenetic and phylogenetic transformations.

The transformations in ontogenetic sequences provide strong evidence for homology among character states, but cannot be used to infer the sequence of phylogenetic transformations. To include the direction of ontogenetic change in the analysis injects assumptions of evolutionary process into the analysis.

D. Relative Value and Utility of Polarity Criteria

The validity and relative value of methods of rooting and polarity in cladistic analysis have been the subjects of considerable debate (e.g., Nelson, 1978, 1985; Brooks and Wiley, 1985; Kluge, 1985; Weston, 1988; Mabee, 1989, 1993,

1996; Bryant, 1992; Patterson, 1996; Meier, 1997). Much of this debate has suffered from the assumption that one method must be inherently superior to the others, the failure to consider the unique auxiliary assumptions of each method (Bryant, 1992), or the fact that the proponents have different theoretical perspectives [e.g., the exchange between Patterson (1996) and Mabee (1996)].

Although much of this debate has been from a theoretical perspective, there have been several attempts to compare criteria, especially outgroup comparison and ontogeny, empirically (e.g., Miyazaki and Mickevich, 1982; Kraus, 1988; Mabee, 1989, 1993; Wheeler, 1990; Meier, 1997). de Pinna (1994) criticized much of this work for simply identifying agreement and disagreement between criteria without providing a means of making an unbiased decision in favor of one over the other. This criticism holds some weight for Mabee's (1989, 1993) analyses that evaluated Nelson's rule by taking the outgroup tree as given; clearly the ontogenetic method can only come out second best in this sort of analysis. However, various of the other studies compared outgroup comparison and the ontogenetic method using independent criteria such as degree of resolution, number of most parsimonious cladograms, amount of homoplasy, number of equally parsimonious roots, and number of characters polarized. Nonetheless, although comparisons using these criteria are informative, it is not clear that any of them provides a valid basis for choosing one method over the other. The use of the amount of homoplasy might seem to have theoretical support; after all, this is the basis for choosing among networks in cladistic analyses. Wheeler (1990) argued that it is homoplasy that causes both outgroup comparison and the ontogenetic criterion to fail. Using this criterion Wheeler (1990) found little difference between these two criteria, whereas in Meier's (1997) analyses outgroup comparison produced less homoplasy. However, homoplasy is associated with the congruence among character distributions, not with the rooting of trees, and there is no clear justification for extending comparisons based on the congruence of character distributions to different data sets. Thus, empirical comparisons have yet to provide a clear rationale for choosing one rooting method over the others. Agreement among rooting methods provides support for the results, but how are disagreements to be resolved?

Methods of polarity determination or rooting fail when their auxiliary methods are invalid. Thus, Bryant (1992) argued that the resolution of disagreements among outgroup comparison, the ontogenetic method, and stratigraphy should be based in each particular instance on a relative assessment of the validity of the auxiliary assumptions of each method. All three methods provide most parsimonious inferences of polarity or rooting given a particular character distribution and specific auxiliary assumptions. The generality criterion has the fewest assumptions, whereas the stratigraphic method has the largest number (Table I). This comparison provides a theoretical basis for preferring the generality criterion, and secondly outgroup comparison, and possibly for avoiding the stratigraphic method. However, one must also consider utility. Although the generality criterion relies on the fewest auxiliary assumptions, it cannot be applied to as many characters as outgroup comparison. The residues in most DNA sequences cannot be polarized using the generality criterion and molecules are rarely preserved in the fossil record; most of these characters must be polarized using outgroup comparison. Many morphological characters can also be polarized using only outgroup comparison.

The rooting of trees using early fossils is problematic because it relies on more auxiliary assumptions than outgroup comparison or the generality criterion, and the validity of these assumptions in a particular instance is often difficult to evaluate. The paleontological method has fallen into disfavor largely because of its assumptions regarding the completeness of the fossil record. There is no question that the fossil record is incomplete and that its use in polarizing a particular character may entail unjustifiable assumptions (Forey, 1992), but is it so incomplete that polarity cannot be reliably determined in enough characters to root a network? Means of assessing the completeness of the fossil record have received considerable attention recently (see references in Bryant, 1992), and correlations between cladistic rank and stratigraphic position (Gauthier *et al.*, 1988; Norell and Novacek, 1992a,b; Benton, 1995; Clyde and Fisher, 1997; Hitchin and Benton, 1997) suggest that the failings of the method are overstated, at least for taxa with relatively good fossil records. Despite its limitations, the complete avoidance of stratigraphic information in the rooting of networks seems extreme (Schoch, 1986; Bryant, 1992) and is not based on extensive empirical comparisons.

CONCLUDING REMARKS

Weston (1994) considered outgroup comparison and the ontogenetic, or generality, criterion as complementary, rather than competing, methods. Certainly, the combined use of the generality criterion, which can root the most inclusive networks, and outgroup comparison, which is applicable to most characters, allows for the rooting of networks at all levels of inclusivity, and based on all types of cladistic data. These two methods also rely on the fewest assumptions, allowing, arguably, for the avoidance of the assumption that evolution is the causal agent of the taxic hierarchy. However, acknowledgment that the relative generality of character distributions provides one line of evidence for evolution does not imply that evolution must be excluded from day to day cladistic analyses. By including evolution among our axiomatic assumptions, additional methods, such as the use of stratigraphy for rooting most parsimonious networks, can be brought to bear on systematic questions. A pluralistic approach also provides the framework for using the results of one rooting method to test the results of another.

Recognition of the generality criterion and the fact that ontogenetic comparisons represent one of several types of data that can be employed to root trees using this method clarify the role of ontogenetic data in polarizing characters. Only the relative generality of character states through the ontogenies of taxa is relevant, and not the direction or other aspects of ontogenetic transformations. Acceptance of these claims should resolve one of the long-standing debates in cladistics, the role of ontogeny in the polarization of character transformations.

ACKOWLEDGMENTS

I thank Günter Wagner for inviting me to contribute to this volume. Peter Weston and Ward Wheeler provided constructive criticism of an earlier version of the manuscript and I feel that their comments improved the final product.

LITERATURE CITED

Ax, P. (1987). "The Phylogenetic System. The Systematization of Organisms on the Basis of their Phylogenesis." John Wiley and Sons, Chichester.

Benton, M. J. (1995). Testing the time axis of phylogenies. *Philos. Trans. R. Soc. Lond. B* **349:** 5-10.

Bishop, M. J. (1982). Criteria for the determination of the direction of character state changes. *Zool. J. Linn. Soc.* **74:**197-206.

Bock, W. J. (1977). Foundations and methods of evolutionary classification. *In* "Major Patterns in Vertebrate Evolution" (M. K. Hecht, P. C. Goody, and B. M. Hecht, eds.), pp. 851-895. NATO Advanced Study Institute Series A 14, Plenum Press, New York.

Bonde, N. (1977). Cladistic classification as applied to vertebrates. *In* "Major Patterns in Vertebrate Evolution" (M. K. Hecht, P. C. Goody, and B. M. Hecht, eds.), pp. 741-804. NATO Advanced Study Institute Series A 14, Plenum Press, New York.

Brady, R. (1985). On the independence of systematics. *Cladistics* **1:**113-126.

Brooks, D. R., and McLennan, D. A. (1991). "Phylogeny, Ecology and Behavior." University of Chicago Press, Chicago.

Brooks, D. R., and Wiley, E. O. (1985). Theories and methods in different approaches to phylogenetic systematics. *Cladistics* **1:**1-11.

Bryant, H. N. (1989). An evaluation of cladistic and character analyses as hypothetico-deductive procedures, and the consequences for character weighting. *Syst. Zool.* **38:**214-227.

Bryant, H. N. (1992). The polarization of character transformations in phylogenetic systematics: role of axiomatic and auxiliary assumptions. *Syst. Zool.* **40:**433-445.

Bryant, H. N. (1997). Hypothetical ancestors and the rooting of cladograms. *Cladistics* **13:**337-348.

Clyde, W. C., and Fisher, D. C. (1997). Comparing the fit of stratigraphic and morphologic data in phylogenetic analysis. *Paleobiology* **23:**1-19.

Colless, D. H. (1969). The interpretation of Hennig's "Phylogenetic Systematics"--a reply to Dr. Schlee. *Syst. Zool.* **18:**134-144.

Crisci, J. V., and Stuessy, T. F. (1980). Determining primitive character states for phylogenetic reconstruction. *Syst. Bot.* **5:**112-135.

de Jong, R. (1980). Some tools for evolutionary and phylogenetic studies. *Z. Zool. Syst. Evolutionsforsch.* **18:**1-23.

de Pinna, M. C. C. (1994). Ontogeny, rooting, and polarity. *In* "Models in Phylogeny Reconstruction" (R. W. Scotland, D. J. Siebert, and D. M. Williams, eds), pp. 157-172. Systematics Association Special Volume No. 52. Clarendon Press, Oxford.

de Queiroz, K. (1985). The ontogenetic method for determining character polarity and its relevance to phylogenetic systematics. *Syst. Zool.* **34:**280-299.

de Queiroz, K., and Donoghue, M. J. (1990). Phylogenetic systematics or Nelson's version of cladistics. *Cladistics* **6**:61-75.

Eldredge, N., and Cracraft, J. (1980). "Phylogenetic Patterns and the Evolutionary Process (Method and Theory in Comparative Biology)." Columbia University Press, New York.

Eldredge, N., and Novacek, M. J. (1985). Systematics and paleobiology. *Paleobiology* **11**:65-74.

Farris, J. S. (1972). Estimating phylogenetic trees from distance matrices. *Am. Nat.* **106**:645-668.

Fink, W. L. (1982). The conceptual relationship between ontogeny and phylogeny. *Paleobiology* **8**:254-264.

Forey, P. L. (1992). Fossils and cladistic analysis. *In* "Cladistics. A Practical Course in Systematics" (P. L. Forey, C. J. Humphries, I L. Kitching, R. W. Scotland, D. J. Siebert, and D. M. Williams, eds.), pp. 124-136. The Systematics Association Publication No. 10, Clarendon Press, Oxford.

Gaffney, E. S. (1979). An introduction to the logic of phylogeny reconstruction. *In* "Phylogenetic Analysis and Paleontology" (J. Cracraft, and N. Eldredge, eds.), pp. 79-111. Columbia University Press, New York.

Gauthier, J., Kluge, A. G., and Rowe, T. (1988). Amniote phylogeny and the importance of fossils. *Cladistics* **4**:105-209.

Gingerich, P. D. (1976). Paleontology and phylogeny: patterns of evolution at the species level in early Tertiary mammals. *Am. J. Sci.* **276**:1-28.

Gutmann, W. F. (1981). Relationships between invertebrate phyla based on functional mechanical analysis of the hydrostatic skeleton. *Am. Zool.* **21**:63-81.

Hennig, W. (1966). "Phylogenetic Systematics." University of Illinois Press, Urbana.

Hitchin, R., and Benton, M. J. (1997). Congruence between parsimony and stratigraphy: comparisons of three indices. *Paleobiology* **23**:20-32.

Kitching, I. L. (1992). The determination of character polarity. *In* "Cladistics. A Practical Course in Systematics" (P. L. Forey, C. J. Humphries, I L. Kitching, R. W. Scotland, D. J. Siebert, and D. M. Williams, eds.), pp. 22-43. The Systematics Association Publication No. 10, Clarendon Press, Oxford.

Kluge, A. G. (1985). Ontogeny and phylogenetic systematics. *Cladistics* **1**:13-27.

Kraus, F. (1988). An empirical evaluation of the use of the ontogeny polarization criterion in phylogenetic inference. *Syst. Zool.* **37**:106-141.

Lakatos, I. (1970). Falsification and the methodology of scientific research programmes. *In* "Criticism and the Growth of Knowledge" (I. Lakatos and A. Musgrave, eds.), pp. 91-196. Cambridge University Press, London.

Lundberg, J. G. (1972). Wagner networks and ancestors. *Syst. Zool.* **21**: 398-413.

Mabee, P. M. (1989). An empirical rejection of the ontogenetic polarity criterion. *Cladistics* **5**: 409-416.

Mabee, P. M. (1993). Phylogenetic interpretation of ontogenetic change: sorting out the actual and artefactual in an empirical case study of centrarchid fishes. *Zool. J. Linn. Soc.* **107**: 175-291.

Mabee, P. M. (1996). Reassessing the ontogenetic criterion: a response to Patterson. *Cladistics* **12**:169-176.

Maddison, W. P., Donoghue, M. J., and Maddison, D. R. (1984). Outgroup analysis and parsimony. *Syst. Zool.* **33**:83-103.

Mayr, E. (1969). "Principles of Systematic Zoology." McGraw-Hill, New York.

Mayr, E. (1974). Cladistic analysis or cladistic classification? *Z. Zool. Syst. Evolutionsforsch.* **12**:94-128.

Meier, R. (1997). A test and review of the empirical performance of the ontogenetic criterion. *Syst. Biol.* **46**:699-721.

Miyazaki, J. M., and Mickevich, M. F. (1982). Evolution of *Chesapecten* (Mollusca: Bivalvia, Miocene-Pliocene) and the biogenetic law. *Evol. Biol.* **15**:369-409.

Neff, N. A. (1986). A rational basis for a priori character weighting. *Syst. Zool.* **35**:110-123.

Nelson, G. J. (1973). The higher-level phylogeny of vertebrates. *Syst. Zool.* **22**:87-91.

Nelson, G. J. (1978). Ontogeny, phylogeny, paleontology, and the biogenetic law. *Syst. Zool.* **27**:324-345.

Nelson, G. (1985). Outgroups and ontogeny. *Cladistics* **1**:29-45.

Nelson, G., and Platnick, N. (1981). "Systematics and Biogeography. Cladistics and Vicariance." Columbia University Press, New York.

Nixon, K. C., and Carpenter, J. M. (1993). On outgroups. *Cladistics* **9**:413-426.

Norell, M. A., and Novacek, M. J. (1992a). The fossil record and evolution: comparing cladistic and paleontologic evidence for vertebrate history. *Science* **255**:1690-1693.

Norell, M. A., and Novacek, M. J. (1992b). Congruence between superpositional and phylogenetic patterns: comparing cladistic patterns with fossil records. *Cladistics* **8**:319-337.

Patterson, C. (1981). Significance of fossils in determining evolutionary relationships. *Annu. Rev. Ecol. Syst.* **12**:195-223.

Patterson, C. (1982). Morphological characters and homology. *In* "Problems of Phylogenetic Reconstruction" (K. A. Joysey and A. E. Friday, eds.), pp. 21-74. Systematics Association Special Volume 21, Academic Press, London.

Patterson, C. (1994). Null or minimal models. *In* "Models in Phylogeny Reconstruction" (R. W. Scotland, D. J. Siebert, and D. M. Williams, eds), pp. 173-192. Systematics Association Special Volume No. 52. Clarendon Press, Oxford.

Patterson, C. (1996). Comments on Mabee's "Empirical rejection of the ontogenetic polarity criterion." *Cladistics* **12**:147-167.

Platnick, N. I. (1979). Philosophy and the transformation of cladistics. *Syst. Zool.* **28**:537-546.

Rieppel, O. (1988). "Fundamentals of Comparative Biology." Birkhauser Verlag, Basel.

Rieppel, O. (1990). Ontogeny - A way forward for systematics, a way backward for phylogeny. *Biol. J. Linn. Soc.* **39**:177-191.

Schaeffer, B., Hecht, M. K., and Eldredge, N. (1972). Phylogeny and paleontology. *Evol. Biol.* **6**:31-46.

Schoch, R. M. (1986). "Phylogeny Reconstruction in Paleontology." Van Nostrand Reinhold Company, New York.

Simpson, G. G. (1961). "Principles of Animal Taxonomy." Columbia University Press, New York.

Stevens, P. F. (1980). Evolutionary polarity of character states. *Annu. Rev. Ecol. Syst.* **11**:333-358.

Szalay F. S. (1977). Phylogenetic relationships and a classification of the eutherian Mammalia. *In* "Major Patterns in Vertebrate Evolution" (M. K. Hecht, P. C. Goody, and B. M. Hecht, eds.), pp. 315-374. NATO Advanced Study Institute Series A 14, Plenum Press, New York.

Watrous, L. E. (1982). Review of "Phylogenetics: the theory and practice of phylogenetic systematics" by E. O. Wiley. *Syst. Zool.* **31**:98-100.

Watrous, L. E., and Wheeler, Q. D. (1981). The out-group comparison method of character analysis. *Syst. Zool.* **30**:1-11.

Weins, J. J. (1998). The accuracy of methods for coding and sampling higher-level taxa for phylogenetic analysis: a simulation study. *Syst. Biol.* **47**:397-413.

Weston, P. H. (1988). Indirect and direct methods in systematics. *In* "Ontogeny and Systematics" (C. J. Humphries, ed.), pp. 27-56. Columbia University Press, New York.

Weston, P. H. (1994). Methods for rooting cladistic trees. *In* "Models in Phylogeny Reconstruction" (R. W. Scotland, D. J. Siebert, and D. M. Williams, eds.), pp. 125-155. Systematics Association Special Volume No. 52. Clarendon Press, Oxford.

Wheeler, Q. D. (1990). Ontogeny and character phylogeny. *Cladistics* **6**:225-268.

Wiley, E. O. (1975). Karl R. Popper, systematics, and classification: a reply to Walter Bock and other evolutionary taxonomists. *Syst. Zool.* **24**:233-243.

Wiley, E. O. (1981). "Phylogenetics. The Theory and Practice of Phylogenetic Systematics." John Wiley and Sons, New York.

Williams, D. M., Scotland, R. W., and Blackmore, S. (1990). Is there a direct ontogenetic criterion in systematics? *Biol. J. Linn. Soc.* **39**:99-108.

THE MECHANISTIC ARCHITECTURE OF CHARACTERS

If a character concept is meaningful it must either refer to a natural unit or at least to a mechanistically meaningful abstraction. In either case, two types of biological questions arise from the idea of a character. First, what are and what cause the relevant biological properties of the characters. For instance, in the case of the adaptive phenotypic characters the relevant biological properties are the heritable phenotypic variations caused by mutations. Hence the question is what mechanisms translate DNA variation into phenotypic variation. The second is, what accounts for the causal homeostasis of character identity. It is an empirical question whether this causal homeostasis exists, but in any case the meaning of any character concept depends on it. Even if most biological knowledge has been gathered by studying individual parts, rather than whole organisms, much of the knowledge is not organized as *knowledge about the architecture of specific characters*. Characters have mostly been studied as difference markers or as manifestations of general biological processes, like inheritance, differentiation, or pattern formation. Nevertheless, with an appropriate reading, some elements of the "architecture" of specific characters can be gained from existing knowledge. Furthermore, the progress in developmental genetics encourages considerations of the developmental and genetic architecture of characters as well as speculations about the origin of characters. Therefore some of the contributions in this section and the next, dedicated to the origin of characters, could be placed in either section.

Some ideas have been put forward regarding the possible architecture of characters, like polygenic inheritance, modularity, quasi-independence, and so on. Some of them are patently wrong, like the one-gene-one-character theory of early genetics, others are simply not sufficiently tested yet, and yet others emerged as recent empirical generalizations, like that of a developmental toolbox of molecular processes used in the development of different characters.

Paul Brakefield reports on one of the characters on which sufficient work has been done to see the contours of the character's architecture, the eye spot patterns of butterflies. Based on the pioneering work of Fred Nijhout, Brakefield and his collaborators have analyzed the eyespot pattern at all levels of organization. He concludes that the unit character is the whole pattern of eyespots rather than individual spots. Much quantitative variability is found in extant populations of *Bicyclus anynana* that explains the interspecific variation in the genus. However, some of the novel phenotypes in contrast seem to be caused by single genes with large effects. This possibly hints at an interesting difference between variation of a character and the origin of new characters discussed further in the next section (see also contributions in section V).

Massimo Pigliucci points out that characters are the product of both genetic and environmental factors. As a consequence the notion of a character meaning a feature or a part of an individual may be misguided. Instead the unit of phenotypic organization is perhaps more abstract, like the developmental reaction norm, of which characters in the traditional sense are just realizations. Pigliucci argues that this is indeed the case and that the traditional view unnecessarily obscures the causal role of the environment in a variety of situations, like the response to novel environments.

Trudy Mackay reviews the molecular methods for and the result from studies into the genetic basis of complex characters, like sensory bristle number. The data reveal a surprising dependency of genetic effects on the sex of the individual and the environment in which the character is observed, supporting Pigliucci's argument for the critical importance of the environment. An interesting observation is also that known developmental genes, discovered because of large effects on development, also supply alleles contributing to quantitative variation and, by extension, to character divergence.

James Cheverud reviews data obtained with the same molecular marker techniques as described by Mackay to test one of the key predictions from the idea of characters as quasi-independent units of the phenotypes. Quasi-independence predicts that features of a character share more genes in common than features from different characters (modularity). He finds that indeed the pleiotropic effects of genes tend to be restricted to functionally and developmentally related traits, making each trait set an individuated character for evolutionary analysis. In addition, there is evidence that pleiotropic effects are genetically variable, which implies that the distribution of pleiotropic effects can evolve by natural selection.

Scott Gilbert and **Jessica Bolker** review and discuss recent evidence revealing a new level of organization: developmental process modules deployed in the development of a variety of morphological characters. They argue that these stable units of developmental interaction are the product of canalization, i.e., the result of natural selection for the stability of the developmental process.

Lisa Nagy and **Terri Williams** review the considerable morphological and developmental literature about the development and evolution of arthropod limbs, another example of a well-known character. Based on this data they raise the important point that modularity found in highly derived species may not imply that the corresponding character in ancestral species is similarly modular. In fact, Nagy and Williams argue that this is positively not the case with arthropod limbs and segments. Modularity is perhaps itself an evolved property. This suggestion connects well with the scenario of the evolution of butterfly wing patterns explained by Nijhout in the next section.

15

THE STRUCTURE OF A CHARACTER AND THE EVOLUTION OF PATTERNS

PAUL M. BRAKEFIELD

Institute of Evolutionary and Ecological Sciences, Leiden University, 2300 RA Leiden,
The Netherlands

INTRODUCTION: DISSECTING A CHARACTER

This essay examines concepts about character evolution in the context of genetical and developmental studies of a specific morphological pattern, that of butterfly eyespots. Although semantics may differ, many of the underlying ideas have already been expressed by Müller and Wagner (1991), Wagner and Altenberg (1996), and others, but here the discussion stems from a single model system. The eyespot pattern of butterflies is composed of a series of individual eyespots or units. Morphological descriptions suggest that the most appropriate and useful application of the term "character" is to the complete eyespot pattern or, perhaps, to the eyespots on a specific wing surface.

I will focus here on several issues based on an analysis of variation in a single species of satyrine butterfly, *Bicyclus anynana*. Our approach centers on the internal genetical and developmental organization of the eyespot pattern (for reviews, see French, 1997; Brakefield, 1998; Brakefield and French, 1999). It has begun to examine how flexibly and rapidly different units or features of the trait may be able to respond to change in the environment and natural selection.

The Character Concept in Evolutionary Biology

Such an integrated study may reveal the extent to which estimating genetic variances and covariances provides a satisfactory description of character structure. An understanding of how genes modulate development and (what genetic variance-covariance matrices reflect in mechanistic terms seen likely to reveal additional insights about character structure and, therefore, about the potential for evolutionary change. A description of the genes and of their developmental roles will also be crucial to formulating predictive models of eyespot formation (see Nijhout and Paulsen, 1997). A knowledge of the internal organization of a character should also reveal whether bias is likely in the production of novel phenotypes, and how such bias relates to the developmental processes. Bias may have resulted in patterns of morphological diversity in eyespot patterns which have been shaped by some balance between natural selection and developmental constraints (see Maynard-Smith et al., 1985; Schlichting and Pigliucci, 1997). Comparisons of such predictions about the likelihood of evolutionary trajectories with observed patterns of change across related taxa should then enable the robustness of ideas about character evolution to be tested.

In investigating the evolution of an eyespot pattern using a multidisciplinary approach we will eventually be able to examine whether the ways in which genes modulate development influence their probability of becoming involved in evolutionary divergence. Putting this point another way, we can examine whether when there are alternative ways in which development can be modulated to produce a given phenotype, can predictions be made about their relative contribution to evolution? Furthermore, studying both the genetics and development of a character should provide insights about whether there is something special about abrupt changes or shifts in phenotype; are such events more likely to involve developmental novelties or additions to the developmental repetoire?

EYESPOTS COMPRISING A MORPHOLOGICAL PATTERN

Ten years ago we began to study the evolution of the wing pattern of African *Bicyclus* butterflies. The most prominent wing pattern element in these butterflies is the marginal eyespot (Fig. 1). These eyespots are frequently exposed while butterflies are at rest. They are likely to function in avoidance of vertebrate predators which hunt their prey by sight. They can act as targets deflecting predator attacks away from the vulnerable body. In this way, they can help the butterfly to escape, albeit having lost part of the outer wing tissue (Wourms and Wasserman, 1985; Brakefield and Reitsma, 1991).

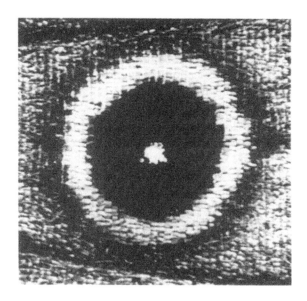

FIGURE 1   A single ventral eyespot of *B. anynana* showing the concentric rings of color. Individual overlapping scale cells can be seen in the gold ring.

Each eyespot consists of a series of annuli arranged around a central white pupil (Fig. 1). Each concentric ring is itself composed of many individual scale cells which contain the same color pigment. The pigments are synthesized toward the end of the pupal stage, shortly before adult eclosion. The different colours of an eyespot are thus produced by sharp transitions in the pigments across the boundaries between the rings. The scale cells are arranged across the wing blades like the tiles on a roof. Each wing surface is effectively a single layer of epidermal cells, some of which are differentiated into the pigmented scale cells (Galant *et al.*, 1998). The two innermost rings of an eyespot in *Bicyclus* are black and gold. Some ventral eyespots may have further rings of different colors (see, e.g., Fig. 1).

Most of the research on *Bicyclus* eyespots has used a laboratory stock *of B. anynana* established from a large number of founders from a single locality in Malawi. The species is easy to rear in large numbers on maize and has a short generation time. *B. anynana* can have eyespots on all four wing surfaces, although they are often absent on the dorsal hindwing (Fig. 2). The wings of a butterfly are bisected by strengthening wing veins or trachea, most of which run proximal to distal. Each wing subdivision, which is bordered by wing veins, is known as a wing cell. The ventral hindwing in *B. anynana* usually shows a complete series of eyespots, one in each wing cell along the wing margins. The eyespots are aligned along the fold or internervule running midway between each pair of veins (Fig. 2A). In contrast, both surfaces of the forewing normally express only two eyespots: one anterior and small and the other posterior and large. Eyespots are usually absent from the two intervening wing cells, as well as from those which lie more anteriorly or posteriorly.

Comparative morphologists in the 1920s and 30s were concerned with describing patterns of wing pattern diversity across species within different major groups of the Lepidoptera, including the Nymphalidae (for review see Nijhout, 1991). Perhaps their most important contribution was the recognition of a so-called prototype or ancestral wing pattern which shows a full ring of marginal eyespot elements on each wing surface. This prototype also incorporates other pattern elements, including medial and parafocal band markings. Some changes in eyespot pattern within the Nymphalidae (which includes the Satyrinae) are then envisaged as involving reductions in eyespot number by their loss in one or more wing cells. Certain extant species appear close to the prototype pattern with a complete complement of eyespots, while many others show no distinct eyespots (see Nijhout, 1991, for illustrations). From the perspective of comparative morphology, the whole eyespot pattern appears to correspond to a single character or module which itself comprises a series of repeated eyespot elements. There is a potential for an eyespot to be expressed in each marginal wing cell. This confers an identity on each eyespot in terms of its position, and enables comparisons of patterns of presence and absence across individuals or taxa.

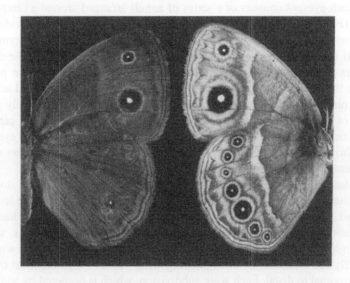

FIGURE 2 The typical eyespot pattern on each wing surface in *B. anynana*. All wings are from the wet season form using representative specimens of the unselected stock reared at 27°C.

This type of morphological organization is supported by a more statistical approach for individual species to examine phenotypic correlations among the eyespots. In another satyrine butterfly, *Maniola jurtina* (Brakefield, 1984), and in the nymphalids, *Precis coenia* and *P. evarete* (Paulsen and Nijhout, 1993), the size of any particular eyespot is positively correlated with the size of other eyespots, especially those on the same wing surface.

If the concept of the prototype is valid it becomes interesting to ask how the pattern of the character it represents can be changed through evolution to yield the present set of states, including those found within *Bicyclus*. Descriptions of genetic variances and covariances, together with information about developmental mechanisms, are also necessary to discover how the eyespot pattern can be decomposed into units or modules which have a (higher) degree of individuality in their development, and thus in their potential for independent evolutionary responses to selection. With such information, we can also eventually match the observed set of patterns to those which our analysis of the internal structure of the character predicts are more likely to be readily generated at the phenotypic level.

DEVELOPMENTAL MECHANISMS AND MORPHOLOGICAL ORGANIZATION

Working with the large posterior eyespot on the forewings of the buckeye butterfly, *Precis coenia,* Nijhout (1980) made the breakthrough which was crucial to revealing the developmental mechanisms of eyespot formation. Butterfly wings develop in the late larva as paired internal epidermal pouches, the imaginal discs. There are two discs on each side of the insect. These protrude at metamorphosis to form the pupal wings, the forewing overlying the hindwing. The cell layer of the dorsal forewing lies immediately under the pupal cuticle, and for a few hours after pupation it is attached to the cuticle. The critical events in pattern determination occur in the early pupae. The trachea of the forewing are visible through the pupal wing case. There are also pupal markings or raised areas of cuticle which overlie the central regions of putative adult eyespots. These provide landmarks for experimental manipulation of eyespot formation.

First, Nijhout (1980) used cautery with fine needles inserted through the pupal cuticle to damage cells of the developing forewing. He found that damage to cells in the central region of the putative posterior eyespot could produce dramatic reductions in the eventual size of the eyespot; the earlier the damage, the smaller the eyespot. More critically, shortly after pupation he was able to transplant the cells of the central region to an adjacent area of the wing where no eyespot pattern was normally observed. An ectopic eyespot formed around the grafted tissue.

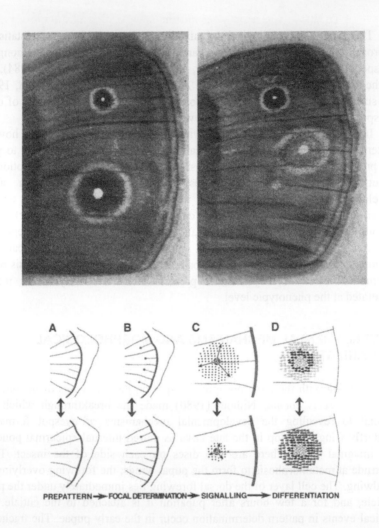

FIGURE 3 A. An ectopic eyespot produced after tranplantation of the focus of the posterior eyespot. B. Diagram of the developmental pathway of eyespot formation.

These results have been fully substantiated by subsequent experiments on *B. anynana* (French and Brakefield, 1992, 1995; Brakefield and French, 1995, 1999). Figure 3 illustrates an example of the result of transplanting an eyespot focus in this species. The overall results are also consistent with Nijhout's (1978) model of eyespot formation in which the focus induces a signal, presumably by diffusion, to surrounding cells. This results in a cone-shaped concentration gradient. At the end of signal induction the surrounding cells "interpret" the gradient, thus gaining information on their position relative to the central focal region and becoming fated to synthesize a particular color pigment.

The later experiments on *B. anynana* have, however, provided several additional insights. First, damage experiments to nonfocal areas of the distal wing blade performed shortly before the end of pattern determination (ca. 24 h after pupation at 27°C) can produce ectopic eyespots which are closely similar to control eyespots except that they have no central, white pupils (cf. in *P. coenia*, Nijhout, 1985). Perhaps, the eyespot foci act as sinks rather than sources. A uniform field of morphogen is then viewed as being initially present across the wing (French and Brakefield, 1992). This is degraded around active foci in early pupae, producing sink-like troughs in the profile. Damage shortly after pupation would then initially yield sinks but later diffusion and healing would tend to recreate the uniform morphogen profile. On the other hand, damage shortly before the end of pattern determination would produce similar sinks but without the time for healing by diffusion from surrounding cells. Results of series of damage experiments, which included the use of different severities of damage, are broadly consistent with this type of interpretation but there are no data which unambiguously distinguish between a source or sink model (French and Brakefield, 1992; Brakefield and French, 1995).

Second, experiments in which cells are damaged in the proximal (inner) region of the forewing do not produce ectopic patterns (Brakefield and French, 1995). Furthermore, focal grafts made into this region of the wing also yield no ectopic patterns. Additional transplantation experiments indicated that while signal transduction can occur through the cell layer of the proximal part of the wing, the cells are not competent to respond, or cannot synthesise a pigment other than that of the background, brown color (French and Brakefield, 1995). Similar distal to proximal grafts performed in *P. coenia* produce a dramatic difference in the ectopic color pattern. While grafts made to distal positions give patterns with a white outer eyespot ring, those to a proximal position yield an outer ring with the same color as two prominent orange bands which occur in that area of the forewing (Brakefield *et al.*, 1996). Variation between scale cells in their competence to produce specific color pigments must be one important mechanism generating spatial color variation in many butterfly wings, as well as differences among individuals and species (see also Koch *et al.*, 1998).

Carroll *et al.* (1994) showed that the developmental gene, *Distal-less*, is expressed in early butterfly wing discs in rays midway between the lacunae that will form wing veins. Later, shortly before pupation, *Distal-less* expression in the rays degrades but their proximal tips become enhanced in small circular groups of cells which correspond to the foci of the future eyespots. The use of *Distal-less* as a molecular probe has enabled a fuller description of eyespot formation in *P. coenia* and *B. anynana* in the form of a developmental pathway (Brakefield *et al.*, 1996). This can be viewed as involving four stages of regulation which are in sequence: (1) establishment of a prepattern with potential foci; (2) determination of the foci; (3) signaling from a focus induces surrounding cells; and (4) induced cells later differentiate into scales of different colors depending on their distance from the focus and their position in the wing. Most recently, Keys *et al.* (1999) have shown that other developmental genes which are involved in wing formation in *Drosophila* have additional components of expression in butterflies that suggest roles in specifying the eyespot pattern. For example, *hedgehog* is expressed strongly and transiently in areas flanking the eyespot foci while the genes *patched, cubitus interruptus,* and *engrailed* become strongly enhanced in the foci themselves, presumably as a result of hedgehog signaling. Further advances in this research at the molecular level are likely to reveal precisely how such signaling pathways with conserved roles in insect wing development are deployed in butterflies in specifying the wing color patterns. We also hope to be able to match some eyespot patterning genes to the regulation of specific signaling pathways.

Thus, this collaborative research effort has provided a basic understanding of eyespot development, although most of the details remain unclear. The involvement of a morphogen diffusion gradient rather than some form of cell-cell relay system has not been demonstrated unambiguously. One important interpretation arising from all these studies is that eyespot formation involves a developmental mechanism which is common to all eyespots on butterfly wings. There is thus convincing support for the concept that the different eyespots are serial repeats and for the view from comparative morphology that the eyespot pattern as a whole represents the most appropriate choice for the unit character. In this context, it then becomes critical to examine potential mechanisms by which differentiation can evolve among the eyespots corresponding to some prototype pattern. We have begun to examine this issue by concentrating on genetic variation for the eyespot pattern in *B. anynana*.

GENETIC VARIATION AND DEVELOPMENT OF FEATURES OF THE EYESPOT PATTERN

High phenotypic correlations are found among the individual eyespots of species including within the genera, *Bicyclus, Maniola,* and *Precis*. These are associated with positive genetic correlations (Brakefield and van Noordwijk, 1985; Holloway *et al.*, 1993; Monteiro *et al.*, 1994; Paulsen, 1994, 1996). This is unsurprising in view of the common developmental mechanism of eyespot formation. The issues of how evolutionary divergence occurs can then be considered in terms of either developmental or genetical mechanisms. Among the specific questions which can be posed from the former perspective are: how can development be uncoupled across eyespots, and how do novelties and modifications of the developmental pathway occur? The parallel questions in terms of genetics then become: how can genetic correlations be changed, and what is the pool of mutants available for genes with localized positional effects and novel phenotypic changes? Linking development and genetics might then indicate how bias in the generation of phenotypic variation is related to developmental organisation. We have tried to integrate the genetics of phenotypic variation in *Bicyclus* eyespots with the developmental insights.

Taking as a character the eyespot pattern based on repeated units of single eyespots, we can then also consider different features or aspects of the pattern and its constituent units. We have recognized the following four features: overall size, color composition, shape, and position. In addition, we have some information on both eyespot number and the presence or absence of particular eyespots. Several years ago we began a series of artificial selection experiments to explore genetic variances for different features of the eyespot pattern of *B. anynana* (Monteiro *et al.*, 1994, 1997a,b,c; Brakefield, 1998). We have also established a series of spontaneous single gene mutants which show a variety of effects on the eyespot pattern (Brakefield *et al.*, 1996; Brakefield, 1998; Brakefield and French, 1999). Our observations indicate that many genes are involved in eyespot formation. Future research will use mutagenesis and QTL gene mapping to more intensively probe the genetics. The best way to illustrate the findings to date and to expand on how such results should yield sounder predictions about evolutionary trajectories for the eyespot pattern is to make specific contrasts among the different features, or among the different mechanisms of producing similar phenotypes.

A. Eyespot Size and Color

Upward and downward artificial selection has been applied to each feature of the large posterior eyespot on the dorsal forewing of *B. anynana*. Lines were obtained with large or small (ultimately absent) eyespots, or with "black" or "gold" eyespots (the outer gold ring either being very narrow or broad relative to the inner part of the eyespot). Fully divergent phenotypic distributions were obtained within five to ten generations (Monteiro *et al.*, 1994, 1997a). These included phenotypes not represented within the unselected stock population (Fig. 4). Although selection was only directed at the posterior eyespot, other eyespots, especially the anterior forewing eyespot on the dorsal surface, tended to show concordant responses to selection. Thus, as mentioned earlier, there are positive genetic correlations among eyespots. Realized heritabilities were slightly higher for eyespot size. At around 50%, they were comparable with morphological characters in *Drosophila*, and they indicate substantial additive genetic variance. The rapid and prolonged responses to selection were consistent with contributions of a number of genes, each of small phenotypic effect. In contrast to the positive genetic correlation for a particular feature among eyespots, correlated responses between eyespot features were minimal (see further later). This is especially striking in specimens from the "black" and "gold" selected lines (Fig. 4).

FIGURE 4 Representative phenotypes from some of the selected lines of B. anynana.

Cautery and transplantation experiments have also been used to examine how the genetic differences between the upward and downward selected lines may influence developmental mechanisms of eyespot formation (Monteiro *et al.*, 1994, 1997a). The results indicate that while eyespot color is regulated through the response to the focal signal, eyespot size is primarily a property of divergence in the activity or strength of the focal signal (although with some contribution of effects traceable to the response). Therefore, although genetic variances are similar for these two eyespot features, the underlying developmental roles of the genes specifying each are very different. In addition, the observations indicate that the developmental options for change are more diverse for eyespot size than for eyespot color. This may suggest that evolution of divergence in size among eyespots may be more readily acheived than for color composition (Brakefield, 1998).

So far we have found no mutant gene for color composition, although a recessive gene, *melanine*, produces a general darkening of all adult cuticle, including the gold eyespot ring (P. M. Brakefield, previously unpublished). On the other hand, we have found several individual genes which influence eyespot size. The best described of these is *Bigeye* which increases the size of all eyespots (Fig. 5). The expression pattern of *Distal-less* indicates that this gene in some way influences the way in which the focal signal is responded to by the surrounding scale-forming cells (Brakefield *et al.*, 1996). This contrasts with the primary developmental effect in the upward selected line at the level of the activity or strength of the focal signal (Monteiro *et al.*, 1994). More recently, we have started to isolate two, apparently single, genes which reduce the size of the dorsal forewing eyespots. The mutant allele at one gene, provisionally called A^-, dramatically reduces the size of the anterior eyespot (which is frequently absent) with no effect on the large posterior eyespot (P. do Ó Beldade and P. M. Brakefield, unpublished data). The other gene, P^-, strongly reduces the posterior eyespot, usually in combination with a less marked effect on the anterior eyespot. The former of these two genes is especially interesting as it demonstrates the existence of single alleles with highly localized positional effects on a subset of eyespots. The *Spotty* allele specifies two novel signaling foci leading to additional eyespots between the anterior and posterior forewing eyespots (Fig. 5). Current crosses will combine the A^- allele with *Spotty* to examine more finely how localized the effect is. It will be especially interesting to discover the developmental basis for such localized effects (see Keys *et al.*, 1999).

I D,

Bigeye. The ventral wing surfaces are shown of female individuals (dorsal forewing patterns tend to follow those of the ventral surface). Each specimen is homozygote for the *Spotty* allele (the wild type is shown in Fig. 2).

Backcrossing the *Bigeye* allele into the downward Low-line for eyespot size for six generations has yielded a line in which *Bigeye* heterozygotes can only be detected by their ventral phenotype (unpublished data). The dorsal eyespots are very small or absent. This is consistent with an effective absence of focal signals in the Low line combined with an effect of *Bigeye* at the level of the response to this signal. In contrast, backcrossing the *Bigeye* allele into the upward High-line appears to yield an additive phenotypic effect in which the resulting eyespots can be larger than in the original *Bigeye* stock. Again, this is predictable from the interpreted changes in development in both the stock and the selected line. Combining *Bigeye* with *Spotty* increases the size of the novel eyespots (Fig. 5).

The occurrence of *Bigeye* and the responses to selection on eyespot size show that similar phenotypes can be produced by very different genetical and developmental changes. A further contrast is illustrated by the lines for eyespot size and color. Although the additive genetic variance is similar for these features, the underlying developmental changes are different. These observations raise a number of issues which are as yet unresolved (Brakefield, 1998). First, for a given change in a feature of a character, does it matter in terms of likelihood of involvement in the evolution of divergence whether a single gene or a number of genes is involved? Second, if two characters or features of a character behave similarly in terms of rate of response to directional selection, does it matter whether they have a similar or dissimilar developmental basis?

B. Eyespot Shape

The eyespots of *B. anynana* have a circular shape. Artificial selection on eyespot shape (toward "fat" or "thin") produced a much slower response than observed for the other features (Monteiro *et al.*, 1997b, c). Realized heritabilities were only about 15%. There were also indications of a rather rapid approach to a limit for the response to selection. Morphometric studies showed that the response to selection was at least partly accounted for by changes in wing shape and in the matrix of scale cells around the forewing eyespot.

Perhaps the options for changing eyespot shape in a bilaterally symmetrical way through modifying the basic components of the developmental pathway of eyespot formation are more limited than for other features (it would be interesting to pursue such ideas by theoretical models, cf. Nijhout and Paulsen, 1997). However, abrupt effects are possible as is shown by two of our mutants, *cyclops* and *comet*, which dramatically change eyespot shape (Fig. 5). The fundamental developmental effect of the former mutant is a change in the venation pattern; one of the major veins is vestigial in each adult wing, leading to a fusion of the flanking eyespots (Brakefield *et al.*, 1996). However, such pleiotropic effects of venation mutants on wing pattern may not be very important in evolution. *Cyclops* is homozygous lethal, and the heterozygotes of mutant phenotype rapidly damage their wings following eclosion, presumably as a direct consequence of the change in venation.

Comet is a fascinating allele for several reasons. Homozygotes not only have the pear-shaped eyespots but the pattern of the parafocal elements (border chevrons and bands) is disrupted producing proximal extensions along the wing veins. Most strikingly, males of this genotype have highly vestigial secondary sexual characters (V. Schneider and P.M. Brakefield, in preparation). Wild-type male *B. anynana* have highly developed androconia in the proximal region of the dorsal hindwing. These comprise two plumes of long hair-like scales overlying regions of other modified scales. The androconia are involved in the production and dissemination of phermones during courtship. These structures are greatly reduced in the *comet* mutant. Although such extreme pleiotropy appears to be the exception (it is not found in any other of our eyespot mutants), it does indicate the potential for effects across many wing characters.

C. Uncoupling of Eyespots

The phenotypic effects of several (but not all) mutants vary among the eyespots. *Spotty* adds eyespots to some wing cells but not to others. The localized effect on a single major vein in *cyclops* produces a more profound effect on the flanking eyespots than on others. Most strikingly, A^- only affects the anterior of the two forewing eyespots. Most recently, butterflies have appeared in a line selected for fast preadult development in which two of the hindwing ventral eyespots (numbers three and four from the top, costal border of the wing) are absent or highly reduced while the other eyespots are unchanged. Early breeding results suggest a simple genetic basis for this pattern change. Intriguingly, one small subgroup of *Bicyclus* shows a very similar change in eyespot pattern (A. F. Monteiro, personal communication). We predict that these types of "uncoupling" genes with major phenotypic effects which are distributed in a nonuniform way across the eyespots have been extremely important in the context of evolutionary novelties and in promoting discrete patterns of morphological diversity. Moreover, the underlying developmental organisation of the eyespot character may generate bias such that certain combinations of eyespots can be readily uncoupled while other cannot (for example, perhaps while eyespots 3 and 4 can "drop out," combinations such as 1 and 2, or 4 and 5 cannot; see Keys *et al.*, 1999).

The effects of uncoupling leading to some degree of independence in development can be dramatic. One specific example is the extreme phenotypic plasticity of the ventral surface eyespots in contrast to an absence of plasticity in those on the dorsal wing surfaces of *B. anynana* and many other satyrine butterflies (Brakefield and Larsen, 1984; Brakefield and French, 1999). The plasticity is induced by temperature and development time during the late larval period (Kooi and Brakefield, 1999). Two seasonal forms are found in nature, one in the wet season with large ventral eyespots and the other in the dry season with small eyespots. This seasonal polyphenism represents an adaptive response to differences in resting background and predator pressure between the seasons (Brakefield and Reitsma, 1991; Brakefield and French, 1999). The plasticity is mediated in the early pupae by the titers of ecdysteroid hormones (Koch *et al.*, 1996). The ventral eyespot foci of the seasonal forms differ in their signaling activity in early pupae (Brakefield *et al.*, 1996). We have recently used results from measurements of selected lines for eyespot size on either dorsal or ventral wing surfaces to suggest that one possible mechanism for this uncoupling between the wing surfaces is the presence of ecdysteroid receptors in cells of the ventral eyespot foci, but not in those of the dorsal wing surface (Brakefield *et al.*, 1998). Whatever the precise mechanism, this example demonstrates that novel means of regulating development of part of the eyespot pattern can be coopted during evolution. In fact, divergence between dorsal and ventral wing surfaces is an almost ubiquitous phenomenon of butterfly wing patterns. This is likely to be driven by butterfly behavior and natural selection but clearly mechanisms have evolved to facilitate independence in development at this level (see also Weatherby *et al.*, 1999).

DISCUSSION

We can now discuss again the most appropriate way of applying the term character to the eyespots of a butterfly such as *B. anynana*. There are several possible choices for the unit character: a single feature of an eyespot, a single eyespot, a subset of eyespots (e.g., those on one wing surface or dorsal/ventral), or the complete pattern. While arguments can be made for each of these from a morphological, genetical, or developmental viewpoint, the best overall choice is probably the complete pattern. Early descriptions of the morphology and variation across species indeed suggested that the unit character is most usefully taken as the whole pattern. This is now strongly supported by our understanding of the developmental mechanisms of eyespot formation from manipulative experiments and molecular research in *P. coenia* and *B. anynana*. All the empirical data are consistent with a single common developmental mechanism underlying the eyespots in these species and probably in all other Lepidoptera.

In our selection experiments on a specific eyespot feature there is a striking lack of substantial correlated responses for other features (cf. "black" and "gold" specimens in Fig. 4). While this has only been quantified for eyespot size and color, it probably holds for other pairs of features with the possible exception of additional "supernumerary" eyespots appearing in upward selected lines for eyespot size. The genetic correlations among eyespot features, even for the same individual eyespot, are thus low. For eyespot size and color, this is presumably because whereas size is specified developmentally primarily by signal strength, the color composition is determined by threshold responses to the signal which correspond to boundaries between eyespot rings. This might suggest that from the perspective of genetics and development the different features should represent the unit characters rather than the complete eyespot pattern. However, it is difficult to imagine that natural selection frequently works in such a tightly targeted manner so that from an integrated evolutionary perspective, the choice of the eyespot pattern, inclusive of the various features, probably remains the most appropriate choice. This discussion also illustrates that more definitive descriptions of character structure, and also of evolutionary constraints, will require specific knowledge about mechanisms of natural selection on the eyespot pattern. We have performed some successful analyses of survival in cohorts of butterflies in the field to examine how visual selection by predators on butterflies at rest influences the size of the ventral eyespots in the seasonal forms (see Brakefield and French, 1999). However, such experiments will have to be greatly refined to produce the necessary detail about natural selection to match our understanding of character structure.

Condamin (1973) recognizes 77 species within the genus *Bicyclus*. These can have highly divergent eyespot patterns. There is also high species richness and diversity in many related genera (including *Mycalesis*). Many differences in the eyespot pattern, especially those which are quantitative and rather uniform across eyespots, are likely to reflect differentiation involving several or many eyespot patterning genes of small effect. There are indeed species of *Bicyclus* which correspond phenotypically to certain of the extreme phenotypes produced by our selection experiments in *B. anynana* (Brakefield, 1998). Such changes may be more characteristic of divergence among the most closely related species toward the tips of the phylogeny.

On the other hand, where more novel phenotypes are concerned, we believe that further study will eventually detect many examples where single genes have played a crucial role in the evolution of morphological diversity and character divergence across taxa. Such genes will be involved in some abrupt changes in features of all the eyespots, as well as in uncoupling certain eyespots leading to independent phenotypic effects across eyespots. *Spotty* may represent a gene of this type (Brakefield *et al.*, 1996). We hope to detect further examples in *B. anynana*. The availability of such genes, and thus the bounds of the developmental repetoire, may have yielded bias in evolutionary trajectories. Perhaps the potential pool of mutants overlaps considerably across species, even rather distantly related ones. If so, then mutagenesis in *B. anynana* may reveal some of the long-term potential for the eyespot character. Given sufficient time and large enough population sizes, absolute constraints in terms of genes or developmental options may have little relevance to the evolution of morphological diversity (although see the discussion of eyespot shape earlier). However, bias in the standing variation within populations is also likely to be critical to the adaptive responses of populations, especially those involving tracking of rapid climate change. If populations on this time horizon tend to follow paths of least resistance in character evolution rather than alternatives, this may also contribute profoundly to shaping long-term patterns of morphological diversity. Further artificial selection experiments are being used to examine these ideas about the relevance of bias introduced through the standing genetic variation within populations (see Brakefield, 1998).

In terms of spectacular visual diversity, the evolution of the wing color pattern of butterflies appears to have behaved in a remarkably unconstrained manner. Although high morphological diversity may suggest extreme freedom in character evolution, such flexibility may be superficial. When different characters are considered, for example, the eyespot pattern as against the pattern of medial bands, there may effectively be few limits to independent evolution in terms of both direction and extent of change (Nijhout, 1991; Paulsen and Nijhout, 1993; Paulsen, 1994, 1996). Such complete independence in genetical, developmental, and evolutionary perspectives is perhaps the best evidence for the existence of different characters. In contrast, if the eyespot pattern indeed reflects a single character in terms of its evolutionary origins, and its genetical and developmental architecture, then evolution may be shaped not only by natural selection but also by bias in the generation of phenotypic variability. Evolution has progressed both by changing the character as a whole and, where possible, by decomposing a single unit into one or more partially independent modules. The type of detailed study at different levels of biological organization which we have begun for the eyespot pattern of *B. anynana* will provide the most detailed understanding of how such a character evolves.

ACKNOWLEDGMENTS

My great debt to all those colleagues and collaborators on the *Bicyclus* project will be clear to all readers of this essay.

LITERATURE CITED

Brakefield, P.M. (1984). The ecological genetics of quantitative characters of *Maniola jurtina* and other butterflies. *In* "The Biology of Butterflies" (R.I. Vane-Wright and P.R. Ackery, eds), pp. 167-190. Academic Press, London.

Brakefield, P.M. (1998). The evolution-development interface and advances with the eyespot patterns of *Bicyclus* butterflies. *Heredity* **80**:265-272.

Brakefield, P.M., and French, V. (1995). Eyespot development on butterfly wings: the epidermal respnse to damage. *Dev. Biol* **168**:98-111.

Brakefield, P.M., and French, V. (1999). Butterfly wings: the evolution of development of colour patterns. *BioEssays* **21**:391-401.

Brakefield, P.M., Gates, J., Keys, D., Kesbeke, F., Wijngaarden, P.J., Monteiro, A., French, V., and Carroll, S.B. (1996). Development, plasticity and evolution of butterfly eyespot patterns. *Nature* **384**:236-242.

Brakefield, P.M., Kesbeke, F., and Koch, P.B. (1998). The regulation of phenotypic plasticity of eyespots in the butterfly *Bicyclus anynana*. *Am Nat* **152**:853-860.

Brakefield, P.M., and Larsen, T.B. (1984). The evolutionary significance of dry and wet season forms in some tropical butterflies. *Biol. J. Linn. Soc.* **22**:1-12.

Brakefield, P.M., and van Noordwijk, A.J. (1985). The genetics of spot pattern characters in the meadow brown butterfly *Maniola jurtina* (Lepidoptera: Satyrinae). *Heredity* **54**:275-284.

Brakefield, P.M., and Reitsma, N. (1991). Phenotypic plasticity, seasonal climate and the population biology of *Bicyclus* butterflies. *Ecol. Entomol.* **16**:291-303.

Carroll, S.B., Gates, J., Keys, D.N., Paddock, S.W., Panganiban, G.E.F., Selegue, J.E., and Williams, J.A. (1994). Pattern formation and eyespot determination in butterfly wings. *Science* **265**:109-114.

Condamin, M. (1973). Monographie du genre *Bicyclus* (Lepidoptera Satyridae). *Mem. Inst. Fond. Afr. Noire* **88**:1-324.

French V. (1997). Pattern formation in colour on butterfly wings. *Curr. Opin. Genet Dev.* **7**:524-529.

French, V., and Brakefield, P.M. (1992). The development of eyespot patterns on butterfly wings: morphogen sources or sinks? *Development* **116**:103-109.

French, V., and Brakefield, P.M. (1995). Eyespot development on butterfly wings: the focal signal. *Dev. Biol.* **168**:112-123.

Galant, R., Skeath, J.B., Paddock, S., Lewis, D.L., and Carroll, S.B. (1998). Expression pattern of a butterfly achaete-scute homolog reveals the homology of butterfly wing scales and insect sensory bristles. *Curr. Biol.* **8**:807-813.

Holloway, G.J., Brakefield, P.M., and Kofman, S. (1993). The genetics of wing pattern elements in the polyphenic butterfly, *Bicyclus anynana*. *Heredity* **70**:179-186.

Keys, D.N., Lewis, D.L., Selegue, J.E., Pearson, B.J., Goodrich, L.V., Johnson, R.L., Gates, J., Scott, M.P., and Carroll, S.B. (1999). Recruitment of a hedgehog regulatory circuit in butterfly eyespot formation. *Science* **283**:532-534.

Koch, P.B., Brakefield, P.M., and Kesbeke, F. (1996). Ecdysteroids control eyespot size and wing color pattern in the polyphenic butterfly, *Bicyclus anynana*. *J. Insect Physiol.* **42**:223-230.

Koch, P.B., Keys, D.N., Rocheleau, T., Aronstein, K., Blackburn, M., Carroll, S.B., and french-Constant, R.H. (1998). Regulation of dopa decarboxylase expression during colour pattern formation in wild-type and melanic tiger swallowtail butterflies. *Development* **125**:2303-2313.

Kooi, R.E., and Brakefield, P.M. (1999). The critical period for wing pattern induction in the polyphenic tropical butterfly *Bicyclus anynana* (Satyrinae). *J Insect Physiol.* **45**:201-212

Maynard Smith, J., Burian, J.R., Kauffman, S., Alberch, P., Campell, J., Goodwin, B., Raup, D., and Wolpert, L. (1985). Developmental constraints and evolution. *Q. Rev. Biol.* **60**:265-287.

Monteiro, A.F., Brakefield, P.M., and French, V. (1994). The evolutionary genetics and developmental basis of wing pattern variation in the butterfly *Bicyclus anynana*. *Evolution* **48**:1147-1157.

Monteiro, A.F., Brakefield, P.M., and French, V. (1997a). Butterfly eyespots: the genetics and development of the color rings. *Evolution* **51**:1207-1216.

Monteiro, A.F., Brakefield, P.M., and French, V. (1997b). The genetics and development of an eyespot pattern in the butterfly *Bicyclus anynana*: response to selection for eyespot shape. *Genetics* **146**:287-294.

Monteiro, A.F., Brakefield, P.M., and French, V. (1997c). The relationship between eyespot shape and wing shape in the butterfly. *Bicyclus anynana*: A genetic and morphometrical approach. *J. Evol. Biol.* **10**:787-802.

Müller, G.B., and Wagner, G.P. (1991). Novelty in evolution: Restructuring the concept. *Annu. Rev .Ecol. Syst.* **22**:229-256.

Nijhout, H.F. (1978). Wing pattern formation in Lepidoptera: a model. *J. Exp. Zool.* **206**:119-136.

Nijhout, H.F. (1980). Pattern formation on Lepidopteran wings: determination of an eyespot. *Dev. Biol.* **80**:267-274.

Nijhout, H.F (1985). Cautery-induced colour patterns in *Precis coenia* (Lepidoptera: Nymphalidae*).*

J. Embryol. Exp. Morphol. **86**:191-203.

Nijhout, H.F. (1991). "The Development and Evolution of Butterfly Wing Patterns." Smithsonian Institution Press, Washington.

Nijhout, H.F., and Paulsen, S.M. (1997). Developmental models and polygenic characters. *Am. Nat.* **149**:394-405.

Paulsen, S.M. (1994). Quantitative genetics of butterfly wing color patterns. *Dev. Genet.* **15**:79-91.

Paulsen, S.M. (1996). Quantitative genetics of the wing color pattern in the buckeye butterfly (*Precis coenia* and *Precis evarete*): evidence against the constancy of G. *Evolution* **50**:1585-1597.

Paulsen, S.M., and Nijhout, H.F. (1993). Phenotypic correlation structure among elements of the color pattern in *Precis coenia* (Lepidoptera: Nymphalidae). *Evolution* **47**:593-618.

Schlichting, C.D., and Pigliucci, M. (1998). "Phenotypic Evolution: A Reaction Norm Perspective." Sinauer Associates, Sunderland, Massachusetts.

Wagner, G.P., and Altenberg, L. (1996). Complex adaptations and the evolution of evolvability. *Evolution* **50**:967-976.

Weatherbee, S.D., Nijhout, H.F., Grunert, L.W., Halder, G., Galant, R., Selegue, J., and Carroll, S. (1999). Ultrabithorax function in butterfly wings and the evolution of insect wing patterns. *Curr. Biol.* **9**:109-115.

Wourms, M.K., and Wasserman, F.E. (1985). Butterfly wing markings are more advantageous during handling than during the initial strike of an avian predator. *Evolution* **39**:845-851.

16

CHARACTERS AND ENVIRONMENTS

MASSIMO PIGLIUCCI

Departments of Botany and of Ecology and Evolutionary Biology, University of Tennessee, Knoxville, TN 37996

CHARACTERS AND ENVIRONMENTS: THE CONCEPT OF REACTION NORMS

When most biologists speak of a given "character," they usually think they know what they mean. A character can be continuous or discrete, for example, the height of a plant or the number of vertebrae in a reptile; but in either case, it is usually obvious how to measure it.[1] Things are only slightly more

---

[1] Even though the science of morphometrics had to become increasingly sophisticated to deal with such an "obvious" problem (Bookstein, 1991; Rohlf and Marcus, 1993; Marcus *et al.*, 1996).

The Character Concept in Evolutionary Biology

complicated by the fact that each character can assume a variety of possible states. We can still identify subspecies of insects characterized by variations in the patterns of their colors, or populations of *Drosophila* with different numbers of bristles. Things become a little trickier once it is realized that the same exact character can assume different forms in organisms with the same genotype if they are raised under different environmental conditions. The common term for this is phenotypic plasticity (Bradshaw,1965; Schlichting, 1986; Sultan, 1987; West-Eberhard, 1989; Scheiner, 1993a; Pigliucci, 1996a), and its impact on the very concepts of character and character states has not so far been investigated in details by evolutionary biologists.

One reason for such a dearth of consideration is that neodarwinists have been busy for a long time simply denying the existence (or at the least the relevance) of phenotypic plasticity (Schlichting and Pigliucci, 1998). The core of the neodarwinian synthesis was the unification of classical natural history (*a la Darwin*) with modern genetics (especially population genetics) (Provine, 1971; Mayr and Provine, 1980; Mayr, 1993). Ecology barely entered the picture [notwithstanding the important work of Clausen, Keck, and Hiesey in the United States: Clausen *et al.* (1940); Clausen and Hiesey, 1960; and of Ford (1931) in England]. The environment was thought of as a "problem" faced by the organism, and a combination of genetics and natural selection would provide the "solution" (Lewontin, 1978; Levins and Lewontin, 1985). An experimental concept developed during that time perhaps best embodied the attitude of an era. Researchers thought that they could minimize the effect of environmental variation on phenotypes by growing different genotypes under one set of standardized conditions. The reasoning behind these "common garden" experiments was that any observed variation would have to be attributed to the genotype (the *important* component from an evolutionary standpoint), while the environment would have been kept in check. That is true enough, except that it was soon realized that the results are going to be dependent on *which* common environment one chooses. Furthermore, many common environments will actually yield misleading outcomes because they represent novel environmental conditions for the genotypes being studied. The phenotype expressed under common garden conditions, therefore, may not reflect what we would observe in the field and in the selective environments historically experienced by a given population (Service and Rose, 1985; Holloway *et al.*, 1990; Joshi and Thompson, 1996; Hawthorne, 1997).

The concept of phenotypic plasticity that emerged in parallel with the neodarwinian synthesis is both simple and intriguing. The most common way to represent the joint effects of genotypes and environments on a given phenotype is to plot the corresponding "norms of reaction" (Fig. 1). These are simply visualized as a graph showing genotypic-specific functions in environment/phenotype space. Each reaction norm is specific to a given genotype for a particular trait in response to a certain set of environmental circumstances. It can be a linear function or it can assume more complex shapes.

In the simplest case of a linear function, it has two fundamental properties (Fig. 1a): the average height of the reaction norm represents the across-environment mean for that genotype and the slope is a measure of plasticity. To be more precise, the slope actually embodies two distinct quantities defining plasticity. The degree of slope measures the intensity of the response to the environment (the *amount* of plasticity). The sign of the slope (positive or negative) represents the *pattern* of plasticity (i.e., whether the character value tends to increase or decrease with an increase in the value of the environmental factor). Of course, a reaction norm can be plastic or nonplastic (Fig. 1b), which means that plasticity and reaction norms *are not* synonyms, contrary to a widespread usage in the literature.

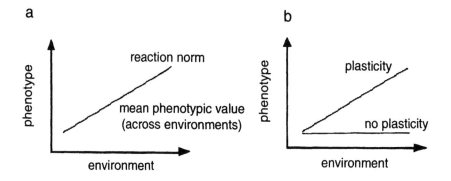

FIGURE 1 The two fundamental properties of reactions norms (mean and plasticity, a) and the difference between a plastic and nonplastic genotype (b).

There are three possible biological phenomena affecting the observed pattern of reaction norms. First, one can have genetic variation for overall trait value if the across-environment means differ significantly (Fig. 2b). Second, each reaction norm may show plasticity if its slope differs significantly from zero. Finally, there can be genetic variation for plasticity (also termed "genotype by environment interaction," following the terminology of standard analyses of variance: Lewontin, 1974) whenever genotypes vary significantly in the slopes of their reaction norms (Fig. 2d,f).

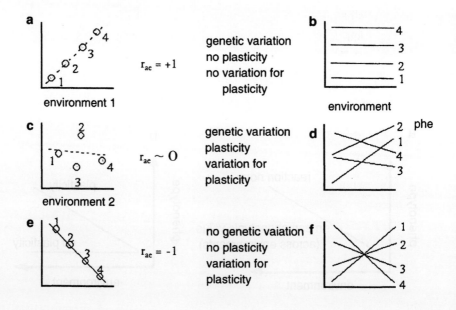

FIGURE 2 A comparison between characters visualized as interenvironment genetic correlations (left: a, c, e) or reaction norms (right: b, d, f). Each line on the right (a genotype's reaction norm) corresponds to a circle on the left. Each number indicates a distinct genotype. After Via, 1987.

Notice that the three phenomena are conceptually independent from each other, and any association among them is a matter of empirical determination. A bundle of reaction norms can show genetic variation without either plasticity or interaction effects if the lines are parallel to each other and to the environmental axis, but separated vertically (Fig. 2b). Alternatively, one can have plasticity with neither genetic variation nor interactions if the norms are parallel to each other, have a slope, but form a tight bundle with no significant differences in across-environment means (not shown). Finally, there can be interaction without genetic variation for the mean or mean plasticity. This latter occurrence is particularly intriguing and counterintuitive. The way such a population would look like in a reaction norm diagram is a series of lines diverging from each other at either extreme of the environmental gradient, while converging toward the center. Furthermore, the slopes of some reaction norms would be negative while some others would be positive (Fig. 2f). While an analysis of variance would detect no genetic variation (for the mean) and no (overall) plasticity, in fact many genotypes would be plastic (but in different fashions), and there would be ample genetic variation for plasticity (because of the divergence of the reaction norms). Obviously, any combination of the three fundamental situations (genetic variation, plasticity, and variation for plasticity) can occur, and has been observed in natural populations of plants and animals.

THE IDEA OF INTERENVIRONMENT GENETIC CORRELATION

There is another way of looking at reaction norms, which makes the relationship between characters and environments even more intuitive. Instead of plotting genotype functions in environment-phenotype space, it is equally possible to represent genotypes as points in environment-environment space (Fig. 2a,c,e). One can think of the expression of the same trait in different environments as two distinct traits (in environment one and environment two), related by a genetic correlation that can assume any value between −1 and +1. If the across-environment genetic correlation (r_{ae}) is close to either positive or negative one, the simplest interpretation is that the same genes affect the trait in both environments (however, if the correlation is negative, those same genes have opposite effects in the two environments). If r_{ae} is close to zero, one can assume that different genes explain the variation for that trait in each environment.[2] This alternative representation has been introduced in evolutionary biology by Via and Lande (1985) as a way of studying the evolution of phenotypic plasticity within the framework of standard quantitative

[2] Of course, there are plenty of other explanations for the observed pattern, for example, that the genes in question are physically linked on a chromosome, as opposed to actually be the same DNA sequence. See Pigliucci and Schlichting (1997) for a discussion of the limits of this approach.

genetics (e.g., Conner and Via, 1993; Andersson and Shaw, 1994). The original idea, however, was proposed in the 1950s by Falconer (1952), and elaborated upon by several authors (Robertson, 1959; Dickerson, 1962; Yamada, 1962; Fernando *et al.*, 1984).

The two ways of viewing phenotypic plasticity, sometimes termed the reaction norm and the character state approaches, are perfectly equivalent from a mathematical viewpoint (Dickerson, 1962; Yamada, 1962; van Tienderen and Koelewijn, 1994; de Jong, 1995). However, they reflect a different biological concept of what a character is, how it is controlled, and how it responds to environmental changes (Scheiner, 1993b; Schlichting and Pigliucci, 1993, 1995a, 1998). Graphically, of course, it becomes difficult to represent reaction norms as character states as soon as the number of environments becomes higher than two or three. More importantly, however, I have argued elsewhere (Pigliucci, 1996b) that both approaches suffer from the simple fact that they are *statistical representations* of the underlying biology. As such, they do not tell us anything about the molecular biology, developmental mechanisms, or physiology that produce these patterns and that, in fact, are the main players that need to be understood in studying the evolution of characters.

MULTIVARIATE DIMENSIONS: CORRELATIONS AMONG PLASTICITIES AND PLASTICITY OF CORRELATIONS

Characters tend to covary across environments. This phenomenon has been termed "plasticity integration" by Schlichting (1986, 1989a), and it has been demonstrated in a number of plant and animal systems (e.g., Marshall *et al.*, 1986; Waitt and Levin, 1993; Newman, 1994). To date, however, few experimental studies have addressed plasticity integration, in part because of the very large size of the necessary experiments. However, plasticity integration is fundamental to our understanding of character evolution. If characters evolve in concert not just within a given environment, but across the range of frequently encountered conditions, this calls for models of phenotypic evolution much more complicated than the ones produced so far (van Tienderen and Jong, 1994; van Tienderen and Koelewijn, 1994).

There are two major hypotheses capable of explaining the existence of patterns of correlations among plasticities[3]: selection and constraints. Existing correlations among plasticities may have been selected because an across-environment integration of the phenotype is adaptive. Alternatively, they may be the result of the incapacity of the genetic system to decouple those reaction norms (for example, because similar regulatory genes and biochemical pathways are responsible for the plasticities of different traits). Schmalhausen (1949)

---

[3] Besides the nondeterministic outcome of random events such as genetic drift.

proposed the first scenario as a general feature of adaptive evolution. The second one is emerging as a likely (but not mutually exclusive) alternative from the massive body of knowledge now accumulated on the redundancy (or lack thereof) of biochemical and developmental pathways (Ingram *et al.*, 1995; Pickett and Meeks-Wagner, 1995). Schmitt and I have suggested that both phenomena may be at work to maintain fairly strict patterns of plasticity integration characterizing the so-called "shade avoidance" response in colonizing plants such as *Arabidopsis thaliana* (Pigliucci and Schmitt, 1999).

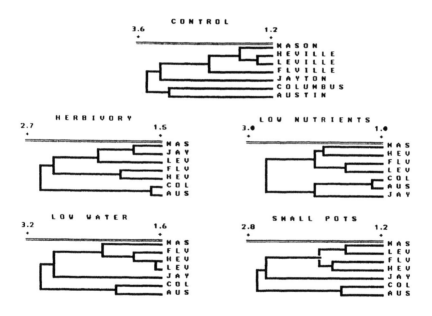

FIGURE 3 Dendrograms showing the similarity of phenotypic correlation matrices among seven populations of *Phlox drummondii*. Each diagram refers to one of five different environmental conditions. Notice how the relative similarity of the correlation matrices changes with the environment, indicating plasticity of phenotypic integration. (From Schlichting and Pigliucci, 1995).

Plasticity integration is not to be confused with *phenotypic integration*, which refers to the degree of correlation among characters within one environment (e.g., Zelditch, 1988; Pigliucci *et al.*, 1991a; Cheverud, 1995, 1996). Of course, things do get confused when one realizes that patterns of phenotypic integration can be different when observed in distinct environments, that is, they themselves are plastic (Lechowicz and Blais, 1988; Schlichting, 1989b; Stearns *et al.*, 1991; Kudoh *et al.*, 1996). An example of the latter is shown in Fig. 3. Schlichting and Pigliucci (1995b) compared the patterns of phenotypic integration (i.e., the character correlations) of several populations of *Phlox drummondii*, finding that populations could be grouped by the similarity of their correlation structures. However, such grouping depends on the particular environment in which the plants were growing (once the genetic background is randomized and accounted for). For example, while populations Austin and Columbus always belong to the same branch on the phenogram, their position relative to Jayton is quite different under the low nutrient and herbivory treatments and when these are compared to the other environments.

This raises not only interesting evolutionary and ecological questions, but it also puts in a new perspective an old problem of systematic studies, at least those conducted at lower taxonomic ranks. In a study on *Ornithogalum montanum* (Liliaceae), for example, I have demonstrated that there was no relationship between the way in which the multivariate phenotype of the plant responded to environmental stress and the subspecific classification of several populations (Pigliucci *et al.*, 1991b). In other words, if one were to proceed to identify the populations as belonging to one subspecies or another, the results would be highly dependent on such a simple parameter as water availability! As mentioned before, a "common garden" approach would not solve the problem either, and in fact it may very well make it worse. Common garden experiments are designed to provide a relatively benign environment to the plants. In the field, however, if definite phenotypic differences evolve they are probably going to be in response to stress or locally different conditions. The benign, controlled conditions of a greenhouse may obliterate any biologically significant difference among genotypes (the "silver spoon effect" quoted in Sultan, 1995). This same problem for taxonomy at the intraspecific level caused by phenotypic plasticity is likely to be important at the interspecific level (certainly for sibling species, for example), and it has been encountered even at the generic level (Trainor and Egan 1991).

Overall, consideration of both plasticity integration and the plasticity of character correlations certainly raises the question of what a character is to begin with. Gould and Lewontin (1979) question what—even within one environmental setting—we can reasonably define as a character. For example, they pointed out that the "chin," long considered a variable trait in human populations, is actually the result of the differential growth of two mandibular fields. We could say that the character "chin" is actually defined by the genetic correlation between the underlying growth fields. As the preceding discussion has showed, this correlation could be altered by environmental conditions (plasticity of integration); alternatively, the reaction norms of the two mandibular fields may be correlated (plasticity integration). The more layers we consider, the more it becomes difficult to agree on what a chin may be. Things turn to be even more intriguing once we take into account the fact that plasticity is not an instantaneous phenomenon, but it develops gradually throughout the ontogeny of an organism.

YET ANOTHER DIMENSION TO THE PROBLEM: THE INTERACTION OF DEVELOPMENT AND PLASTICITY

Phenotypic plasticity is not just an environmental phenomenon. It is the result of complex genotype–environment interactions. However, these interactions do not occur at one point in time. Rather, phenotypic plasticity is a developmental process, and the reaction norms that we usually measure in the adult stage, or at reproductive maturity, are in fact the result of a positive feedback between environments and genes throughout the ontogeny of an organism (Schmalhausen, 1949; Smith-Gill, 1983; Pigliucci *et al.*, 1996; Schlichting and Pigliucci, 1998).

The literature on developmental plasticity is assuming an increasingly important role in shaping our thinking about plasticity in particular and development in general. Clear empirical examples of how ontogenetic trajectories are shaped by the interaction between development, genotype, and environment have been published especially, but not uniquely, in plant biology (Dong and Pierdominici ,1995; Martin-Mora and James, 1995; Pigliucci and Schlichting, 1995; Brakefield *et al.*, 1996; Bruni *et al.*, 1996; Gedroc *et al.*, 1996; Pigliucci *et al.*, 1997).

The emerging picture from all these studies is that adult characters are shaped gradually by the way genetic instructions are expressed in an ever-changing environmental milieu. This is far from being a vague statement. We can now track the differential expression of genes at different times during ontogeny, in different tissues, and in response to distinct environmental conditions (Parsons and Mattoo, 1991; Nelson and Langdale, 1992; Wei and Deng, 1992; Estevez *et al.*, 1993; Neyfakh and Hartl, 1993; Crews *et al.*, 1994; Prandl *et al.*, 1995). This promises to open the way to the construction of a long-awaited bridge between organismal and molecular biology, with epigenetics providing the link between the two.

There is another sense in which plasticity is relevant to the formation of characters through development. Schlichting and Pigliucci (1998) have argued that the internal conditions of an organism are somewhat analogous to the external environment. To be more precise, one can see the epigenetic processes occurring throughout development as an interaction of genes with *two* kinds of environments: the classical external environment comprising biotic and abiotic factors, and the internal environment, including diffusing chemicals, cell-cell interactions, and so on. While this perspective may be pushing the concept of "environment" too far for some people, it is still useful to think about how genetic instructions give origin to epigenetic processes. Considering internal and external environments as comparable interactors with the genes may shed some light onto the evolution of epigenetic systems themselves.

This suggestion of treating internal and external environments as analogous entities raises the question of what exactly we consider an "environment" and how do we measure it. This is one of the most important and deceptively simple questions related to the problem of how phenotypic plasticity changes our way of looking at characters.

BUT WHAT IS THE ENVIRONMENT?

Perhaps one of the most fundamental problems with ecology is that, even though it is supposed to study the effects of environments on organisms, we are still at a loss when it comes down to pinpoint the biologically relevant components of an organism's milieu. This is true notwithstanding innumerable studies quantifying physical and biological aspects of the environment. An empirical approach to answer this question has been taken, for example, by Bell and Lechowicz (Bell and Lechowicz, 1991; Lechowicz and Bell, 1991). In the first paper of that series, these authors argued that there are fundamentally two approaches that can be used to study environmental heterogeneity. The indirect method is the most intuitive and is based on actual measurements of the physical environment in which plants and animals live. The problem is, as any ecologist knows, that the amount of variance detectable even on small spatial scales in factors such as temperature, humidity, and nutrient availability is simply mind-boggling. To this, one has to add that such measurements are also usually not consistent in time because of both seasonal and short time fluctuations. The second approach is what Bell and Lechowicz term direct. This consists of the idea of exploiting the organisms themselves as indicators of the quality of the environment through the use of bioassays. Unfortunately, there are several problems with this approach too. First, it is actually feasible only for sedentary or easy to track organisms (mostly plants and sedentary animals such as corals). Second, while one gets a fairly accurate idea of how the organism perceives environmental quality, the direct method does not provide the investigator with any clue about which aspects of the environment are actually causing the observed response. The latter is a potentially devastating limitation, since— unless coupled with other approaches—it confines ecology to the study of correlations and variances, precluding an understanding of the causal mechanisms lurking behind the observed statistical patterns.

These limitations can be overcome, as Bell and Lechowicz showed in applying both approaches to their experimental systems, with very interesting results. The direct method was used to study *Arabidopsis thaliana* and *Hordeum vulgare* in an undisturbed forest near Montreal (Bell and Lechowicz, 1991). Nested sampling of a 50 x 50-m area showed a steady decrease in environmental variance from scales of 10 to 0.1 m (the regression of log variance vs. log distance was linear). While the correlation between sites decreases with distance, this is true at all spatial scales examined (and it occurs at the same rate), suggesting that the environments considered in the study are equally complex at all spatial scales (i.e., they are fractal). The indirect method was applied by sampling 555 points in the same grid and measuring soil pH and availability of K^+ and NO_3^- ions (Lechowicz and Bell, 1991). The results demonstrated that all three edaphic measures are predictably similar up to 2 m, but that at larger spatial scales the autocorrelation is negligible. Bell and Lechowicz concluded that the environment varies at scales that are relevant to seed dispersal and genetic neighborhood size of typical understory herbs such as *Impatiens*. This finding affects our understanding of the relationship between the grain of the environment as perceived by the organism and the maintenance of genetic variation and, therefore, the response to selection in natural populations.

Stratton (1994) and Stratton and Bennington (1996) have further investigated the relationship between character expression, fitness, and the degree and pattern of environmental heterogeneity. In a study on *Erigeron annuus* (Stratton, 1994, 1995) Stratton planted plants at 630 locations within a 0.5-ha field to directly measure environmental effects. The experiment included three genotypes with clearly identifiable phenotypic markers so that he could ascertain the genotype of the plant by sight. The results indicated that almost all the genotype by environment interaction is observable at the smallest spatial scale (10 cm), with reversals in the relative fitness of different genotypes occurring at this same scale. Stratton therefore concluded that the environmentally induced pattern of spatial heterogeneity in relative fitness affects the fate of the next generation of plants, which mostly disperse over the same range of environments.[4] However, when he attempted to combine this direct method with indirect measurements of soil nutrients and percent cover of surrounding vegetation, Stratton found that only 12% of the genotype-environment variance could be explained by heterogeneity in the indirect measurements. Of course, one can always argue that some critical environmental factor was not measured in that particular instance, but this does illustrate the difficulty of combining approaches to obtain a more complete picture of the ecological genetics of the species studied. Unfortunately, to the best of my knowledge no better alternative has yet been proposed.

---

[4] Stratton also found a positive autocorrelation between *absolute* fitness and spatial distance, indicating a larger scale variation in habitat quality.

It seems clear that evolutionary ecologists must combine different approaches to the study of the environmental component of the genotype-environment interaction. Field-conducted bioassays provide the only reliable indication that organisms are indeed experiencing environmental heterogeneity, and that the observed variation in character expression is in fact germane to natural conditions. Measurements of physical and biotic parameters of the environment can provide preliminary, correlational, evidence of the causal mechanisms underlying the observed patterns of character variation. On the other hand, controlled laboratory studies focusing on environmental manipulation of the most likely candidate factors are the only method to directly address the causality question. However, manipulation experiments need to be guided by field findings in order to make them relevant to natural conditions and to narrow the range and number of possible environmental variables.

Bell (1992) elegantly summarized what we know of environments and how they elicit phenotypic responses in his list of five attributes of environments. First, environmental variance is relatively large. By "relatively," Bell means when compared to either genetic or genotype-by-environment variance. In fact, studies of a variety of traits in crop plants (our largest database to date) clearly show that the environmental variance explains close to 80% of the total phenotypic variance of a given character. Therefore, geneticists and evolutionists are advised to pay attention to what was once considered environmental "noise" (Sultan, 1992). Second, the environment is complex at all scales in space and time. Bell and Lechowicz's own data discussed earlier clearly bear out this point. As a consequence, Bell argues, the reductionist approach that has been so successful in genetics may turn out not to be applicable in ecology (contra some current trends in the field). Third, the response of organisms to the environment is indefinitely inconsistent. By this Bell means that the ranking of performance of different organisms varies among environments. This is another factor decoupling ecology from genetics, since genetic (and phylogenetic) diversity cannot be explained by (i.e., they do not covary with) environmental diversity. This decoupling applies to macroevolution as well, once we consider that genetic information varies continuously throughout a phylogeny (because new species are genetically related to their immediate ancestors or sister groups), while environmental information is phylogenetically patchy, since a new species can colonize an entirely different environmental milieu from its closest relatives. Fourth, environmental variation is largely self-regulated, that is, organisms act as environmental factors. This property of the organism-environment system has played an increasingly important part in ecology (shaping, for example, the concept of Hutchinsonian niche) and, more recently, in evolutionary biology (Lewontin, 1978; Levins and Lewontin, 1985). The implication drawn by Bell is that physical and biotic factors are therefore fundamentally distinct. However, we have to consider that biotic factors such as plant density will inevitably influence physical factors such as light and nutrient availability, thereby causally connecting the two types of factors. The major point is that we cannot consider environments and organismal traits as distinct entities: they influence each other. Finally, in Bell's classification scheme, environments tend continually to deteriorate. This can be rephrased by suggesting that environments change as a result of the presence of organisms, and that therefore the adaptive landscape for a given group of organisms changes literally under their feet (or roots) (Lewontin, 1978). The outcome of this was canonized by Van Valen with his famous "Red Queen hypothesis" (van Valen, 1973), according to which most evolutionary "energy" is spent by a species simply to maintain its position on an adaptive peek that keeps shifting away.

There is another component to environmental variation and its relationship to organismal characteristics that is not part of Bell's classification. When we refer to the reaction of an organism to an environmental factor we usually mean the whole biologically relevant range of that factor. For example, people study reaction norms to "temperature" or "water," meaning the range covering cold to hot (David et al., 1990; Huey et al., 1991; Dahlhoff and Somero, 1993; Berrigan and Charnov, 1994; McMichael and Burke, 1994; Schrag et al., 1994; Gilchrist, 1995; Bennett and Lenski, 1997; Brakefield and Kesbeke, 1997; Sgro` and Hoffmann, 1998), or drought to flood (Pigliucci et al., 1991b; Sultan and Bazzaz, 1993; Voesenek and Veen, 1994; Bruni et al., 1996; Keeland and Sharitz, 1997). Of course, some of these studies are conducted at only one extreme of the environmental gradient, but still most of the current discussion is framed in terms of response to the whole gradient (as in "the reaction norms of *Drosophila melanogaster* to temperature...")). I make the suggestion here that the extremes of the gradient are likely to be more different from each other (in terms of organismal response or perception) than either is to the extremes of a completely different gradient. In other words, it may be that the response to high temperature is more similar to the response to low water (since the two are ecologically coupled) then the reaction to high temperature is to the reaction to low temperature.

This is born out by the molecular literature. Evidence is accumulating that the molecular machinery necessary to respond to drought (Good and Zaplachinski, 1994; Welin et al., 1994; Jagtap and Bhargava, 1995; Mantyla et al., 1995; Yamaguchi-Shinozaki et al., 1995) is different from the one employed in response to flooding (Armstrong et al., 1994; Hurng and Kao, 1994; Blom and Voesenek, 1996). The same seems to be true for heat shock (Rickey and Belknap, 1991; Hubel and Schoffl, 1994; Loeschcke and Krebs, 1994; Prandl et al., 1995; Krebs and Feder, 1997; Downs et al., 1998) vs. cold (Koga-Ban et al., 1991; Welin et al., 1994; Mantyla et al., 1995; Mizoguchi et al., 1996). Furthermore, we have evidence of common molecular machinery coupling one extreme of distinct environmental factors (for example, cold and drought: Welin et al., 1994; Mizoguchi et al., 1996). This seems to refute models of "generalized stress response" (Chapin, 1991), or at least to limit their scope and universality. Future studies of the way in which characters react to environments, therefore, will have to take into account the ecological and (related) molecular commonalities between different aspects of the environment, rather than follow a human-based intuitive model of what an environmental factor is.

INFORMATION, THE GENOTYPE-ENVIRONMENT DETERMINATION OF TRAITS, AND THE CONCEPT OF INTERENVIRONMENTAL HOMOLOGY

The existence of phenotypic plasticity affects another major area of inquiry in the evolutionary biology of characters: the idea of homology between traits. The concept of homology is older than evolutionary biology itself, and it has undergone several dramatic redefinitions in recent times (Wagner, 1989). Some authors distinguish between interorganismic and intraorganismic homologies (McKitrick, 1994). The first is found when we compare across species structures with similar evolutionary histories, such as the bird's wings and the forelimbs of other vertebrates (Gatesy and Dial, 1996a). The second type deals with the duplication and specialization of the same ancestral structure (as in the case of body segments in insects: Carroll et al., 1995; Osorio et al., 1995). I suggest that the property that genotypes have to produce different character states in different environments may constitute a third category of homology, interenvironmental homology.

Obviously, interenvironmental homology is related to the other two, as those may in fact be related to each other. It is very reasonable to think of interorganismal homology as related to intraorganismal homology, as different taxa inherit in part or modify through evolutionary time the repertoire available to their ancestors. For example, intraorganismal homologous structures such as distinct body segments in insects and their relatives are clearly related to the interorganismal homology of the same body segment across insect species. Similarly, an intraorganismal homology can be related to an interenvironmental one (and possibly vice versa, see later). A particularly clear example of the relationship between intraorganismal and interenvironmental homology is provided by heterophylly, the production of entire or dissected leaves in response to environmental conditions such as water level fluctuation in plants (Cook and Johnson, 1968; Deschamp and Cooke, 1985; Bruni et al., 1996). Heterophylly is one of the few clear examples of how phenotypic plasticity can be adaptive, and we are in the process of better understanding not only the several cues that trigger the response under field conditions, but also of elucidating the hormonal and eventually genetic pathways leading to it (Kane and Albert, 1982, 1987). Structures involved in heterophylly (i.e., entire or dissected leaves) are clearly homologues to each other at the intraorganismic level, yet they are expressed in different environments. A similar example from the animal literature is provided by wing dimorphism in insects (Novotny, 1994;

Roff 1994a, 1994b), and in particular by the environmentally determined dichotomy between macropterous and brachypterous animals. Macroptery (fully developed wings) is favored when food scarcity or crowding requires the animals to move away from their current local environment. However, brachyptery (rudimental wings) are favored under the opposite circumstances because of the cost of maintaining flight-capable wings when they are not necessary.

A better understanding of the genetic and developmental basis of heterophylly and wing dimorphism would directly provide insights into intraorganismal homology.[5] In fact, Winn (1996a,b) has emphasized the connection between phenotypic plasticity and environment-independent intraindividual variation in the morphology of the same structure. Wagner (1989) concluded his review of the homology concept invoking comparative and experimental approaches to unravel the problem of how the individuality of characters is realized developmentally and how it emerges during phylogeny. The possible role of phenotypic plasticity in the evolution of phenotypic novelties (West-Eberhard, 1989; Schlichting and Pigliucci, 1998; see later) may offer an excellent system in which to apply both the experimental and the comparative methods to tackle the intimately related problems of homology and the origin of new traits.

Studies of phenotypic plasticity, and in particular the concept of interenvironment genetic correlations cited earlier, can also shed light onto the very definition of homology. According to both Van Valen (1982) and Roth (1988) homology is a "correspondence caused by a continuity of information." This implies that the homology of two structures is an unbroken connection between them throughout evolutionary time. According to Van Valen, the existence of this connection does not have to be based on the similarity of the genetic basis for the two traits. What Roth later termed "genetic piracy" can come into play as well. Sometimes clearly homologous structures from an evolutionary standpoint are actually under the influence of different genes in different organisms. Van Valen would still consider these traits homologous, and Roth would suggest that this is an example of genetic piracy, in that a new gene (or set of genes) takes over the role previously held by a different DNA sequence (see Roth, 1988 for a discussion of specific examples). The important point is that this takeover at the genetic level does not interfere with a continuity of structures at the phenotypic level. The interenvironment genetic correlation measures a similar phenomenon. If the absolute value of the correlation is less

---

[5] Because both leaves and wings are hypothesized to be homologues to other fundamental structures such as flower organs and other body segments, respectively, which are also being studied intensively from a molecular developmental level (Fenster et al., 1995; Weigel and Nilsson, 1995; Sessions, 1997; Carroll et al., 1995; Osorio et al., 1995) – the connection among all three types of homology should prove fascinating for probing a variety of evolutionary questions.

than one, this indicates that different genes are at least partially affecting the expression of the same trait in distinct environments. In a sense, the strength of the interenvironment genetic correlation measures the degree of "genetic piracy" underlying the system.

Finally, the study of interenvironment homology can yield insights into another fundamental question asked by Wagner (1989): does it make any sense to talk of "degrees" of homology, or is homology an all or nothing property of biological systems? Everything changes throughout evolutionary time, although admittedly at different paces and in different manners. How, then, can anything characterizing living organisms not be a matter of degree?[6] In the specific case of homology, the phenomena of genetic piracy and interenvironmental correlations make it clear that the information underlying homologous structures changes through time, culminating in the possibility of no similarity of genetic information whatsoever between two homologues characters. By extension of the same logic, we can tell that two extant structures are definitely homologues (e.g., the linear and dissected leaves of a heterophyllous plant, or the two sets of wings in some insects), or probably homologues (e.g., the leaves and the flower parts of the same plant, or wing- and leg-bearing segments in insects). However, it is easy to imagine that at one point both the informational basis and the physical appearance of two characters will diverge enough through evolutionary time to make it difficult for us to recognize their ancestral homologous state. Since even characters that are still clearly homologues evolve phenotypic novelties, which are an integral part of them but were not present in the ancestral trait, should we not consider these as examples of partial homology?[7]

Our discussion so far has moved from a consideration of the environmental influences on character expression in extant populations (a microevolutionary level of analysis) to broader implications on homology and character evolution (at the macroevolutionary level). The last, but certainly not least, aspect of considering the relationship between environments and characters implies an alternative (albeit not at all new) way of looking at macroevolution itself.

ENVIRONMENTS, PHENOTYPIC NOVELTIES, AND GENETIC ASSIMILATION

Perhaps the single most difficult problem in evolutionary biology is the one posed by the appearance of the so-called "phenotypic novelties." A precise definition of phenotypic novelty depends on the kind of organism and what level

---

[6] Even the very property of "living" systems may actually have emerged by steps, although not necessarily gradual ones (Farmer et al., 1986; Casti, 1989; Doolittle, 1994).

[7] For example, the stamen as a whole may be homologous to a leaf, but what about the specific parts that make up the stamen itself, such as the filament and the anther?

of macroevolution one is considering. Several authors have addressed the problem (Levin, 1983; Agur and Kerszberg, 1987; Muller, 1990; Jablonski, 1993; Brush, 1996; Gatesy and Dial, 1996b; Swalla and Jeffery, 1996; Toquenaga and Wade, 1996; Hunter, 1998). At least one of these attempts has directly involved phenotypic plasticity (West-Eberhard, 1989).

West-Eberhard's argument is that an environmentally induced, nongenetic, change in phenotype can yield dramatic alterations in morphology that would be perceived as novel phenotypes. If the environmental change is semipermanent, the new morphology requires no initial change in the genetic system. However, if the new phenotype happens to be advantageous, selection will increase the frequency of whatever genetic modifiers stabilize the phenotype regardless of environmental fluctuations. This idea is not new. As Schlichting and Pigliucci (1998) have pointed out, it is a reformulation of Waddington's genetic assimilation (Waddington 1942, 1952, 1960), in itself not different from Schmalhausen's (1949) "stabilizing selection," or even from Baldwin's (1896) "new factor in evolution," and closely related to Goldschmidt's (1940) concept of phenocopy.

To some extent, the reason why the role of phenotypic plasticity (which is a prerequisite for genetic assimilation) has not been taken seriously ever since the neo-Darwinian synthesis is that environmental influences have been historically considered a "nuisance" to be dealt with in practice, but certainly not relevant in theory (Sultan, 1992). Therefore, classical evolution by gradual gene substitution has for all effective purposes been the only game in town. However, it has been a particularly inefficient explanatory frame at best, relying on such poorly substantiated phenomena as "preadaptation" (Futuyma, 1998). Modern evolutionary biologists are faced with increasing pressure to admit the possible role of at least two more mechanisms. On the one hand, the effect of mutations with major developmental effects (Gottlieb, 1984; Doebley et al., 1990; Orr and Coyne, 1992; Dorweiler et al., 1993; Wagner et al., 1994; Walsh, 1995; Brakefield et al., 1996; Schluter, 1996; Sordino and Duboule, 1996; Dorweiler and Doebley, 1997). On the other hand, the contribution of phenotypic plasticity (Levin, 1988; West-Eberhard, 1989; Smith, 1990; Schlichting and Pigliucci, 1993; Crews et al., 1994; Whiteman, 1994; Janzen, 1995; Brakefield et al., 1996). The role of mutations with major developmental effects (not dissimilar from Goldschmidt's infamous "hopeful monsters") is beyond the scope of this chapter. I will therefore concentrate on the second phenomenon.

One example of the potential contribution of plasticity to macroevolution concerns the evolution of two sympatric species of three-spine sticklebacks, *Gasterosteus* sp. (Day et al., 1994) (Fig. 4). Day et al. hypothesized that plasticity of the skull morphology in the ancestor of these taxa may have played a role in their current separation in morphospace, as well as in the observable niche partitioning. One species lives in a limnetic environment and feeds mainly on plankton. Accordingly, it is characterized by reduced gape width, longer gill rakers, less deep head, and longer snout. The second species, on the other hand, lives in a benthic environment, feeding upon worms. Consequently, its gape is wider, gill rakers shorter, head deeper, and snout shorter.

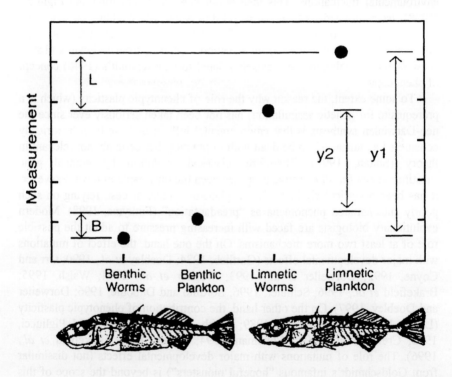

FIGURE 4 Reaction norms of two species of three-spine sicklebacks exposed to their natural and each other's diet. Notice how each species displays a pattern of plasticity that partially bridges the morphological gap between the two. From Day *et al.* (1994).

Day and collaborators then fed fishes of each species their natural diet and that of the other species. Their prediction was that each species would demonstrate adaptive plasticity by a shift of character means (in response to the "wrong" diet) in the direction of the other species' morphology. This is exactly what happened, although the reaction of one species was more pronounced than the other and the intensity of the response varied among characters (Fig. 4). As pointed out by Schlichting and Pigliucci (1998), this is what we would expect if adaptive plasticity had provided the initial steps toward the evolution of the new morphology. The hypothesis is that a simple change in diet, forced by environmental circumstances, created a partially new morphology from the existing plasticity of the bone developmental system. This allowed the species to survive, albeit under suboptimal conditions, in the new environment. At this point classical selection for gene substitutions would gradually increase the fit between the phenotype and the environment. The advantage of this scenario is that it does away with the idea of preadaptation. No preexisting genetic condition (other than the existence of a plastic, but not necessarily adaptive, reaction norm) is necessary because the new morphology is a by-product of the existence of a reaction norm. Not even mosaic evolution is necessary (although it may occur in parallel) because the plasticity of the developmental system will change several characters simultaneously, not one at a time [see West-Eberhard (1989) for a splendid example of this].

Unfortunately, examples like these are hard to find in the literature. Part of the reason is the previously mentioned lack of expectation that phenotypic plasticity may play a major role in macroevolution. In part, however, this is certainly due to the experimental difficulty of investigating the first steps of genetic assimilation. After all, it may take very few generations of strong selection for modifiers to bridge the gap between one stable morphology and the other. Following that, the "signature" of genetic assimilation would be much harder to detect. Of course, this is equally true of the classical concepts of mosaic evolution and especially of preadaptation. However, such is the power of an established scientific paradigm (Kuhn, 1970). As in the case of other macroevolutionary theories, this one also needs to be investigated by the use of the comparative method (Harvey and Purvis, 1991). However, unlike the traditional gene substitution-only theory, evolution by phenotypic plasticity and genetic assimilation is actually testable under laboratory conditions, as the case of the sticklebacks and a few other published to date (Meyer, 1987; van Tienderen, 1990; Wainwright, 1994) clearly demonstrate.

Overall, the very concepts of character definition, character measurement, homology, and evolution of morphological novelties need to be reexamined in light of recent empirical and theoretical research on phenotypic plasticity. A full integration of genetically mediated environmental effects on the phenotype will not be a panacea to solve every problem in evolutionary biology, but it will surely help to consider the environment as an equal player in the evolutionary game.

ACKNOWLEDGMENTS

I wish to thank Hilary Callahan, Mark Camara, Mitch Cruzan, Heidi Pollard, and Anna Maleszyk for comments on this manuscript. I am very much indebted to Cindy Jones, Carl Schlichting, Kurt Schwenk, and Gunter Wagner for discussions on homology, constraints, and evolution. This research was partly supported by NSF grant DEB-9527551.

LITERATURE CITED

Agur, Z., and Kerszberg, M. (1987). The emergence of phenotypic novelties through progressive genetic change. *Am. Nat.* **129**:862-875.

Andersson, S., and Shaw, R. G. (1994). Phenotypic plasticity in *Crepis tectorum* (Asteraceae): genetic correlations across light regimens. *Heredity* **72**:113-125.

Armstrong, W., Brandle, R., and Jackson, M. B. (1994). Mechanisms of flood tolerance in plants. *Acta Bot. Neer.* **43**:307-358.

Baldwin, J. M. (1896). A new factor in evolution. *Am. Nat.* **30**:354-451.

Bell, G., (1992). Five properties of environments. *In* "Molds, Molecules, and Metazoa" (P. R. Grant and H. S. Horn, eds.), pp. 33-56. Princeton University Press, Princeton.

Bell, G., and Lechowicz, M. J. (1991). The ecology and genetics of fitness in forest plants. I. Environmental heterogeneity measured by explant trials. *J. Ecol.* **79**:663-685.

Bennett, A. F., and Lenski, R. E. (1997). Evolutionary adaptation to temperature. VI. Phenotypic acclimation and its evolution in *Escherichia coli. Evolution* **51**:36-44.

Berrigan, D., and Charnov, E. L. (1994). Reaction norms for age and size at maturity in response to temperature: a puzzle for life historians. *Oikos* **70**:474-478.

Blom, C. W. P. M., and Voesenek, L. A. C. J. (1996). Flooding: the survival strategies of plants. *Trends Ecol. Evol.* **11**:290-295.

Bookstein, F.L. (1991). "Morphometric Tools for Landmark Data: Geometry and Biology." Cambridge University Press, Cambridge.

Bradshaw, A. D. (1965). Evolutionary significance of phenotypic plasticity in plants. *Adv. Genet.* **13**:115-155.

Brakefield, P. M., Gates, J., Keys, D., Kesbeke, F., Wijngaarden, P. J., Monteiro, A., French, V., and Carroll, S. B. (1996). Development, plasticity and evolution of butterfly eyespot patterns. *Nature* **384**:236-242.

Brakefield, P. M., and Kesbeke, F. (1997). Genotype-environment interactions for insect growth in constant and fluctuating temperature regimes. *Proc. R. Soc. Lond.* **264**:717-723.

Bruni, N. C., Young, J. P., and Dengler, N. G. (1996). Leaf developmental plasticity of *Ranunculus flabellaris* in response to terrestrial and submerged environments. *Can. J. Bot.* **74**:823-837.

Brush, A. H. (1996). On the origin of feathers. *J. Evol. Biol.* **9**:131-142.

Carroll, S. B., Weatherbee, S. D., and Langeland, J. A. (1995). Homeotic genes and the regulation and evolution of insect wing number. *Nature* **375**:58-61.

Casti, J. L. (1989). "Paradigms Lost." Avon Books, New York.

Chapin III, F. S. (1991). Integrated responses of plants to stress. *Bioscience* **41**:29-36.

Cheverud, J. M. (1995). Morphological integration in the saddle-back tamarin (*Saguinus fuscicollins*) cranium. *Am. Nat.* **145**:63-89.

Cheverud, J. M. (1996). Quantitative genetic analysis of cranial morphology in the cotton-top (*Sanguinus oedipus*) and saddle-back (*S. fuscicolis*) tamarins. *J. Evol. Biol.* **9**:5-42.

Clausen, J., and Hiesey, W. M. (1960). The balance between coherence and variation in evolution. *Proc. Natl. Acad. Sci.U S A* **46**:494-506.

Clausen, J., Keck, D., and Hiesey, W. M. (1940). "Experimental Studies on the Nature of Species. I. Effect of Varied Environments on Western North American Plants." Carnegie Institution of Washington, Washington, DC.

Conner, J., and Via, S. (1993). Patterns of phenotypic and genetic correlations among morphological and life-history traits in wild radish, *Raphanus raphanistrum. Evolution* **47**:704-711.

Cook, S. A., and Johnson, M. P. (1968). Adaptation to heterogenous environments. I. Variation in heterophylly in *Ranunculus flammula* L. *Evolution* **22**:496-516.

Crews, D., Bergeron, J. M., Bull, J.J., Flores, D., Tousignant, A., Skipper J. K., and Wibbels, T. (1994). Temperature-dependent sex determination in reptiles: proximate mechanisms, ultimate outcomes, and practical applications. *Dev.Genet.* **15**:297-312.

Dahlhoff, E., and Somero, G. N. (1993). Effects of temperature on mitochrondria from Abalone (genus *Haliotis*): adaptive plasticity and its limits. *J. Exp. Biol.***185**:151-168.

David, J. R., Capy, P., and Gauthier, J. P. (1990). Abdominal pigmentation and growth temperature in *Drosophila melanogaster*: similarities and differences in the norms of reaction of successive segments. *J. Evol. Biol.* **3**:429-445.

Day, T., Pritchard, J., and Schluter, D. (1994). Ecology and genetics of phenotypic plasticity: a comparison of two sticklebacks. *Evolution* **48**:1723-1734.

de Jong, G. (1995). Phenotypic plasticity as a product of selection in a variable environment. *Am. Nat.* **145**:493-512.

Deschamp, P. A., and Cooke, T. J. (1985). Leaf dimorphism in the aquatic Angiosperm *Callitriche heterophylla. Am. J. Bot.* **72**:1377-1387.

Dickerson, G. E. (1962). Implications of genetic environmental interaction in animal breeding. *Anim. Prod.* **4**:47-64.

Doebley, J., Stec, A., Wendel, J., and Edwards, M. (1990). Genetic and morphological analysis of a maize-teosinte F2 population: implications for the origin of maize. *Proc. Natl. Acad. Sci. USA* **87**:9888-9892.

Dong, M., and Pierdominici , M. G. (1995). Morphology and growth of stolons and rhizomes in three clonal grasses, as affected by different light supply. *Vegetatio* **116**:25-32.

Doolittle, R. F. (1994). The early evolutionary history of proteins. *Boll.Zool.* **61**:99-103.

Dorweiler, J., Stec, A., Kermicle, J., and Doebley, J. (1993). *Teosinte glume architecture 1*: a genetic locus controlling a key step in maize evolution. *Science* **262**:233-235.

Dorweiler, J. E. and Doebley, J. (1997). Developmental analysis of *Teosinte glume architecture 1*: a key locus in the evolution of maize (Poaceae). *Am. J. Bot.* **84**:1313-1322.

Downs, C. A., Heckathorn, S. A., Bryan, J. K., and Coleman, J. S. (1998). The methionine-rich low-molecular-weight chloroplast heat-shock protein: evolutionary conservation and accumulation in relation to thermotolerance. *Am. J. Bot.* **85**:175-183.

Estevez, M., Attisano, L., Wrana, J. L., Albert, P. S., Massague`, J., and Riddle, D. L. 1993. The *daf-4* gene encodes a bone morphogenetic protein receptor controlling *C. elegans* dauer larva development. *Nature* **365**:644-649.

Falconer, D. S. (1952). The problem of environment and selection. *Am. Nat.* **86**:293-298.

Farmer, J. D., Kauffman, S. A., and Packard N. H. (1986). Autocatalytic replication of polymers. *Physica* **22D**:50-67.

Fenster, C. B., Diggle, P. K., Barrett, S. C. H., and Ritland, K. (1995). The genetics of floral development differentiating two species of *Mimulus* (Scophulariaceae). *Heredity* **74**:258-266.

Fernando, R. L., Knights, S. A., and Gianola, D. (1984). On a method of estimating the genetic correlation between characters measured in different experimental units. *Theor. Appl. Genet.* **67**:175-178.

Ford, E. B. (1931). "Mendelism and Evolution." Methuen, London.

Futuyma, D. (1998). "Evolutionary Biology." Sinauer, Sunderland, MA.

Gatesy, S. M., and Dial, K. P. (1996a). From frond to fan: *Archaeopteryx* and the evolution of short-tailed birds. *Evolution* **50**:2037-2048.

Gatesy, S. M., and Dial, K. P. (1996b). Locomotor modules and the evolution of avian flight. *Evolution* **50**:331-340.

Gedroc, J. J., McConnaughay, K. D. M., and Coleman, J. S. (1996). Plasticity in root/shoot partitioning: optimal, ontogenetic, or both? *Funct. Ecol.* **10**:44-50.

Gilchrist, G. W. (1995). Specialists and generalists in changing environments. I. Fitness landscapes of thermal sensitivity. *Am. Nat.* **146**:252-270.

Goldschmidt, R. (1940). "The Material Basis of Evolution." Yale University Press, New Haven, CT.

Good, A. G., and Zaplachinski, S. T. (1994). The effects of drought stress on free amino acid accumulation and protein synthesis in *Brassica napus*. *Physiol. Plant.* **90**:9-14.

Gottlieb, L. D. (1984). Genetics and morphological evolution in plants. *Am.Nat.***123**:681-709.

Gould, S. J., and Lewontin, R. C. (1979). The spandrels of San Marco and the Panglossian paradigm: a critique of the adaptationist programme. *Proc. R. Soc. Lond. B* **205**:581-598.

Harvey, P. H., and Purvis, A. (1991). Comparative methods for explaining adaptations. *Nature* **351**:619-624.

Hawthorne, D. J. (1997). Ecological history and evolution in a novel environment: habitat heterogeneity and insect adaptation to a new host plant. *Evolution* **51**:153-162.

Holloway, G. J., Povey, S. R., and Sibly, R. M. (1990). The effect of new environment on adapted genetic architecture. *Heredity* **64**:323-330.

Hubel, A., and Schoffl, F. (1994). *Arabidopsis* heat shock factor: isolation and characterization of the gene and the recombinant protein. *Plant Mol. Biol.* **26**:353-362.

Huey, R. B., Partridge, L. and Fowler, K. (1991). Thermal sensitivity of *Drosophila melanogaster* responds rapidly to laboratory natural selection. *Evolution* **45**:751-756.

Hunter, J. P. (1998). Key innovations and the ecology of macroevolution. *Trends Ecol. Evol.* **13**:31-35.

Hurng, W. P., and Kao, C. H. (1994). Effect of flooding on the activities of some enzymes of activated oxygen metabolism, the levels of antioxidants, and lipid peroxidation in senescing tobacco leaves. *Plant Growth Regul.* **14**:37-44.

Ingram, G. C., Goodrich, J., Wilkinson, M. D., Simon, R., Haughn G. W., and Coen E.S. (1995). Parallels between *UNUSUAL FLORAL ORGANS* and *FIMBRIATA*, genes controlling flower development in *Arabidopsis* and *Antirrhinum*. *Plant Cell* **7**:1501-1510.

Jablonski, D. (1993). The tropics as a source of evolutionary novelty through geological time. *Nature* **364**:142-144.

Jagtap, V., and Bhargava, S. (1995). Variation in the antioxidant metabolism of drought tolerant and drought susceptible varieties of *Sorghum bicolor* (L.) Moench. exposed to high light, low water and high temperature stress. J. Plant Physiol. **145**:195-197.

Janzen, F. J. (1995). Experimental evidence for the evolutionary significance of temperature-dependent sex determination. *Evolution* **49**:864-873.

Joshi, A., and Thompson, J. N. (1996). Evolution of broad and specific competitive ability in novel versus familiar environments in *Drosophila* species. *Evolution* **50**:188-194.

Kane, M. E., and Albert, L. S. (1982). Environmental and growth regulator effects on heterophylly and growth of *Proserpinaca intermedia* (Haloragaceae). *Aquat. Bot.* **13**:73-85.

Kane, M. E., and Albert, L. S. (1987). Integrative regulation of leaf morphogenesis by gibberellic and abscisic acids in the aquatic angiosperm *Proserpinaca palustris* L. *Aquat. Bot.* **28**:89-96.

Keeland, B. D., and Sharitz, R. R. (1997). The effects of water-level fluctuation on weekly tree growth in a southeastern USA swamp. *Am. J. Bot.* **84**:131-139.

Koga-Ban, Y., Abe, M., and Kitagawa, Y. (1991). Alteration in gene expression during cold treatment of rice plant. *Plant Cell Physiol.* **32**:901-905.

Krebs, R. A., and Feder, M. E. (1997). Deleterious consequences of Hsp70 overexpression in *Drosophila melanogaster* larvae. *Cell Stress Chaperones* **2**:60-71.

Kudoh, H., Ishiguri, Y., and Kawano, S. (1996). Phenotypic plasticity in age and size at maturity and its effects on the integrated phenotypic expressions of life history traits of *Cardamine flexuosa* (Cruciferae). *J. Evol. Biol.* **9**:541-570.

Kuhn, T. (1970). "The Structure of Scientific Revolutions." University of Chicago Press, Chicago.

Lechowicz, M. J., and Bell, G. (1991). The ecology and genetic of fitness in forest plants. II. Microspatial heterogeneity of the edaphic environment. *J. Ecol.* **79**:687-696.

Lechowicz, M. J., and Blais, P .A. (1988). Assessing the contributions of multiple interacting traits to plant reproductive success: environmental dependence. *J. Evol. Biol.* **1**:255-273.

Levin, D. A., (1983). Polyploidy and novelty in flowering plants. *Am. Nat.* **122**:1-25.

Levin, D. A. (1988). Plasticity, canalization and evolutionary stasis in plants. *In* "Plant Population Biology." pp. 35-45. Blackwell, Oxford.

Levins, R., and Lewontin, R. C. (1985). "The Dialectical Biologist." Harvard University Press, Cambridge, MA.

Lewontin, R. C. (1974). The analysis of variance and the analysis of causes. *Am. J. Hum.Genet.* **26**:400-411.

Lewontin, R.C. (1978). Adaptation. *Sci. Am.* 213-230.

Loeschcke, V., and Krebs, R. A. (1994). Genetic variation for resistance and acclimation to high temperature stress in *Drosophila buzzatii*. *Biol. J. Linn. Soc.* **52**:83-92.

Mantyla, E., Lang, V., and Palva, E. T. (1995). Role of abscisic acid in drought-induced freezing tolerance, cold acclimation, and accumulation of LTI78 and RAB18 proteins in *Arabidopsis thaliana*. *Plant Physiol.* **107**:141-148.

Marcus, L. F., Corti, M., Loy, A., Naylor, G. J. P., and Slice, D. E. (1996). "Advances in morphometrics." Plenum Press, New York.

Marshall, D. L., Levin, D. A., and Fowler, N. L. (1986). Plasticity of yield components in response to stress in *Sesbania macrocarpa* and *Sesbania vesicaria* (Leguminosae). *Am. Nat.* **127**:508-521.

Martin-Mora, E., and James, F. C. (1995). Developmental plasticity in the shell of the queen conch *Strombus gigas*. *Ecology* **76**:981-994.

Mayr, E. (1993). What was the evolutionary synthesis? *Trends Ecol. Evol.* **8**:31-33.

Mayr, E., and Provine, W. B. (1980). "The Evolutionary Synthesis. Perspectives on the Unification of Biology." Harvard University Press, Cambridge, MA.

McKitrick, M. C. (1994). On homology and the ontological relationship of parts. *Syst. Biol.* **43**:1-10.

McMichael, B. L., and Burke, J. J. (1994). Metabolic activity of cotton roots in response to temperature. *Environ. Exp. Bot.* **34**:201-206.

Meyer, A. (1987). Phenotypic plasticity and heterochrony in *Cichlasoma managuense* (Pisces, Cichlidae) and their implications for speciation in Cichlid fishes. *Evolution* **41**:1357-1369.

Mizoguchi, T., Irie, K., Hirayama, T., Hayashida, N., Yamaguchi-Shinozaki, K., Matsumoto, K., and Shinozaki, K. (1996). A gene encoding a mitogen-activated protein kinase kinase kinase is induced simultaneously with genes for a mitogen-activated protein kinase and an S6 ribosomal protein kinase by touch, cold, and water stress in *Arabidopsis thaliana*. *Proc. Natl. Acad. Sci. USA* **93**:765-769.

Muller, G. B. (1990). Developmental mechanisms at the origin of morphological novelty: a side-effect hypothesis. *In* "Evolutionary Innovations," pp. 99-130. Chicago University Press, Chicago.

Nelson, T., and Langdale, J. A. (1992). Developmental genetics of C_4 photosynthesis. *Annu. Rev. Plant Physiol. Plant Mol. Biol.* **43**:25-47.

Newman, R. A. (1994). Genetic variation for phenotypic plasticity in the larval life history of spadefoot toads (*Scaphiopus couchii*). *Evolution* **48**:1773-1785.

Neyfakh, A. A., and Hartl , D. L. (1993). Genetic control of the rate of embryonic development: selection for faster development at elevated temperatures. *Evolution* **47**:1625-1631.

Novotny, V. (1994). Relation between temporal persistence of host plants and wing length in leafhoppers (Hemiptera: Auchenorrhyncha). *Ecol. Entomol.* **19**:168-176.

Orr, H. A., and Coyne, J. A. (1992). The genetics of adaptation: a reassessment. *Am. Nat.* **140**:725-742.

Osorio, D., Averof, M., and Bacon, J. P. (1995). Arthropod evolution: great brains, beautiful bodies. *Trends Ecol. Evol.* **10**:449-454.

Parsons, B. L., and Mattoo, A. K. (1991). Wound-regulated accumulation of specific transcripts in tomato fruit: interactions with fruit development, ethylene and light. *Plant Mol. Biol.* **17**:453-464.

Pickett, F. B., and Meeks-Wagner, D. R. (1995). Seeing double: appreciating genetic redundancy. *Plant Cell* **7**:1347-1356.

Pigliucci, M. (1996a). How organisms respond to environmental changes: from phenotypes to molecules (and vice versa). *Trends Ecol. Evol.* **11**:168-173.

Pigliucci, M. (1996b). Modelling phenotypic plasticity. II. Do genetic correlations matter? *Heredity* **77**:453-460.

Pigliucci, M., diIorio, P., and Schlichting, C. D. (1997). Phenotypic plasticity of growth trajectories in two species of *Lobelia* in response to nutrient availability. *J. Ecol.* **85**:265-276.

Pigliucci, M., Paoletti, C., Fineschi, S., and Malvolti, M. E. (1991a). Phenotypic integration in chestnut (*Castanea sativa* Mill.): leaves versus fruits. *Bot. Gaz.* **152**:514-521.

Pigliucci, M., Politi, M. G., and Bellincampi, D. (1991b). Implications of phenotypic plasticity for numerical taxonomy of *Ornithogalum montanum* (Liliaceae*). Can. J. Bot.* **69**:34-38.

Pigliucci, M., and Schlichting, C. D. (1995). Ontogenetic reaction norms in *Lobelia siphilitica* (Lobeliaceae): response to shading. *Ecology* **76**:2134-2144.

Pigliucci, M., and Schlichting, C. D. (1997). On the limits of quantitative genetics for the study of phenotypic evolution. *Acta Biotheor.* **45**:143-160.

Pigliucci, M., Schlichting, C. D., Jones, C. S., and Schwenk, K. (1996). Developmental reaction norms: the interactions among allometry, ontogeny and plasticity. *Plant Species Biol.* **11**:69-85.

Pigliucci, M., and Schmitt, J. (1999). Genes affecting phenotypic plasticity in *Arabidopsis*: pleiotropic effects and reproductive fitness of photomorphogenic mutants. *J. Evol. Biol.* **12**:551-562.

Prandl, R., Kloske, E., and Schoffl, F. (1995). Developmental regulation and tissue-specific differences of heat shock gene expression in transgenic tobacco and *Arabidopsis* plants. *Plant Mol. Biol.* **28**:73-82.

Provine, W. B. (1971). Population genetics: the synthesis of Mendelism, Darwinism, and Biometry. *In* "The Origin of Theoretical Population Genetics," pp. 130-178. Chicago University Press, Chicago.

Rickey, T. M., and Belknap, W. R. (1991). Comparison of the expression of several stress-responsive genes in potato tubers. *Plant Mol. Biol.* **16**:1009-1018.

Robertson, A. (1959). The sampling variance of the genetic correlation coefficient. *Biometrics* **15**:469-485.

Roff, D. A. (1994a). Habitat persistence and the evolution of wing dimorphism in insects. *Am. Nat.* **144**:772-798.

Roff, D. A. (1994b). Why is there so much genetic variation for wing dimorphism? *Res. Pop. Ecol.* **36**:145-150.

Rohlf, F. J., and Marcus, L. F. (1993). A revolution in morphometrics. *Trends Ecol. Evol.* **8**:129-132.

Roth, L. (1988). The biological basis of homology. *In* "Ontogeny and Systematics" (C. J. Humphries, ed.), pp. 1-26. Columbia University Press, New York.

Scheiner, S. M. (1993a). Genetics and evolution of phenotypic plasticity. *Annu. Rev. Ecol. Syst.* **24**:35-68.

Scheiner, S. M. (1993b). Plasticity as a selectable trait: reply to Via. Am. Nat. **142**:371-373.

Schlichting, C. D. (1986). The evolution of phenotypic plasticity in plants. *Annu. Rev. Ecol. Syst.* **17**:667-693.

Schlichting, C. D. (1989a). Phenotypic integration and environmental change. *Bioscience* **39**:460-464.

Schlichting, C. D. (1989b). Phenotypic plasticity in *Phlox*. II. Plasticity of character correlations. *Oecologia* **78**:496-501.

Schlichting, C. D., and Pigliucci, M. (1993). Evolution of phenotypic plasticity via regulatory genes. *Am. Nat.* **142**:366-370.

Schlichting, C. D., and Pigliucci, M. (1995a). Gene regulation, quantitative genetics and the evolution of reaction norms. *Evol. Ecol.* **9**:154-168.

Schlichting, C. D., and Pigliucci, M. (1995b). Lost in phenotypic space: environment-dependent morphology in *Phlox drummondii* (Polemoniaceae). *Int. J. Plant Sci.* **156**:542-546.

Schlichting, C. D., and Pigliucci, M. (1998). "Phenotypic Evolution. A Reaction Norm Perspective." Sinauer, Sunderland, MA.

Schluter, D. (1996). Adaptive radiation along genetic lines of least resistance. *Evolution* **50**:1766-1774.

Schmalhausen, I. I. (1949). "Factors of Evolution. The Theory of Stabilizing Selection." University of Chicago Press, Chicago.

Schrag, S. J., Ndifon, G. T., and Read, A. F. (1994). Temperature-determined outcrossing ability in wild populations of a simultaneous hermaphrodite snail. *Ecology* **75**:2066-2077.

Service, P. M., and Rose, M. R. (1985). Genetic covariation among life-history components: the effect of novel environments. *Evolution* **39**:943-945.

Sessions, R. A. (1997). *Arabidopsis* (Brassicaceae) flower development and gynoecium patterning in wild type and *Ettin* mutants. *Am. J. Bot.* **84**:1179-1191.

Sgro`, C. M., and Hoffmann, A. A. (1998). Effects of temperature extremes on genetic variances for life history traits in *Drosophila melanogaster* as determined from parent-offspring comparisons. *J. Evol. Biol.* **11**:1-20.

Smith, H. (1990). Signal perception, differential expression within multigene families and the molecular basis of phenotypic plasticity. *Plant Cell Environ.* **13**:585-594.

Smith-Gill, S. J. (1983). Developmental plasticity: developmental conversion versus phenotypic modulation. *Am. Zool.* **23**:47-55.

Sordino, P., and Duboule, D. (1996). A molecular approach to the evolution of vertebrate paired appendages. *Trends Ecol. Evol.* **11**:114-118.

Stearns, S., de Jong, G., and Newman, B. (1991). The effects of phenotypic plasticity on genetic correlations. *Trends Ecol. Evol.* **6**:122-126.

Stratton, D. A. (1994). Genotype-by-environment interactions for fitness of *Erigeron annuus* show fine-scale selective heterogeneity. *Evolution* **48**:1607-1618.

Stratton, D. A. (1995). Spatial scale of variation in fitness of *Erigeron annuus*. *Am. Nat.* **146**:608-624.

Stratton, D. A., and Bennington, C.C. (1996). Measuring spatial variation in natural selection using randomly-sown seeds of *Arabidopsis thaliana*. *J. Evol. Biol.* **9**:215-228.

Sultan, S. E. (1987). Evolutionary implications of phenotypic plasticity in plants. *Evol. Biol.* **21**:127-178.

Sultan, S. E. (1992). What has survived of Darwin's theory? Phenotypic plasticity and the neo-darwinian legacy. *Evol. Trends Plants* **6**:61-71.

Sultan, S. E. (1995). Phenotypic plasticity and plant adaptation. *Acta Bot. Neerl.* **44**:363-383.

Sultan, S. E., and Bazzaz, F. A. (1993). Phenotypic plasticity in *Polygonum persicaria*. II. Norms of reaction to soil moisture and the mainteinance of genetic diversity. *Evolution* **47**:1032-1049.

Swalla, B. J., and Jeffery, W. R. (1996). Requirement of the *Manx* gene for expression of chordate features in a tailless ascidian larva. *Science* **274**:1205-1208.

Toquenaga, Y., and Wade, M. J. (1996). Sewall Wright meets Artificial Life: the origin and maintenance of evolutionary novelty. *Trends Ecol. Evol.* **11**:478-482.

Trainor, F. R., and Egan, P. F. (1991). Discovering the various ecomorphs of *Scenedesmus*: the end of a taxonomic era. *Arch. Protistenkd.* **139**:125-132.

van Tienderen, P. H. (1990). Morphological variation in *Plantago lanceolata*: limits of plasticity. *Evol. Trends Plants* **4**:35-43.

van Tienderen, P. H., and de Jong, G. (1994). A general model of the relation between phenotypic selection and genetic response. *J. Evol. Biol.* **7**:1-12.

van Tienderen, P. H., and Koelewijn, H. P. (1994). Selection on reaction norms, genetic correlations and constraints. *Genet. Res.* **64**:115-125.

van Valen, L. (1973). A new evolutionary law. *Evol. Theory* **1**:1-30.

van Valen, L. M. (1982). Homology and causes. *J. Morphol.* **173**:305-312.

Via, S. (1987). Genetic constraints on the evolution of phenotypic plasticity. *In* "Genetic Constraints and Adaptive Evolution" (V. Loeschcke, ed.), pp. 47-71. Springer-Verlag, Berlin.

Via, S., and Lande, R. (1985). Genotype-environment interaction and the evolution of phenotypic plasticity. *Evolution* **39**:505-522.

Voesenek, L. A. C. J., and van der Veen, R. (1994). The role of phytohormones in plant stress: too much or too little water. *Acta Bot. Neerl.* **43**:91-127.

Waddington, C. H. (1942). Canalization of development and the inheritance of acquired characters. *Nature* **150**:563-565.

Waddington, C. H. (1952). Selection of the genetic basis for an acquired character. *Nature* **169**:278.

Waddington, C. H. (1960). Experiments on canalizing selection. *Genet.Res.* **1**:140-150.

Wagner, A., Wagner, G. P., and Similion, P. (1994). Epistasis can facilitate the evolution of reproductive isolation by peak shifts: a two-locus two-allele model. *Genetics* **138**:533-545.

Wagner, G. P. (1989). The biological homology concept. *Annu.Rev. Ecol. Syst.* **20**:51-69.

Wainwright, P. C. (1994). Functional morphology as a tool in ecological research. *In* "Ecological Morphology: Integrative Organismal Biology" (P. C. Wainwright and S. M. Reilly, eds.), pp. 42-59. University of Chicago Press, Chicago.

Waitt, D. E., and Levin, D. A. (1993). Phenotypic integration and plastic correlations in *Phlox drummondii* (Polemoniaceae). *Am. J. Bot.* **80**:1224-1233.

Walsh, J. B. (1995). How often do duplicated genes evolve new functions? *Genetics* **139**:421-428.

Wei, N., and Deng, X.-W. (1992). *COP9*: a new genetic locus involved in light-regulated development and gene expression in *Arabidopsis. Plant Cell* **4**:1507-1518.

Weigel, D., and Nilsson, O. (1995). A developmental switch sufficient for flower initiation in diverse plants. *Nature* **377**:495-500.

Welin, B. V., Olson, A., Nylander, M., and Palva, E. T. (1994). Characterization and differential expression of *dhn/lea/rab*-like genes during cold acclimation and drought stress in *Arabidopsis thaliana. Plant Mol. Biol.* **26**:131-144.

West-Eberhard, M. J. (1989). Phenotypic plasticity and the origins of diversity. *Annu. Rev. Ecol. Syst.* **20**:249-278.

Whiteman, H. H. (1994). Evolution of facultative paedomorphosis in salamanders. *Q. Rev. Biol.* **69**:205-221.

Winn, A. A. (1996a). Adaptation to fine-grained environmental variation: an analysis of within-individual leaf variation in an annual plant. *Evolution* **50**:1111-1118.

Winn, A. A. (1996b). The contributions of programmed developmental change and phenotypic plasticity to within-individual variation in leaf traits in *Dicerandra linearifolia. J. Evol. Biol.* **9**:737-752.

Yamada, U. (1962). Genotype by environment interaction and genetic correlation of the same trait under different environments. *Jpn. J. Genet.* **37**:498-509.

Yamaguchi-Shinozaki, K., Urao T., and Shinozaki, K. (1995). Regulation of genes that are induced by drought stress in *Arabidopsis thaliana. J. Plant Res.* **108**:127-136.

Zelditch, M. L. (1988). Ontogenetic variation in patterns of phenotypic integration in the laboratory rat. *Evolution* **42**:28-41.

17

THE GENETIC ARCHITECTURE OF QUANTITATIVE TRAITS

TRUDY F. C. MACKAY

Department of Genetics, North Carolina State University, Raleigh, NC 27695

INTRODUCTION

Most morphological characters used in taxonomy are likely to have a complex genetic architecture, with many potentially interacting genes contributing to the ultimate phenotype. A full understanding of the biology of characters cannot neglect the underlying genes; further, resolution of differing phylogenetic

The Character Concept in Evolutionary Biology
Copyright © 2001 by Academic Press. All right of reproduction in any form reserved.

relationships inferred from phenotypic and molecular data cannot be achieved until we determine the phenotypic effects of molecular variation at the actual loci affecting the phenotype. The same morphological characters that discriminate among taxa often vary continuously within taxa, albeit over a much narrower range. The continuous variation in phenotype of these "quantitative" characters is attributable to the simultaneous segregation of alleles at multiple loci affecting the trait, called Quantitative Trait Loci, or QTL. Allelic effects at QTL are small relative to the background noise from the segregation of the other QTL, and sensitivity to environmental variation further blurs the distinction between genotypes (Falconer and Mackay, 1996). To the extent that the QTL at which variation occurs within a species are the same as those that cause divergence between species, we can make progress toward understanding the genetic basis of character *divergence* by studying character *variation* within a population.

Although simpler than determining the genetic basis of phenotypic divergence between taxa, a full understanding of the nature of genetic variation for quantitative traits is by no means trivial. Such an understanding requires that we enumerate the loci at which mutational and segregating variation affecting the trait occurs, estimate mutation rates and allele frequencies, and determine the distribution of effects of mutations and segregating alleles at each locus. That is, we need to know the homozygous, heterozygous, epistatic, and pleiotropic effects of QTL alleles, in a typical range of environments, and the nature of the molecular polymorphisms responsible for the difference in phenotype among QTL alleles. Until recently, such genetic and molecular dissection has not been possible – the impediments have been technical, not conceptual. Here, I review what has been learned recently about the genetic architecture of the model quantitative traits, numbers of *Drosophila* sternopleural and abdominal bristles. The problems encountered and solutions adopted will apply in general to efforts to genetically dissect other *Drosophila* quantitative traits, and to other traits and species. Further, the properties of bristle number QTL in terms of distributions of additive, epistatic, and pleiotropic effects and the nature of the genetic loci giving rise to naturally occurring variation for bristle number may hold in general for other morphological characters, including those that are divergent between taxa.

DROSOPHILA SENSORY BRISTLE NUMBER AS A MODEL SYSTEM

Understanding the genetic and molecular basis of variation for a quantitative trait has the highest probability of success in a genetically well-characterized model organism, such as *Drosophila melanogaster*. In addition to its relatively short life cycle, which facilitates rapid production of inbred lines and strains selected for divergent values of the character of interest, it is possible to construct "designer genotypes," using "balancer" chromosomes to substitute whole chromosomes into a specified background genotype in a few generations.

Decades of research with this organism has provided us with a library of stocks bearing mutations of most, if not all, visible phenotypes, and deficiency chromosomes covering approximately 70% of the genome (Lindsley and Zimm, 1992). P transposable element mutagenesis is a convenient tool for generating additional mutations affecting quantitative traits and for characterizing the loci disrupted by P element insertions at the molecular level. P elements have also been harnessed as efficient transformation vectors. There is an active *Drosophila* genome project, which has made available P1 and cosmid clones, and homozygous lethal P element insertion lines spanning the entire genome. *Drosophila* is highly polymorphic for markers that can be used for mapping QTL, on both coarse (centi-Morgan, cM) and fine (kilobase, kb) scales. Thus, this organism provides us with all the tools necessary for mapping QTL and characterizing them at the level of genetic locus [see Ashburner (1989) for a comprehensive discussion of *Drosophila* genetic analysis].

Drosophila bristles are external mechano- and chemosensory structures. Although flies are covered in bristles and hairs, the positions of most of these characters are invariant. However, the numbers of mechanosensory bristles on the sternopleural plates and on the abdominal sternites are phenotypically and genetically variable, and have a long history of use as model quantitative characters. Abdominal and sternopleural bristle numbers have high heritabilities—approximately 0.5—in natural populations (Falconer and Mackay, 1996). They have been used to elucidate basic quantitative genetic principles such as short- (Clayton et al., 1957a; Frankham et al., 1968) and long-term (Clayton and Robertson, 1957; Jones et al., 1968) response to selection; selection limits (Yoo, 1980); spontaneous mutation rates (reviewed by Keightley et al., 1993; Houle et al., 1996); and the effects of recombination (McPhee and Robertson, 1970), population structure (Madalena and Robertson, 1975), and environmental heterogeneity (Mackay, 1981) on genetic variation and response to selection. Like many quantitative traits, bristle number is thought to be under stabilizing natural selection, since mean bristle numbers are relatively constant across natural populations (López-Fanjul and Hill, 1973a, b). Alleles with extreme bristle number effects most likely have deleterious pleiotropic effects on larval competitive ability (Linney et al., 1971) or some other aspect of fitness (Nuzhdin et al., 1995). Bristle numbers were the first *Drosophila* traits for which the chromosomal locations, numbers, and effects of QTL contributing to selection response were estimated (Breese and Mather, 1957; Thoday, 1979; Shrimpton and Robertson, 1988a, b). Finally, many genes are known that affect bristle development (Jan and Jan, 1993; Kania et al., 1995; Salzberg et al., 1997) and/or that have major mutant effects on bristle number (Lindsley and Zimm, 1992; Lyman et al., 1996). These genes are candidate bristle number QTL.

DROSOPHILA BRISTLE NUMBER QTL

The first step in the elucidation of the genetic architecture of bristle number, or indeed any quantitative trait, is to determine the map positions and effects of the QTL at which naturally occurring variation in bristle number occurs. Since multiple loci with segregating allelic effects that are too small to be perceived over and above segregation at other loci and superimposed environmental variation are expected, QTL are mapped by linkage to marker loci whose genotype can be determined unambiguously. In species like *Drosophila* that are amenable to inbreeding, the most efficient way to map QTL is to cross two lines that are fixed for alternate alleles at QTL and marker loci, and to score individuals from backcross, F2, or other segregating generations for their quantitative trait phenotype and marker genotype. The presence of a QTL linked to a marker locus (or between adjacent marker loci) is inferred if there is a difference in quantitative trait phenotype between marker genotype classes.

Although conceptually simple, there are many statistical and genetical issues to consider when designing and interpreting such experiments. The first is genetic sampling. While any pair of inbred lines is likely to be fixed for alleles at QTL affecting the desired trait, whether or not they differ in the mean value of the trait, experiments that utilize inbred derivatives of strains that have been divergently selected for the character of interest, from a large base population, will contain a better sample of the relevant QTL. The second and third issues, statistical power and experimental design, are related. Very large experiments are necessary to detect QTL of even moderate effect (Falconer and Mackay, 1996; Lynch and Walsh, 1997). The effect of a QTL depends on the mean difference between homozygous genotypes and the phenotypic standard deviation, and any design that minimizes the latter will have more power to detect QTL with small effects. One way to reduce the within-marker-class phenotypic standard deviation is to construct a mapping population of recombinant inbred (RI) lines, which enables measurement of the character on multiple individuals of the same genotype. RI lines also offer the advantage of permanency (apart from the accumulation of spontaneous mutations over time), and are thus useful for assessing environmental sensitivities of QTL alleles. The second way to reduce the within-marker-class phenotypic standard deviation is to control for the effects of other segregating QTL. This can be done genetically, by constructing introgression or chromosomal congenic stocks in which QTL are mapped on each chromosome in turn, in a homozygous background genotype; and/or statistically, using multiple regression methods to control for the effects of QTL a specified distance away from the test marker (or interval between adjacent markers) (Zeng, 1993, 1994; Jansen and Stam, 1994). Fourth, the precision in estimating the map positions of the QTL depends on the availability of a dense polymorphic marker map, and the opportunity for recombination between closely linked markers. The latter requirement can be met by very large mapping populations or by using mapping populations in which multiple rounds of recombination occur (for example, RI lines) (Darvasi, 1998). Finally, QTL mapping entails a large number of statistical tests, and care must be taken to set an appropriate experiment-wise false-positive error rate.

Recently, we have completed a series of four experiments to map and determine the properties of QTL affecting variation in abdominal and sternopleural bristle number. The first three experiments used lines derived from the Raleigh, NC natural population and were designed to enable inferences about numbers and effects of QTL alleles segregating in nature. The fourth experiment used two laboratory inbred lines and was designed to determine the extent to which QTL alleles exhibit genotype x environment interaction (GEI).

The experiments to infer properties of naturally occurring bristle number QTL were begun by obtaining a large sample of flies from a single collection site (Raleigh Farmer's Market), and deriving from this base population lines selected for 25 generations for high and low abdominal bristle number (Long *et al.*, 1995), and high and low sternopleural bristle number (Gurganus *et al.*, 1999). Given the size of the base population and of the selection lines, one expects that most loci at which alleles segregated at intermediate frequency in the wild will have been sampled. The lines were selected for 25 generations to minimize the chance that new mutations affecting bristle number contribute to selection response (Hill, 1982), while bringing to high frequency alleles affecting the traits. As the selection lines were not highly inbred, homozygous derivatives of the selection lines were constructed. Next, the X and third chromosomes of each homozygous high selection line were substituted into the homozygous low line backgrounds. Four RI mapping populations were generated by a single generation of recombination between high and low selection line chromosomes, all in a low line background: X and chromosome 3 populations from the abdominal bristle selection lines, and X and chromosome 3 from the sternopleural bristle selection lines. The X and third chromosomes comprise approximately 60% of the *Drosophila* genome. As noted earlier, the RI line design combined with considering one chromosome at a time in a common background is expected to increase the power to detect QTL.

These experiments utilized only one round of recombination and mapped the divergence between high and low QTL alleles. To obtain finer-scale localization of QTL and to infer whether the selection response was attributable to high or low, or both, alleles at each QTL, mapping populations of RI lines were constructed from the same high and low X and third chromosomes backcrossed for three generations to an unselected inbred strain. The background genotype of the mapping populations was that of the unselected strain (Nuzhdin *et al.*, 1999). This experiment is thus an advanced generation intercross, RI, chromosomal congenic design, and should yield accurate estimates of QTL positions and effects. In all of these experiments, both bristle traits were recorded on a sample of 10 males and 10 females from each of two replicate vials, for each RI line. A total of 640 RI lines were constructed and scored in the three experiments.

The fourth experiment utilized two homozygous strains that had not been selected for bristle number, Oregon and 2b. The F1 progeny were backcrossed to 2b, and the backcross progeny were mated at random for four generations. A total of 98 RI lines were derived from the advanced intercross population by 25 generations of strict full-sib inbreeding (Nuzhdin *et al.* 1997). All 98 RI lines were reared in three temperature environments, and abdominal and sternopleural bristle numbers were recorded on a sample of 10 males and 10 females from each of two replicate vials, for each RI line, in each environment (Gurganus *et al.*, 1998).

Recombination breakpoints were inferred for all of the RI lines using the highly polymorphic cytological insertion sites of *roo* retrotransposable element markers. These markers provide a dense (3-4 cM) informative marker map, where the order and physical location of markers are known. Composite interval mapping (Zeng, 1993; 1994) and/or single marker step-wise multiple regression (Churchill and Doerge, 1994; Doerge and Churchill, 1996) were used to map bristle number QTL. The results of the two methods were nearly identical, and both reduce within-marker-class variation by controlling for the segregation of other QTL. The *P* value for declaring a significant QTL in the analyses of Long *et al.* (1995) was taken as $0.05/n$, where *n* was the number of markers tested (a Bonferroni correction for multiple tests). In the analyses of Gurganus *et al.* (1998, 1999) and Nuzhdin *et al.* (1999), empirical significance thresholds under the null hypothesis were determined by randomly permuting the trait data among the genotype data 1000 times and recording the most significant test statistic for each permuted data set. Real test statistics exceeding the 5% tail of the permutation distribution were considered significant.

At this point it might be noted that QTL mapping is hardly a trivial enterprise, even using fruit flies. The results of these recent studies, discussed later, entailed creating and genotyping 738 RI lines, and scoring a total of 37,360 flies for each of two bristle traits!

The most precise estimates of chromosomal locations and effects of bristle number QTL are from the analysis of Nuzhdin *et al.* (1999). The cytogenetic map positions corresponding to the highest test statistic for each significant QTL are given in Fig. 1. To date, 28 QTL associated with response and correlated response to selection for bristle number have been detected on the first and third chromosomes. This is a minimum estimate of the total number of QTL on these chromosomes contributing to variation in the Raleigh population, since further recombination may reveal more QTL, larger samples of the initial base population and during selection may have succeeded in capturing additional alleles, and existing alleles with smaller effects could have been detected with larger numbers of RI lines. There are a number of additional features of these results that are worth noting.

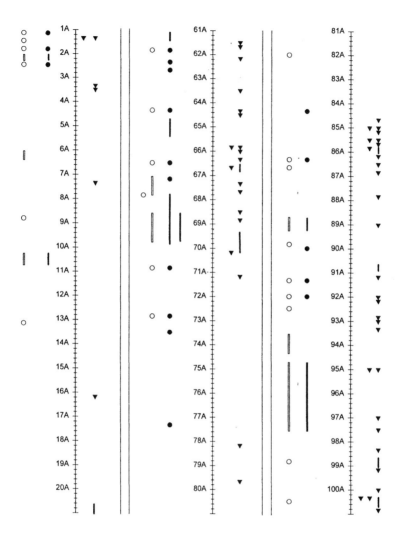

Figure 1 Correspondence between map positions of QTL and of candidate genes affecting abdominal and sternopleural bristle number. The physical maps of the *X* chromosome and chromosomes *3L* and *3R* are shown with the 20 numbered major cytological divisions per chromosome arm, each with six lettered (A-F) subdivisions. The estimates of abdominal (open circles and lines) and sternopleural (solid circles and lines) bristle number QTL map positions, shown to the left of each chromosome, were taken from Nuzhdin *et al.* (1999). Circles are used to indicate QTL corresponding to a single marker locus and lines if the map position covers two or more markers. The candidate gene map positions (solid triangles) shown to the right of each chromosome were taken from Lindsley and Zimm (1992); Jan and Jan (1993); Salzberg *et al.* (1994, 1997); Kania *et al.* (1995); Lyman *et al.* (1996) and Lai *et al.* (1998). Single triangles indicate candidate genes with map positions accurate to a cytological subdivision, and vertical bars depict candidate genes with map positions that are less precisely estimated.

(1) The same high and low selected chromosomes were used by Long *et al.* (1995) and Gurganus *et al.* (1999), respectively, to map the QTL responsible for divergent responses to selection for abdominal and sternopleural bristle number, and correlated responses in sternopleural and abdominal bristle number. Although fewer QTL covering larger genomic regions were detected in these studies than that of Nuzhdin *et al.* (1999), the chromosomal regions to which the QTL mapped were in excellent agreement for all three studies, attributing to the accuracy of the mapping methods. Interestingly, the QTL map locations inferred from the unrelated Oregon and 2b strains (Gurganus *et al.*, 1998) also overlap those of the QTL detected by Nuzhdin *et al.* (1999), as do those of much earlier studies using further unrelated strains, morphological markers, and progeny testing to estimate the map position of the QTL within an interval (Wolstenholme and Thoday, 1963; Thoday *et al.*, 1964; Spickett and Thoday, 1966; Piper, 1972; Shrimpton and Robertson, 1988b). Finding the same genomic regions containing bristle number QTL recurring in studies that span a period of 35 years, for populations of diverse origin, and using different mapping methodologies, strongly suggests that many of the same loci cause variation in bristle number in different geographical populations, an inference supported by classical quantitative genetic analyses (López-Fanjul and Hill, 1973a,b).

(2) True *et al.* (1997) have mapped QTL affecting morphological divergence in bristle characters between two sibling species, *Drosophila simulans* and *D. mauritiana*. One QTL was found to affect the divergence in abdominal bristle number, and it mapped to the tip of chromosome *3L*. Although the map position is approximate, this QTL maps to the same region as QTL, affecting variation in abdominal bristle number within *D. melanogaster*. This observation is consistent with the hypothesis that QTL affecting variation in a character within a species also contribute to divergence of that character between species.

(3) The QTL detected in the studies of Long *et al.* (1995) and Gurganus *et al.* (1999) had large effects and accounted for the majority of the difference between the selected chromosomes in bristle number. Although a larger number of QTL was detected by Nuzhdin *et al.* (1999), the effects remained as large, if not larger, than those estimated for a smaller number of factors. One explanation for this paradox is that many of the factors mapped by Nuzhdin *et al.* (1999) had effects in the opposite direction to selection and were presumably on the selected chromosome as a consequence of hitchhiking along with the selected loci. In addition, the estimates of effects by Long *et al.* (1995) and Gurganus *et al.* (1999) were in a low selection line background, leaving open the possibility that epistasis between high and low alleles at interacting loci may have decreased the estimate of the effect of the high allele. Although bristle number QTL effects were large on average, they are by no means equal. Some QTL have very large effects and account for the majority of the difference between selected chromosomes. Such a leptokurtic distribution of QTL effects is in accord with

the hypothesis first proposed by Robertson (1967), and substantiated as well by the mapping studies of Shrimpton and Robertson (1988b) and Gurganus et al. (1998).

(4) In no case did the QTL associated with response to selection for high bristle number map to the same position as those associated with response to selection for low bristle number. As analyses of limits to selection from single-pair bottleneck and from larger base populations suggest that many of the alleles affecting selection response are at intermediate frequency in nature (Robertson, 1968; Frankham, 1980), this result is consistent with an asymmetrical distribution of allelic effects at the majority of loci contributing to selection response; i.e., at most loci a common allele exists with either a high or a low effect on bristle number relative to other alleles segregating at the locus. The result is also consistent with a symmetrical distribution of QTL allelic effects in nature, but with rare alleles affecting selection response, such that the initial sample from nature was unlikely to contain both a high and a low allele at any given locus.

(5) Two-thirds of the QTL associated with correlated selection responses were loosely linked to the QTL associated with direct selection response, and one-third were associated with the same markers as those affecting direct responses. Pleiotropy or close linkage can be presumed as the cause of the latter correlated responses pending further finer scale mapping. This could partly explain why correlated selection responses are notoriously unpredictable from parameters estimated in the base population (Clayton et al., 1957b), as much will depend on linkage disequilibrium between bristle number QTL in the sample used to initiate the selection lines.

(6) Epistatic interactions between bristle number QTL are not uncommon (Shrimpton and Robertson, 1988a, b; Long et al., 1995; Gurganus et al., 1999). Further, epistatic effects can be as large, or larger, than main effects.

(7) QTL effects are often both sex and environment specific. Although sexual dimorphism for bristle number has long been known, sex-specific bristle number QTL effects (i.e., variation in the contribution of each QTL to the overall difference in bristle number between the sexes) were unexpected. Of the 22 QTL alleles with effects on sternopleural bristle number mapped by Nuzhdin et al. (1999), 13 had sex-specific effects; similarly, of the 21 alleles with effects on abdominal bristle number, 14 were sex specific. Sex-specific effects are also commonly observed for spontaneous (Mackay et al., 1995) and P-element-induced (Lyman et al., 1996) mutations affecting bristle number. Gurganus et al. (1998) mapped 9 QTL affecting sternopleural bristle number and 11 QTL affecting abdominal bristle number, 14 of which exhibited GEI with respect to temperature. To further complicate the picture, both epistatic (Long et al., 1995) and GEI interaction effects (Gurganus et al., 1998) are often sex specific. Sex and environment x QTL interactions could lead to the maintenance of quantitative genetic variation for bristle number (Gillespie and Turelli, 1989).

Furthermore, sex-specific QTL effects can facilitate the rapid evolution of sexual dimorphism.

FROM QTL TO GENE

Even the most precisely determined QTL positions can cover rather large genomic regions. This is because an accurate map position depends on recombination events close to the QTL, which in turn requires large numbers of meioses and consequently very large sample sizes and/or multiple-generation experimental designs. This problem is particularly acute in regions of restricted recombination, such as pericentromeric regions of *Drosophila*. One way to accurately map a QTL to the level of resolution that is required for subsequent identification by sequencing a cosmid clone is by selecting for recombinants using polymorphic markers closely linked to the QTL. While feasible, this is a rather laborious procedure in *Drosophila*, as the marker genotypes cannot be determined until the individuals have been bred. Marker assessment is destructive, and replicate measurements of the same genotype are necessary to ascertain phenotypes with quantitative effects. A more feasible method may be to use overlapping deficiencies, if they are available for the regions of interest, to more accurately estimate QTL map positions.

Candidate Genes

Currently, the best-case scenario for going from mapped factor to genetic locus is the existence of a candidate locus affecting the trait in the interval to which the QTL maps. Candidate loci are genes in the metabolic or developmental pathways leading to the trait phenotype. Putative positional candidate loci may be identified through mutational analysis of the trait or, for organisms with complete genome sequences in the QTL region, by postulating a functional relationship between particular coding sequences in the region and the quantitative trait phenotype. Underlying the candidate gene approach is the assumption that the variation in quantitative traits is caused by the segregation of "isoalleles" with quantitative effects at loci previously identified by alleles with major, qualitative effects (Thompson, 1975; Mackay 1985; Robertson, 1985). A limitation of the candidate locus approach is that all the loci leading to the trait phenotype are not usually known, and loci that are not obvious candidates may have undescribed and unexpected pleiotropic effects on the trait. Traits with a rich source of potential candidate loci will, therefore, be those for which comprehensive screens for mutations affecting the trait have been conducted. Further, screens for mutations with quantitative effects are essential, as it is possible that the spectrum of mutational effects at some loci does not include alleles with major effects, perhaps through functional redundancy. *Drosophila* bristle numbers are traits for which such comprehensive mutant screens have been conducted, in the context of searching for essential genes affecting peripheral nervous system development (reviewed by Jan and Jan, 1993; Kania *et al.*, 1995; Salzberg *et al.*, 1997) and for mutations with quantitative effects on adult bristle number (Lyman *et al.*, 1996).

Figure 1 shows the map positions of the candidate loci affecting *Drosophila* bristle number on the *X* and third chromosomes, and the relative map positions of the candidate genes and bristle number QTL. There are a total of 61 candidate genes listed. From this figure, we can see that there are a few QTL for which there are no candidate genes. For these QTL, there is no option but very fine-scale mapping for understanding to what loci the QTL correspond. In this case, quantitative genetic analysis of naturally occurring variation could reveal novel loci essential for bristle formation. However, for most QTL there is at least one and often there are several potential candidates mapping to the same region. How can we implicate the candidate gene and the QTL as one and the same entity? We have addressed this problem for bristle number QTL using three kinds of genetic tests: (1) introgression of naturally occurring candidate gene alleles into a standard background; (2) tests for quantitative failure of QTL alleles to complement mutations of the candidate gene; and (3) tests for association of molecular variation at the candidate gene with phenotypic variation for the quantitative trait in a sample of alleles from nature.

Introgression of Candidate Gene Alleles

We have constructed "near-isoallelic" lines, in which a sample of approximately 50 alleles from the Raleigh natural population at each of six genomic regions containing candidate loci for bristle number QTL are introgressed into a common homozygous background. The gene regions are those including the *achaete-scute* (*ASC*) and *Notch* (*N*) loci at the tip of the *X* chromosome; *scabrous* (*sca*) and *daughterless* (*da*) on chromosome *2*; *extramacrochaetae (emc), hairy (h)*, and the *Delta (Dl) – Hairless (H)* region on chromosome *3*. The introgression for each gene region was accomplished using visible mutations for the candidate bristle loci. First, the bristle mutations were backcrossed for 20 generations to the standard homozygous line. Then, 50 homozygous *X*, second and third chromosomes from nature were substituted into the same standard background. Finally, the region containing the naturally occurring candidate gene allele was introgressed into the standard background by crossing each of 50 alleles to the stock containing the mutation of the candidate gene and selecting for heterozygous backcross progeny for 10 generations, then breeding the wild-derived introgressed segment to homozygosity. After 10 backcross generations, one expects the candidate gene allele plus a 10 cM linked chromosomal fragment originating from the wild-derived chromosome on either side of the candidate gene in an otherwise homozygous background (Crow and Kimura, 1970). Two independent backcrosses were done for each of the 50 alleles to control for variation in the size of the introgressed fragment. In total, 6 gene regions x 50 alleles each x 2 backcross replicates per allele = 600 isoallelic lines (24,000 flies) were evaluated for sternopleural and abdominal bristle number. There was significant variation among lines of all gene regions, for both bristle traits. One cannot infer the extent to which the variation in bristle number is attributable to allelic variation at the candidate locus from these results, however, and can only conclude that there is a gene or genes in the region affecting quantitative genetic variation in these traits.

Quantitative Complementation Tests

To test for a genetic interaction between a recessive mutation, m^*, and a candidate gene defined by a recessive mutation, m, affecting the same phenotype as m^*, one crosses the two strains and determines whether the heterozygote displays the mutant phenotype (m^* fails to complement, or interacts, with m) or is wild type (m^* complements, or does not interact, with m). The logic of complementation testing can be extended to alleles with additive, quantitative effects by recognizing that the traditional test is for a difference in heterozygous effects of m^* in a wild type and mutant background. A quantitative complementation test is a test for interaction between at least two QTL alleles (e.g., the high, H and low, L alleles from selection lines) and mutant (Tester, T) and wild type (Control, C) alleles at the candidate gene. If the difference between T and C alleles is the same in the H and L QTL allele heterozygotes, there is no interaction, and the QTL alleles complement the candidate gene alleles. If the difference between the T and C alleles depends on the QTL allele, there is an interaction and failure of the QTL alleles to complement the candidate gene alleles. This principle is illustrated in Fig. 2. Quantitative failure to complement is detected formally by a significant interaction term in a two-way analysis of variance.

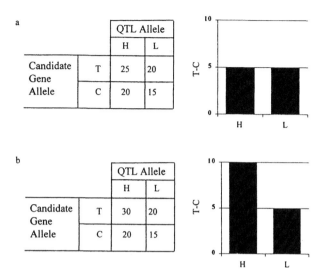

Figure 2 Quantitative complementation tests. A minimum of four alleles are required: two QTL alleles with different effects (here labeled high, H, and low, L), and a mutation (Tester, T) and wild type (Control, C) allele at the candidate gene. The phenotype of the quantitative trait is measured on the four (H/T, H/C, L/T and L/C) heterozygous genotypes. *(a)* If the difference between the T and C alleles at the candidate gene is the same in the high and low QTL allele backgrounds (H/T H/C = L/T L/C), there is no interaction between alleles at the candidate gene and QTL alleles; i.e., complementation. *(b)* If the difference between the T and C alleles at the candidate gene is not the same in the high and low QTL allele backgrounds (H/T H/C L/T L/C), there is an interaction between alleles at the candidate gene and QTL alleles; i.e., failure to complement.

QTL alleles on the Raleigh population high and low selected chromosomes failed to complement mutations at several candidate bristle number loci (Long *et al.*, 1996; Gurganus *et al.*, 1999), as did QTL alleles that had accumulated spontaneous mutations affecting bristle number (Mackay and Fry, 1996). However, interpretation of these results is ambiguous. As in traditional complementation tests, quantitative failure to complement can be attributable to allelic or epistatic interactions of the QTL and candidate gene alleles. It could be argued that epistatic interactions are more likely if the chromosomes used in the test contain high (or low) alleles at multiple QTL in association, as was the case for the experiments described earlier. Further, these studies have shown that the results of these tests can be highly dependent on the tester mutation used, and thus observed complementation for one allele does not necessarily mean that QTL alleles will not interact with other mutations at the candidate gene.

Allelic Association Tests

Mapping QTL in crosses between homozygous lines is possible because of the maximum linkage disequilibrium generated between QTL and marker locus alleles. The magnitude of the disequilibrium is expected to decay in subsequent generations of random mating as a function of the map distance between the quantitative trait and marker loci and the number of random mating generations (Falconer and Mackay, 1996; Weir 1996). Thus, as the number of generations of random mating (the number of opportunities for recombination) increases, the map distance between the QTL and markers in strong linkage disequilibrium with it decreases. Over time, only markers very closely linked to the QTL will be in linkage disequilibrium with the QTL alleles. This is the population genetic explanation for why multiple-generation mapping methods will give more precise localizations of QTL positions. The flip side of the argument is that only very tightly linked polymorphisms are expected to be in linkage disequilibrium in random mating populations. One can thus use this population genetic expectation as a basis for ultra-high resolution mapping of quantitative traits. If there is no association (i.e., linkage equilibrium) between the quantitative trait phenotype and molecular polymorphism within a candidate gene, then one can conclude that variation at the candidate gene is not a cause of variation in the trait. If, on the other hand, such an association is found, then one has the opportunity to identify the actual molecular polymorphisms responsible for variation in quantitative trait phenotypes (QTNs, or quantitative trait nucleotides).

The logic of linkage disequilibrium mapping is analogous to mapping QTL in line crosses, but on the scale of a single gene, rather than the whole genome. The experimental design requires a sample of alleles of a candidate gene from nature, the haplotype of each allele with respect to molecular polymorphisms at the candidate gene, and an estimate of the quantitative trait phenotype for each allele. One then assesses whether there is a difference in the mean value of the quantitative trait between the marker allele classes; such a difference indicates there is linkage disequilibrium between the marker and the QTL alleles. This kind of data is easier to obtain in *Drosophila* than for other diploid, outcrossing organisms, because of the ability to make whole chromosomes sampled from nature homozygous, in a standard homozygous background. The linkage phase of the polymorphic sites is thus unambiguous, and the phenotypic variation observed is attributable to the QTL on the single chromosome examined.

This experimental design was used to associate phenotypic variation in bristle number with molecular variation in the *achaete-scute* complex (*ASC*) at the tip of the *X* chromosome (Mackay and Langley, 1990) and at *scabrous* (*sca*) on chromosome *2* (Lai *et al.*, 1994). There was a highly significant association between the presence of large insertions at *ASC* and bristle number, with insertion-bearing haplotypes averaging 1.6 fewer abdominal and 1.2 fewer sternopleural bristles than alleles without large insertions. At *sca*, more polymorphic sites were associated with bristle number than expected by chance, sternopleural bristles than alleles without large insertions. At *sca*, more polymorphic sites were associated with bristle number than expected by chance, indicating that some molecular polymorphisms in the *sca* gene region are significantly associated with variation in bristle number. While these analyses impute these loci as genes at which alleles affecting quantitative variation segregate, the actual polymorphisms associated with the effects on the traits have not been identified. Further, the observed associations could be caused by linkage disequilibrium between the molecular polymorphisms at the candidate genes and QTL alleles at other bristle number loci on the same chromosome. Finally, it is important to recognize that a significant association can arise if a molecular polymorphism actually causes the quantitative variation (i.e., the polymorphism is the QTN), but also as the consequence of population admixture or other events in the history of the population.

A Combined Approach

While introgression, quantitative complementation and association tests are individually not sufficient to prove that a candidate gene is the genetic locus corresponding to a QTL, the combination of all of these methods considerably strengthens the case. We have recently completed such combined analyses for two candidate bristle number loci, *Delta* (*Dl*) and *scabrous*. Forty-seven second and 55 third chromosomes containing *sca* and *Dl* alleles, respectively, were sampled from the Raleigh natural population. Molecular variation was assessed in the sample of alleles using six-base restriction site polymorphism and single-stranded conformation polymorphism (SSCP) for the *sca* alleles (Lai *et al.*, 1994) and four- and six- base restriction site polymorphism at *Dl* (Long *et al.*, 1998). There were a total of 27 polymorphic sites in the 45-kb region including the *sca* locus and 53 polymorphic sites in the 57-kb region including *Dl*. The pattern of molecular polymorphism was such that there was little linkage disequilibrium between polymorphic sites, indicative of much historical recombination. Each allele represented a unique haplotype. Phenotypic variation was assessed for each chromosome in several ways: as a whole chromosome homozygote in a standard background, as a homozygous introgression line, and as heterozygous introgression lines against wild type and mutant alleles of the candidate genes. The quantitative complementation test results indicate that naturally occurring *sca* alleles fail to complement the abdominal bristle effect, but not the sternopleural bristle effect, of a *sca* mutant allele (Lyman *et al.*, 1999). Naturally occurring *Dl* alleles fail to complement the sternopleural bristle number effect of a *Dl* mutation in both sexes, and the abdominal bristle effect of the tester mutation in females only (Lyman and Mackay, 1998).

The analysis of associations between bristle number phenotypes and marker genotypes for these data poses some statistical problems. There are many tests for association performed on each data set, and the tests are not independent because both the markers and the phenotypes are partly correlated. The solution proposed by Churchill and Doerge (1994; Doerge and Churchill, 1996) in the context of genome-wide scans for QTL was applied to these candidate gene scans for QTNs. The threshold for significance under the null hypothesis of no association was evaluated empirically, by permuting the trait data at random among the marker haplotypes and determining the distribution of test scores in the permuted data set. The site with the highest significant observed test statistic was then fitted in a multiple regression model, and the procedure repeated on the residuals until no further significant sites were detected. The multiple correlated phenotypic data were accommodated by doing the association tests on the first principal component score of each line. The outcome of these analyses was amazingly clear, and is illustrated in Fig. 3. At *sca*, an SSCP "allele" in the third intron has an effect of 4.4 abdominal bristles (Lyman *et al.* 1999). At *Dl*, a *Hae*III site in Intron 2 with a frequency of $p = 0.60$ has an effect of 0.65 sternopleural bristles in both sexes; and an *Scr*F1 site ($p = 0.58$) in Intron 5 has an effect of 1.1 abdominal bristles, in females only (Long *et al.*, 1998). There are several interesting features of these results and points to consider in designing future experiments:

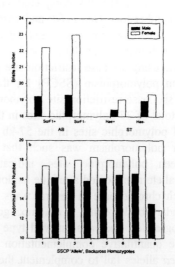

FIGURE 3 Allelic association studies at two candidate bristle number genes, *Dl* (Long *et al.*, 1998) and *sca* (Lyman *et al.*, 1999). Panel *(a)* Effects of two polymorphic restriction sites at *Dl* on bristle number in homozygous *Dl* region introgression lines (see text for explanation). An *Scr*F1 site in Intron 5 has an effect of 1.1 abdominal bristles, in females only, and an *Hae*III site in Intron 2 has an effect of 0.65 sternopleural bristles in both sexes. Panel *(b)* An SSCP "allele" in Intron 3 of *sca* has an effect of 4.4 abdominal bristles in homozygous *sca* region introgression lines.

(1) Molecular variation at candidate QTL can be associated with phenotypic variation in quantitative traits.

(2) Although an accurate estimate of QTL map position and evidence for genetic interaction of the QTL and candidate gene are not necessary prior to conducting association studies, this evidence is desirable as it serves to whittle down the large number of potential candidate genes to be evaluated. For *sca* and *Dl*, the complementation tests on introgression lines led to the same conclusions as the association tests.

(3) The molecular polymorphisms associated with bristle number variation are at intermediate frequency. These results are consistent with the existence of QTL alleles with moderately large effects at intermediate frequency in nature, or with rare QTL alleles with large effects.

(4) Abdominal bristle number effects at *Dl* were strongly sex-specific, confirming at the molecular level our previous observation of sex-specific QTL.

(5) The molecular polymorphisms associated with bristle number are in introns, suggesting that variation in non-coding regions contributes to quantitative variation in phenotype.

(6) It is interesting to speculate on the nature of pleiotropy. Two independent polymorphisms at *Dl* were associated with two bristle traits. Different *Dl* alleles thus vary as to whether sternopleural and abdominal bristle number are positively, negatively, or uncorrelated. If generally true, this may partly explain our inability to accurately predict correlated responses to selection (Clayton *et al.*, 1957b).

(7) Efforts to map the actual QTNs will require a higher density of polymorphic markers than used in these studies. In order to detect linkage disequilibrium between the QTN and marker alleles, adjacent marker alleles must be themselves in linkage disequilibrium, and disequilibrium falls off rapidly with physical distance (e.g., Aquadro *et al.*, 1986; Miyashita and Langley, 1988). The markers used in *sca* and *Dl* were not evenly spaced, and the significant associations were found in regions where the marker density was greatest. Thus, it is not inconceivable that there are additional polymorphisms associated with bristle number variation in these genes that have gone undetected.

(8) Since population history (e.g., recent admixture) can generate disequilibria between polymorphisms at the candidate gene and the quantitative trait phenotype, independent replication in other populations is necessary to confirm whether observed associations are causal or artefactual.

It has been over 30 years since Alan Robertson (1967) despaired that the statistical fog surrounding the study of quantitative traits could ever be lifted, enabling the description of these traits in terms of complex genetics rather than complex statistics. Some of that fog is now dissipating for the model quantitative traits, abdominal and sternopleural bristle numbers of *Drosophila*. A clear path has been revealed toward the ultimate goal of understanding how genetic variation at the molecular level accounts for variation at the level of phenotype, how this variation is maintained within populations, and what is the molecular basis of morphological divergence between species. Traversing this path should prove to be quite stimulating, given recent results illustrating the necessity of incorporating nonadditive interactions between QTL alleles and sex and environment by QTL interactions in our descriptions of the genetics and evolution of quantitative traits.

ACKNOWLEDGMENTS

I thank the National Institutes of Health grants GM 45344 and GM 45146 for financial support, Richard Lyman for preparing the figures, and Robert Anholt for comments on the manuscript.

LITERATURE CITED

Ashburner, M. (1989). *"Drosophila,* a Laboratory Handbook." Cold Spring Harbor Laboratory Press, Cold Spring Harbor.

Aquadro, C. F., Deese, S. F., Bland, M. M., Langley, C. H., and Laurie-Ahlberg, C. C. (1986). Molecular population genetics of the *alcohol dehydrogenase* gene region of *Drosophila melanogaster. Genetics* 114:1165-1190.

Breese, E. L., and Mather K. (1957). The organization of polygenic activity within a chromosome in *Drosophila.* I. Hair characters. *Heredity* 11:373-395.

Churchill, G. A., and Doerge, R. W. (1994). Empirical threshold values for quantitative trait mapping. *Genetics* 138:963-971.

Clayton, G. A., Knight, G. R., Morris, J. A., and Robertson, A. (1957b). An experimental check on quantitative genetical theory. III. Correlated responses. *J. Genet.* 55:171-180.

Clayton, G. A., Morris, J. A., and Robertson, A. (1957a). An experimental check on quantitative genetical theory. I. Short-term responses to selection. *J. Genet.* 55:131-151.

Clayton, G. A., and Robertson, A. (1957). An experimental check on quantitative genetical theory. II. The long-term effects of selection. *J. Genet.* 55:152-170.

Crow, J. F., and Kimura, M. (1970). "An Introduction to Population Genetics Theory." Harper and Row, New York.

Darvasi, A. (1998). Experimental strategies for the genetic dissection of complex traits in animal models. *Nat. Genet.* 18:19-24.

Doerge, R. W., and Churchill, G. A. (1996). Permutation tests for multiple loci affecting a quantitative character. *Genetics* 142:285-294.

Falconer, D. S., and Mackay, T. F. C. (1996). "Introduction to Quantitative Genetics, 4/e." Addison Wesley Longman, Harlow, Essex.

Frankham, R. (1980). The founder effect and response to artificial selection in *Drosophila. In* "Selection Experiments in Laboratory and Domestic Animals" (A. Robertson, ed.), pp. 87-90. Commonwealth Agricultural Bureaux, Slough, UK.

Frankham, R., Jones, L. P., and Barker, J. S. F. (1968). The effects of population size and selection intensity in selection for a quantitative character in *Drosophila.* III. Analysis of the lines. *Genet. Res.* 12:267-283.

Gillespie, J. H., and Turelli, M. (1989). Genotype-environment interactions and the maintenance of polygenic variation. *Genetics* 121:129-138.

Gurganus, M. C., Fry, J. D., Nuzhdin, S. V., Pasyukova, E. G., Lyman, R. F., and Mackay, T. F. C. (1998). Genotype-environment interaction for quantitative trait loci affecting sensory bristle number in *Drosophila melanogaster. Genetics* 149:1883-1898.

Gurganus, M. C., Nuzhdin, S. V., and Mackay, T. F. C. (1999). High resolution mapping of quantitative trait loci for sternopleural bristle number in *Drosophila melanogaster*. *Genetics* 152:1585-1604.

Hill, W. G. (1982). Predictions of response to artificial selection from new mutations. *Genet. Res.* 40:255-278.

Houle, D., Morikawa, B., and Lynch, M. (1996). Comparing mutational variabilities. *Genetics* 143:1467-1483.

Jan, Y. N., and Jan, L. Y. (1993). The peripheral nervous system. In "The Development of *Drosophila melanogaster*" (M. Bate and A. Martinez Arias, eds.), Vol. 2, pp. 1207-1244. Cold Spring Harbor Laboratory Press, Plainview, New York

Jansen, R. C., and Stam, P. (1994). High resolution of quantitative traits into multiple loci via interval mapping. Genetics 136:1447-1455.

Jones, L. P., Frankham, R., and Barker, J. S. F. (1968). The effects of population size and selection intensity in selection for a quantitative character in *Drosophila*. II. Long-term response to selection. *Genet. Res.* 12:249-266.

Kania, A., Salzberg, A., Bhat, M., D'Evelyn, D., He, Y. *et al.* (1995). P-element mutations affecting embryonic peripheral nervous system development in *Drosophila melanogaster*. *Genetics* 139:1663-1678.

Keightley, P. D., Mackay, T. F. C., and Caballero, A. (1993). Accounting for bias in estimates of the rate of polygenic mutation. *Proc. R. Soc. Lond. Ser. B* 253:291-296.

Lai, C., Lyman, R. F., Long, A. D., Langley, C. H., and Mackay, T. F. C. (1994). Naturally occurring variation in bristle number and DNA polymorphisms at the *scabrous* locus in *Drosophila melanogaster*. *Science* 266:1697-1702.

Lai, C., McMahon, R., Young, C., Mackay, T. F. C., and Langley, C. H. (1998). *quemao*, a *Drosophila* bristle locus, encodes geranygerany pyrophosphate synthase. *Genetics* 149:1051-1061.

Lindsley, D. L., and Zimm, G. G. (1992). "The Genome of *Drosophila melanogaster*." Academic Press, San Diego.

Linney, R., Barnes, B. W., and Kearsey, M. J. (1971). Variation for metrical characters in *Drosophila* populations. III. The nature of selection. *Heredity* 27:163-174.

Long, A. D., Lyman, R. F., Langley C. H., and Mackay, T. F. C. (1998). Two sites in the *Delta* gene region contribute to naturally occurring variation in bristle number in *Drosophila melanogaster*. *Genetics* 149:999-1017.

Long, A. D., Mullaney, S. L., Reid, L. A., Fry, J. D., Langley, C. H. *et al.* (1995). High resolution mapping of genetic factors affecting abdominal bristle number in *Drosophila melanogaster*. *Genetics* 139:1273-1291.

Long, A. D., Mullaney, S. L., Mackay, T. F. C., and Langley, C. H. (1996). Genetic interactions between naturally occurring alleles at quantitative trait loci and mutant alleles at candidate loci affecting bristle number in *Drosophila melanogaster*. *Genetics* 144:1497-1518.

López-Fanjul, C., and Hill, W. G. (1973a). Genetic differences between populations of *Drosophila melanogaster* for a quantitative trait. I. Laboratory populations. *Genet. Res.* 22:51-68.

López-Fanjul, C., and Hill, W. G. (1973b). Genetic differences between populations of *Drosophila melanogaster* for a quantitative trait. II. Wild and laboratory populations. *Genet. Res.* 22:69-78.

Lyman, R. F., Lawrence, F., Nuzhdin, S. V., and Mackay, T. F. C. (1996). Effects of single *P* element insertions on bristle number and viability in *Drosophila melanogaster*. *Genetics* 143:277-292.

Lyman, R. F., Lai, C., and Mackay, T. F. C. (1999). Linkage disequilibrium mapping of molecular polymorphisms at the *scabrous* locus associated with naturally occurring bristle number in *Drosophila melanogaster*. Genet. Res. 74:301-311.

Lyman, R. F., and Mackay, T. F. C. (1998). Candidate quantitative trait loci and naturally occurring phenotypic variation for bristle number in *Drosophila melanogaster*: The *Delta-Hairless* gene region. *Genetics* 149:983-998.

Lynch, M. and Walsh, B. (1997). "Genetics and Analysis of Quantitative Traits." Sinauer Associates, Inc., Sunderland, MA.

Mackay, T. F. C. (1981). Genetic variation in varying environments. *Genet. Res.* **37**:79-93.

Mackay, T. F. C. (1985). Transposable element-induced response to artificial selection in *Drosophila melanogaster. Genetics* **111**:351-374.

Mackay, T. F. C., and Fry, J. D. (1996) Polygenic mutation in *Drosophila melanogaster*: genetic interactions between selection lines and candidate quantitative trait loci. *Genetics* **144**:671-688.

Mackay, T. F. C., and Langley, C. H. (1990). Molecular and phenotypic variation in the *achaete-scute* region of *Drosophila melanogaster. Nature* **348**:64-66.

Mackay, T. F., CLyman, R. F., and Hill, W. G. (1995). Polygenic mutation in *Drosophila melanogaster*: non-linear divergence among unselected strains. *Genetics* **139**:849-859.

Madalena, F. E., and Robertson, A. (1975). Population structure in artificial selection: studies with *Drosophila melanogaster. Genet. Res.* **24**:113-126.

McPhee, C. P., and Robertson, A. (1970). The effect of suppressing crossing-over on the response to selection in *Drosophila melanogaster. Genet. Res.* **16**:1-16.

Miyashita, N., and Langley, C. H. (1988). Molecular and phenotypic variation of the *white* locus region in *Drosophila melanogaster. Genetics* **120**:199-212.

Nuzhdin, S. V., Dilda, C. L., and Mackay, T. F. C. (1999). The genetic architecture of selection response: inferences from fine-scale mapping of bristle number quantitative trait loci in *Drosophila melanogaster. Genetics* **153**:1317-1331.

Nuzhdin, S. V., Fry, J. D., and Mackay, T. F. C. (1995). Polygenic mutation in *Drosophila melanogaster*: the causal relationship of bristle number to fitness. *Genetics* **139**:861-872.

Nuzhdin, S. V., Pasyukova, E. G., Dilda, C. L., Zeng Z.-B., and Mackay, T. F. C. (1997). Sex-specific quantitative trait loci affecting longevity in *Drosophila melanogaster. Proc. Natl. Acad. Sci. USA* **94**:9734-9739.

Piper, L. R. (1972). "The Isolation of Genes Underlying Continuous Variation." Ph.D. thesis, University of Edinburgh.

Robertson, A. (1967). The nature of quantitative genetic variation. *In* "Heritage from Mendel" (A. Brink, ed.), pp. 265-280. The University of Wisconsin Press, Madison, WI.

Robertson, A. (1968). The spectrum of genetic variation. *In* "Population Biology and Evolution" (R. C. Lewontin, ed.), pp. 5-16. Syracuse University Press, Syracuse, NY.

Robertson, D. S. (1985). A possible technique for isolating genic DNA for quantitative traits in plants. *J. Theor. Biol.* **117**:1-10.

Salzberg, A., D'Evelyn, D., Schulze, K. L., Lee, J.-K., Strumpf, D. *et al.* (1994). Mutations affecting the pattern of the PNS in *Drosophila* reveal novel aspects of neuronal development. *Neuron* **13**:269-287.

Salzberg, A., Prokopenko, S. N., He, Y., Tsai, P. *et al.* (1997). *P*-element insertion alleles of essential genes on the third chromosome of *Drosophila melanogaster*: mutations affecting embryonic PNS development. *Genetics* **147**:1723-1741.

Shrimpton, A. E., and Robertson, A. (1988a). The isolation of polygenic factors controlling bristle score in *Drosophila melanogaster*. I. Allocation of third chromosome bristle effects to chromosome sections. *Genetics* **118**:437-443.

Shrimpton, A. E., and Robertson, A. (1988b). The isolation of polygenic factors controlling bristle score in *Drosophila melanogaster*. II. Distribution of third chromosome bristle effects within chromosome sections. *Genetics* **118**:445-459.

Spickett, S. G., and Thoday, J. M. (1966). Regular responses to selection. 3. Interactions between located polygenes. *Genet. Res.* **7**:96-121.

Thoday, J. M. (1979). Polygene mapping: uses and limitations. *In* "Quantitative Genetic Variation" (J. N. Thompson, Jr., and J. M. Thoday, eds.), pp. 220-233 Academic Press, New York.

Thoday, J. M., Gibson, J. B., and Spickett, S. G. (1964). Regular responses to selection. 2. Recombinations and accelerated response. *Genet. Res.* **5**:1-19.

Thompson, J. N., Jr. (1975). A test of the influence of isoallelic variation upon a quantitative character. *Heredity* **35**:401-406.

True, J. R., Liu, J., Stam L. F., Zeng, Z.-B., and Laurie, C. C. (1997). Quantitative genetic analysis of divergence in male secondary sexual traits between *Drosophila simulans* and *D. mauritiana*. *Evolution* **51**: 816-832.

Weir, B. S. (1996). "Genetic Data Analysis II." Sinauer Associates, Inc., Sunderland, MA.

Wolstenholme, D. R., and Thoday, J. M. (1963). Effects of disruptive selection. VII. A third chromosome polymorphism. *Heredity* **10**:413-431.

Yoo, B. H. (1980). Long-term selection for a quantitative character in large replicate populations of *Drosophila melanogaster*. I. Response to selection. *Genet. Res.* **35**:1-17.

Zeng, Z.-B. (1993). Theoretical basis of precision mapping of quantitative trait loci. *Proc. Natl. Acad. Sci. USA* **90**:10972-10976.

Zeng, Z.-B. (1994). Precision mapping of quantitative trait loci. *Genetics* **136**:1457-1468.

18

THE GENETIC ARCHITECTURE OF PLEIOTROPIC RELATIONS AND DIFFERENTIAL EPISTASIS

JAMES M. CHEVERUD

Department of Anatomy and Neurobiology, Washington University School of Medicine, St. Louis, MO, 63110

INTRODUCTION

Patterns of morphological integration follow patterns of developmental and functional relationship among traits. However, the genetic architecture underlying morphological integration is unknown. Are functionally and developmentally related traits affected by specific gene loci, with different loci affecting different trait complexes, or are the levels of correlation between traits controlled by a balance of positive and negative pleiotropy? Furthermore, if the genetic architecture itself can evolve, there should be genetic variation in pleiotropic patterns due to differential epistasis.

A quantitative trait locus (QTL) study of the mouse mandible is undertaken to address these questions in a F2 intercross of the LG/J and SM/J mouse strains. Mandibular measurements were collected for 480 individuals and measurements were classified into subregions based on their developmental and functional relationships. Seventy-six polymorphic microsatellite loci were scored on these same individuals. Multivariate interval mapping was used to identify QTLs for mandibular subregions. Variation in pleiotropy was detected with an analysis of covariance between mandibular subregions and the interaction of genotypic scores with total mandibular length.

The Character Concept in Evolutionary Biology

In general, pleiotropic effects of genes seem to be restricted to functionally and developmentally related trait sets, making each trait set an individuated character for evolutionary analysis. We also found variation in pleiotropy between local mandibular lengths and total mandibular length. This indicates that the pleiotropic effects of a locus can vary due to differential epistasis, where epistasis nullifies the effect of a target locus on some characters, but not on others. This implies that the nature of individuated characters displayed by a species may evolve.

The theory of evolutionary morphological integration (Olson and Miller, 1958; Cheverud 1982, 1984, 1995, 1996a, b) states that functionally and/or developmentally interacting traits will experience stabilizing selection favoring functionally compatible trait values. Given the presence of genetic variation, this stabilizing selection results in the evolution of strong cohesive genetic correlations among functionally related traits and relatively low genetic correlations among unrelated traits (Lande, 1980; Cheverud, 1982, 1984, 1996). The "modularization" of the phenotype into relatively independent functionally interacting trait groups (Wagner, 1996; Wagner and Altenberg, 1996) produces individuated character complexes. The members of these character complexes vary together within a species and evolve together in a coordinated fashion because they are genetically correlated. The integration of traits within a character complex requires that they be considered as a linked evolutionary entity, or character, in evolutionary analysis.

GENETIC CORRELATION AND PATTERNS OF PLEIOTROPY

Genetic correlations among traits arise from the pleiotropic effects of genes on multiple traits and/or linkage disequilibrium among distinct loci, each affecting a single member of the character complex (Falconer and Mackay, 1996). Linkage disequilibrium is not usually considered as of long-term importance in trait coevolution. It arises from the evolutionary history of a population and linkage relationships among the loci affecting the traits. Genetic correlations due to linkage disequilibrium will occur preferentially among functionally interacting traits if linkage relationships evolve so that distinct genes affecting interacting traits tend to lie near one another in the genome. While linkage relationships among loci may evolve, studies of genes affecting character complexes tend to indicate that such genes are spread throughout the genome, rather than being restricted to a specific chromosomal location (Cheverud et al., 1996, 1997; Tanksley et al., 1982; Weller et al., 1988; Paterson et al., 1988; Martin et al., 1989; Edwards et al., 1987). While there is certainly much variability in genome structure among species, syntenic relationships indicate that genome structure is conserved over long periods of evolutionary time on a fine scale, limiting the evolution of integrated character complexes through genome rearrangements. For these reasons, pleiotropy is usually considered as the most important source for genetic correlation.

Cheverud (1984) showed that a population's genetic correlation due to pleiotropy is a weighted average of positive pleiotropy at loci segregating (++) and (--) alleles, negative pleiotropy at loci segregating (+-) and (-+) alleles, and non pleiotropy at loci segregating (+0) and (-0) or (0+) and (0-) alleles. The weight of a locus' contribution to the correlation is determined by the strength of its effects on the traits. While this simple theory is straightforward, we know little about the underlying genetic architecture responsible for genetic correlation, i.e., the number of loci affecting the trait, dominance interactions between alleles at these loci, epistatic interactions among different loci, and patterns of pleiotropic effects produced by a locus. Different genetic architectures for genetic correlation can lead to different evolutionary expectations (Gromko, 1995; Lascoux, 1997).

Riedl (1978) predicted that the genetic architecture evolves to mimic the functional/developmental system. According to Riedl (1978; Wagner, 1996; Wagner and Altenberg, 1996), the pleiotropic effects of genes will evolve so that functionally and developmentally related traits will be affected by the same set of loci, distinct from loci affecting other character complexes. One then expects a nested hierarchy of effects that conforms to a nested hierarchy of functional and/or developmental relationship. As with morphological integration, this hypothesis predicts relatively high genetic correlations among functionally and developmentally related traits. However, it goes beyond predicting correlations by specifying the underlying genetic architecture for pleiotropic relations. Pleiotropic relations at individual loci will evolve so that different character complexes are affected by different sets of genes while traits within a complex will be affected by the same set of genes. This is in contrast to the possibility that the balance of positive and negative pleiotropy plays an important role in determining genetic correlation.

TABLE I Differential epistasis for two traits, A and B, at focal locus, X, and modifier locus, Y

	Trait A					Trait B		
	$\underline{Yy}$	$\underline{Yy}$	$\underline{YY}$			$\underline{yy}$	$\underline{Yy}$	$\underline{YY}$
xx |	-2	-1	0		xx |	-2	-1	0
Xx |	-1	0	1		Xx |	-1	0	0
XX |	0	1	2		XX |	0	1	0

DIFFERENTIAL EPISTASIS

If, as Riedl (1978) suggests, patterns of pleiotropy evolve so that functionally and developmentally related traits share a common set of genes distinct from genes affecting other character complexes, there must be genetic variation in the pleiotropic effects associated with a locus. Variation in pleiotropy can arise through differential epistasis which occurs when variability in a subset of the affected traits is suppressed by a modifier locus, the modifier locus having no effect on the remaining traits. For example, in Table I the cell values represent two-locus genotypic values for traits A and B at loci X and Y. Loci X and Y contribute additively to trait A, without epistasis. However, loci X and Y have a strong epistatic contribution to trait B. While locus X affects trait A regardless of the genotype present at locus Y, locus X only affects trait B in the presence of the yy and Yy genotypes. When the YY genotype is present, trait B is not affected by locus X. This is differential epistasis because epistasis occurs for trait B but not for trait A. The X locus has pleiotropic effects on traits A and B when the yy and Yy genotypes are present but locus X is not pleiotropic when the YY genotype is present. The Y locus genotype controls the pleiotropic pattern displayed by the X locus. This is only one of many potential scenarios in which differential epistasis results in variation in pleiotropic effects. With differential epistasis, changes in genotype frequency at the modifier locus change the pleiotropic pattern displayed by the focal locus.

I will first summarize recent results bearing on patterns of pleiotropic effects produced by quantitative trait loci (QTLs) on mandibular morphology. Is the relative lack of association between traits in different character complexes due to a balance of positive and negative pleiotropy [(++)(--) vs (+-)(-+) combinations at various loci]? Or is it due to modularization of the genetic architecture, so that traits participating in different character complexes are affected by different sets of genes? I will then present preliminary data on potential genetic variability in pleiotropy at the loci found to affect mandibular morphology. Specifically, I examine the possibility for variability in mandibular allometry due to differential epistasis.

MAMMALIAN MANDIBLE

The mammalian mandible is a developmentally complex organ, arising from several different tissues and their interactions, making it a suitable model system for studying the relationship among function, development, and genetic architecture (Atchley and Hall, 1991; Atchley, 1993). The bony mandible forms by membranous ossification from several distinct mesenchymal concentrations.

The cells in these condensations arise from cranial neural crest cells that have migrated from the region surrounding the midbrain down into the mandibular division of the first pharyngeal arch. There are several separate mandibular condensations in the mouse, two alveolar concentrations related to the incisor and the molars and a series of condensations for muscle attachment on the ascending ramus (see Fig. 1). The alveolar condensations participate with the overlying epidermis in a series of inductive interactions (Hall, 1994) responsible for the formation of teeth and the alveolar bone supporting them. Even after bone formation is complete, alveolar bone will resorb when the teeth are removed. Thus the presence and size of the alveolar regions are regulated primarily by the size of the associated teeth. The ascending ramus is composed of four regions for muscle attachment: (1) the body of the ascending ramus to which the masseter and medial pterygoid muscles attach; (2) the coronoid process to which the temporalis muscle attaches; (3) the angular process to which an extension of the masseter muscle attaches; (4) and the condyloid process to which the lateral pterygoid muscle attaches and through which the mandible articulates with the cranium. The ablation of a functioning muscle results in the resorption of the associated muscular process. Condensations form without muscular action but the mature form and maintenance of the ascending ramus and its components depend on the activity of the attached muscles (Herring and Lakers, 1981; Moore, 1981; Atchley *et al.*, 1984).

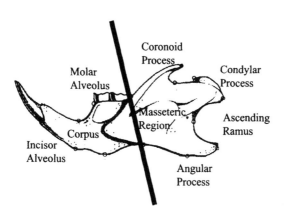

FIGURE 1   Functional and developmental units of the mandible (after Hall, 1994).

These mandibular regions have separate developmental and functional identities in that they arise from different mesenchymal condensations and relate to different nonbony tissues. Therefore, each can be considered as a character complex or morphological module distinct from the other modules of the mandible. This has been borne out in a series of studies of mandibular morphology in which functionally and developmentally related traits have a relatively high genetic correlation (Bailey, 1956; Atchley *et al.*, 1985a,b; Cheverud *et al.*, 1991). Furthermore, as predicted by Riedl (1978), Bailey (1985, 1986) and Cheverud *et al.* (1997) found that restricted chromosomal regions affect localized regions or modules of the mandible.

In this chapter, I examine the pleiotropic effects of individual gene loci on mandibular morphology. Under Riedl's (1979) hypothesis, I expect most genes to have effects restricted to a character complex with a deficit of genes displaying antagonistic pleiotropy. The earlier analysis of Cheverud *et al.* (1997) is modified to take advantage of new multivariate modes of analysis and formal tests for pleiotropy.

MATERIALS AND METHODS

Study Population

A F_2 generation produced by intercrossing two inbred mouse strains, Small (SM/J) and Large (LG/J), provides the study population. These two strains differ dramatically in overall size (Chai, 1956), having been derived from separate selection experiments for small (MacArthur, 1944) and large (Goodale, 1938, 1941) 60-day body weight, respectively. Loci carrying different alleles in the parental inbred strains will be heterozygous in the F_1 hybrid offspring. Intercrossing F_1 hybrid animals results in Mendelian segregation of alleles at varying loci in the F_2 generation.

TABLE II   Mandibular measurements as shown in Fig. 2. Measurements are classified as belonging to the alveolar process (AL) and its subdivisions, the incisor (IN) or molar (MO) alveolus, or to the ascending ramus (MU) and its subdivisions, the coronoid process (CR), the condyloid process (CN), the angular process (AG), or the body of the ascending ramus (MS).

Trait Number	Trait Name	Region
1	Coronoid height	MU, CR
2	Superior condylar length	MU, CN
3	Condylar width	MU, CN
4	Inferior condylar length	MU, CN
5	Condylar base length	MU, CN, MS
6	Posterior angular height	MU, AG
7	Posterior angular length	MU, AG
8	Anterior angular length	MU, AG
9	Superior angular length	MU, AG, MS
10	Posterior corpus height	AL, IN, MU, MS
11	Coronoid base length	MU, CR, Ms
12	Posterior-inferior basal length	AL, IN
13	Anterior-inferior basal length	AL, IN
14	Inferior incisor alveolus length	AL, IN
15	Incisor alveolus width	AL, IN
16	Superior incisor alveolus length	AL, IN
17	Anterior corpus height	AL, IN
18	Molar alveolus height	AL, MO
19	Superior molar alveolus length	AL, MO
20	Inferior molar alveolus length	AL, MO
21	Superior coronoid length	MU, CN

Ten SM/J males were mated with 10 LG/J females producing 41 F_1 hybrid animals. The F_1 hybrids were intercrossed to produce 535 F_2 progeny. Individual F_1 dams produced several litters of varying size. Mated pairs were housed together until the dams were deemed pregnant, at which time the male was removed from the cage. Progeny were housed with their dam for 3 weeks and then were weaned to standard laboratory rodent chow and randomly allocated to single-sex cages of 5 animals each. The animals were maintained until 10 weeks of age at which time they were sacrificed and necropsied. Organs were harvested and saved for DNA extraction. Carcasses were macerated by dermestid beetles and the right and left side mandibles were separated in preparation for measurement. Complete mandibles were available for 480 animals. Further information on animal husbandry is reported in Cheverud *et al.* (1996).

Mandibular and Molecular Measurements

The two-dimensional coordinates of 15 mandibular landmarks were obtained from lateral views of the right mandible using a digital-video data-collection system. A series of 21 linear distances were calculated from the landmark coordinate data (see Fig. 2). These distances were chosen to delineate the size and shape of individual mandibular components and provide complete coverage of the mandible. For the most part, measurements sharing a landmark in common lie at right or obtuse angles to one another, minimizing positive correlation due to morphological redundancy (Cheverud and Richtsmeier, 1986). The attribution of measurements to functional and developmental mandibular units is given in Table II and follows from discussions in Atchley and Hall (1991) and Atchley (1993), as summarized earlier. Traits were placed in specific mandibular units (see Fig. 1) when they crossed or formed a boundary of that unit.

Prior to genetic analysis, the effects of dam, litter size, experimental block, and sex were removed from the data, as described in Cheverud *et al.* (1996). These corrections resulted in a reduction of only 12% of the total variance in mandibular traits, but reducing the variance due to nongenetic factors enhances our ability to detect individual gene effects.

TABLE III   Positions of microsatellite markers along the mouse autosomes. Distances are given in Haldane's centi-Morgans.

Marker	Distance (cM)	Marker	Distance (cM)	Marker	Distance (cM)
D1Mit3	0	*D7Mit21*	0	*D15Mit13*	0
D1Mit20	8	*D7Nds1*	52	*D15Mit5*	24
D1Mit7	44	*D7Mit17*	66	*D15Mit2*	54
D1Mit11	56	*D7Mit9*	80	*D15Mit42*	80
D1Mit14	82	*D7Nds4*	114		
D1Mit17	122			*D16Mit2*	0
		D8Mit8	0	*D16Mit5*	30
D2Mit1	0	*D8Mit56*	82		
D2Mit17	~80			*D17Mit46*	0
D2Mit28	88	*D9Mit2*	0	*D17Mit16*	10
D2Mit22	98	*D9Mit4*	14	*D17Mit39*	46
		D9Mit8	28		
D3Mit54	0	*D9Mit19*	72	*D18Mit12*	0
D3Mit3	22			*D18Mit17*	4
D3Mit22	38	*D10Mit2*	0	*D18Mit8*	48
D3Mit12	60	*D10Mit20*	44		
D3Mit32	130	*D10Mit10*	64	*D19Mit16*	0
		D10Mit14	82	*D19Mit14*	6
D4Mit2	0			*D19Mit2*	56
D4Mit17	30	*D11Mit62*	0		
D4Mit45	42	*D11Mit64*	46		
D4Mit16	62	*D11Mit15*	56		
D4Mit13	84	*D11Mit14*	78		
		D11Mit48	108		
D5Mit47	0				
D5Mit6	26	*D12Mit37*	0		
D5Mit61	92	*D12Mit2*	20		
D5Mit26	102	*D12Mit5*	40		
D5Mit32	134	*D12Mit6*	50		
D5Mit43	144	*D12Nds2*	70		
D6Mit1	0	*D13Mit1*	0		
D6Mit9	52	*D13Mit9*	56		
D6Nds5	68	*D13Mit35*	90		
D6Mit15	96				
		D14Nds1	0		
		D14Mit5	44		
		D14Mit7	62		

DNA was extracted from the spleens of individual mice using protocols described in Routman and Cheverud (1994, 1995). Seventy-six microsatellite markers were scored on the 535 F_2 hybrid mice. PCR amplification of microsatellite loci followed the protocol suggested by Dietrich *et al.* (1992, 1996) with minor modifications (Routman and Cheverud, 1994, 1995).

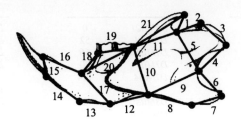

FIGURE 2 Linear measurements describing mandibular morphology as listed in Table II.

The marker loci cover all 19 autosomes. The X chromosome was not analyzed at this time because of insufficient molecular variability. The markers define a total of 55 intervals covering 1500 cM, for an average interval length of 27.5 cM. The specific loci scored and map locations are given in Fig. 3 and Table III (Cheverud *et al.*, 1996). The map distances were obtained from MAPMAKER 3.0b (Lander *et al.*, 1987; Lincoln *et al.*, 1992).

QTL Detection and Pleiotropic Patterns

A multivariate interval mapping approach was used to detect quantitative trait loci affecting mandibular morphology (Lander and Botstein, 1989; Haley and Knott, 1992). Each functionally and developmentally related set of traits [coronoid process (CR), condyloid process (CN), masseteric region (MS), angular process (AG), incisor alveolus (IN), and molar alveolus (MO)] was analyzed separately. Additionally, models combining the four muscle-attachment regions and both alveolar regions were performed along with a whole-mandible analysis. The specific algorithms used follow those described by Haley and Knott (1992) in the context of single trait analysis. The marker genotypes and their levels of recombination are used to obtain the probability that an arbitrary position lying between flanking markers is homozygous SM/J, heterozygous, or homozygous LG/J. These probabilities are then multiplied by (-1), (0), and (+1), respectively, and summed to obtain an additive genotypic score (a). Likewise, the probability of heterozygosity at the arbitrary intermediate location is calculated and used as a dominance genotypic score (d). The members of each character complex are then jointly regressed onto the additive and dominance genotypic scores to obtain the probability of a gene affecting the character set at the specified location. This probability can be obtained by canonical correlation analysis (Blackith and Reyment, 1971) in SAS (Leamy *et al.*, 1998) or by set correlation methods (Cohen, 1982) in SYSTAT (Cohen and Wilkinson, 1997). The set correlation procedure in SYSTAT provides F and X^2 statistics and associated probabilities for the full multivariate trait set and single F values and associated probabilities for each trait. Calculations were performed every 2 cM in each of the 55 intervals of our mouse genetic map. Locations of QTLs are specified by the chromosomal position with the locally lowest probability of indicating a QTL by chance, when no true QTL exists. Chromosomal positions are not directly fitted in the statistical model. Instead, an exhaustive genome scan (every 2 cM) is performed. Confidence limits for these positions are determined by the interval encompassing a chromosomal region for which the increase in probability of a false positive result is less than an order of magnitude (Lander and Botstein, 1989).

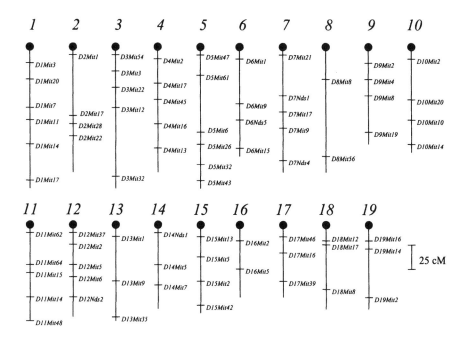

FIGURE 3   Map of microsatellite markers scored in the LG/J by SM/J cross.

The tests described earlier are for a single QTL on each chromosome. If it appeared that an additional QTL for the same traits may also reside on the chromosome, specific two-QTL models were fit to the data using the additive and dominance genotypic scores for two chromosomal positions at a time (Haley and Knott, 1992). If the two-QTL model fit significantly better than the one-QTL model at the 5% level, the two-QTL model was accepted. Significance was determined using a X^2 test with 2 degrees of freedom on twice the difference in natural logarithms of the likelihood of each model.

When QTLs for two or more separately analyzed trait sets map to the same chromosome, it is necessary to consider whether they map to the same position, indicating pleiotropy, or to different positions. Pleiotropy was tested by first identifying the most likely chromosomal position for each trait set considered separately (sets A and B) and then for the combination of both trait sets (set AB). A X^2 value for model fit was obtained for trait set A at its most likely position and at the most likely combined-trait position (set AB), controlling for variation in set B variables (Cohen, 1982; Cohen and Wilkinson, 1997). Likewise, a X^2 value for model fit was obtained for trait set B at its most likely position and at the most likely combined trait position (set AB), controlling for variation in set A variables. Variation in alternate trait sets is controlled for in the model so that the correlation structure of the variables is consistent in the dual location and single location models. The differences in X^2 values between the separate position model and the combined position model for trait sets A and B were added together to produce a X^2 test for pleiotropy. This test is considered as having 1 degree of freedom because in the separate position model two positions are allowed while in the single position model only one is allowed. Lynch and Walsh (1998) point out that the sum of two X^2 random variables is also X^2 distributed. A significant result indicates that the two trait sets are likely to be affected by two different QTLs. A nonsignificant result indicates that a single QTL affecting both trait sets may be present.

This approach to testing for pleiotropy was modeled using simulated data to determine whether the test statistic follows a chi-square distribution with 1 degree of freedom. A QTL affecting two traits, A and B, was placed in the center of a 20-cM interval. The traits are positively correlated due to pleiotropy at the QTL. Each individual was assigned a genotypic value based on the QTL alleles present. A residual term accounting for the effects of all unlinked loci and environmental factors was added to the genotypic value to produce the phenotypic value. The residuals for traits A and B were uncorrelated so that only the QTL causes intertrait correlation. Each trait has 5% of its variance due to the QTL. QTL locations for traits A, B, and their combination were obtained using standard interval mapping methods. The X^2 values for the separate locus versus single pleiotropic locus model were compared, as described earlier, using the difference in chi-square between the separate gene and pleiotropy models. The average absolute distance between the map positions of traits A and B was 4.24 cM. The expected and observed cumulative frequency distributions are provided in Fig. 4. If the observed cumulative frequency distribution matches a chi-square distribution with 1 degree of freedom, the cumulative frequencies should lie along a straight line with a slope of 1.0 and an intercept of zero. This appears to be the case for our simulation (see Fig. 4). Deviations from a 1 degree of freedom chi-square distribution are not statistically significant using a X^2 goodness-of-fit test ($X^2 = 4.48$, 6 df, Prob. = 0.387). Therefore, this procedure provides an appropriate test of the null hypothesis that both traits are affected by a single pleiotropic QTL.

It is expected that some false positive results will be obtained given the large number of significance tests performed for this study. With 19 chromosomes and 21 traits, one expects about one false positive result per trait at the 5% level, or 21 false attributions of traits to chromosomal locations (QTLs) scattered randomly across the genome. The hypothesis tests performed for this study do not depend on the positive nature of any single chromosomal location-trait association observed. Instead, the tests depend on the pattern of traits associated with each QTL. We do not expect false positive results to produce clusters of developmentally and functionally related traits in restricted chromosomal regions. There are 55 regions over which 21 false positive results would be spread. Assuming each interval is of equal length, there is a 38% chance of a false positive result in any given interval. The possibility of two such false positives in a single interval is about 14%. The probability that these two false positives are from the same portion of the mandible is 7%. Thus, to the extent that false positive results are produced in this analysis, they will only rarely give rise to false pleiotropic relations among unrelated traits. It is more likely that false positives would include inappropriate traits for genes with pleiotropic effects.

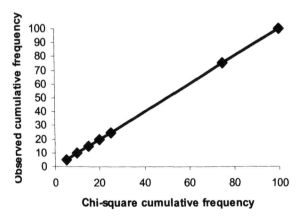

FIGURE 4 Probability plot for the observed cumulative frequency distribution in relation to the cumulative frequency distribution expected under X^2 with 1 degree of freedom.

Given this approach, it is possible that closely linked QTL could be responsible for effects attributed here to a single locus. Our analysis is limited by the resolution provided by recombination in the gametes that produced the 535 F_2 animals raised for this study. Fine-mapping genomic regions of interest in later intercross generations (Darvasi and Soller, 1995) may lead to finer resolution. However, over-inclusive aggregation of QTL-effect locations into a single QTL when the effects are actually due to multiple linked QTLs will bias the results against the hypothesis of genetically separate mandibular modules, unless it is supposed that closely linked genes tend to affect closely related morphological traits.

TABLE IV   Positions of mandibular QTLs affecting functionally and developmentally related trait sets. Loci affecting both MU and AL affect the entire mandible. The position given is relative to the markers in Table I, followed by 1 LOD score support intervals.

QTL	Trait Sets	Position	Centromeric	Telomeric	Probability
QTMAN1-1	CR	28	4	52	0.0093
QTMAN1-2	CN, IN	46	46	54	4.41×10^{-7}
QTMAN1-3	MO	122	68	122	0.0190
QTMAN2-1		0	-	-	0.0050
QTMAN2-2	MU		80	98	0.0096
QTMAN3-1	IN	14	0	36	0.0006
QTMAN3-2	MU	44	32	50	1.35×10^{-4}
QTMAN3-3	IN	114	92	(130)	0.0031
QTMAN4-1	MU,AL	30	18	40	1.30×10^{-6}
QTMAN4-2	MU	68	60	74	3.64×10^{-7}
QTMAN5-1	MU	102	100	142	0.0016
QTMAN5-2	CN	54	34	96	9.57×10^{-4}
QTMAN5-3	IN	132	118	(144)	0.0083
QTMAN6-1	MU, AL	20	6	34	8.30×10^{-6}
QTMAN6-2	MU, AL	92	84	(96)	2.60×10^{-8}
QTMAN7-1	AG, IN	62	58	66	2.73×10^{-11}
QTMAN7-2	CR, MO	78	74	88	0.0002
QTMAN8-1	MU	54	34	78	2.02×10^{-6}
QTMAN9-1	CN, IN	26	18	36	1.66×10^{-5}
QTMAN10-1	AL	24	12	38	2.65×10^{-7}
QTMAN10-2	MU	70	62	78	2.76×10^{-7}
QTMAN10-3	AL	56	52	68	4.40×10^{-8}
QTMAN11-2	MU, MO	52	46	64	1.00×10^{-15}
QTMAN11-3	IN	70	64	84	2.30×10^{-8}

QTL	Trait Sets	Position	Centromeric	Telomeric	Probability
QTMAN11-4	MO	108	78	(108)	0.0048
QTMAN12-1	AL	12	8	16	8.73×10^{-14}
QTMAN12-2	MU	28	22	38	4.76×10^{-7}
QTMAN12-3	MU, AL	70	50	(70)	4.42×10^{-5}
QTMAN13-1	MU	18	(0)	32	2.50×10^{-5}
QTMAN13-2	AL	72	64	78	4.37×10^{-12}
QTMAN14-1	AL	0	(0)	14	0.0127
QTMAN14-2	MU, IN	56	48	(62)	8.00×10^{-9}
QTMAN15-1	AG	12	2	42	4.72×10^{-4}
QTMAN15-2	AL	36	26	46	5.15×10^{-7}
QTMAN15-3	CN	50	42	52	3.00×10^{-9}
QTMAN15-4	CR	80	70	(80)	0.0029
QTMAN16-1	MU, AL	30	18	(30)	1.41×10^{-6}
QTMAN17-1	AL	8	(0)	20	0.0290
QTMAN17-2	MU	20	10	(46)	0.0057
QTMAN18-1	MU	30	16	46	3.09×10^{-5}
QTMAN19-1	MU, AL	0	(0)	4	0.0020

Differential epistasis

Multivariate analyses of covariance were performed as a preliminary test of differential epistasis. The mandible measures from each subregion were jointly regressed on the additive and dominance genotype scores (described earlier; Haley and Knott, 1992), total mandible length (from the condyloid process to the incisor alveolus), and the interaction of mandible length with each genotypic score. Significant interactions indicate that the regression, and hence allometry, of mandible measurements on total mandible length varies with QTL genotype. The variability in correlation across QTL genotypes indicates variation in pleiotropy. In this preliminary design the variation in pleiotropy is examined phenotypically. The observed interaction may be due to differential epistasis or differential genotype-environment interaction. These two possibilities can be resolved by examining the variation in allometry across QTL genotypes in variable genetic backgrounds, as in recombinant inbred strains, multiple measured genotype analyses, or family structured analyses.

RESULTS

Pleiotropic Patterns

The locations and affected mandibular regions for each of the 41 identified quantitative trait loci are presented in Table IV and Figure 5. Each chromosome carries at least one QTL affecting mandibular morphology. Multivariate analysis discovered four more QTLs than the 37 identified in Cheverud *et al.* (1997). These include a third locus at the telomeric end of chromosome 1 (*QTMAN1-3*), two additional loci on chromosome 5 (*QTMAN5-2, QTMAN5-3*), one locus on chromosome 10 (*QTMAN10-3*), and one locus on chromosome 15 (*QTMAN15-4*).

TABLE V   QTLs showing differential epistasis between mandibular length and mandibular regions.   Interaction type indicates whether the trait on mandible length regression and dominance).   When two regions are affected, separate values are given for each region.

QTL	Region	Interaction Type	Probability
QTMAN3-1	CR	additive	0.041
QTMAN3-2	CR	additive and dominance	0.012
QTMAN3-3	CR, CN	additive	0.015
QTMAN4-2	AG	additive	0.044
QTMAN7-1	IN	additive	0.023
QTMAN9-1	MS, IN	additive, additive	0.035, 0.030
QTMAN11-3	MS	dominance	0.031
QTMAN12-1	IN	dominance	0.019
QTMAN12-2	IN	additive and dominance	0.037
QTMAN13-1	CN	dominance	0.031
QTMAN14-1	CR, IN	additive, dominance	0.012, 0.042
QTMAN15-4	IN	additive	0.033
QTMAN17-1	IN	additive	0.002
QTMAN18-1	MS	dominance	0.041
QTMAN19-1	MU	dominance	0.018

Also, one previously identified QTL *(QTMAN 11-1)* was not separately resolved by the multivariate analyses. This unresolved QTL affected only two mandibular traits from different subregions (Cheverud *et al.*, 1997).

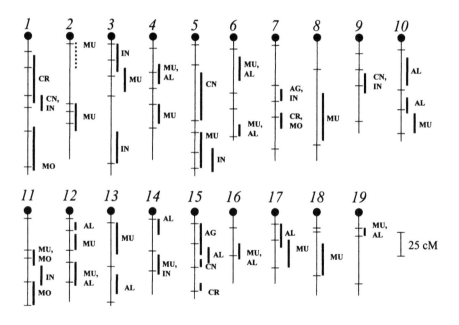

FIGURE 5 QTL locations and confidence limits (|) on maps of autosomal mouse chromosomes. Character complexes mapping to a location are indicated with the symbols defined in Table II. A dashed line indicates uncertain interval length because the flanking region was not mapped.

TABLE VI   An illustration of differential epistasis.   Regression and correlation of coronoid measurements with mandible length for the three genotypes at _QTMAN3-2_.

Measurement	SM homozygote		Heterozygote		LG homozygote	
	ß	r	ß	r	ß	r
posterior coronoid (1)	0.10*	0.28	0.03^{ns}	0.08	0.02^{ns}	0.06
basal coronoid (11)	0.30*	0.50	0.30*	0.56	0.16*	0.39
anterior coronoid (21)	0.38*	0.51	0.32*	0.47	0.17*	0.29

An asterisk (*) indicates significantly different from zero at the 0.01 level while "ns" indicates not significant.

The pattern of results strongly suggests that morphological integration occurs because of nonpleiotropic gene effects rather than negative pleiotropy. Of the 41 QTLs, 27% affected only individual subregions of the mandible (CR = 2, CN = 2, AG = 1, MO = 2, IN = 4), 44% had their effects restricted to either the alveolus (AL = 7) or muscle-associated regions of the ascending ramus (MU = 11), and 29% had effects across the whole mandible (MU and AL = 12). QTLs affecting the whole mandible were relatively homogenous and of the same sign without significant evidence of antagonistic pleiotropy. If part or all of the ascending ramus is larger in animals homozygous for the LG/J allele, the alveolus would also be larger.

Relative to the earlier study (Cheverud _et al.,_ 1997), much stronger support for QTLs was found using multivariate interval mapping methods. Most of the probabilities of no QTL effect associated with individual traits were in the range of 0.01 to 0.001 for the single trait analyses. However, nearly all the multivariate probabilities of no QTL effect were less than 0.001 and ranged lower by 12 orders of magnitude. Clearly, much statistical power can be gained through multivariate QTL analysis (Jiang and Zeng, 1995). When pleiotropy is present, grouping related traits dramatically enhances the power of the analysis.

Differential Epistasis

Many of the QTLs affecting mandibular morphology also cause variable allometry and correlation between mandibular traits and total mandibular length (see Table V). Variability in the allometry of mandible measures was typically restricted to functionally and developmentally related subsets of mandibular traits. Furthermore, two-thirds of these instances of differential epistasis were due to interactions with the additive genotypic value.

As an example of the kinds of variability in allometry represented in these results, the regression (ß) and correlation (r) of the three coronoid measurements, posterior coronoid length (1), basal coronoid length (11) and anterior coronoid length (21), on total mandible length are given in Table VI for each genotype at *QMAN3-2*. Note that posterior coronoid length (1) is only correlated with total length in the SM homozygotes. For the other two measurements, the regression of coronoid length on total length is lower in the LG homozygote than in the heterozygote or SM homozygote.

The other QTLs listed in Table V show the same kind of variation in allometry across QTL genotypes. Variability in allometry across genotypes is due to differential epistasis between the QTL and other loci affecting the mandible and/or to genotypic interaction with unspecified environmental factors. Further research in appropriately structured populations will allow these possibilities to be distinguished.

DISCUSSION

As in the previous analysis of this data (Cheverud *et al.*, 1997), there is strong support for the idea that the genetic basis of morphological integration lies in phenotypic modularization (Wagner, 1996; Wagner and Altenberg, 1996). Specific genes affect specific character complexes while different character complexes are affected by different sets of genes. Similar findings were made with respect to early and late growth (Cheverud *et al.*, 1996) and cranial morphology (Leamy *et al.*, 1998). For the most part, pleiotropic relations are restricted to functionally and developmentally related traits. The results can be distinguished from a random pleiotropy model because the expectation of such a model is that traits mapping together would not necessarily be functionally and/or developmentally related. Instead, unrelated trait sets would be grouped together, increasing the frequency of whole mandible QTLs. This was demonstrated in Cheverud *et al.* (1997), where the functional and/or developmental distribution of traits for each QTL was tested against a random pleiotropy model. The tests typically resulted in rejection of the random pleiotropy model.

However, the separate genetic basis for different character complexes in the mandible is not absolute. There is a hierarchy of results corresponding to a hierarchy of functional and developmental relationship. At the lowest level of the hierarchy there are individual traits. There are no instances of single trait mapping in this data set. Stepping up the hierarchy to the next level, 27% of the QTLs had their effects restricted to a single mandibular character complex (CR, CN, AG, MS, IN, MO). At the next level, mandibular character complexes are combined into alveolar (AL) and muscle attachment (MU) regions. Forty-four percent of the loci had effects restricted to traits participating in these more general character complexes (MU or AL). Only 29% of the QTLs affected traits across the whole mandible (MU and AL), our highest hierarchical level in this analysis. At even higher levels of the hierarchy of morphological units, it is still possible that some subset of these whole mandible QTLs only affect the mandible, and not functionally independent features in the head such as the braincase, or body size as a whole (Cheverud, 1982; Leamy et al., 1998).

The strong interdependence within sets of related traits and relative pleiotropic independence of the sets themselves fits the prediction made by Riedl (1978). Patterns of pleiotropy are associated with patterns of functional and developmental relationship in that pleiotropic gene effects are restricted to mandibular subregions and fail to indicate negative pleiotropy. Most gene effects are localized at the alveolar versus muscular process level, or even lower, while relatively few QTLs affect the whole mandible. Had whole mandible QTLs been common and region-specific QTLs rare or there was substantial negative pleiotropy, Riedl's (1978) hypothesis would not have been supported.

We also discovered that pleiotropic effects of loci on character complexes were variable and can potentially evolve, as predicted by Riedl (1978) and Wagner (1996; Wagner and Altenberg, 1996). Variation in pleiotropic relations may result from differential epistasis, where epistatic interactions between loci are different for different subsets of traits, opening a window on the evolution of genetic architecture itself. As the underlying genetic basis for trait variation plays an important role in evolutionary change, the underlying architecture itself is subject to evolution under epistasis. Our recent studies of epistasis for adult body size (Routman and Cheverud, 1997) indicate that epistasis among QTLs is likely to be common and similar in magnitude to direct, single locus effects. The evolutionary implications of epistasis are likely to be profound since they concern the evolution of the underlying genetic basis for morphological variation. However, as yet we have little guidance from theory regarding the consequences of epistasis, except that it is likely to affect the stability of the genetic variance/covariance patterns which link selection to evolutionary response. It seems likely that epistasis may make genetic variance/covariance patterns either more or less stable depending on the forms of epistasis present. Riedl's (1978) intuition that genetic architecture evolves to fit the functional and developmental relations among traits remains without explicit theoretical foundation.

Since traits in character complexes share a unique genetic basis, the evolution of functionally and developmentally related traits should be considered jointly rather than independently. This implies that character complexes can be considered as individual characters with a unique genetic underpinning. Individuation of a character complex is realized in the patterns of pleiotropic relationships. It remains to be seen whether selection modifies pleiotropic effects so that patterns of pleiotropy reflect patterns of functional and developmental relationship. However, genetic variations in pleiotropic patterns caused by differential epistasis have been identified here for the mammalian mandible and such variation is necessary if pleiotropic patterns are to evolve. Given that an individuated character is composed of a set of traits under the control of a common set of gene loci, selection on variation in pleiotropic patterns will alter the nature of the individuated characters comprising an organism.

ACKNOWLEDGMENTS

I thank E. Routman and D. Irschick for their work in preparing our mandibular gene mapping project and L. Leamy and G. Wagner for interesting conversations on the subject of this paper. T. Vaughn and B. Ackermann provided helpful comments on the manuscript. The work reported on here was supported by NSF grants DEB-9419992 and DEB-9726433.

LITERATURE CITED

Atchley, W. R. (1993). Genetic and developmental aspects of variability in the mammalian mandible. *In* "The Skull" (J. Hanken and B. K. Hall, eds.), Vol. 1, pp. 207-247, University of Chicago Press, Chicago.

Atchley, W. R., and Hall, B. (1991). A model for development and evolution of complex morphological structures. *Biol. Rev. Cambridge* **66**:101-157.

Atchley, W. A., Herring, S. W., Riska, B., and Plummer, A. (1984). Effects of the muscular dysgenesis gene on developmental stability in the mouse mandible. *Craniofacial Genet Dev. Biol.* **4**:179-189.

Atchley, W. R., Plummer, A., and Riska, B. (1985a). Genetics of mandible form in the mouse. *Genetics* **111**:555-577.

Atchley, W. R., Plummer, A., and Riska, B. (1985b). Genetic analysis of size-scaling patterns and the mouse mandible. *Genetics* **111**:579-595.

Bailey, D. (1956). A comparison of genetic and environmental principal components of morphogenesis in mice. *Growth* **20**:63-74.

Bailey, D. (1985). Genes that affect the shape of the murine mandible: congenic strain analysis. *J. Hered.* **76**:107-114.

Bailey, D. (1986). Genes that affect morphogenesis of the murine mandible: recombinant-inbred strain analysis. *J. Hered.* **77**:17-25.

Blackith, R. E., and Reyment, R. A. (1971). "Multivariate Morphometrics." Academic Press, New York.

Chai, C. (1956). Analysis of quantitative inheritance of body size in mice. I. Hybridization and maternal influence. *Genetics* **41**:157-164.

Cheverud, J. (1982). Phenotypic, genetic, and environmental morphological integration in the cranium. *Evolution* **36**:499-516.

Cheverud, J. (1984). Quantitative genetics and developmental constraints on evolution by selection. *J. Theor. Biol.* **110**:155-172.

Cheverud, J. (1995). Morphological integration in the saddle-back tamarin (*Saguinus fuscicollis*) cranium. *Am. Nat.* **145**:63-89.

Cheverud, J. (1996a). Quantitative genetic analysis of cranial morphology in the cotton-top (*Saguinus oedipus*) and saddle-back (*S. fuscicollis*) tamarins. *J. Evol. Biol.* **9**:5-42.

Cheverud, J. (1996b). Developmental integration and the evolution of pleiotropy. *Am. Zool.* **36**:44-50.

Cheverud, J., Hartman, S., Richtsmeier, J., and Atchley, W. R. (1991). A quantitative genetic analysis of localized morphology in mandibles of inbred mice using finite element scaling analysis. *J. Craniofacial Genet. Dev. Biol.* **11**:122-137.

Cheverud, J., and Richtsmeier, J. (1986). Finite-element scaling applied to sexual dimorphism in rhesus macaque (*Macaca mulatta*) facial growth. *Syst. Zool.* **35**:381-399.

Cheverud, J., Routman, E., Duarte, F. M., van Swinderen, B., Cothran, K., and Perel, C. (1996). Quantitative trait loci for murine growth. *Genetics* **142**:1305-1319.

Cheverud, J. M., Routman, E. J., and Irschick, D. K. (1997). Pleiotropic effects of individual gene loci on mandibular morphology. *Evolution* **51**:2004-2014.

Cohen, J. (1982). Set correlation as a general multivariate data-analytic method. *Multivar. Behav. Res.* **17**:301-341.

Cohen, J., and Wilkinson, L. (1997). Set and canonical correlation. *In* "SYSTAT 7.0: New Statistics" (L. Wilkinson, ed.), pp. 169-190. SPSS Inc., Chicago.

Darvasi, A., and Soller, M. 1995. Advanced intercross lines, an experimental population for fine genetic mapping. *Genetics* **141**:1199-1207.

Dietrich, W., Katz, H., Lincoln, S., Shin, H.-S., Friedman, J., Dracopoli, N., and Lander, E. S. (1992). A genetic map of the mouse suitable for typing intraspecific crosses. *Genetics* **131**:423-447.

Dietrich, W., Miller, J., Steen, R., Merchant, M., Damron-Boles, D., Husain, Z., Dredge, R., Daly, M., Ingalls, K., O'Connor, T., Evans, C., DeAngelis, M., Levison, D., Kruglyak, L., Goodman, N., Copeland, N., Jenkins, N., Hawkins, T., Stein, L., Page, D., and Lander, E. (1996). A comprehensive genetic map of the mouse genome. *Nature* **380**:149-152.

Edwards, M., Stuber, C., and Wendel, J. (1987). Molecular-marker-facilitated investigations of quantitative-trait loci in maize. I. Numbers, genomic distribution and types of gene action. *Genetics* **116**:113-125.

Falconer, D. S., and Mackay, T. F. C. (1996). "Introduction to Quantitative Genetics." Longman Press, New York.

Goodale, H. (1938). A study of the inheritance of body weight in the albino mouse by selection. *J. Hered.* **29**:101-112.

Goodale, H. (1941). Progress report on possibilities in progeny test breeding. *Science* **94**:442-443.

Gromko, M. H. (1995). Unpredictability of correlated response to selection: pleiotropy and sampling interact. *Evolution* **49**:685-693.

Haley, C. S., and Knott, S. A. (1992). A simple regression method for mapping quantitative trait loci in line crosses using flanking markers. *Heredity* **69**:315-324.

Hall, B. K. (1994). "Evolutionary Developmental Biology." Chapman & Hall, London.

Herring, S. W., and Lakars, T. C. (1981). Craniofacial development in the absence of muscle contraction. *J. Craniofacial Genet. Dev. Biol.* **1**:341-357.

Jiang, C., and Zeng, Z.-B. (1995). Multiple trait analysis of genetic mapping for quantitative trait loci. *Genetics* **140**:1111-1127.

Lande, R. (1980). The genetic covariance between characters maintained by pleiotropic mutations. *Genetics* **94**:203-215.

Lander, E. S., and Botstein, D. (1989). Mapping Mendelian factors underlying quantitative traits using RFLP linkage maps. *Genetics* **121**:185-199.

Lander, E. S., Green, P., Abrahamson, J., Barlow, A., Daley, M., Lincoln, S., and Newburg, L. (1987). MAPMAKER: an interactive computer package for constructing primary genetic linkage maps of experimental and natural populations. *Genomics* **1**:174-181.

Lascoux, M. (1997). Unpredictability of correlated response to selection: linkage and initial frequency also matter. *Evolution* **51**:1394-1400.

Leamy, L., Routman, E., and Cheverud, J. (1998). Quantitative trait loci for early and late developing skull characters in mice: a test of the genetic independence model of morphological integration. *Am. Nat.*

Lincoln, S., Daly, M., and Lander, E. (1992). "Constructing Genetic Maps with MAPMAKER/EXP 3.0." 3rd Edition. Whitehead Institute Technical Report, Boston.

Lynch, M., and Walsh, B. (1998). "Genetics and Analysis of Quantitative Traits." Sinauer Associates, Sunderland, Massachusetts.

MacArthur, J. (1944). Genetics of body size and related characters. I. Selection of small and large races of the laboratory mouse. *Am. Nat.* **78**:142-157.

Martin, B., Nienhuis, J., King, G., and Schaefer, A. (1989). Restriction fragment length polymorphisms associated with water use efficiency in tomato. *Science* **243**:1725-1728.

Moore, W. (1981). "The Mammalian Skull." Cambridge University Press, Cambridge.

Olson, E., and Miller, R. (1958). "Morphological Integration." University of Chicago Press, Chicago.

Paterson, A., Lander, E., Hewitt, J., Peterson, S., Lincoln, S., and Tanksley, S. (1988). Resolution of quantitative traits into Mendelian factors by using a complete linkage map of restriction length polymorphisms. *Nature* **335**:721-726.

Riedl, R. (1978). "Order in Living Organisms." Wiley Press, New York.

Routman, E., and Cheverud, J. (1994). A rapid method of scoring simple sequence repeat polymorphisms with agarose gel electrophoresis. *Mamm. Genome* **5**:187-188.

Routman, E., and Cheverud, J. (1995). Polymorphism for PCR-analyzed microsatellites: data for two additional inbred mouse strains and the utility of agarose gel electrophoresis. *Mamm. Genome* **6**:401-404.

Routman, E. J., and Cheverud, J. M. (1997). Gene effects on a quantitative trait: two-locus epistatic effects measured at microsatellite markers and at estimated QTL. *Evolution* **51**:1654-1662.

Tanksley, S., Medina-Filho, H., and Rick, C. (1982). Use of naturally-occurring enzyme variation to detect and map genes controlling quantitative traits in the interspecific backcross of tomato. *Heredity* **49**:11-25.

Wagner, G. (1996). Homology, natural kinds, and the evolution of modularity. *Am. Zool.* **36**:36-43.

Wagner, G., and Altenberg, L. (1996). Perspective: complex adaptations and the evolution of evolvability. *Evolution* **50**:967-976.

Weller, J., Soller, M., and Brody, T. (1988). Linkage analysis of quantitative traits in an interspecific cross of tomato (*Lycopersicon esculentum* X *Lycopersicon pimpinellifolium*) by means of genetic markers. *Genetics* **118**:329-339.

19

HOMOLOGIES OF PROCESS AND MODULAR ELEMENTS OF EMBRYONIC CONSTRUCTION

SCOTT F. GILBERT[1] AND JESSICA A. BOLKER[2]

[1]*Martin Biological Laboratories, Swarthmore College, Swarthmore, PA 19081*
[2]*Department of Zoology, University of New Hampshire, Durham, NH 03824*

INTRODUCTION

A. From Pattern to Process

Biologists have been arguing about homology since the 1830s, and despite enormous advances in molecular, evolutionary, and developmental biology, the debate continues. In his aptly titled pamphlet *Homology: An Unsolved Problem*, Gavin de Beer (1971) drew a clear line between structure, which he viewed as the only appropriate thing to be homologized, and function: "An organ is homologous with another because of what it is, not because of what it does." Distinguishing structure from function was straightforward when we knew little about the developmental mechanisms responsible for the emergence of anatomical structures. Recent progress in developmental genetics, however, has given us remarkable insights into the molecular mechanisms of morphogenesis, but has at the same time blurred the clear divide between structure and function.

At the genetic or molecular level, it is difficult to tell where one ends and the other begins. One scientist's "cause" is another scientist's "phenotype."

De Beer's analyses of structural homologies excluded not only function, but also generative processes. In justifying this approach, he cited several classic cases of homologous structures that develop in different ways. In salamanders, lenses derived from the dorsal iris and those derived from the surface ectoderm were homologous, although they arose from different tissues. Similarly, vertebrate neural tubes produced by primary neurulation were homologous to those produced by secondary neurulation. Both examples demonstrate the independence of structural homology from developmental processes.

The entry of developmental genetics into the realm of morphogenesis has changed this situation. The original aim of developmental genetics had been to explain the problem of cell differentiation (see Gilbert, 1997): How does one cell become different from other cell types and from its progenitors; how does an epidermal skin cell become different from a nerve cell and from the ectodermal cell that had the potency to become either? However, by the early 1990s, genes were found whose expression determined not merely cells but entire body units: the homeotic gene complex in *Drosophila* was shown to specify the identity of body segments, and mutations of similar genes in mice caused region-specific (rather than cell-type specific) defects (Nüsslein-Volhard *et al.*, 1987; Chisaka and Capecchi, 1991). At the same time, the intercellular and intracellular pathways active in forming organs were being elucidated. This program began with studies of genes whose misregulation produced cancers, but these oncogenic pathways were soon found to be the same ones embryonic cells use to regulate their growth and coordinated differentiation (see Gilbert, 1996). The similarities of these developmental genes and pathways between evolutionarily separated organisms and between different organs within the same organism have reopened the debate about the relevance of development to homology and to evolution. One issue concerns the role of "process" characters in assessing the homology of the structures they help build. We are not going to deal with that in this chapter (see Bolker and Raff, 1996 for discussion). Instead, we will focus on a second, more fundamental question: whether and how we can homologize the processes themselves.

This is an important exercise. Homologies of processes become critical to the discussion of evolution and development when we consider (1) that evolution depends on hereditable changes in development, (2) that development is modular such that different modules can change without affecting other modules, (3) that modules can be coopted into new functions, and (4) that modules depend on intercellular communication transduced by families of paracrine factors, signaling cascades, and transcription factors. Thus, the signal transduction pathways, themselves, can constitute modules that can be used for several different developmental and physiological functions. Like intracellular biochemical pathways, they can be conserved through evolution and modified both within and between species. However, unlike the metabolic biochemical modules, those pathways of intercellular communication can act to create new morphological entities during development.

B. The Embryo and Homology

The notion that homologies were best studied in embryos can be traced back to Martin Barry (1837) who claimed that structural homologies were best recognized without the "embarrassment of function," and, moreover, that the "manner of development" was similar between different animal groups. Later, structural embryonic homologies were considered some of the most important evidence for evolutionary theory: the larval tunicate showed the ascidians to be chordates, the larval barnacle demonstrated that the cirripedes were modified crustaceans (Kowalewski, 1871; Müller, 1864), and the embryonic mammalian jaw, not its adult structure, showed the affinity of mammals to the reptiles (Reichert, 1837). Wilson (1898) used homologous cleavage and cell partitioning patterns to show the affinity of mollusc, annelid, and flatworm phyla. Such homologies indicated common descent. They were *evidence* of evolution, but shed no light on evolutionary *mechanisms*—natural and sexual selection. In the early 20th century, developmental mechanics separated embryology from evolutionary biology, genetics replaced embryology as the perceived motor of evolution, and arguments over homology and homoplasy frustrated attempts to create phylogenetic trees (Bowler, 1996; Gilbert *et al.*, 1996). From this point, embryonic homologies were little studied.

One reason for the recent revival of interest in developmental homology is its central importance to the contemporary (re)synthesis of developmental and evolutionary biology. Three research programs have converged to make developmental homology a critical issue. First, paleontology's punctuated equilibrium model of evolution requires rapid morphological change, and its supporters have proposed developmental phenomena such as allometry, heterochrony, and homeosis as candidate mechanisms (Gould, 1977). Homologous structures in different lineages became the evidence for the historical occurrence of these processes, so criteria for recognizing homologies are central to this program.

In molecular biology, nucleic acid and protein sequencing have shown that gene structures could be homologous (Britten, 1967; Fitch, 1970). While this program originally centered on proteins such as globins, whose structures were homologous both within the organism and between species, the most interesting homologies came from a third program, the analysis of developmental regulatory genes. E. B. Lewis and colleagues pioneered the genetic analysis of a region on *Drosophila* chromosome 3 that contained several homeotic genes, and in the 1980s the 180 base-pair homeobox was found in these genes. These fly "Hom-C" genes were thus identified as serially homologous. Shortly thereafter, molecular hybridization studies showed that these genes were also specially homologous: the same genes were found in vertebrates (Carrasco *et al.*, 1984; for review, see Gilbert *et al.*,1996). Not only did the genes appear to be homologous within a species, but the same genes appeared to reside on the vertebrate chromosomes in the same order as they did on the fly chromosomes. The complexes were, themselves, homologous.

Other highly conserved developmental regulatory genes have since been identified, and it is clear that analogous structures are often made from homologous genes. (We do not subscribe to the view that the expression of homologous genes renders the structures themselves homologous; Dickinson, 1995; Bolker and Raff, 1996.) The *PAX6* (*eyeless*) gene is expressed during the development of photoreceptors throughout the animal kingdom, the *Tinman* (*Csh*) gene appears to be the basis for both the *Drosophila* and the vertebrate hearts, and the expression of the *fringe* gene apparently defines the region of limb outgrowth in both flies and birds, even though the mechanisms of limb formation in arthropods and vertebrates are completely different. In none of these cases does the expression of a highly conserved gene confer homology on the resulting structures.

A final reason to reconsider homologies of process is the wealth of new data concerning the importance of modularity in both evolution and development. The notion of modularity is attracting increasing interest from developmental and evolutionary biologists as well as from philosophers (Raff, 1996; Wagner, 1996; proceedings of the Friday Harbor Workshop on Modularity in Development and Evolution, 1997). Biologists have long recognized that structures (or parts of structures) can act as modules, distinct but potentially interacting developmental or evolutionary units. We are now realizing that dynamic entities such as morphogenetic fields (Gilbert *et al.*, 1996; Raff, 1996), developmental equivalence groups, or signal transduction pathways may also act as modules.

WHAT ARE HOMOLOGIES OF PROCESS?

A. Extending the Concept of Homology

Extending the concept of homology to cover genetic characters as well as aspects of morphology has been straightforward. Genes or DNA sequences are still physical entities, with a defined structure, and the original homology criteria derived from comparative anatomy (relative position, transformation, special quality) readily apply. Recognizing "homologies of process," however, represents a qualitative change. If we are to homologize processes—as indeed is already common practice—we must consider whether we can apply the original criteria of homology, and what complications arise from applying them to a new sort of character.

Process homology confronts head-on two central issues of homology debates. The first is the problem of function, and the importance of excluding functional criteria from homology definitions or assessments. It is not at all easy to define a process independent of its function. The second is that although hypotheses of homology must specify a particular level of biological organization (Striedter and Northcutt, 1991; Bolker and Raff, 1996; Abouheif, 1997; Dickinson, 1995), gene expression bridges two such levels, the genetic and the phenotypic. These are nontrivial problems. Before we go on to address the difficulties with homologies of process, however, we need to explain what we mean by the term.

The concept of homologies of process has its origin in Howard Schneiderman's observations of "homologous specification" in *Drosophila*. Here, transdifferentiating cells or the tissue resulting from homeotic gene mutations retain positional information that reflects their position in the imaginal disc (Postlethwait and Schneiderman, 1971). Antennal imaginal disc tissue misdirected to differentiate as leg tissue will differentiate according to its location in the antennal disc. Thus, if the disc region destined to become the most distal portion of the antenna forms leg structures instead, it becomes a claw (not a coxa or trochanter).

Schneiderman's observations had three central implications for the homology problem. First, they reinforced the idea that homology had to be defined at a particular level. Second, they showed that homology at the structural level was separable from that of process. Third, and most importantly, they demonstrated that structures with no *anatomical* homology—the eye and the leg—could nevertheless have an underlying homology of *process* in their construction.

B. Recent Examples

Since Schneiderman's initial description of "homologous specification," molecular biologists and developmental geneticists have discovered many additional examples that illustrate these principles.   We now use the term "process homology" (or, more formally, "homology of process") to describe the relationship between pathways that are composed of homologous proteins and that are related by common ancestry.   These homologous pathways need not form anatomical homologs.

Along with applying a new label—"process homology"—to the relationship uniting the eye, wing, and leg discs, we can now identify its molecular basis: the interaction of the hedgehog and Wnt pathways.   These two paracrine factors interact within the disc to specify the proximal/distal axes of the respective organ (see Ingham, 1994). The same molecules that specify these axes in the eye also specify them in the leg and wing discs, constituting a serial process homology.

Moreover, the same pathways exist in vertebrates. For instance, in the Wnt signaling cascade, every element in the insect pathway has a homologue in the vertebrate embryo, and the same interactions that transmit the *Drosophila* Wingless signal to the nucleus occur in vertebrates. The genes and the protein interactions are the same: only the "readout" (the target genes) is changed between tissues and species (Cadigan and Nusse, 1997; Fig. 1). The Wnt pathways in vertebrates, nematodes, and arthropods comprise homologous proteins arranged in similar fashions: we therefore consider them homologous pathways. The Wnt pathway (or the set of homologous Wnt pathways), then, is a homologous cassette of information, a homologous module of gene-protein interactions—hence, a homology of process.

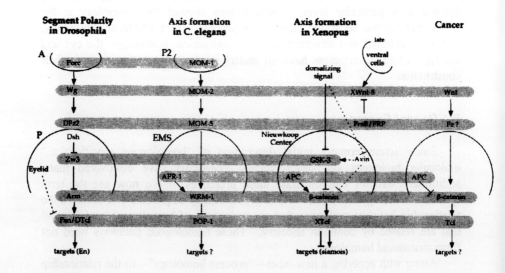

FIGURE 1 Comparison of Wnt pathways in embryogenesis and mammalian carcinogenesis. Related genes are highlighted across the different systems. Potential differences in the pathways are shown shaded. Broken lines indicate alternative pathways (after Cadigan and Nusse, 1997).

A second example of pathway and process homology is the Rel protein pathway. In the *Drosophila* blastoderm, the binding of ligand (Spätzle protein) from the ventral follicle cells to its receptor (Toll protein) transduces a phosphorylation signal that separates a potential transcription factor (Dorsal protein) from its cytoplasmic inhibitor (the cactus protein). Once released, the factor enters the nucleus and regulates the transcription of particular genes, a critical step in forming the ventral portion of the fly embryo. This *Drosophila* pathway corresponds protein for protein with the pathway that liberates NF-kB from its inhibitor, IkB, when interleukin-1 coactivates mammalian B lymphocytes through its cell surface receptor (Shelton and Wasserman, 1993). This pathway is also used in the formation of the fly's immune system and for the vertebrate limb (Lemaitre *et al.*, 1996; Bushdid *et al.*, 1998; Kanegae *et al.*, 1998; Wu and Anderson, 1998). While there is no homology being postulated among the mammalian B lymphocyte, the fly fat body, the insect blastoderm, and the avian limb, there is a remarkable homology in the pathways that help construct these three structures.

GENERAL PRINCIPLES

A. Processes and the Nature of Homology

We are used to thinking of pathways or signal transduction systems as temporal series; indeed, we usually write them with arrows showing the sequential activation of each part. The molecules of a pathway, however, are also physically linked: the surface of one protein fits into the surface of the next. In either the anatomical or the process case, homologous structures (the proteins in a pathway, or the bones in a skeleton) are linked together in the same physical organization. Thus, the standard criteria for anatomical homology remain relevant to homologies of process. Just as "the forelimb" is a homologous entity (between species) because it is composed of parts that are themselves homologous, "the Wnt pathway" is a unit of anatomical homology. However, because pathways have temporal as well as physical organization, "the Wnt pathway" is more than a unit of structure: it is also a modular unit of process.

We recognize homologies of process in the common cassettes or modules that occur in different lineages, and these can be used in different ways. The structure and expression patterns of one set of genes may not indicate homology. The coordinated assembly of several genes and gene products into a functional module, and the sharing of such modules between species or between tissues of an organism, is more significant. Such shared modules embody homologies of process.

They need not, however, imply homologies of resulting structures. For example, the presence of the Rel pathway in both the *Drosophila* and the mammalian immune systems does not mean that the immune systems are homologous (i.e., derived from a common ancestral immune system). Rather, it means that both immune systems use a common modular subunit in their respective constructions. The Rel pathway is also used in *Drosophila* segmentation and in vertebrate limb development, and these developmental events are certainly not homologous.

B. Processes as Characters

Processes, or assemblies of processes into pathways or cassettes, are themselves characters. This idea has not penetrated discussions of homology in the evolutionary literature, perhaps because evolutionary biology and systematics have traditionally focused on adult structures. It is, however, essential for evolutionary analyses of development: many of the key characters of embryos are processes. Moreover, many embryonic structures are themselves transient (i.e., time dependent). Thus if we seek homologies in embryos, they should be homologies of processes rather than of structures. This is not a new conclusion, at least among those who have sought to integrate developmental and evolutionary biology.

Waddington (1975) emphasized the importance of processes as characters, citing Whitehead's (1933)"...replacement of 'things' by processes which have an individual character which depends upon the 'concrescence' into a unity of very many relations with other processes." Waddington envisioned these "concrescences" in embryos as assemblies of many genes and their products into stabilized pathways or modules. Bonner (1988) has associated modularity with "gene nets" that can participate in many different aspects of development, and Wagner (1996) identifies homologs as units of evolutionary transformation.

Embryonic induction, a central mechanism of metazoan development, neatly illustrates the usefulness of considering pathways as characters. As Waddington pointed out, a cell's competence to respond to a particular inducer depends on the prior existence of a response pathway in that cell. Altering the receptor allows the cell to respond to a new inducer, using the same internal pathway. Similarly, altering the transcription factors activated by the pathway could cause a cell to express a new set of genes.

This seems to have occurred in the classic locus of induction, the amphibian organizer. In *Xenopus*, the organizer-inducing activity of the Nieuwkoop Center appears to result from the activation of ß-catenin proteins (Brannon *et al.*, 1997; Larabell *et al.*, 1997). The normal activator of ß-catenin is a Wnt protein, but some other protein apparently activates the distal portion of the Wnt pathway in the dorsal vegetal cells of the *Xenopus* blastula (Yost *et al.*, 1998; Miller *et al.*, 1999; Fig. 1). Similarly, the Rel cascade initiated by the binding of Toll and

Spätzle during *Drosophila* development is also used by the fly's immune cells to activate genes necessary for fighting fungal infections (Lemaitre *et al.*, 1996). In both these cases, a conserved pathway serves different functions as it is initiated by different activators in distinct contexts.

These two examples also illustrate other properties of homologies of process. As mentioned earlier, homologous pathways are composed of the products of homologous genes. We expect such homologous pathways to represent variations on a common theme. Figure 1 shows some of the variations played upon the Wnt pathway. Homology is not identity, and the existence of pathways allows for evolutionary modulation at each step.

C. What Process Homology Is Not: Function as Criterion

Two of the major, and most common, misuses of developmental processes in homology assessments are (1) the use of processes as criteria for homology, rather than as characters, and (2) the conflation of processes with function. First, in contrast to the common practice of using processes to help assess the homologies of genes or structures, we ask whether we can establish homologies between the processes themselves, and what such homologous relationships might reveal about the evolution of developmental pathways. In our view, developmental processes are *characters in and of themselves*, not merely criteria for assessing structural homology.

Second, process homology has nothing to do with so-called "functional homology." Process homology treats processes as characters, that is, as parts of organisms that can be compared and homologized, just as the classical homology concept treats morphological characters. The term "functional homology," on the other hand, represents a fundamental misunderstanding of the central concept of homology. Homology is a relationship of "sameness" due to descent from a common ancestral precursor. Similarity of function carries no implication of evolutionary relationship (indeed it can arise by numerous means other than phylogenetic continuity), and relationships based on functional resemblance are analogies, not homologies. It is therefore critical in discussing process homology to distinguish process from function as clearly as possible.

By a "process" we refer to an action (*what* happens), not to its functional outcome (*why* it happens). For instance, making a phone call is a process that can be described entirely without reference to why one might make a phone call, or the effect the call might produce. The process of making the call is the same whether its function is to remind one's spouse to buy milk, or to accept a job offer. The result of the call may be to ensure that there is milk for breakfast, awaken someone from a nap, or set paperwork in motion, but it is in any case entirely distinct from the process of telephoning (or the equipment used to transmit one's voice).

Function can be an organismal trait (forming a leg, inducing a neural tube, etc.) or it can be a molecular trait (such as phosphorylating a particular protein, transcribing a particular gene). Both are causal, but at different levels. Homologies of process are seen at the molecular level. This distinction between levels is essential when analyzing homologous pathways with respect to function.[1]

D. The Function Problem

Objections, explanations, and pleas notwithstanding (e.g., de Beer, 1971; Dickinson, 1995; Bolker and Raff, 1996; Abouheif et al., 1997), the notion that function is linked to homology refuses to die. One reason for its persistence is the common use of "homology" as an impressive-sounding synonym for "similarity" in the molecular literature (Hillis, 1994); a second is that many developmental geneticists are unaware of the history and established meanings of the term. A third reason, however, may be that evolutionists have failed to explain clearly why the notion of "functional homology" is problematic. This flat refusal to consider function in any way when assessing homologies has been particularly unconvincing to biologists working at the molecular level, where structure and function are so intimately intertwined.

The ultimate problem with including function as a criterion (or definition) of homology is that function is not evolutionarily transmitted in the way that structure—or process—is. There is no genetic basis for the function of a structure, or a gene, or a signal cascade, only for the phenotypic character itself. The ends for which the character is used, its functions, are not heritable, and evolutionary homology is a relationship based on inheritance of a character (whether morphological or process) from a common ancestor.

We believe that there are important insights to be gained into development, evolution, and their relationship from examining homologous processes. At the same time, we contend that it is essential to exclude function from assessments or assignments of homology. So how can we include processes among the

---

[1] In this regard, we should distinguish homologies of process from the concept of homodynamy. Homodynamy (Baltzer, 1950, 1952) concerns developmental inductions that are followed in the same manner between two organisms. For instance, homodynamy is invoked when embryonic frog epidermis, transplanted into that region of the newt embryo destined to become jaw, produces a frog jaw in the newt head (see Riedl, 1978). There is a similarity in the ability to respond to the same inducer to "form a jaw." It indicates that the same "readout system" can respond to the same inducer even in diverse species. Thus, homodynamy is a phenomenon at the organismal level, not at the molecular level. Moreover, it can easily be counterfeited by other organs producing the same signal. The newt otic vesicle or nose rudiment will cause the formation of ectopic limbs when transplanted beneath the flank ectoderm of newt embryos (Balinsky, 1933). This is probably because they secrete FGF8, an excellent inducer of limb formation, although probably not the actual inducer used during embryogenesis to form limbs (Crossley and Martin, 1995; Ohuchi et al., 1997).

developmental "characters" we compare without conflating process and function, or homology and analogy?

One answer is simply to be careful. The current trend in developmental and molecular genetic studies is to focus on apparently universal and deeply conserved features of organisms. It is essential that we remember that even seemingly universal similarities can arise in more than one way. Our first, and often correct, assumption is that such similarities represent synapomorphies. However, extraordinarily similar structures and processes, even with identical functions, can also arise through convergence. [This has happened, for example, in lysozyme sequences (Stewart and Wilson, 1987; Swanson *et al.*, *1991*).]

A second approach is to exploit the concept of modules to help distinguish processes as such from their results or functions. Homologous pathways, which act as modules, are composed of homologous proteins arranged in a homologous manner. These modules can function in several capacities during development. For instance, the RTK-RAS pathway can be used as a module for the morphogenesis of a *C. elegans* vulva or for the determination of a *Drosophila* photoreceptor cell. The Wnt pathway can be used for the development of fly wings or mouse kidneys. Gene level homologies cannot be determined solely by patterns of gene expression. The case for homology at the gene level can, however, be bolstered by the gene product's similarity to the product of another (putatively homologous) gene that belongs to a homologous pathway. As with organismal-level homologies, evidence for the homology of the whole strengthens the evidence for the homologies of the corresponding parts.

Tricky as it may be, careful discrimination between processes and their results can yield crucial evolutionary (and even therapeutic) insights. In their analysis of recent biomineralization research involving combinations of bone and nacre, Atlan *et al.* (997; Westbroek and Marin, 1998) conclude that

> although nacre and bone are not homologous as such, parts of the complex machinery that directs their formation may be. Logically, this would imply that, before the Cambrian diversification, these homologous parts served some function other than biomineralization and that they were coopted for the latter purpose when the different metazoan stocks began producing skeletons. (p. 862)

In this case, widely shared—and homologous—elements of developmental "machinery" or processes have a longer evolutionary history than the tissues they now help to build.

E. Implications of Process Homology for the Levels Issue

The importance of specifying the level at which homologies are assigned has been discussed at length elsewhere (e.g., Bolker and Raff, 1996; Dickinson,

1995). In short, hypotheses of homology relate characters at the same level of the biological hierarchy, and homologies at one level—for example, that of gene sequences—do not automatically imply homologies at another level—for example, the structures in which the genes are expressed. (This is especially true when the expression pattern has already been used to classify the two similar genes as homologous.)

Homologies of process form an intermediate level between gene homologies and morphological homologies. Identifying process homologies can therefore help us assess (a) whether genetic pathways themselves are homologous, (b) whether their component parts are homologous, and (c) whether the structures in which they are expressed are homologous. Homologies of process are thus critical connections between the genetic and the morphological levels of organization.

In theory, one should keep different levels entirely distinct. In fact, life is more complicated: gene expression patterns represent an "intrusion" of the genetic level onto the morphological (or, more generally, the phenotypic) level. This complexity is reflected in the different ways evolutionary and developmental biologists describe genes. Evolutionary biologists and population geneticists treat genes as markers or patterns analogous to morphological characters—effectively, as modules of structure. In contrast, developmental geneticists and molecular biologists see genes as causal agents, the basis of specific functions, or elements in networks of functionally interconnected units: modules of process or of function.

This distinction is more than semantic: it represents a fundamental difference in the type of character one considers a gene to be. One can classify the spatial and temporal pattern of gene expression as a quasi-morphological character. Traditional homology concepts, and classical criteria, work fairly well for this. Alternatively, one can consider the gene's action as an element of a process. In this latter case, the concepts of modularity and process homology offer a way to homologize dynamic aspects of gene expression and interactions, while avoiding the morass of "functional" homology.

F. Expression Data and Gene Homologies

Molecular data—sequences and/or gene expression patterns—have been used in two ways to indicate homology, and the two uses are often confused. The traditional approach is to cite expression patterns to support hypotheses of *homology between gene sequences*. The second approach is to treat similar expression patterns of homologous genes as evidence for *morphological homologies*. While we believe the former strategy is appropriate (if used carefully), the second is more problematic because homologous genes can be used in many different structures and contexts.

The designation of two genes as homologous has often been based on combinations of structural and functional properties, a procedure that risks conflating homology and analogy. Still, from a less purist and more practical perspective, similar expression patterns and/or functions usually tip the probabilistic argument in favor of a pair of genes' having a common ancestor rather than a fortuitous homoplasy. For example, the *chordin and short-gastrulation* genes, whose products have only 29% of their amino acid sequences in common, are considered homologs. These two genes are expressed in the same (i.e., presumably homologous) places in vertebrate and fly embryos. Moreover, their products have similar functions: binding BMP4/DPP proteins, thereby allowing neuralization of the ectoderm (Hawley *et al.*, 1995; Sasai *et al.*, 1995).

Ultimately, neither structure nor expression pattern alone is a good criterion for gene-level homology. Sequence similarity can be caused not only by divergence from a common sequence but also by (1) convergence to a shared sequence due to selection for a common function such as DNA binding (molecular homoplasy) or (2) the accumulation of exons from unrelated genes. Just because ankyrin has EGF-like domains does not mean that it evolved from a member of the EGF family.

When homologous developmental genes show similar expression patterns, it is often proposed that the structures in which they are expressed are homologous. Thus, the observation of *Distal-less* expression in *Drosophila* leg primordia and in the maxilla was widely seen as proof that the maxillary appendages were homologous to the legs (Panganiban *et al.*, 1994). This particular scenario is credible on paleontological grounds as well, but the expression data alone are not sufficient evidence for homology. This was underscored by later research (Panganiban *et al.*, 1997) demonstrating *Distal-less* expression not only in the feet of insects, but also in the tube feet of sea urchins, and in just about any other structure that everts from the main body axis. A proposed synapomorphy of insect legs became a symplesiomorphy characteristic of any "stickem-outy" from the developing animal body. The conserved genes, per se, tell us nothing about the evolutionary relationships between particular appendages.

G. Deep Homology and Its Limitations

In some cases, gene-level homologies, such as the presence of the Hox cluster, can represent fundamental synapomorphies that link widely divergent groups. Such "deep homology" may exist between processes as well. Two processes have been cited as examples of deep homologies. De Robertis and Sasai (1996) suggested that the character "formation of neural ectoderm" is one such deep homology, and that every neural cord uses the same pathway. Similarly, Shubin and colleagues (1997) proposed that limb formation evolved only once, and that despite the enormous divergence between insect and vertebrate limbs, they are constructed according to the same rules and use the same homologous processes.

The chordin/BMP4 pathway that triggers neural ectoderm formation in both the vertebrate and fly ectoderms is an excellent example of a homology of process. In *Xenopus*, the paracrine factor BMP4 is secreted by ventral mesodermal cells and is activated by the combined action of BMP1 and BMP7. Subsequent binding of BMP4 to ectodermal cells causes these cells to activate transcription factors that in turn cause the ectoderm to become epidermal (i.e., skin). However, the organizer region (underlying dorsal mesoderm) secretes chordin, which binds to and blocks the action of BMP-4. This prevents the ectoderm from expressing genes specifying an epidermal fate, so that it instead becomes neural. Activation of this pathway on the dorsal side of *Xenopus* and zebrafish embryos results in the formation of a dorsal neural tube. In the fly *Drosophila*, a BMP4 homolog (the decapentaplegic protein) is activated by BMP1 and BMP7 homologs (the tolloid and screw proteins, respectively) and instructs the ectoderm to produce skin. As in vertebrates, an arthropod homologue of chordin (the short-gastrulation protein, sog) blocks the effects of decapentaplegic, causing the ectoderm to become neural where sog is secreted. In contrast to the vertebrate pattern, this takes place on the ventral surface of the fly, producing a ventral neural cord.

Despite major differences between the neural cords of flies and vertebrates, the instructions telling them where to form are remarkably similar. This similarity has been confirmed by reciprocal substitution experiments (Holley *et al.*, 1995; Schmidt *et al.*, 1995; Marqués *et al.*, 1997). Chordin mRNA will cause neural formation in flies; injection of sog causes ectopic neural tube formation in the frog. BMP4 will prevent fly neural cord formation, and decapentaplegic protein will block the formation of the frog neural tube. There seems to be only one set of instructions for forming neural ectoderm, though they can be followed in different regions of the embryo. The BMP/chordin interaction specifies neural ectoderm, whether dorsally in the frog or ventrally in the fly (for review and references, see DeRobertis and Sasai, 1996; Gilbert, 1997).

Although the insect leg and the vertebrate leg are obviously analogous structures, they appear to share a deep homology in the way they develop, and thus serve as a second example of deep homology (Shubin *et al.*, 1997). Development of vertebrate and invertebrate limbs uses homologous sets of genetic instructions, in largely—though not entirely—homologous ways. Both express the fringe protein at the distalmost end of the proximal-distal axis and use boundary conditions to activate their respective *fringe* homologs. Both use hedgehog homologues to establish the anterior-posterior axis, and ectopic expression of *hedgehog* in the posterior section of the limb disc or field causes the formation of mirror-image ectopic limbs. In both vertebrate and arthropod limbs, the dorsal-ventral axis is marked by the activation of *apterous* homologs by Wingless homologs.

However, there are significant differences as well. Insects have no equivalent of the vertebrate apical ectodermal ridge (AER), nor do they use fibroblast growth factors to mediate cell proliferation or to maintain hedgehog expression. Moreover, even though insect and vertebrate limb primordia express many similar genes, the functions of those genes can vary. While *apterous* is critical in activating *fringe* expression in the fly limb disc, its homolog, *lmx*, is not important for *fringe* homolog expression in the vertebrate limb field. The Engrailed protein confers posterior identity in the fly limb, but ventral identity in the vertebrate limb. Such complications undermine the hypothesis of deep homology between limbs. Much of the "toolbox" for making limbs is also shared with nonhomologous structures such as vertebrate teeth (Vaahtokaari *et al.*, 1996), which strengthens the argument for convergence rather than true homology between developmental processes in insect and vertebrate limbs. The deep homology of limbs therefore remains controversial.

The argument from parsimony (i.e., the "small toolbox argument") must be kept in mind whenever processes are claimed as homologous, since the smallness of the toolbox implies the distinct possibility of convergence. For instance, if certain fibroblast growth factors generally maintain cell division, certain bone morphogenetic proteins generally initiate apoptosis, and certain transforming growth factors generally stimulate extracellular matrix production, one would expect these proteins to be utilized in different embryos whenever processes of cell division, apoptosis, and matrix formation are required. These commonalities of process, however, do not render the various resulting structures homologous, any more than their common expression of Pax6 makes a mouse's nose homologous to its eyes.

H. Rewards of Thinking about Processes

Despite its potential for misuse, we believe the concept of process homology is a powerful tool. Extending the concept of homology from morphological to molecular characters has already shed new light on the relationships and evolution of organisms and the genes they carry. Further extension of the concept of homology to include developmental processes will illuminate relationships between different ontogenies. Ultimately it is ontogenies, not individual organisms or their genes, that evolve. Homology was initially about phenotypes, and the concept was successfully expanded to include genotypes as well; now it is time to apply it to the ontogenetic processes that link these two levels at which evolution is studied.

A carefully defined and applied concept of process homology will refine comparisons of developmental processes between divergent model species and provide a framework for integrating data from nonmodel organisms as well.

For example, recent work by Minsuk (1996, 1997) has shed new light on the long-studied problem of amphibian mesoderm morphogenesis by going beyond classical anatomical comparisons in traditional "model" species. Minsuk's comparative morphogenetic and experimental approach has broadened our understanding of morphogenetic mechanisms and their evolution within amphibians, and also deepened our understanding of the gastrulation processes of *Xenopus*, the best-known species.

Another merit of process homology for developmental biologists is its inclusion of a time axis, which traditional homology concepts lack. Development is ultimately a process, and its central "characters" are as much temporal as spatial. One difficulty of applying a traditional (structural) homology concept to embryos is that many of their structures (e.g., somites) are transient or are made up of continually changing populations of cells (e.g., the dorsal blastopore lip, or Hensen's node). The most significant features of embryos are not structures. Rather, they are the processes and changes embryos undergo, and the mechanisms by which those changes occur—whether changes in gene regulation or the mechanical interactions among different regions of the embryo. Moreover, evolutionary-scale comparisons between ontogenies often describe differences in terms of heterochronies, or shifts in developmental timing. Neither sort of character—transient structures, or timing differences— can be addressed satisfactorily by either a structural or a genetic homology concept. Carefully defining and identifying process homologies will give us a conceptual tool for comparing dynamic developmental characters and looking at how they have evolved in different lineages.

The concept of process homology is useful for examining not only the evolution of development, but also evolution more generally, including the role of genes. Classical evolutionary theory considered evolution a subset of population genetics and explained natural selection in terms of changes in gene frequencies. Each gene was seen as an independent, autonomous unit within the genome and could be transmitted independently of any other. Natural selection occurred via the differential transmission of independent alleles through adult organisms that competed for reproductive advantage.

The present synthesis of evolutionary biology and developmental genetics provides a different, and complementary, view of the roles of genes in evolution. First, the new view focuses on the construction of embryos rather than on competition between adults, with an eye toward the kinds of developmental differences that can generate new structures. Second, the developmental synthesis highlights genetic regulatory regions rather than exonic differences (such as changes that make one enzyme variant more efficient than another). Third, this developmental approach focuses on the similarities between genes rather than on their differences.

Most significantly, we now look at genes as parts of a pathway or field, rather than as autonomous agents. By describing homologies of process between whole pathways, we begin to see genes integrated into new, larger evolutionary units. Recognizing developmental genes, processes, and pathways as modules, and as evolutionary characters in their own right, is a critical step toward a new synthesis of evolution and development. Evolution ultimately depends on heritable changes in development, and the modular organization of development facilitates evolutionary change in several ways.

First, this organization allows different modules to change without affecting others. Second, modules can be coopted into new functions. Third, since modules depend on the intercellular communication network of paracrine factors, signaling cascades, and transcription factors, these signal transduction pathways, themselves, can constitute modules that can be used for several different developmental and physiological functions. They can be conserved through evolution and modified both within and between species.

We repeatedly find homologies of process among the signal transduction pathways that operate between embryonic cells. What we now call paracrine factors and ligands turn out to be the long-sought inducers of classical experimental embryology. Embryologists now recognize receptors and signal transducing molecules as components of the competence apparatus that enables certain cells to respond to specific inducers. These signaling pathways are the bases of embryonic induction, which is in turn the core of organogenesis. If macroevolution involves changing morphological features, then the alteration of signal transduction pathways becomes critical for any discussion of large scale evolution.

CONCLUSIONS: PROCESS HOMOLOGY, DEVELOPMENT, AND VIEWS OF EVOLUTION

Identifying the ways in which homologous processes are regulated, replicated, and changed over time will enable us to better understand how changes in development generate changes in morphology and, ultimately, the evolution of new groups of animals.

Waddington (1953) noted that natural selection worked in two distinct modes. The traditional one concerned the elimination of adult phenotypes ("normative selection"), while the less studied mode ("stabilizing selection") eliminated individuals with unstable systems of epigenetic interactions. Concepts of modularity and process homology give us a way to begin exploring the latter mode, which focuses on developmental processes rather than adult phenotypes. Such exploration opens up valuable ways of looking not only at the history of phylogenesis, but also at its mechanisms.

The developmental approach to homology and evolution complements the classical genetic approach. Both are needed. Before this decade, we did not know enough about the mechanisms of development to understand how changes in developmental processes could explain evolution. We are now beginning to acquire that knowledge through our understanding of homologous genes and processes. Looking at the Modern Synthesis in 1953, J.B.S. Haldane used a developmental analogy to express a hopeful sense of things to come: "The current instar of the evolutionary theory may be defined by such books as those of Huxley, Simpson, Dobzhansky, Mayr, and Stebbins. We are certainly not ready for a new moult, but signs of new organs are perhaps visible." The structure of a more inclusive, developmentally oriented evolutionary synthesis is beginning to emerge, and we echo Haldane's conviction that the current diversity of views on the form it should take "...is not a bad sign. It points forward to a broader synthesis in the future."

ACKNOWLEDGMENTS

The authors wish to thank G. Wagner for his comments on the draft of this manuscript. JB thanks P. Bixby of the UNH Writing Center for expert editorial assistance.

LITERATURE CITED

Abouheif, E. (1997). Developmental genetics and homology: a hierarchical approach. *Trends Ecol. Evol.* 12 (10):405-408.

Atlan, G., Balmain, N., Berland, S., Vidal, B., and Lopez, E. (1997). Reconstruction of human maxillary defects with nacre powder: histological evidence for bone regeneration. *C.R. Acad. Sci. Paris/Life Sci.* 320:253-258.

Balinsky, B. I. (1933). Das Extremitätenseitenfeld, seine ausdehnung und bescaffenheit. *Roux's Arch. Ent. Org.* 130:704 -736.

Baltzer, F. (1950). Entwicklungsphysiologische Betrachtungun über Probleme der Homologie und Evolution. *Rev. Suisse Zool.* 57:451 - 477.

Baltzer, F. (1952). Expreimentelle Beiräge zur Frage der Homologie. *Experientia* 8:285 - 297.

Barry, M. (1837). On the unity of structure in the animal kingdom. *Edinb. New Phil. J.* 22:116-141.

Bolker, J. A., and Raff, R. A. (1996). Developmental genetics and traditional homology. *Bioessays* 18(6):489-494.

Bowler, P. J. (1996) "Life's Splendid Drama." University of Chicago Press, Chicago.

Brannon, M., Gomperts, M., Sumoy, L., Moon, R. T., and Kimelman, D. (1997). A beta-catenin/Xtf-3 complex binds to the *Siamois* promoter to regulate dorsal axis specification in *Xenopus*. *Genes Dev.* 11:2359-2370.

Britten, R. (1967). *Carnegie Inst. Wash. Year Book.* 66:68.

Bushdid, P. B., Brantley, D. M., Yull, F., Blaeuer, G. L., Hoffman, L. H., Niswander, L., and Kerr, L. D. (1998). Inhibition of NF-kB activity results in disruption of the apical ectodermal ridge and aberrant limb morphogenesis. *Nature* 392:615-618.

Cadigan, K. M., and Nusse, R. (1997). Wnt signaling: a common theme in animal development. *Genes Dev.* 1:3286-3305.

Carrasco, A. E., McGinnis, W., Gehring, W. J., and De Robertis, E. M. (1984). Cloning of an *X. laevis* gene expressed during embryogenesis coding for a peptide region homologous to *Drosophila* homeotic genes. *Cell* 37:409-414.

Chisaka, O., and Capecchi, M. R. (1991). Regionally restricted developmental defects resulting from targeted disruption of the homeobox gene hox 1.5. *Nature* 350:473-479.

Crossley, P. H., and Martin, G. R. (1995). The mouse Fgf8 gene encodes a family of polypeptides and is expressed in regions that direct outgrowth and patterning of the developing embryo. *Development* 121:439-451.

de Beer, G. R. (1971). "Homology: An Unsolved Problem."   Oxford Biol. Readers 11. Oxford University Press, London.

De Robertis, E. M., and Sasai, Y. (1996). A common plan for dorsoventral patterning in Bilateria. *Nature* **380**:37-40.

Dickinson, W. J. (1995). Molecules and morphology: where's the homology? *Trends. Genet.* **11**:119-121.

Fitch, W. M. (1970). Distinguishing homologous from analogous proteins. *Syst. Zool.* **19**:99-113.

Gilbert, S. F., Opitz, J., and Raff, R. A. (1996). Resynthesizing evolutionary and developmental biology. *Dev. Biol.* **173**:357-372.

Gilbert, S. F. (1996). Cellular dialogues during development. *In* "Gene Regulation and Fetal Development" (G. Martini and H. Neri, eds.), March of Dimes Foundation Birth Defects: Original Article Series Vol. 30 (1), pp. 1-12. Wiley-Liss, New York.

Gilbert, S. F. (1997). Enzymatic adaptation and the entrance of molecular biology into embryology. *In* "The Philosophy and History of Molecular Biology: New Perspectives" (S. Sarkar, ed.), pp. 101-123. Kluwer Academic Publishers, Dordrecht.

Gould, S. J. (1977). "Ontogeny and Phylogeny." Harvard University Press, Cambridge.

Hawley, S. H. B., Wünnenberg-Stapleton, K., Hashimoto, C., Laurent, M. N., Watabe, T., Blumberg, B. W., and Cho, K. W. Y. (1995). Disruption of BMP signals in embryonic *Xenopus* ectoderm leads to direct neural induction. *Genes Dev.* **9**:2923-2935.

Hillis, D. M. (1994). Homology in molecular biology. *In* "Homology: The Hierarchical Basis of Comparative Biology" (B. K. Hall, ed.), pp. 339-368. Academic Press, New York.

Holley, S. A., Jackson, P. D., Sasai, Y., Lu, B., De Robertis, E. M., Hoffmann, F. M., and Ferguson, E. L. (1995). A conserved system for dorsal-ventral patterning in insects and vertebrates involving sog and chordin.*Nature* **376**:249-253.

Ingham, P. W. (1994). Hedgehog points the way.*Curr. Biol.* **4**:347-350.

Kanegae, Y., Tavares, A. T., Izpisúa-Belmonte, J. C., and Verma, I. M. (1998). Role of Rel/NF-kB transcription factors during outgrowth of the vertebrate limb. *Nature* **392**:611-614.

Kaufman, T. C., Seeger, M. A., and Olsen, G. (1990). Molecular and genetic organization of the *Antennapedia* gene complex of *Drosophila melanogaster. Adv. Genet.* **27**:309-362.

Kowalewski, A. (1871). Die Entwicklung der einfaschen Ascidien. *Arch. Micr. Anat.* **7**:101-130.

Larabell, C. A., Torres, M., Rowning, B. A., Yost, C., Miller, J. R., Wu, M., Kimelman, D., and Moon, R. T. (1997). Establishment of the dorsoventral axis in Xenopus embryos is presaged by early asymmetries in beta-catenin that are modulated by the Wnt signaling pathway. *J. Cell. Biol.* **136**:1123-1136.

Lemaitre, B., Nicolas, E., Michaut, L., Reichart, J.-M., and Hoffman, J. A. (1996). The dorsoventral regulatory gene cassette spaetzle/Toll/cactus controls the epotent antifungal response in *Drosophila* adults. *Cell* **86**:973-983.

Marqués, G., Mussacchio, M., Shimell, M. J., Wünnenberg-Stapleton, K., Cho, K. W. Y., and O'Connor, M. B. (1997). Production of a Dpp activity gradient in the early Drosophila embryo through the opposing actions of the sog and tld proteins. *Cell* **91**:417-426.

Medawar, P. B., and Medawar, J. S. (1977). "The Life Science." Harper & Row, New York.

Miller, J. R., Rowning, B. A., Larabell, C. A., Yang-Snyder, J. A., Bates, R. L., and Moon, R. T. (1999). Establishment of the dorsal-ventral axis is *Xenopus* embryos coincides with the dorsal enrichment of Dishevelled that is dependent on cortical rotation. *J. Cell Biol.* **146**:427-437.

Minsuk, S. B., and Keller, R.E. (1996). Dorsal mesoderm has a dual origin and forms by a novel mechanism in *Hymenochirus*, a relative of *Xenopus. Dev. Biol.* **174**:92-103.

Minsuk, S. B., and Keller, R.E. (1997). Surface mesoderm in *Xenopus*: a revision of the stage-10 fate map. *Dev. Genes Evol.* **207**:389-401.

Nüsslein-Volhard, C., Frönhofer, H. G., and Lehmann, R. (1987). Determination of anterioposterior polarity in *Drosophila. Science* **238**:1675-1681.

Ohuchi, H. *et al.* (1997). The mesenchymal factor, FGF10, initiates and maintains the outgrowth of the chick limb bud through interaction with FGF8, and apical ectodermal factor. *Development* **124:** 2235-2244.

Postlethwait, J. H., and Schneiderman, H. A. (1971). Pattern formation and determination in the antenna of the homeotic mutant Antenapedia of *Drosophila melanogaster. Dev. Biol.* **25:**606-640.

Raff, R.A. (1996). "The Shape of Life." Chicago University Press, Chicago.

Riedl, R. (1978). "Order in Living Organisms: A Systems Analysis of Evolution." (R. P. S. Jefferies, trans.), John Wiley, New York.

Sasai, Y., Lu, B., Steinbeisser, H., and De Robertis, E. M. (1995). Regulation of neural induction by the Chd and BMP4 antagonist patterning signals in *Xenopus. Nature* **376:**333-336.

Schmidt, J., Francois, V., Bier, E., and Kimelman, D. (1995). Drosophila short gastrulation induces an ectopic axis in Xenopus: evidence for conserved mechanisms of dorsal-ventral patterning. *Development* **121:**4319-4328.

Shelton, C. A., and Wasserman, S. A. (1993). pelle encodes a protein kinase required to establish dorsoventral polarity in the Drosophila embryo. *Cell* **72:**515-525.

Stewart, C.-B., and Wilson, A.C. (1987). Sequence convergence and functional adaptation of stomach lysozymes from foregut fermenters. *Cold Spring Harbor Symp. Quant. Biol.* **52:**891-899.

Striedter, G. F., and Northcullt, R. G. (1991). Biological hierarchies and the concept of homology. *Brain Behav. Evol* **38:**177-189.

Swanson, K. W., Irwin, D. M., and Wilson, A. C. (1991). Stomach lysozyme gene of the langur monkey: Tests for convergence and positive selection. *J. Mol. Evol.* **33:**418-425.

Waddington, C. H. (1953). Epigenetics and evolution. *In* "Soc. Exper. Biol. Symposium 7: Evolution" (R. Brown and J. F. Danielli, eds.), pp. 186-199. Cambridge University Press, Cambridge.

Waddington, C. H. (1975). "The Evolution of an Evolutionist." Cornell University Press, Ithaca, New York.

Wagner, G. P. (1996). Homologues, natural kinds, and the evolution of modularity. *Amer. Zool.* **36:**36-43.

Warren, R. W., Nagy, L., Selegue, J., Gates, J., and Carroll, S. (1994). Evolution of homeotic gene regulation and function in flies and butterflies. *Nature* **372:**458 - 461.

Westbroek, P., and Marin, F., (1998). A marriage of bone and nacre. *Nature* **392:**861-862.

Whitehead, A. N. (1933). "Adventures of Ideas," pp. 274 – 276. Macmillan, New York.

Wilson, E. B. (1898).Cell lineages and ancestral reminiscences. *Biological Lectures from the Marine Biological Laboratories, Woods Hole, Masssachusetts.* Ginn, Boston, pp. 21-42.

Wu, L. P., and Anderson, K. V. (1998). Regulated nuclear import of Rel proteins in the *Drosophila* immune response. *Nature* **392:**93-97.

Yost, C., Farr, G. H., Pierce, S. B., Ferkey, D. M., Chen, M. M., and Kimelman, D. (1998). GBP, an inhibitor of GSK-3, is implicated in *Xenopus* development and oncogenesis. *Cell* **93:**1031-1041.

20

COMPARATIVE LIMB DEVELOPMENT AS A TOOL FOR UNDERSTANDING THE EVOLUTIONARY DIVERSIFICATION OF LIMBS IN ARTHROPODS: CHALLENGING THE MODULARITY PARADIGM

LISA M. NAGY[1] AND TERRI A. WILLIAMS[2]

[1]Department of Molecular and Cellular Biology, University of Arizona, Tucson, AZ 85721
[2]Department of Ecology and Evolutionary Biology, Yale University, New Haven CT 06520

INTRODUCTION

In this chapter we compare arthropod appendages by analyzing similarities and differences in their development. The comparison of arthropod limbs, even between species, is a comparison of serially repeated structures. How such reiterated structures are to be interpreted is far from obvious. One common view is that serial repetition is evidence of a modular organization, i.e., a body built out of identical subunits. From this perspective, we look for mechanisms of limb development that are semiautonomous in function and conserved between species. In addition, the assumption of a modular body plan implies a complete and continuous identity of structural modules. In this chapter we evaluate both these assumptions from a developmental perspective and ask (1) is the arthropod body plan modular in organization and (2) to what degree do all arthropod appendages share a common identity that can be used as a basis for homology? We conclude that arthropods are not simply built from identical repeated units but that modularity is exhibited in different degrees in different lineages and is thus an evolutionarily variable character. Similarly, although limbs show widespread conservation in some patterning processes that position the limb primordia, much of limb development is not conserved. Developmental criteria can clarify problems in establishing homologies between structurally distinct limbs. Furthermore, understanding how arthropod limbs have evolved may require reanalyzing them within the context of patterns of axial diversification and the mechanisms of segmentation that establish them.

MODULARITY, IDENTITY, AND MECHANISMS OF DEVELOPMENT

The fact that arthropods are made up of serially repeated structures—segments and limbs—is one of the most salient characteristics of the phylum (Fig. 1). The diversification of these repeated structures has been the basis of arthropod evolution (Brusca and Brusca, 1990). Because segments and limbs are repeated along the body axis, they have been considered a series of structural modules that originally repeated identically along the body axis and then differentiated through the course of evolution (Raff and Kaufmann, 1983; Gerhart and Kirschner, 1997). The very idea of modularity rests on the assumption that organisms can be decomposed into separable elements (Raff, 1996, Wagner, 1996). This separability is due in part to an independence in the developmental processes that produce different modules. For example, in vertebrates, limb buds transplanted to novel positions on the flank develop normally (see Gilbert, 1997). This kind of manipulation demonstrates the relative independence of limb development from its normal context. Thus, vertebrate limbs can be considered modular. Similarly in arthropods, isolated *Drosophila* limb discs cultured within the body of another insect develop to proper mature form (Schneiderman and Bryant, 1971). However, arthropods have been considered exemplars of modular organization not only because limbs can develop independently, but also because the body plan is composed of many pairs of limbs, attached in a one to one fashion to the repeated segments. Arthropod limbs and segments are thought to be reiterated, independent units out of which the body plan is constructed.

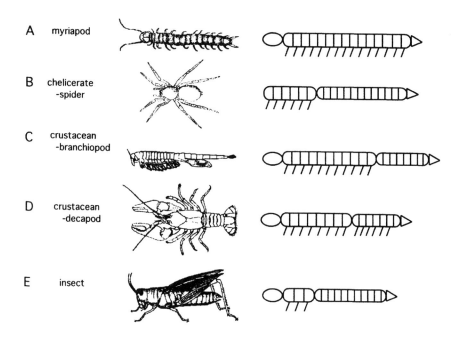

FIGURE 1   The modularity of arthropod body plans. The segmental body plan of some representative arthropods is schematized by the icons to the right of each of each animal. These icons portray the number and position of reiterated subunits, i.e., segments and limbs, and make obvious the diversity in segmental organization. Note also that the segments in different taxa are clustered into groups called tagma.

The assumption that arthropod limbs have diversified in a semi-autonomous or modular fashion is supported by both dissociability in developmental timing and divergence of adjacent limbs. Segments immediately adjacent to each other can be functionally specialized for widely different roles, and adjacent limbs can have highly divergent forms and functions, e.g., tiny mouthparts, large pincers, and walking legs (Fig. 2). Modularity of limb development is also suggested by the fact that, despite generally being coupled, the timing of limb development relative to segment development can vary in both insect and crustacean lineages. For example, in metamorphic insects, limb primordia form during embryogenesis, but then invaginate to form discs that are not expressed until later when the animal metamorphoses into the adult (Snodgrass, 1954; Anderson, 1972). Many crustaceans pass through some form of metamorphosis as well (Snodgrass, 1956; Williamson, 1982) and exhibit temporal dissociation between limb and segment development. Some decapod crustaceans (e.g., larval lobsters) delay the formation of the abdominal swimming appendages until the juvenile phase, even though the larvae already have developed abdominal segments (Gurney, 1942). Individual segments can be gained or lost from the body plan within particular lineages (Brusca and Brusca, 1990). All of these patterns support the idea of modular organization.

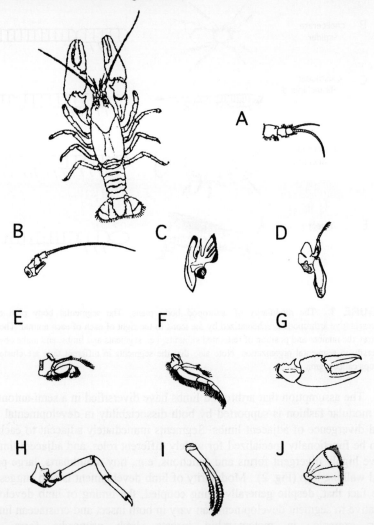

Figure 2 Tagmosis and limb diversity in a representative arthropod. Diagram of a crayfish with some representative limbs (after Snodgrass, 1965). Note the distinct body regions of the crayfish as well as the abrupt transitions in limb forms along the body axis. Limbs: (A) First antenna; (B) second antenna; (C) second maxilla; (D) first maxilliped; (E) second maxilliped; (F) third maxilliped; (G) chela; (H) third walking leg; (I) pleopods; and (J) uropods.

Not only do segments and limbs appear as modules within arthropods, limbs themselves are composed of subunits that potentially could be modular. One type of subunit is the limb segment that is separated by joints and has discrete musculature and nerves permitting independent movement and control. Another subunit within limbs are the branches that comprise the whole limb (see diversity of branches in Fig. 2). The presence of branches themselves provides material for diversification since branches of the same limb can be quite distinct in their form and function. For example, a single limb can have one branch specialized for walking, another for swimming, and another for respiration (McLaughlin, 1982). This independence of branches is also evident in the separate neural control of each branch. For example, the inner branches in a series of limbs can be more tightly coordinated along the body than an inner branch is with the lateral branch of the same limb (Laverak *et al.*, 1977). Thus, it has been relatively easy to characterize arthropod morphology as modular.

Viewing the reiterated structures of arthropod morphology as modular has profound consequences for the comparison of those structures. This is because modules, by definition, are fundamentally identical. If arthropod limbs and segments are modular then all segments are fundamentally identical and all limbs are fundamentally identical. The differences between segments or between limbs are only superficial or somehow peripheral to this fundamental identity. Another way to say this is that limbs (or segments) are equivalently individualized characters throughout the arthropods. Being equivalently individualized characters implies that they share the same degree of structural complexity and are delimited by a similar set of underlying developmental constraints (Wagner, 1989a; Wagner, 1989b). In this sense, limbs would have direct homology as appendages, either serially along the body axis or between species.

IS THE APPARENT MODULAR MORPHOLOGY OF ARTHROPODS REFLECTED IN UNDERLYING DEVELOPMENTAL MECHANISMS?

A natural starting point to answer whether modular morphology is reflected in underlying developmental mechanisms is an analysis of *Drosophila* development since there, as in no other arthropod, the developmental mechanisms responsible for generating body parts are well explored. During embryonic development in *Drosophila*, the body is subdivided into reiterated units by the gradual refinement of spatial gradients (reviewed in Akam, 1987; Ingham, 1988; Martinez-Arias, 1993). Similarly, limb primordia are set up along the body axis using regulatory signals that are reiterated in each segment (reviewed in Cohen *et al.*, 1993). After a number of segments and limbs have been similarly specified, they diverge in character. Axial diversification, resulting in differences to both segment and limb character, is produced in part by the differential activity of "selector" genes—the Hox genes—imposed on a series of identical segments (reviewed in Lewis, 1964; McGinnis and Krumlauf, 1992; Carroll, 1995).

Superficially, *Drosophila* seems to provide a tidy developmental rationale for the patterns of diversification among arthropods. Morphological units—limbs and segments—could be produced by a conserved set of developmental mechanisms reiterated along the body axis. Varying the regulation of Hox genes expressed in different segments could initiate the divergence in character between different segments and different limbs. Through tinkering with the developmental toolbox, evolution could produce endless variants of segments or limbs on the anterior/posterior (A/P) axis. Unfortunately, this interpretation, though tantalizing at first glance, does not seem born out, at least in its simple version. As elaborated later, many regulatory mechanisms controlling pattern in *Drosophila* do not map simply onto morphological characters, either within *Drosophila*, or when viewed comparatively.

To evaluate how understanding mechanisms of limb patterning can help us understand limb diversification, we compare these mechanisms across arthropods. After an overview of arthropod morphogenesis, we provide a brief summary of the regulatory mechanisms that position and elaborate limbs in *Drosophila*. We focus on the degree to which limbs are set up by reiterated regulatory networks indicative of developmental modularity. Subsequently, we compare limb development in *Drosophila* with other arthropods to evaluate whether limb patterning has a shared developmental basis throughout the arthropods. We focus on the developmental basis for drawing homologies between appendages and parts of appendages. Then, because limbs and segments have a shared developmental basis, we discuss axial patterning among arthropods. In discussing both limbs and segments, we ask what evidence supports the view that the appearance of modular subunits in adult morphology represents a causal underpinning of modular development mechanisms. If limbs are reiterated modules, this should be reflected in their development. We ask whether arthropod appendages are produced by a set of developmental mechanisms with a defined system of regulation and constraint that is common to all arthropods.

OVERVIEW OF APPENDAGE PATTERNING IN *DROSOPHILA*

Because *Drosophila* will serve as the basis for the comparative analysis, it is important to recognize that its mode of morphogenesis is highly derived and specialized. Most arthropods do not show a radical metamorphosis during their life history. Indeed, for the vast majority of arthropods segmentation, limb development, and growth are concurrent processes. Unlike vertebrate limbs, paired appendages in arthropods can be found in every segment. Indeed, the temporal growth and differentiation of the limbs are inextricably linked to the growth and differentiation of segments. Primitively, arthropods hatched as free-living larvae that basically consisted of a recognizable anterior end with head structures followed by a posterior end, or telson (Fig. 3; Anderson, 1973, 1982; Brusca and Brusca, 1990). Between the head and the telson is a morphologically undifferentiated field of cells within which segments and limbs develop in an anterior to posterior progression. As described later, limb primordia are positioned with regard to segmental boundaries and show a one-to-one correspondence with the segments. At any given larval stage, the A/P axis displays segments and limbs in successive stages of maturity. This mode of development is found in extant groups as well as in fossils arthropods from the Cambrian (Walossek, 1993). The same general pattern, i.e., the sequential addition of segments and limbs, can be found in many arthropods that develop the adult morphology in the embryo, although in many species that have a metamorphic lifestyle, this pattern is lost (Sander, 1976). *Drosophila*, like other holometabolous insects, undergoes a radical metamorphosis from larva to adult; it does not add segments or limbs in a gradual and sequential manner (Fig. 3).[1] Instead, the full complement of adult segment primordia are set up at once along the body axis and, subsequently, limb primordia are formed in all the limb-bearing segments. The fact that the derived mode of morphogenesis displayed by *Drosophila* is so distinct from typical arthropod morphogenesis makes the direct comparison of developmental patterning signals from *Drosophila* to other arthropods problematic.

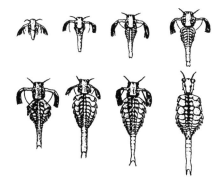

FIGURE 3 Arthropods show distinct patterns of morphogenesis. Most arthropods reach maturity through a series of gradual changes during morphogenesis, represented here by the larval development in the branchiopod crustacean, *Artemia* (after Heath, 1924). *Artemia* passes through a series of larval stages in which segments are added sequentially to the posterior and limbs concurrently develop on those segments. By contrast, in the insect *Drosophila*, cells that will become the adult structures are set aside during embryology and develop autonomously inside the larvae only to emerge at the time of metamorphosis to the adult.

---

[1] The variation in segmentation that we refer to here precedes the conserved function of the A/P patterning genes *engrailed* and *wingless* described later. For reviews on variation in mechanisms of arthropod segmentation, see Sander (1976), Patel (1994), and Nagy (1994).

IN *DROSOPHILA* THE EARLY DEVELOPMENT OF LIMB PRIMORDIA IS LINKED TO SEGMENT DEVELOPMENT, LATER LIMB PRIMORDIA DEVELOP DISTINCT, AUTONOMOUS FATES

Limb patterning in *Drosophila* occurs in two temporally distinct phases. The early positioning of the limb primordia takes place during embryogenesis and involves gene circuits that concurrently establish segmental positional information along the A/P and dorsal/ventral (D/V) axes. Once appendage primordia have been established within the segmental patterning system, they have the potential to launch on distinct developmental pathways to form distinct adult appendages (reviewed in Cohen *et al.*, 1993; Blair, 1999; Brook *et al.*, 1998). This second phase, proximal/distal (P/D) limb outgrowth and elaboration, occurs during larval development. We begin with a review of these two phases of limb development in *Drosophila* and then compare these patterning steps to what is known in other arthropods.

HOW ARE THE EARLY LIMB PRIMORDIA POSITIONED WITHIN THE SEGMENT AND DO ALL SEGMENTS HAVE THE POTENTIAL TO MAKE LIMBS?

In *Drosophila*, early appendage patterning uses the same regulatory molecules responsible for patterning the A/P and D/V axes as those which set up the segmented body plan. Along both A/P and D/V axes of the embryo, boundaries between populations of cells are established in response to thresholds of maternally provided morphogens that supply graded signals of positional information (reviewed in Lawrence and Struhl, 1996). In each appendage-bearing segment, these boundaries function to set aside a bilateral cluster of cells that will later become the paired segmental appendages (Cohen, 1990; Cohen *et al.*, 1993).

Along the A/P axis, *engrailed (en)* -expressing cells come to mark the posterior portion of each segment in arthropods. Just anterior to these cells lie cells which express the *wingless (wg)* gene. The boundary between *en*- and *wg*-expressing cells in each segment serves to position the limb primordia, or imaginal discs, along the A/P axis of that segment (Fig. 4A; reviewed in Cohen *et al.*, 1993). Similarly, the D/V limb "address" of the limb primordia is provided by the combined activity of some of the genes that pattern the D/V axis of the embryo. The epidermal growth factor (EGF) receptor homologue, *DER* (Raz and Shilo, 1993), is essential for restricting the limb primordia ventrally, and the *decapentaplegic (dpp)* gene acts to inhibit limb development dorsally (Goto and Hayashi, 1997). A cluster of cells at the boundary between *wg* and *en* expression and at the edge of the D/V extent of the *dpp* expression begin expressing *Distalless (Dll)* (Fig. 4A). *Dll* is known to be essential for outgrowth of limbs in *Drosophila*. Null mutants show loss of the distal limb structures up to the proximal limb segment, the coxa (Cohen and Jürgens, 1989).

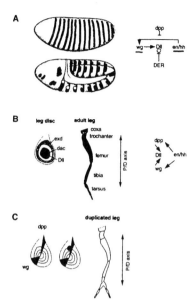

FIGURE 4　　Model for *Drosophila* appendage development. A. Early appendage patterning: Early in development the *Drosophila* embryo becomes divided into segmental units. The maintenance of each segment depends on the reiterated expression of *wg* (grey) and *en /hh* (black) expression. Mutant analyses have shown that appendage primordia, represented by the expression of the *Dll* gene (black circles), are positioned at the intersection of the expression of the *wg* and *en* genes along the A/P axis and the boundaries of the expression of the *dpp* and *DER* genes along the D/V axes. All of these growth factors shown in the schematic on the right are presumed to activate transcription factors that directly regulate the *Dll* promoter, although this has not yet been directly shown. B. Diagram of a limb disc and leg of *Drosophila* illustrating the patterning of the disc from periphery to center which, during the eversion to metamorphosis, translates into proximal to distal of the adult leg. The genes involved in specifying the P/D axis are indicated on the disc and the adult leg. In the leg disc, Wg-Dpp create a distal organizer responsible for patterning the D/V and P/D axes of the leg. Different levels of the growth factors Wg and Dpp directly activate and maintain cell fates: cells fated to form the distal tip of the leg receive the highest levels of Dpp and Wg and activate the expression of the transcription factor *Dll* (black). Slightly more proximal cells express the transcription factor *dachshund* (*dac*, dark grey), which is repressed in more distal cells by the higher levels of Dpp and Wg (Lecuit and Cohen, 1997). While the coxal leg segment is outside the domain effected by Wg and Dpp, it is nonetheless patterned indirectly by these two growth factors. *homothorax* , another target of Wg-Dpp in the distal domain, represses the function of the transcription factor *extradenticle* (*exd*, light grey) The absence of homothorax in the proximal domain allows for the translocation of exd to the nucleus, where it directs proximal limb development. (Mann and Abu-Shaar, 1996; Aspland and White, 1997; Rieckhof *et al.*, 1997). Most of the genes that are involved in positioning the appendage primordia (A) function in patterning the axes of the developing leg, although their regulatory interactions are modified (compare schematics on the right in B with A). C. Leg disc showing normal Wg-Dpp expression (left). Loss of these growth factors can result in loss of distal leg structures; by contrast ectopic expression (right disc), represented by the circle of Wg expression (black) in the disc, which creates a new boundary between these growth factors, also creates new distal outgrowths or forked legs with a superficially branched appearance (Campbell *et al.*, 1993; Diaz-Benjumea *et al.*, 1994; Campbell and Tomlinson, 1995; Held, 1995).

The homeotic genes, the well-known "selector" genes postulated to function as "on-off" switches that determine segment identity, surprisingly have no function in setting up the limb primordia in each segment (reviewed in Carroll, 1995). Because the homeotics regulate the unique identities of *Drosophila* segments, as judged by the presence or character of the limbs, it was originally supposed that they would be responsible for the earliest stages of limb development in flies by promoting both appendage development and subsequent differentiation. However, *Drosophila* appendages are not actively promoted by the Hox genes. For example, in the absence of the abdominal homeotic genes *Ultrabithorax* (*Ubx*), *abdominal-A* (*abd-A*), and *Abdominal-B* (*Abd-B*), identical limb primordia develop in every segment, including the normally limbless abdominal segments (Lewis, 1978; Vachon *et al.*, 1992). Each trunk segment can form an appendage without any homeotic input. Hox genes simply modify the response of each segment to limb-inducing information present in every trunk segment, either by repressing limb development or by modulating the character of the limb. Thus, in *Drosophila*, the information required to pattern the body axis is sufficient to pattern the limb primordia without other input. This information is inherent and identical in every trunk segment. The same patterning circuits used for establishing the A/P and D/V address of the limb primordia within the segments are used to position all appendages along the body axis, despite subsequent differences in morphology.

HOW IS THE SEMI-AUTONOMY OF LATER LIMB DEVELOPMENT AND SPECIFICATION CONTROLLED?

Although the limb primordia are established in the embryo, they consist of only 20-25 cells (see Cohen, 1993). Growth and development of the primordia into a limb is not seen until later, during larval development. Later development of the *Drosophila* leg depends on some of the same signaling molecules that positioned the primordia in the embryo—the secreted signaling molecules Wg and Dpp. These two proteins act as concentration-dependent morphogens to specify both the D/V and P/D axes of the leg (Fig. 4B, reviewed in Campbell and Tomlinson, 1995; Held, 1995; Brook *et al.*, 1998; Blair, 1999). Wg expression is restricted to a ventrolateral stripe; Wg specifies ventral cell fates and represses Dpp. Dpp expression is strongest in a dorsolateral stripe; Dpp specifies dorsal cell fates and represses Wg. Thus, Wg and Dpp define the D/V axis through a mutual antagonism. During larval development, the Wg and Dpp expression domains meet at the center of the imaginal disc, which corresponds to the presumptive distal tip of the leg. Cooperatively, Wg and Dpp activate, in a concentration-dependent manner, target genes which are expressed in discrete domains along the P/D axis of the limb (*Distalless, dachsund*, Lecuit and Cohen, 1997; Fig. 4B, for more details see legend to Fig. 4). These targets function initially to divide the limb into proximal and distal domains and subsequently to pattern the distal portion of the leg, from the claw to the tibial segment. Thus, while Wg and Dpp act in opposition to define D/V positions along the leg, they act cooperatively to define the P/D axis. Together, Wg, Dpp and their targets create a regulatory module that patterns the D/V and P/D axes of the adult limb.

As discussed earlier, the apparently modular morphological units within limbs are branches and individual segments separated by joints. While *Drosophila* legs are not themselves branched, experimental manipulations that create new juxtapositions of Wg and Dpp expression can create new branches (reviewed in Campbell and Tomlinson, 1995; Held, 1995; Brook *et al.*, 1998), leading to the idea that the diversification in branching patterns exhibited by arthropods might be the result of reiterations of this basic patterning module (Fig. 4C, and comparative discussion later). However, exactly how joints and repeated limb segments are generated is not yet well understood. While there are some genes which are expressed in segmentally reiterated patterns, e.g., *Notch and annulin* (de Celis *et al.*, 1998; Singer *et al.*, 1992; Bastiani *et al.*, 1992), the relationship between these genes and the Wg and Dpp morphogens is not known.

As described, all limbs appear to initiate adult development using a common patterning module, suggesting a common identity of all limbs. Unfortunately, the global patterning system provides no direct developmental basis for the formation of specific subparts of the appendages, features easily recognized by comparative morphology. For example, genes that control different proximal/distal regions of the limb do not map to specific limb segments or branches. Rather, differentiation of limbs appears to rest on a different suite of patterning genes, whose direct relationship to the underlying global patterning system is not yet well described. The genes responsible for limb differentiation do not alter the global patterning system that initiates limb development in each appendage.[2] This is apparent from the analysis of the phenotypes of homeotic mutants, which modify the character of each limb (reviewed in Lawrence and Morata, 1994; Biggin and McGinnis, 1997). Homeotic transformations illustrate the global nature of the limb patterning system. On any segment, a homeotic transformation—leg to antennae, for example—does not modify positional information: although cells take on a new fate—leg vs. antenna—they still use the same limb coordinate system to read out that fate.

---

[2] There are two recent and notable exceptions to this idea. It has been recently suggested that *Dll*, defined here as part of the global patterning system, should also be considered a "selector" or master regulatory control gene for the leg. Misexpression of *Dll* in the wing can result in transformations to leg (Gorfinkiel *et al.*, 1997). It has also recently been shown that differences between the *Drosophila* wing and haltere can indeed be attributed, among many things, to differences in the pattern of *wingless* expression (Weatherbee *et al.*, 1998). The implication of this finding is that the global patterning system shared between appendages may also be subject to evolutionary modifications.

SUMMARY OF *DROSOPHILA* MODEL

Thus, in *Drosophila*, some but not all developmental mechanisms do support the modular view of morphology. In early development, similar segmental anlagen are patterned by regulatory mechanisms that are reiterated along the body axis. All segments have the potential to make limbs and position the early primordia using the same genes that pattern the segmental anlagen. Once positioned, limb primordia use reiterated patterning systems for subsequent elongation. The A/P, D/V, and P/D coordinate system of the limb can be read by cells fated to produce limbs of very distinct morphology. However, the parts of limbs, despite their suggestive reiterations or functional divisions, do not appear to be controlled by regulatory mechanisms that map directly onto morphology or could be used equally in different limbs. The limb-patterning system functions to pattern the limb as a whole, rather than individual genes encoding individual limb segments. This does not mean that modifications to this global patterning system do not underlie evolutionary diversification in limbs, but that limb segments, unlike body segments, are not patterned by any known mechanism that divides the limb into developmental units that directly correspond to obvious morphological units.

To help us resolve difficulties in defining biological homologues, we elaborate comparisons of limb development among arthropods, using the *Drosophila* model as a point of departure. Such a comparison is inherently difficult because morphogenesis in *Drosophila* is derived. Since *Drosophila* morphogenesis is geared to orchestrate the radical metamorphosis from larva to adult, many regulatory features in *Drosophila* show a decoupling in the production of larval and adult structures. Indeed, the early and late phases in limb patterning are a reflection of the temporal decoupling of segmentation and limb development found in metamorphic insects. In this case, the temporal decoupling depends upon the capacity for the same genes, i.e., *wg* and *dpp*, to function differentially and dissociably at two separate times, in two unique expression domains. This ability depends on modular elements in their promoters. For example, although both early segmentation and later wing development require the *wg* gene, there are *wg* mutant alleles that effect the wing exclusively and leave segmentation unaffected (Baker, 1988). As discussed earlier, many arthropods go through a more gradual morphogenesis and show no temporal decoupling between larval and adult structures. Therefore, it is difficult to know what aspects of patterning in *Drosophila* should be widely applicable. Would we expect the limb patterning pathway that functions in the *Drosophila* embryo or larvae (or neither) to be used in other arthropods?

A COMPARATIVE ANALYSIS OF APPENDAGE DEVELOPMENT

In this section, we compare the model of early limb positioning and later outgrowth in *Drosophila* to other arthropod species and ask whether this reflects a conserved, constrained patterning system for setting up all arthropod limbs, despite great morphological diversity. This analysis is necessarily confounded by the degree of morphological variation within arthropods.   Therefore, we subsequently discuss the uses and limitations of such molecular comparisons as applied to morphological variation.   Another important caveat is that the following discussion of the comparative aspects of development in arthropods is based on those species for which either an adequate description of limb morphogenesis or some comparative gene expression data exists. These species are few in number and highly divergent taxonomically. This inherent sampling error has the potential to highly skew our conclusions. Only data on additional species can rectify this problem.

REGULATION OF EARLY POSITIONING OF LIMBS: THE A/P CIRCUITRY IS WIDELY CONSERVED BUT THE D/V CIRCUITRY IS NOT

The developmental analysis of *Drosophila* led us to conclude that segments all have inherent information to position limb primordia. Do the limbs of other arthropods initiate development in a similar way? Two lines of evidence suggest that they do: conservation in genetic regulatory circuits and patterns of morphogenesis.   During morphogenesis, limb positioning in arthropods is consistent with the idea that a boundary acts to position limbs along the A/P axis. Limb buds arise in the middle of the segment, regardless of their subsequent shape and form (Dohle and Scholtz, 1988; Manton, 1928, 1934; Williams and Muller, 1996), and the relative timing at which the limb primordia are initially established appears conserved.   Molecular data corroborate the data from morphogenesis. Recall that A/P positioning of primordia in *Drosophila* depends on the interaction between *en* and *wg* and that this boundary is critical in activating *Dll* and thus promoting limb development.   *en* is expressed in a segmentally reiterated pattern and with a similar time course during development in every arthropod examined to date (Brown *et al.*, 1994;   Campbell and Caveney, 1989; Fleig, 1990; Patel *et al.*, 1989a, b; Scholtz *et al.*, 1993, 1994; Scholtz, 1995; Manzanares, 1993; Sommer and Tautz, 1991). This suggests it serves a homologous function. The other partner in the A/P positioning circuit, *wg*, has been less widely analyzed. The expected pattern is seen in several insects (Kraft and Jäckle, 1994; Nagy and Carroll, 1994), but, in the one crustacean examined, the *wg* ortholog displays a segmentally reiterated pattern only on the ventral body wall (*Triops*, Nulsen and Nagy, 1999). Nonetheless, it is predicted that this expression is positioned adjacent to *en* and functions homologously. Thus, some regulatory control of A/P positioning of limb primordia seems unambiguously conserved, given both the similarities in morphogenesis and the conservation of the regulatory circuit for A/P positioning.

However, this conservation does not extend to D/V positioning of primordia. Among arthropods, gross variability in adult limb structure, like the number and position of branches, exists along the D/V axis and there appears to be a corresponding variability in underlying developmental mechanisms. During morphogenesis, the position and extent of the limb primordia along the D/V axis is variable between species. In many arthropods, the limb buds develop as a small circular group of cells within a restricted region of the ventral body wall and most of the cells composing the ventral part of the segment become body wall proper. However, in branchiopod crustaceans like *Artemia* and *Triops*, virtually the entire ventral to ventrolateral region of each segment forms the limb primordia (Anderson, 1967; Benesch, 1969; Williams and Müller, 1996). Molecular data corroborate variability in D/V patterning. In the beetle *Tribolium castaneum*, the dynamics of embryonic *dpp* expression are quite different than in *Drosophila* (Sanchez-Salazar *et al.*, 1997) although the expression of *wg*, *Dll*, and *dpp* is consistent with these genes interacting to position limb primordia. However, in the development of branched limbs in branchiopods and malacostracans, *Dll* appears as multiple spots of expression along the D/V axis on the ventral body wall, rather than a single primordium observed in the development of uniramous insects (Panganiban *et al.*, 1995). This appearance of multiple sites of *Dll* expression cannot be explained as arising from a single *wg/dpp* interaction like that which regulates *Dll* in *Drosophila* (Williams, 1998; Nulsen and Nagy, 1999). Thus, based on patterns of morphogenesis and a presumed lack of a shared D/V network, only the A/P portion of the genetic circuit that positions limb primordia in *Drosophila* appears to be conserved.

LATER LIMB DEVELOPMENT IS NOT WIDELY CONSERVED

Because arthropod limbs are so diverse, comparisons of later limb development involve comparisons between structurally similar or dissimilar limbs. The main structural distinction we focus on here involves branching. Understanding mechanisms of branching is important because arthropod limbs show great structural diversity in the presence and arrangement of branches or outgrowths. Some regularities exist—insects have an unbranched walking leg, crustaceans typically have a two branched leg, chelicerates have unbranched walking legs and lobate gills—although only as generalities in the midst of great variation (Brusca and Brusca, 1990). Our ability to define homologues should rest, in part, on a detailed structural similarity between limbs; this may or may not be well established. For example, the inner branch, or endopod, in crustaceans is a multijointed rod-shaped branch used for walking and is often identified with the unbranched walking leg in insects (Snodgrass, 1935). However, because they often lack a central axial branch, multibranched, phyllopodous leaf-like legs in crustaceans have long been difficult to compare to uni- or biramous limbs (Hessler, 1982). Thus, identifying developmental criteria for comparing limb branches across taxa would be useful.

In later limb patterning in *Drosophila*, a global P/D patterning system is read by cells fated to form very different limbs. Is this patterning system found in the unbranched limbs in other insects? Dll expression was examined in the thoracic legs of a butterfly (Panganiban *et al.*, 1994) and cricket (Niwa *et al.*, 1997) and, in both cases, was found to be similar to the expression of Dll in *Drosophila*. This suggests that all insects—whether their legs developed from imaginal discs in metamorphosing insects or by a gradual outpocketing of the body wall as in crickets—share aspects of a leg P/D patterning system.

By definition, branched limbs have more than one point of distalization and outgrowth. Are all branches equivalent in terms of patterning, i.e., do all arise using the same P/D patterning circuit found in insect legs? Adult morphology suggests not, since branches typically originate from what appears to be the main axis of the limb. However, experimental perturbations to the developing leg discs in *Drosophila* produced forked structures superficially resembling branched limbs (Fig 4C; reviewed in Campbell and Tomlinson, 1995), leading to the expectation that naturally occurring branched limbs are patterned via reiteration of the P/D patterning module that controls outgrowth of the leg in *Drosophila* (reviewed in Williams and Nagy, 1996)

Does limb branching arise by simply reiterating a semiautonomous regulatory process that can be plugged in at any time in development and thus represent a developmentally constrained feature in arthropod limb evolution? Wg and Dpp together create a signaling circuit that activates *Dll* expression and is the major determinant of proximo-distal limb patterning in *Drosophila*. When Dll was first examined in crustaceans, species with both biramous and multiramous limbs showed Dll expression in all limb branches (Panganiban *et al.*, 1995). This seemed to support the hypothesis based on patterning the *Drosophila* leg that branches would be formed simply by using reiterations of the P/D leg patterning circuit. However, this conclusion was based on the expression of a single gene. Subsequent evidence from comparison of other genes yields a less straightforward model for limb branching. This is clearly illustrated by the *wg* expression pattern in the developing *Triops* limb (Fig. 5). The *Triops* limb has eight branches. The *wg-Dll* expression pattern in five of these branches is consistent with a simple reiteration of a *Drosophila*-like P/D gene network; yet the other three branches have *wg* expression patterns that imply P/D outgrowth and *Dll* activation can be regulated by other mechanisms (Nulsen and Nagy, 1999; Williams, 1998). This means that, at the simplest level, we cannot assign a kind of general identity to all limb branches. This is not to say that homologies cannot be drawn between branches but that developmental genetics cannot provide a rationale for inferences between broad taxonomic distances.

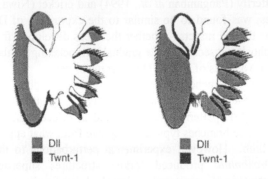

FIGURE 5 The development of branched limbs. Expression of *Dll* and *wg* in the multi-branched
limb of the crustacean, *Triops*. A. A predicted pattern of expression if the signals were expressed as
they are in the *Drosophila* limb. B. The actual pattern of expression of *Dll* and *wg* found in the
developing *Triops* limb. Note that *wg* expression is reversed from ventral to dorsal in two branches
and completely encompasses the margin of the dorsalmost branch. Similarly, *Dll* is not expressed in
the dorsalmost branch and is expressed more strongly along the margins of the branches than at the
tips. *Dll* is also found in the limb stem between two distal branches.

The benefits of comparing more closely related species within a
phylogenetic framework are illustrated by an analysis of limb branching in two
independently evolved lineages within crustaceans (Fig. 6; Williams, 1999).
Both branchiopods and phyllocarids (basal malacostracans) have multibranched,
phyllopodous limbs. However, limb morphogenesis in phyllocarids follows the
basic development of the biramous limb found in other malacostracans with the
late addition of a lateral branch onto a biramous limb bud. Thus, homologies
between the two main limb branches in phyllocarids and other malacostracans are
supported by developmental data. By contrast, branchiopod limb morphogenesis
shows no trace of a biramous limb bud. Instead, all the limb branches present in
the adult form at the earliest stage of limb morphogenesis, completely
obliterating the biramous pattern and providing no developmental support for

drawing homologies between branchiopod limb branches and the typical biramous crustacean limb (Fig. 6; Williams,1999). Thus, while limb branches are clear homologues in closely related taxa, there are not obvious shared developmental underpinnings responsible for producing branches in every type of arthropod limb. However, what does appear conserved is the earliest expression of Dll within limb buds. This expression could be set up using the same regulatory circuit found in *Drosophila* and could function to promote generic outgrowth of limb buds. If this were true then limbs as a whole would share a widespread patterning mechanism while the question would remain as to what patterning events control branching and other details of diverse limb morphologies.

Development of typical biramous limb

Development of multibranched malacostracan limb, *Nebalia*

Development of multibranched branchiopod limb; *Triops*

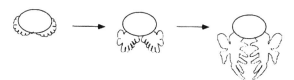

FIGURE 6 A comparison of morphogenesis in branched limbs in two independently evolved crustacean lineages. A. Morphogenesis in a typical biramous crustacean limb. Note the early bifurcation of the limb bud and the subsequent outgrowth and elaboration of the two branches. B. Morphogenesis in a phyllocarid (*Nebalia*), a basal malacostracan. Limbs pass through a biramous stage and then, late in morphogenesis, form an addition lateral limb. C. Morphogenesis in a branchiopod, *Triops*, in which no trace of the biramous limb development remains. Instead, all limb branches form early in morphogenesis. B and C represent distinct evolutionary routes for producing multibranched, phyllopodous limbs. Because limb morphogenesis in branchiopods has been so highly modified, homologies to limb branches in other species are difficult to discern.

SUMMARY

The establishment of A/P domains in the ventral body wall, which in *Drosophila* provides a boundary for positioning limb primordia, appears widely conserved among arthropods. Branching diversity arises ventro-medially or dorso-laterally along a regular A/P boundary. These branches do not appear to share any generic patterning system. Instead, comparisons of both morphogenesis and gene expression indicate numerous pathways produce branching. Since variation has not been documented in the regulation of a single P/D axis to promote generic limb outgrowth, this may also be widely conserved. However, the details of diverse morphologies—between different kinds of limbs—are subsequently controlled by mechanisms not yet understood. This means that although arthropod appendages share some deep similarities, or a basic individualization, the comparability of limb parts remains an open question. Based on developmental criteria, homologies may be found in certain cases but not others. Thus, genetic regulation of appendage patterning does not provide support for a view that diverse limbs have been constructed through evolution from a few, simpler underlying building blocks. Although appendage primordia, positioned along the body using segmental A/P patterning coordinates, may be comparable modules, the relationships and homologies of subparts of the limbs are not, as yet, clarified by developmental genetic data, and all limb branches do not appear to arise from the iteration of a basic P/D patterning module.

USING COMPARATIVE PATTERNING MECHANISMS TO UNDERSTAND MORPHOLOGICAL VARIATION

In this section we further explore the uses of development for interpreting identity in arthropod limbs. In particular, we examine cases of using regulatory mechanisms as a means of establishing morphological homology; in the first case, the commonalties of limb patterning can be used as criteria to identify limb homologues; in the second case, molecular criteria from later limb development have been used unconvincingly to establish homologies between limb parts. We also discuss in more detail limb branches as reiterated structures and ask if all limb branches can be said to be equally individualized during development.

CATEGORIZING ALL BODY WALL EXTENSIONS IS DIFFICULT; CONSERVED A/P CIRCUITRY MAY PROVIDE AN ADDITIONAL CRITERION

The conservation of A/P positioning bears on the issue of comparing whole appendages, that is, which of the many body wall extensions in arthropods

The conservation of A/P positioning bears on the issue of comparing whole appendages, that is, which of the many body wall extensions in arthropods constitute appendages? Defining appendages is nontrivial because of the great morphological diversity in the extensions arising from the body wall in arthropods. It is easy to classify antennae, mouthparts, and most ventral trunk limbs as appendages—all are associated with a single segment and have independent, intrinsic musculature for control. However, many structures found on arthropods—gills, genitalia, labrum, cerci, larval insect prolegs, carapaces, dorsal spines, dorsal armor in malacostracans—do not exactly fit a straightforward definition of an appendage in terms of position or morphology and yet on occasion have been argued to be appendage homologues (Sharov, 1966; Rempel, 1975; Scholtz, 1998; Popadic *et al.*, 1998). In addition, some adult structures that seem unambiguous appendages become problematic in light of their developmental history (see discussion of wing primordia, later). In view of these ambiguities, one developmental criterion for defining appendages, in addition to descriptive embryology, might be those extensions of the body wall that use the conserved circuitry for positioning relative to the A/P axis of the segment. Thus, genitalia, cerci, prolegs, and the labrum[3] may be limb homologues while carapaces, dorsal spines, and armor would not. This developmental criterion could add novel characters to long-standing morphological debates. In crustaceans, controversy exists as to whether the terminal region, which is generally called the telson and bears paired outgrowths (furcal rami), is in fact an anal somite with true appendages (Bowman, 1971; Schminke 1976). If the latter were true then we would predict that the furcal rami would not simply express *Dll* but *Dll* would be activated by the *wg* /*en* boundary as in all well-defined segments.

DELINEATING BOUNDARIES OF LIMBS AND PARTS OF LIMBS

Another difficult issue raised by comparing late limb development in structurally diverse limbs is locating corresponding morphological boundaries in appendages of different species. Identifying such boundaries between species has been attempted recently using genetic markers and illustrates the pitfalls of using gene markers of later appendage development to establish identity.

The first difficulty in identifying morphological boundaries across limbs is in defining what constitutes body wall and what constitutes limb. The difficulties may arise because of lack of morphological landmarks. For example, branchiopod or cephalocarid crustacean limbs are broadly inserted into the body

---

[3] In crustaceans, contradictory data exist as to the status of the labrum as an appendage. Popadic *et al.* (1998) report *Distalless* in the labrum of a variety of arthropods, including crustaceans. However, Schlotz (1995) found no *engrailed* expression in the labrum of four species of malacostracan crustaceans.

wall and lack joints or other external features that might define a boundary between limb and body wall. In addition, delineation may be problematic during development if the adult location of an outgrowth differs from its developmental origin. Decapod crustaceans have a number of gills in the thoracic region located around the dorsal leg margin. Although these are designated by their attachment point in the adult, e.g., body wall or leg (pleurobranch or podobranch), this does not strictly correspond to the developmental position of the outgrowths; pleurobranchs can originate on the leg but "migrate" during the course of development to a position on the body wall (Hong, 1988). Similarly, the fly wing primordium originates from the same cluster of cells as the fly leg primordium, but the wing primordium migrates dorsally, potentially obscuring its developmental and evolutionary origin (Cohen *et al.*, 1993). On an evolutionary time scale, this ambiguity of limb vs. body wall can be even more problematic since it is common for hypotheses concerning the evolution of arthropod appendages to assume transformation of ancestral leg parts into descendant body wall parts or visa versa. For example, currently there are two opposing theories concerning the evolutionary origin of the wing. In one scenario, the wing is a novel morphological structure that arose from small dorsal flaps of body wall present on primitive wingless insects (Snodgrass, 1935). In another, the wing is homologous to a dorsal branch, or exite, of the leg of an ancestral arthropod that has migrated to a more dorsal position in the insects (Wigglesworth, 1973; Kukalova-Peck, 1978).

In *Drosophila*, each wing imaginal disc is part of a single primordium with a leg disc in the second and third thoracic segments of the fly, identified by the expression of *Dll*. A cluster of cells from each primordium will migrate dorsally, repress *Dll* expression, and shortly thereafter begin expressing wing-specific genes. Several of these genes, including *apterous* and *nubbin*, have unique wing-specific functions (Cohen *et al.*, 1992; Cifuentes and Garcia-Bellido, 1997). Thus, in the adult fly, there are two separate appendages that have arisen from one primordium separated by extensive intervening body wall, each with autonomous late development regulated by separate genetic circuits. In an analysis of wing-specific genes in crustaceans, Averof and Cohen (1997) found *apterous* and *nubbin* expressed within dorsal limb branches in a branchiopod and a malacostracan. They claim that these genes identify homologous structures across taxa and that the dorsal limb branches (epipods) in crustaceans are homologous to the wing of a fly. From a topological viewpoint, this idea gains general support from the morphogenesis of the wing primordia from the dorsal region of the leg primordia. On the other hand, topological comparisons of corresponding adult structures between crustaceans and flies are not at all obvious. In *Drosophila*, apterous specifies dorsal fate in the wing and is expressed only in the dorsal surface of the wing. The structural differences between the crustacean limb branches and the fly wing are so great that the corresponding surface cannot be identified in crustaceans. In *Artemia*, the epipod, instead of being flattened dorsoventrally as is the wing in *Drosophila*, is flattened anteroposteriorly. Thus, the dorsal surface, strictly speaking, is merely the ridge of cells where the front and back epithelial sheets are joined. However, in *Artemia*, apterous protein is expressed throughout the entire branch. Despite the extreme geometric dissimilarity, does the whole epipod thus represent the dorsal wing surface? If so, where is the ventral surface? One general problem in using regulatory genes to identify corresponding adult structures is simply that the modifications to development that permit evolutionary change are not in any way guaranteed to somehow recapitulate the course of that evolutionary change directly (see, for example, Alberch and Gale, 1985). Given its spatial expression and specific regulatory role in *Drosophila*, it is difficult with the data at hand to formulate a hypothesis as to the degree of homology that the shared *apterous* expression might represent.

The difficulties of defining boundaries within the limb are further illustrated by the hypothesis that all arthropod limbs have two basic regions: a proximal and a distal region. These regions are supposed to reflect the evolutionary history of arthropods limbs. This hypothesis originated with Snodgrass (1935) and was based on both comparative morphology and assumptions about the mechanics of locomotion. He speculated that the arthropod limb was originally an unjointed, unbranched limb and evolved joints via a series of small, functionally relevant steps. The first step was the formation of a joint at the body wall allowing the limb to swing forward and back. The next step formed a joint which subdivided the limb into a proximal domain (coxapodite) and a distal domain (telopodite) and allowed the telopodite to move up and down. The final step was the formation of a "knee" joint more distally allowing the distal part of the telopodite to project vertically toward the ground. From this origin, both the coxapodite and telopodite underwent further subdivisions via jointing and fusions in a lineage dependent fashion.[4]

Studies of regulation of limb development in *Drosophila* have explicitly claimed to support Snodgrass' hypothesis. It has been argued that the first developmental subdivision during the development of the *Drosophila* leg corresponds to this original evolutionary subdivision into coxopodite and telopodite. The first targets of the Wg-Dpp signaling center that patterns the leg divide the leg into proximal and distal domains, marked by *exd* and *Dll* expression (Gonzalez-Crespo and Morata, 1996). These domains are further subdivided by later regulatory events (Lecuit and Cohen, 1997; Abu-Shaar and Mann, 1998). Thus, the sequence of gene regulatory events during *Drosophila* limb development might be seen as recapitulating evolutionary modifications of limbs. If this were the case, we would expect to find these mutually exclusive

---

[4] Contrary to published reports (Cohen and Jürgens, 1989; Gonzalez-Crespo and Morata, 1996), Snodgrass (1935) does not claim that the coxapodite is evolutionarily derived from body wall and the telopodite is a novel structure. He claims that both are part of the primitive limb but that secondarily the coxapodite can be subdivided into subcoxa—next to the body wall—and coxa; the subcoxa may be subsequently incorporated into the pleural body wall.

domains in all limbs. As discussed earlier, *Dll* is expressed in a distal domain in all crustacean limbs examined. While it has been reported that *Artemia Dll* and *exd* are expressed in mutually exclusive domains within the limb (Gonzalez-Crespo and Morata, 1996), the domains do not appear as well-defined proximal and distal domains and are in fact overlapping (Nulsen and Nagy, pers. obs.). This fact is perhaps not surprising given the structural and morphogenetic differences in a branchiopod leg and a fly leg (see Figs. 4 and 5 and discussion earlier). Again this is a case where genetic markers remain ambiguous due, in part, to the ease with which one can imagine that some kind of co-option of genes has occurred. This is especially true in the case of *Dll* since *Dll* is highly promiscuous and appears in extensions of the body wall throughout the metazoa, including echinoderm tube feet, ascidian ampullae, and annelid parapodia. It seems much more likely to be performing some very general shared function, e.g., regulating cellular proliferation, throughout disparate organisms, than defining the distal domains of homologous structures.

SERIALLY REITERATED STRUCTURES, LIKE ARTHROPOD LIMB BRANCHES, POSE SPECIAL PROBLEMS OF HOMOLOGY

The comparison of patterning mechanisms yielded no support for the idea that branches are modules. Indeed, because in part of their iterated nature they can be subject to special problems of homology. For example, drawing homologies between limbs with different numbers of branches can be inherently problematic as has been demonstrated in assigning homology to different digits in the vertebrate limb. The difficulty in both cases arises from assigning and comparing individuality of elements within sets that contain different numbers. This question of a stereotyped individuality of reiterated elements has been explicitly linked to developmental processes within vertebrate limbs. Differing perspectives address whether homology can be claimed only at the level of the whole limb or down to parts of the limb, i.e., the digits themselves. In the structuralist perspective, structures that result from global patterning mechanisms (e.g., digits) lack individuality since variation in their final form results from variation in the initial conditions of the global patterning mechanisms (Goodwin and Trainor, 1983). Parts generated in this manner simply represent different solutions to a common set of generative rules and thus do not have the attribute of individual homology. However, a similar emphasis on developmental mechanisms led Shubin and Alberch (1986) to claim that homologies could be found between digits since limb development displays repeatable chondrogenic interactions that produce identifiable local patterns within the limb. In a more complex view of digit identity, Wagner and Gauthier (1999) argue that the identity of developmental condensations vs. adult digits can undergo a frameshift during development, i.e., the individuality of digits defined by early chondrogenesis in the bird (digits 2-3-4) changes during development such that the individuality found in the adult is shifted (digits 1-2-3). This frameshift illustrates the flexibility of the developmental system and the fact that identity is not established at a single time point.

While much less is known about the comparable tissue level dynamics of limb morphogenesis in arthropods, one can nevertheless ask whether all arthropod limbs display repeatable features of morphogenesis that permit assigning identity to particular limb branches. Primitively and in most extant arthropods, limb morphogenesis begins as a simple outpocketing of the segmental body wall. Schram (1986) suggested that all crustacean limbs subsequently pass through a biramous stage during early morphogenesis. While this indeed may be true for many species of crustaceans, the analysis of morphogenesis in branchiopod crustaceans discussed earlier demonstrates that there is no developmental stage that is similar to the two-branched limb bud. Instead, all limb branches arise simultaneously by subdivision of the early limb bud, and early alterations to limb morphogenesis are responsible for the formation of branchiopod multibranched limbs. The difficulties in drawing homologies between branchiopod limbs and other branched limbs may be due to a lack of stereotyped individuality between their branches vs. branches found in the biramous limbs of other crustacean species (Williams, 1999). Thus, as with vertebrate digits, arthropod limb branches can elude straightforward assignments of homology simply because they are serially repeated structures (see general discussions in Bateson, 1894; Van Valen, 1993; Wagner, 1989b).

SUMMARY

Despite the desire to have molecular markers for morphological homology, arthropod limbs are not clearly characterized in such a fashion. In general, it is clear that using expression patterns of single genes is at best a weak criterion of homology (see also, Bolker and Raff, 1996; Jockusch and Nagy, 1996). Conserved gene circuits or networks, in which a number of regulatory interactions are maintained, provide a stronger criterion than single genes that an integrated developmental process has been conserved. Thus, when expression patterns of a number of interacting genes appear in the same developmental sequence in many different species, conservation is more likely. The use of the *wg, en* boundary along the A/P axis to activate *Dll* and establish limb primordia is an example of such conservation. However, as with single gene comparisons, the actual functional role of the genes is simply inferred in nonmodel systems. Given the pitfalls of interpreting expression patterns in animals where gene function remains unknown, homology of genes is best viewed as specific to the genetic level and not, by itself, as a sign of homology on a morphological level. A more practical and ultimately fruitful approach is exemplified by recent work on the detailed expression of *Ubx* in which different levels of expression are responsible for differences in trichrome patterns on the second thoracic legs in three closely related *Drosophila* species (Stern, 1998).

Furthermore, general problems exist in using genetic markers to define homologous structures. One particularly clear way to visualize such pitfalls is to ask how the presence of certain genes could be unambiguously interpreted to signify identity of adult structure vs. simply co-option into a novel pathway. This is a common conceptual difficulty when trying to interpret the developmental basis for evolutionary change. If an evolutionary transformation is posed, e.g., fusion of leg segments to the body wall, it is necessary to ask: is there some way that the novel body wall in descendants would retain, either through morphological or genetic markers, a recognizable "limb identity." The developmental basis for an evolutionary fusion might involve the respecification of a field of cells so that a population of cells in the ancestor becomes fated to be a body wall in the descendent. In this kind of early differential allocation of cells, one would not expect to see either genetic or morphological remnants of leg character in the body wall of the descendent. In this case, given their complexity, the behaviors of cells and tissues during morphogenesis may provide a better guide to homology than underlying the regulatory information. It is always possible that orthologous genes have, in these cases, simply been co-opted into distinct networks. This is particularly likely when the genes in question do not provide a very clear correspondence with resulting adult morphology. The redeployment of conserved regulatory genes via evolutionary tinkering is simply too widespread to assume conservation of gene function, particularly across relative large phylogenetic distances (Duboule and Wilkins, 1998).

Finally, reiterated structures like limb branches can elude simple comparison since such structures may or may not be equally individualized in different species. Indeed, it is important to remember that limbs themselves are reiterated along the body axis. Whereas to this point we have compared limbs as if they existed in isolation—a fly leg, an *Artemia* leg—we now turn to segmentation to consider how this may affect our comparative analysis.

SEGMENTATION IN ARTHROPODS: AXIAL PATTERNING AND LIMB DIVERSIFICATION

Why is the axial patterning of segments important to understanding limb diversification? First, as described earlier, positioning of limb primordia occurs within the A/P framework of a segment and thus limb and segment development are inextricably linked. Second, limbs never occur in isolation. Like the segments that bear them, limbs develop as reiterated structures along the body axis. Therefore we briefly discuss patterns of segmentation in arthropods and point out some specific ways that differences in segmental character may impact evolutionary diversification of limbs.

Although variation in patterns of segmentation along the body axis in arthropods is enormously diverse, it can be characterized by two extremes: homonomy, in which segment and appendage character is basically similar along the A/P axis, and heteronomy, in which differences from one segment to the next can be sharp and abrupt. Extant groups that display homonomy include branchiopods, cephalocarids, and remipedes, among crustaceans, and millipedes and centipedes, among myriapods. Although homonomous groups show relatively little morphological differentiation among limbs, such differentiation does exist as graded changes in morphology along the A/P axis. *Drosophila* and other metamorphic insects are good examples of heteronomous taxa. In these groups, not only can adjacent segments differ widely in character from one another but also segments of similar character are clustered together into functional units called tagma (see Fig. 1).

An increase in axial diversity is commonly held to be one of the prominent trends in arthropod evolution (Cisne, 1974; Wills *et al.*, 1997). The traditional view is that the most ancestral arthropods exhibited a high degree of homonomy and that adaptation and specialization to particular ecological niches led to axial diversification of limbs and segments and, thus, heteronomy (Manton, 1977). In other words, heteronomy evolved from homonomy. In simplified terms, this has been construed to mean that the axially differentiated segments of heteronomy evolved from a series of identical segments. This idea gained credence from studies of the development of segments in *Drosophila* in which a series of reiterated identical segmental primordia diverge through the differential activity of Hox genes that appear to act as "selector" genes to specify segment character. Thus, segments have been assumed to be modules, i.e., semiautonomous, identical building blocks, within which character can differentiate during development or diversify during evolution. We question this assumption later. We propose that not all arthropod segments exhibit the same degree of developmental modularity. In our view, segment modularity is simply another character that has varied during the course of arthropod evolution.

The idea that segments are developmental modules rests on oversimplified assumptions concerning both Hox gene function and the processes of segmentation within arthropods. Far from being simple selectors of segment character, it is known that, even within *Drosophila*, the spatial and temporal details of Hox expression within a segment are crucial in determining the exact role they will play in segment character (Castelli-Gair and Akam, 1995). It has even been argued recently that the entire perspective of Hox genes as selector genes is a misconception (Akam, 1998; Castelli-Gair, 1998). Although changes in Hox gene regulation will likely have influenced many aspects of arthropod evolution, differences in axial patterning are not likely to be explained by such changes (see also Budd, 1999). Similarly, the segmental primordia in different species—upon which Hox genes ultimately act—may not be directly comparable. As described previously, patterns of segment morphogenesis are varied among arthropods. *Drosophila* establishes all its body segments by subdivision of the cells in the posterior (e.g., Dohle and Scholtz, 1988) or by unknown mechanisms. In these cases, the timing and mechanisms by which segment character is established are not yet known. In general, homonomous groups show sequential segmentation. Developmental differences between segmentation in homonomous and heteronomous arthropods may be significant and could play a role in the subsequent differentiation of limbs.

The developmental coupling of limbs and segments differs in heteronomous vs. homonomous groups. In general, limb differentiation appears to be more readily decoupled from segmentation in heteronomous taxa. In these groups, differences in limb morphology often arise early in limb development soon after the limb primordia are formed, suggesting an early dissociation between segment-forming and limb-forming processes. In the extreme case of metamorphic taxa, this is manifest as a complete temporal dissociation in the timing of limb and segment development. By contrast, in homonomous groups, the early stages of limb development are often quite similar for the whole series of limbs. Limb differentiation occurs late in development, e.g., via the progressive reduction or amplification of limb branches. For example, in the branchiopod *Triops*, a long series of similar thoracic appendages develop early in larval life (Williams and Müller, 1996). Later in development, the first thoracic appendage gradually but dramatically changes its shape by virtually losing one branch and elaborating another into a long sensory process (Williams, pers. obs.). These differences in the timing of limb differentiation suggest that homonomous and heteronomous groups differ fundamentally in the regulatory circuitry that links segments and limbs. This may result from the lack of temporal dissociation in homonomous groups between the gradual and sequential addition of segments and the progressive development of limbs on those segments. Limb morphogenesis may be more constrained along the body axis in homonomously segmented arthropods compared to those groups where limb morphogenesis is temporally dissociated from segment development. Indeed, specification of both limbs and segments may be more autonomous along the A/P body axis in some heteronomous groups than in homonomous groups, suggesting that modularity itself evolves. This general idea is reflected even within *Drosophila* tagma by contrasting the relatively homonomous abdominal segments to the thorax. The thoracic segments are highly differentiated and also more developmentally partitioned or modular with respect to one another (as evidenced by mutations that affect the character of any one thoracic segment). By contrast, mutations affecting abdominal segment character tend to disrupt clusters of segments anteriorly or posteriorly and not change the identity of a single segment, in particular in the middle of the abdomen. This kind of linkage, we suggest, may be more profound in groups with completely homonomous segmentation. For example, in the homonomous cephalocarid crustaceans, the trunk limbs, although basically similar in each segment, show gradations in morphology from anterior to posterior. This consists of sequential reduction in medial elements and, finally, in the most posterior trunk limb, a loss of the entire medial branch to posterior. This consists of sequential reduction in medial elements and, finally, in the most posterior trunk limb, a loss of the entire medial branch (endopod; Sanders, 1957). This suggests an axial differentiation in limbs arising simply from differential growth.

Another pattern that suggests the interdependence of axial neighbors in homonomous groups occurs in some branchiopod crustaceans. Notostracans exhibit a mismatch between dorsal and ventral segmentation. A canonical segment has, among other things, only a single ganglion and a single pair of appendages. However, within notostracans, posterior to the eleventh trunk segment, there is a mismatch between dorsally reiterated segmental structures and ventrally reiterated segmental structures. Ventral structures are more numerous and, as a result, four, five, or more ventral pairs of appendages are associated with each apparent dorsal segment (Linder, 1952).

What impact does the view of variability of segment modularity have on our understanding of limbs as characters and of the diversification of limbs within arthropods? In taxa whose segments show a high degree of modularity, axial diversification of limb types may result from changes to *both* early limb morphogenesis (e.g., in the D/V allocation of primordia) and later limb outgrowth. This is due to the temporal dissociation between limb and segment development discussed earlier. By contrast, groups without a high degree of segment modularity, which we argue are represented by some extant homonomous taxa, may generate axial diversity in limbs through late changes of differential growth. Furthermore, we hypothesize that although early changes to morphogenesis could occur in these groups, such changes would necessarily affect the entire axial limb series. Because of these differences in developmental constraints, we predict that the classes of variants in limb morphology produced in each case would be different.

AXIAL DIFFERENTIATION, MODULARITY, AND INTEGRATION

Homonomy is simply a distinct expression of axial patterning from heteronomy. Whether homonomy should be considered primitive, however, is not clear. The earliest arthropod fossils show both heteronomous and homonomous body plans, and there is certainly no evidence that an arthropod with a series of identical segments ever existed. Patterns of axial diversification, both homonomous and heteronomous, have evolved. Segment modularity has evolved separately and, in some cases, modularity and axial diversification may have become linked. For example, the evolution of segment modularity may be a preadaptation for the temporal dissociation of limb and segment development. Such temporal dissociation permits the evolution of highly divergent limb morphologies even on adjacent segments. However, modularity in segmentation is not a causal prerequisite to heteronomy of axial patterns. Thus, we treat the relationship between segmentation and patterns of axial diversity as open to empirical exploration (Van Valen, 1993). To evaluate transformations in the axial body plan of arthropods involves considering three *separate* issues: the nature and degree of axial differentiation, the degree of modularity of the segments, and the integration of the processes that generate differentiation and modularity. In no arthropod are all three issues fully addressed. Even in *Drosophila*, we know relatively little about mechanisms of integration because developmental studies rarely take this perspective.

How does integration play a role in the evolution of modularity? We suggest that homonomous arthropods exhibit tightly integrated limb and segment development, constraining early limb development to be similar along the body axis. Axial change of limb morphology is restricted to later limb development or to very graded axial changes. A good example of the latter is the progressive reduction and loss of one of the limb branches in the posterior segments of cephalocarids mentioned earlier. Such axial change might form the first selective basis for differences along the body, leading to a selective pressure to decouple limbs and segments both spatially and temporally and, thus, the potential for tagmosis. However, any such changes that arose would still have to produce functionally coherent organisms. This is permitted by the fact that mechanisms of integration coexist with mechanisms of reiteration.

Such coupling clearly must be a part of the development of any reiterated structures. Even *Drosophila*, which appears to have modular segments, displays various mechanisms that ensure the integration of the developmental pattern *as a whole*. For example, the maternal effect gene, *bicoid*, acts in a concentration-dependent manner to set up the A/P patterning and ultimately segments within the embryo. Experimentally changing the bicoid dosage leads to alterations in the cascade of regulatory genes that position the segments such that the anterior segments are abnormally expanded (Fröhnhoffer and Nüsslein-Volhard, 1986; Berleth *et al.*, 1988). However, normal animals are nonetheless produced as a result of increased cell death in the expanded anterior of the embryo and decreased cell death in the contracted posterior (Namba *et al.*, 1997). Thus, embryos are able to regulate such that the overall pattern of normal segments is established.

Understanding developmental integration is likely crucial in interpreting morphological transformations in evolution. In analyzing morphogenesis, for example, one could ask by what means are transformations in limbs permitted by features of the developmental system that are flexible instead of hardwired in the face of perturbations? For example, it is known in *Drosophila* and other insects that muscles and neural elements follow the lead of the epidermis to establish a coordinated limb pattern. Instead of being hardwired, muscle and nerve precursors use landmarks in the ectoderm as guideposts for outgrowth as well as a process of reciprocal feedback to establish correct patterning. This system can be robust to change since a modification in patterning the ectoderm would not require a coordinated modification to other tissue to persist as a potential variant. Whether this behavior, known mainly from *Drosophila* and crickets, holds true throughout the arthropods is unexplored.   More importantly, whether this behavior is the result of a shared set of underlying mechanisms is also unknown.

SUMMARY

We argue that the assumption that arthropods are modular in their basic character is not clearly supported outside the model system, *Drosophila melanogaster.* A global coordinate system exists in *Drosophila* to position all limb primordia along the body. This appears to be partially conserved among arthropods with the conservation in A/P positioning perhaps reflecting the deep link between segment and limb patterning. Although many of the same individual genes are shared between *Drosophila* and other arthropods, mechanisms controlling segmentation and later limb development remain unknown. Thus, it is difficult to assess the degree of modularity other arthropod possess. Other aspects of development, in extant homonomous groups in particular, suggest that their segmentation does not have an equivalently modular basis. This points to the need to carefully distinguish between present-day modularity—like that we find in the well-studied model system—and the assumption that homologous morphologies developed via homologous modular mechanisms in the past. A difficulty in using the notion of modularity in an evolutionary analysis is that even if we uncover evidence of modularity in a highly differentiated model system, there is no reason to believe that feature of its development is not also highly specialized. The degree of modularity can be an axis of evolutionary change permitting new types of specialization.

Finally, we stress the importance of viewing characters like limbs within the context of the whole body axis. This has led to the view that homonomy and heteronomy may be expressions of differential modularity. Furthermore, we hypothesize that the potential for modularity to evolve rests on another feature fundamental to arthropod axial organization: integration. Whether modular in nature or not, arthropods are manifestly made of reiterated structures along the body axis. In order to successfully generate behaviors these structures have to be coordinated into a functioning whole.

ACKNOWLEDGMENTS

We thank K. Dunlap, T. Hansen, E. Jockusch, and G. Wagner for critical readings of the text.

LITERATURE CITED

Abu-Shaar, M., and Mann, R. S. (1998). Generation of multiple antagonistic domains along the proximodistal axis during *Drosophila* leg development. *Development* **125**:3821-3830.

Akam, M. (1987). The molecular basis for metameric pattern in the *Drosophila* embryo. *Development* **101**:1-22.

Akam, M. (1998). Hox genes, homeosis, and the evolution of segment identity: no need for hopeless monsters. *Int. J. Dev. Biol.* **42**:445-451.

Alberch, P., and Gale, E. (1985). A developmental analysis of an evolutionary trend: digital reduction in amphibians. *Evolution* **39**:8-23.

Anderson, D. T. (1967). Larval development and segment formation in the branchiopod crustaceans *Limnadia stanleyana* King (Conchostraca) and *Artemia salina* (L.) (Anostraca). *Aust. J. Zool.* **15**:47-91.

Anderson, D. T. (1972). The development of holometabolous insects. *In* "Developmental Systems: Insects" (S. Counce and C. H. Waddington, eds.), pp. 165-242. Academic Press, New York.

Anderson, D. T. (1973). "Embryology and Phylogeny in Annelids and Arthropods." Pergamon Press, Oxford.

Anderson, D. T. (1982). Embryology. *In* "The Biology of Crustacea: Embryology, Morphology, and Genetics" (L. G. Abele, ed.), Vol. 2, pp. 1-42. Academic Press, New York.

Averof, M., and Cohen, S. (1997). Evolutionary origin of insect wings from ancestral gills. *Nature* **385**:627-630.

Baker, N. (1988). Embryonic and imaginal requirements for *wingless*, a segment-polarity gene in *Drosophila*. *Dev. Biol.* **125**:96-108.

Bastiani, M. J., de Couet, H. G., Quinn, J. M., Karlstrom, R. O., Kotrla, K., Goodman, C. S., and Ball, E. E. (1992). Position-specific expression of the annulin protein during grasshopper embryogenesis. *Dev. Biol.* **154**:129-42.

Bateson, W. (1894). "Materials for the Study of Variation." McMillan and Co., London.

Benesch, R. (1969). Zur Ontogenie und Morphologie von *Artemia salina* L. *Zool.Jahrb (Anat. Ontog. Tiere)* **86**:307-458.

Berleth, T. M., Burri, G., Thoma, D., Bopp, S., Richstein, S., Frigerio, G., Nolland, M., and Nusslein-Volhard, C. (1988). The role of localization of *bicoid* RNA in organizing the anterior pattern of the *Drosophila* embryo. *EMBO (Eur. Mol. Biol. Organ.) J.* **7**:1749-1756.

Biggin, M. D., and McGinnis, W. (1997). Regulation of segmentation and segmental identity by *Drosophila* homeoproteins: the role of DNA binding in functional activity and specificity. *Development* **124**:4425-4433.

Blair, S. S. (1999). *Drosophila* imaginal disc development: patterning the adult fly. *In* "Development-Genetics, Epigenetics and Environmental Regulation" (V. E. A. Russo, D. Cove, L. Edgar, R. Jaenisch, and F. Salamini, eds.). Springer, Heidelberg.

Bolker, J. A., and Raff, R. A. (1996). Developmental genetics and traditional homology. *Bioessays* **18**:489-494.

Bowman, T. E. (1971). The case of the nonubiquitous telson and the fraudulent furca. *Crustaceana* **21**:165-175.

Brook, W. J., Diza-Benjumea, F. J., and Cohen, S. M. (1998). Organizing spatial pattern in limb development. *Annu. Rev. Cell. Dev. Biol.* **12**:161-80.

Brown, S. J., Patel, N. H., *et al.* (1994). Embryonic expresion of the single *Tribolium engrailed* homolog. *Dev. Genet.* **15**:7-18.

Brusca, R. C., and Brusca, G. J. (1990). "Invertebrates." Sinauer Associates, New York.

Budd, G. E. (1999). Does evolution in body patterning genes drive morphological change—or vise versa. *Bioessays* **21**:326-332.

Campbell, G., and Tomlinson, A. (1995). Initiation of proximodistal axis in insect legs. *Development* **121**:619-628.

Campbell, G., Weaver, T., and Tomlinson, A. (1993). Axis specification in the developing *Drosophila* appendage: the role of *wingless*, *dpp* and the homeobox gene *aristaless*. *Cell* **74**:1113-1123.

Campbell, G. L., and Caveney, S. (1989). *engrailed* gene expression in the abdominal segment of *Oncopeltus*: gradients and cell states in the insect segment. *Development* **106**:727-737.

Carroll, S. B. (1995). Homeotic genes and the evolution of arthropods and chordates. *Nature* **376**:479-485.

Castelli-Gair, J. (1998). Implications of the spatial and temporal regulation of Hox genes on development and evolution. *Int. J. Dev. Biol.* **42**:437-444.

Castelli-Gair, J., and Akam, M. (1995). How the Hox gene *Ultrabithorax* specifies two different segments: the significance of spatial and temporal regulation within metameres. *Development* **121**:2973-2982.

Cifuentes, F. J., and Garcia-Bellido, A. (1997). Proximo-distal specification in the wing disc of Drosophila by the *nubbin* gene. *Proc. Natl. Acad. Sci. USA* **94**:11405-10.

Cisne, J. L. (1974). Evolution of the world fauna of aquatic free-living arthropods. *Evolution* **28**:337-366.

Cohen, B., Simcox, A. A., and Cohen, S. M. (1993). Allocation of the imaginal primordia in the *Drosophila* embryo. *Development* **117**:597-608.

Cohen, S. M. (1990). Specification of limb development in the *Drosophila* embryo by positional cues from the segmentation genes. *Nature* **343**:173-177.

Cohen, S. M. (1993). Imaginal disc development. *In* "The Development of *Drosophila melanogaster*" (M. Bate and A. Martinez-Arias, eds.), pp. 747-842. CSHL Press, New York.

Cohen, S. M., and Jürgens, G. (1989). Proximal-distal pattern formation in *Drosophila*: graded requirement for *Distal-less* gene activity during limb development. *Roux's Arch. Dev. Biol.* **198**:157-169.

de Celis, J. F., Tyler, D. M., de Celis, J., and Bray, S. J. (1998). Notch signalling mediates segmentation of the *Drosophila* leg. *Development* **125**:4617-4628.

Diaz-Benjumea, F. J., Cohen, B., and Cohen, S. M. (1994). Cell interations between compartments establishes the proximal-distal axis of *Drosophila* legs. *Nature* **372**:175-179.

Dohle, W., and Scholtz, G. (1988). Clonal analysis of the crustacean segment: the discordance between genealogical and segmental borders. *Development (Suppl.)* **104**:147-160.

Duboule, D., and Wilkins, A. S. (1998). The evolution of "bricolage." *Trends Genet.* **14**:54-59.

Evoy, W. H., and Ayers, J. (1982). Locomotion and control of limb movements. *In* "Neural Integration and Behavior" (D. C. Sandeman and H. L. Atwood, eds.), pp. 62-106. Academic Press, New York.

Fleig, R. (1990). *engrailed* expression and body segmentation in the honeybee *Apis mellifera*. *Roux's Arch. Dev. Biol.* **198**:467-473.

Fröhnhofer, H. G., and Nüsslein-Volhard, C. (1986). Organization of anterior pattern in the *Drosophila* embryo by the maternal gene *bicoid*. *Nature* **324**:120-125.

Gerhart, J., and Kirschner, M. (1997). "Cells, Embryos, and Evolution." Blackwell, Oxford.

Gilbert, S. (1997). "Developmental Biology." Sinauer Associates, Inc., Sunderland, MA.

Gonzalez-Crespo, S., and Morata, G. (1996). Genetic evidence for the subdivision of the arthropod limb into coxopodite and telopodite. *Development* **122**:3921-3928.

Goodwin, B. C., and Trainor, L. E. H. (1983). The ontogeny and phylogeny of the pentadactyl limb. *In* "Development and Evolution" (B. C. Goodwin, N. Holder, and C. C. Wylie, eds.), pp. 75-98. Cambridge University Press, Cambridge.

Gorfinkiel, N., Morata, G., and Guerrero, I. (1997). The homeobox gene *Distal-less* induces ventral appendage development in *Drosophla*. *Genes Dev.* **11**:2259-2271.

Goto, S., and Hayashi, S. (1997). Specification of the embryonic limb primordium by graded activity of Decapentaplegic. *Development* **124**:125-132.

Gurney, R. (1942). "Larvae of Decapod Crustacea." Ray Society, London.

Heath, H. (1924). The external development of certain phyllopods. *J. Morphol.* **38**:453-476.

Held, L. I. (1995). Axes, boundaries and coordinates: the ABCs of fly leg development. *Bioessays* **17**:721-732.

Hessler, R. H. (1982). Evolution within the crustacea. *In* "Biology of the Crustacea, Systematics, the Fossil Record and Biogeography" (L. Abele, ed.), Vol. 1, pp. 149-185. Academic Press, New York.

Hong, S. Y. (1988). Development of epipods and gills in some pagurids and brachyurans. *J. Nat. Hist.* **22**:1005-1040.

Ingham, P. (1988). The molecular genetics of embryonic pattern formation in *Drosophila*. *Nature* **335**:25-34.

Jockusch, E. L., and Nagy, L. M. (1997). Insect evolution: how did insect wings originate? *Curr. Biol.* **7**:R358- R361.

Kraft, R., and Jackle, H. (1994). *Drosophila* mode of metamerization in the embryogenesis of the lepidopteran insect *Manduca sexta*. *PNAS* **91**:6634-6638.

Kukalova-Peck, J. (1978). Origin and evolution of insect wings and their relation to metamorphosis, as documented by the fossil record. *J. Morphol.* **156**:53-126.

Laverak, M. S., Neil, D. M., and Robertson, R. M. (1977). Metachronal exopodite beating in the mysid *Praunus flexuosus*: a quantitative analysis. *Proc. R. Soc Lond. B* **198**:139-154.

Lawrence, P. A., and Morata, G. (1994). Homeobox genes: their function in *Drosophila* segmentation and pattern formation. *Cell* **78**:181-189.

Lawrence, P. A., and Struhl, G. (1996). Morphogens, compartments, and pattern: lessons from *Drosophila*? *Cell* **85**:951-961.

Lecuit, T., and Cohen, S. M. (1997). Proximal-distal axis formation in the *Drosophila* leg. *Nature* **388**:139-145.

Lewis, E. B. (1964). Genetic control and regulation of developmental pathways. *In* "The Chromosomes in Development" (M. Loke, ed.). Academic Press, New York.

Lewis, E. B. (1978). A gene complex controlling segmentation in *Drosophila*. *Nature* **276**:565-570.

Linder, F. (1952). Contributions to the morphology and taxonomy of the Branchiopoda Notostraca, with special reference to the North American species. *Proc. U. S. Natl. Mus.* **102**:1-69.

Manton, S. M. (1928). On the embryology of the mysid crustacean, *Hemimysis lamornae*. *Philos. Trans. R. Soc. Lond. B* **216**:363-463.

Manton, S. M. (1934). On the embryology of the crustacean *Nebalia bipes*. *Philos. Trans. R.. Soc. Lond. B* **223**:163-238.

Manton, S. M. (1977). "The Arthropoda." Clarendon Press, Oxford.

Manzanares, M., Marco, R., and Garesse. R. (1993). Genomic organization and the developmental pattern of expression of the *engrailed* gene from the brine shrimp *Artemia*. *Development* **118**:1209-1219.

Manzanares, M., Williams, T. A., Marco, R., and Garesse, R. (1996). Segmentation in the crustacean *Artemia* as revealed by *engrailed* protein distribution. *Roux's. Arch. Dev. Biol.* **205**:424-431.

Martinez-Arias, A. (1993). Larval epidermis of *Drosophila*. *In* "The Development of *Drosophila melanogaster*" (M. Bate and A. Martinez-Arias, eds.), pp. 517- 608. CSHL Press, New York.

McGinnis, W., and Krumlauf, R. (1992). Homeobox genes and axial patterning. *Cell* **68**:283-302.

McLaughlin, P. A. (1982). Comparative morphology of crustacean appendages. *In* "Embryology, Morphology, and Genetics" (L. G. Abele, ed.), pp. 197-256. Academic Press, New York.

Nagy, L. M. (1994). A glance posterior. *Curr. Biol.* **9**:811-814.

Nagy, L. M., and Carroll, S. (1994). Conservation of *wingless* patterning functions in the short-germ embryos of *Tribolium castaneum*. *Nature* **367**:460-463.

Namba, R., Pazdera, T. M., Cerrone, R. L., and Minden, J. S. (1997). *Drosophila* embryonic pattern repair: how embryos respond to *bicoid* dosage alteration. *Development* **124**:1393-1403.

Niwa, N., Saitoh, M., Ohuchi, H., Yoshioka, H., and Noji, S. (1997). Correlation between *Distal-less* expression patterns and structures of appendages in development of the two-spotted cricket, *Gryllus bimaculatus*. *Zool. Sci.* **14**:115-125.

Nulsen, C., and Nagy, L. M. (1999). The role of *wingless* in the development of multi-branched crustacean limbs. *Dev. Genes Evol.* **209**:340-348.

Panganiban, G., Nagy, L., and Carroll, S. (1994). The role of the *Distal-less* gene in the development and evolution of insect limbs. *Curr. Biol.* **4**:671-675.

Panganiban, G., Sebring, A., Nagy, L., and Carroll, S. (1995). The development of crustacean limbs and the evolution of arthropods. *Science* **270**:1363-1366.

Patel, N. H. (1994). Developmental evolution: insights from studies of insect segmentation. *Science* **266**:581-590.

Patel, N. H., Kornberg, T. B., and Goodman, C. S. (1989a). Expression of *engrailed* during segmentation in grasshopper and crayfish. *Development* **107**:201-212.

Patel, N. H., Martin-Blanco, E., Coleman, K. G., Poole, S. J., Ellis, M. C., Kornberg, T. B., and Goodman, C. S. (1989b). Expression of *engrailed* proteins in arthropods, annelids, and chordates. *Cell* **58**:955-968.

Popadic, A., Panganiban, G., Rusch, D., Shear, W. A., and Kaufman, T. C. (1998). Molecular evidence for the gnathobasic derivation of arthropod mandibles and for the appendicular origin of the labrum and other structures. *Dev. Genes Evol.* **208**:142-150.

Raff, R. (1996). "The Shape of Life: Genes, Development, and the Evolution of Animal Form." University of Chicago Press, Chicago.

Raff, R. A., and Kaufmann, T. C. (1983). "Embyros, Genes, and Evolution." Macmillan, New York.

Raz, E., and Shilo, B. (1993). Establishment of ventral cell fates in the *Drosophila* embryonic ectoderm requires DER, the EGF receptor homolog. *Genes Dev.* **7**:1937-1948.

Reese, E. S. (1983). Evolution, neuroethology, and behavioral adaptations of crustacean appendages. *In* "Studies in Adaptation, the Behavior of Higher Crustacea" (S. Rebach and D. W. Dunham, eds.), pp. 57-84. John Wiley and Sons, New York.

Rempel, J. G. (1975). The evolution of the insect head: the endless dispute. *Quaest Entomol.* **11**:7-25.

Rieckhof, G. E., Casares, F., Ryoo, H. D., Abu-Shaar, M., and Mann, R. S. (1997). Nuclear translocation of *extradenticle* requires *homothorax*, which encodes an *extradenticle*-related homeodomain protein. *Cell* **91**:171-183.

Sanchez-Salazar, J., Pletcher, M. T., Bennett, R. L., Dandamudi, T. J., Denell, R. E., and Doctor, J. S. (1996). The *Tribolium decapentaplegic* gene is similar in sequence, structure, and expression to the *Drosophila dpp* gene. *Dev. Genes Evol.* **206**:237-246.

Sander, K. (1976). Specification of the basic body pattern in insect embryogenesis. *Adv. Insect Physiol.* **12**:125-238.

Sanders, H. L. (1957). The Cephalocarida and crustacean phylogeny. *Syst. Zool.* **6**:112-129.

Schminke, H. K. (1976). The ubiquitous telson and the deceptive furca. *Crustaceana* **30**:292-300.

Schneiderman, H. A., and Bryant, P. (1971). Genetic analysis of developmental mechanisms in *Drosophila*. *Nature* **234**:187-194.

Scholtz, G. (1995). Expression of the *engrailed* gene reveals nine putative segment-anlagen in the embryonic pleon of the freshwater crayfish *Cherax destructor* (Crustacea, Malacostraca, Decapoda). *Biol. Bull.* **188**:157-165.

Scholtz, G. (1998). Cleavage, germ band formation and head segmentation: the ground pattern of the Euarthropoda. *In* "Arthropod Relationships" (R. A. Fortey and R. H. Thomas, eds.), pp. 317-332. Chapman and Hall, London.

Scholtz, G., Dohle, W., Sandeman, R. E., and Richter, S. (1993). Expression of *engrailed* can be lost and regained in cells of one clone in crustacean embryos. *Int. J. Dev. Biol.* **37**:299-304.

Scholtz, G., Patal, N. H., and Dohle, W. (1994). Serially homologous *engrailed* stripes are generated via different cell lineages in the germ band of amphipod crustaceans (Malacostraca, Peracarida). *Int. J. Dev. Biol.* **38**:471-478.

Schram, F. R. (1986). "Crustacea." Oxford University Press, Oxford.

Sharov, A. G. (1966). "Basic Arthropodan Stock, with Special Reference to Insects." Oxford University Press, Oxford.

Shubin, N. H., and Alberch, P. (1986). A morphogenetic approach to the origin and basic organization of the tetrapod limb. *Evol. Biol.* **20**:319-387.

Singer, M. A., Hortsch, M., Goodman, C. S., and Bentley, D. (1992). *Annulin*, a protein expressed at limb segment boundaries in the grasshopper embryo, is homologous to protein cross-linking transglutaminases. *Dev. Biol.* **154**:143-159.

Snodgrass, R. E. (1935). "Principles of Insect Morphology." MacGraw Hill Book Co., New York.

Snodgrass, R. E. (1954). Insect metamorphosis. *Smithson. Misc. Collect.* **122**:1-124.

Snodgrass, R. E. (1956). Crustacean metamorphoses. *Smithson. Misc. Collect.* **131**:1-78.

Snodgrass, R. E. (1965). "A Textbook of Arthropod Anatomy." Hafner Publishing Co., New York.

Sommer, R., and Tautz, D. (1991). Segmentation gene expression in the housefly *Musca domestica*. *Development* **113**:419-430.

Stern, D. (1998). A role of *Ultrabithorax* in morphological differences between *Drosophila* species. *Nature* **396**:463-466.

Vachon, G., Cohen, B., Pfeifle, C., McGuffin, M. E., Botas, J., and Cohen, S. (1992). Homeotic genes of the Bithorax complex repress limb development in the abdomen of the *Drosophila* embryo through the target gene *Distal-less*. *Cell* **71**:437-450.

Van Valen, L. M. (1993). Serial homology: the crests and cusps of mammalian teeth. *Acta Palaeontol. Polon.* **38**:145-158.

Wagner, G. (1996). Homologues, natural kinds and the evolution of modularity. *Am. Zool.* **36**:36-43.

Wagner, G. P. (1989). The biological homology concept. *Annu. Rev. Ecol. Syst.* **20**:51-69.

Wagner, G. P. (1989). The origin of morphological characters and the biological basis of homology. *Evolution* **43**:1157-1171.

Wagner, G. P., and Gauthier, J. A. (1999). 1,2,3 = 2,3,4: A solution to the problem of the homology of the digits in the avian hand. *Proc. Natl. Acad. Sci. USA* **96**:5111-5116.

Walossek, D. (1993). The Upper Cambrian *Rehbachiella* and the phylogeny of Branchiopoda and Crustacea. *Fossils Strata* **32**:1-202.

Weatherbee, S. D., Halder, G., Kim, J., Hudson, A., and Carroll, S. (1998). Ultrabithorax regulates genes at several levels of the wing-patterning hierarchy to shape the development of the *Drosophila* haltere. *Genes Dev.* **12**:1474-1482.

Wigglesworth, V. B. (1973). Evolution of insect wings and flight. *Nature* **246**:127-129.

Williams, T. A. (1998). *Distalless* expression in crustaceans and the patterning of branched limbs. *Dev. Genes Evol.* **207**:427-434.

Williams, T. A. (1999). Morphogenesis and homology in arthropod limbs. *Am. Zool.* **39**:664-675.

Williams, T. A., and Muller, G. B. (1996). Limb development in a primitive crustacean, *Triops longicaudatus*: subdivision of the early limb bud gives rise to multibranched limbs. *Dev. Genes Evol.* **206**:161-168.

Williams, T. A., and Nagy, L. M. (1996). Comparative limb development in insects and crustaceans. *Semin. Cell Dev. Biol.* **7**:615-628.

Williamson, D. I. (1982). Larval morphology and diversity. *In* "Embryology, Morphology, and Genetics" (L. G. Abele, ed.), pp. 43-110. Academic Press, New York.

Wills, M. A., Briggs, D. E. G., and Fortey, R. A. (1997). Evolutionary correlates of arthropod tagmosis: scrambled legs. *In* "Arthropod Relationships" (R. A. Fortey and R. H. Thomas, eds.), pp. 57-65. Chapman and Hall, London.

THE EVOLUTIONARY ORIGIN
OF CHARACTERS

Major steps in evolution often, if not always, involve the origin of new characters which then become characteristic of many descendent species. Examples are the origin of feathers among the ancestors of birds and the origin of the autopodium (hand and foot), which leads to the origin of the tetrapod limb. Whether the origin of new characters is dynamically different from the adaptive modification of existing characters by mutation and selection is an open question. As explained by Paul Brakefield in the previous section, however, the genetics of butterfly eyespot patterns suggests that modification of the pattern is due to quantitative polygenic variation, while changes in the kind of pattern appears to be caused by single mutations with large effects. This may hint at the possibility that innovations involve different developmental genetic pathways than quantitative variation (although not necessarily different genes, rather different kinds of alleles, as shown by Trudy Mackay).

The origin of new characters is the area in which any character concept will meet its most stringent test of utility. The question is: can an idea about what characters *are* guide research into the causes of character origination? When we find out what the mechanistic basis of character identity is, then we can investigate how these mechanisms arose in evolution. As mentioned in the introduction to Section IV, the heuristic value of the biological species concept is that it implies that the origin of a species is the origin of isolation mechanisms, i.e., the origin of mechanisms maintaining species identity. This helped to focus research on relevant questions. We are far from a character concept that could do the same for the origin of characters, but this has to be the expectation for any serious character concept. It is interesting to compare what the authors of this section have to say about character origination with the character concepts proposed by the authors of Section II. I leave this exercise to the reader.

The first three chapters in this section reflect upon the origin of specific characters. **Peter Endress** writes about the origin of flowers. He offers a phenomenological definition of a key innovation, as a synapomophy that has not significantly reverted to the previous states among the descendent species. Using this definition he examines the evolution of flower elements and concludes that there are two stages in the evolution of a new character: (1) the occurrence of a new feature, and (2) the consolidation in the genetic apparatus to provide stability to the new character. This resonates well with the findings about the origin of butterfly wing pattern elements (Nijhout, this volume) as well as the abstract ideas put forward by Newman and Muller in this section.

Fred Nijhout reviews the evidence pertaining to the origin of the nymphalid butterfly wing pattern. He concludes that the pattern elements arose from a simple global patterning mechanism creating circular pigment patterns. The asymmetries of the wing, in which these patterns develop, provided a blueprint for the individuation of the different parts of the pattern. Finally, the compartmentalization of the wing surface led to the individuation of smaller parts of the pattern to allow the evolution of individual morphologies. In contrast to the flower example, which is a composite structure, where the emphasis is on the addition and stabilization of new elements, the wing patterns arise from the differentiation of a global unitary pattern by individuation.

Javier Capdevila and **Juan Carlos Belmonte** compare the scenarios for the origin of the tetrapod limb based on developmental genetic evidence. The innovation associated with the fin-limb transition is the origin of fingers. Two kinds of genetic changes are currently considered to be involved in the origin of fingers: a change in the pattern of the 5' Hoxd genes in the developing autopodium and a distalization of the expression of Shh, the molecular correlate of the ZPA. Both scenarios are consistent with the idea that innovations may have a simple genetic basis, as hinted at by Brakefield. In fact the regulation of all Hoxd genes in the hand seems to be due to a single enhancer site and is thus a simple molecular trait.

Stuart Newman and **Gerd Müller** propose a scenario for the origin of new characters which assumes that the strict mapping of genetic variation to phenotypic variation is the result of evolution rather than a precondition. According to their scenario, new characters first arose from spontaneous self-organization of the developing tissues and are secondarily stabilized by genetic changes. Only then has a heritable canonical character arisen. This scenario is roughly consistent with the one painted by Nijhout for the evolution of butterfly wing patterns, which derive from global pattern formation processes and the subsequent individualization of parts of the pattern. It is presumable that the various parts of the pattern are initially not under direct genetic control, while in the derived state pattern elements do have partially independent genetic control.

As mentioned earlier, innovations are interesting because they are assumed to play a more important role in large-scale patterns of evolution than simple adaptive modifications of existing characters. This idea finds its expression in the concept of key innovations (see also Endress), but the concept has experienced vigorous opposition from some evolutionary biologists. **Frietson Galis** reexamines the notion of key innovations and proposes a redefinition of the term that is intended to overcome these difficulties. The important feature of her definition is that she defines key innovations with respect to their effect on character evolution, rather than on the rate of diversification, which was the main problem with previous ideas about key innovations. Of course innovations may have consequences on diversification, but this is no longer the basis of the definition. Her notion of key innovations is consistent with but goes beyond the working definition used by Endress as it assumes a specific role of key innovations in opening up opportunities for phenotypic evolution.

21

ORIGINS OF FLOWER MORPHOLOGY

PETER K. ENDRESS

Institute of Systematic Botany, University of Zurich, CH-8008 Zurich, Switzerland

FLORAL STRUCTURAL ELEMENTS AS THE RESULT OF KEY INNOVATIONS

Flowers are morphologically the most complicated parts of plants. Instead of being single organs, such as leaves, roots, and stems, flowers are composite structures composed of a number of organs that form an ordered pattern. It is this complexity that makes flowers so aesthetically highly attractive.

In an evolutionary context, floral structural elements may be defined as the result of key innovations. Key innovations are intriguing steps in the phylogeny of organisms (e.g., Lauder and Liem, 1989; Cheverud, 1990; Cracraft, 1990; Müller, 1990; Nitecki, 1990; Seilacher, 1990; Baum and Larson, 1991; Müller and Wagner, 1991; Thomson, 1992; Sanderson and Donoghue, 1994; Heard and Hauser 1995; Smith and Szathmáry, 1995; Szathmáry and Smith, 1995; Berenbaum *et al.*, 1996; Raff, 1996; Rohde, 1996; Crane and Kenrick, 1997;

The Character Concept in Evolutionary Biology

Baum, 1998; Hunter, 1998; Masters *et al.*, 1998). At present they find increasing attention due to new avenues of evolutionary biology using molecular techniques. Key innovations are new characters that acquire an indispensable biological role (developmental or ecological) that causes them to be conserved despite changing adaptive pressures (Müller and Wagner, 1991). The definitions given in the literature differ depending on the level of their focus: (1) process of evolution, which leads to a key innovation, thus emphasizing the starting point of a key innovation vs (2) pattern of distribution of characters in large clades, thus emphasizing the result of a key innovation.

How do key innovations manifest themselves in macroevolution? A key innovation is a new feature that is stable in a larger clade and did not or did not significantly revert to the previous state; the less a synapomorphy of a clade is reversed and the larger this clade is, the more this synapomorphy is a key innovation. This definition becomes practically applicable as more robust phylogenetic trees of larger clades become available, the number of which is rapidly increasing (e.g., APG, 1998). Such stabilized patterns in flowers seem especially interesting because most floral features are not stable at high levels. Stability may be interpreted as that the feature became so essential that it was never again reverted.

FLORAL ORGANS AND OTHER STABILIZED FEATURES AS STRUCTURAL ELEMENTS

What are structural elements of flower morphology? In the first place certainly those parts that are commonly called floral organs. They are sepals, petals, stamens, and carpels. These organs are present in flowers in a certain number and arrangement. It is especially apt to call them elements or building blocks as each organ commonly occurs in more than one copy in each flower. Thus they are modules of the modular structural complexes, which are represented by the flowers. Flowers, in turn, commonly occur in small or large numbers on an individual plant. They form inflorescences. Thus, flowers are by themselves elements or modules of a higher order. This may also be repeated at still higher orders: inflorescences may be elements or modules of compound inflorescences.

It is generally agreed that all angiosperm flowers are homologous in being uniaxial structures and not branching systems (for discussion see Friis and Endress, 1990). However some authors (Burger 1977; Meeuse, 1992; Leroy, 1993) hypothesized that conventional flowers evolved by aggregation of several primitively simple flowers of basal angiosperms. Thus, according to them not all angiosperm flowers are homologous: those of some basal angiosperms are uniaxial, whereas those of most extant angiosperms are biaxial. However, comparative developmental and paleobotanical studies as well as recent molecular developmental studies do not show evidence in favor of these latter hypotheses.

Are there elements other than floral organs as such in floral morphology? If elements are taken in a broader sense, in evolutionary terms as relatively stable structures of some complexity, of homologues, certainly there are. Thus, we may consider as elements not only organs by themselves but also any character states of organs that are stabilized to some degree or evolutionarily stable organ complexes (e.g., postgenital fusion, thecal organization of anthers, syncarpy, gynostemium; see later). Organisms are organized in "stratified complexity patterns," which was elegantly visualized in the parable of the two watchmakers (Simon, 1962). Two watchmakers, Tempus and Hora, made watches out of 1000 pieces. Tempus operated with single pieces. Whenever distracted, the unfinished watch fell into pieces. Hora, on the contrary, first put together small units that consisted of 10 pieces, and those again into larger units that consisted of 10 small units each. At each distraction his loss was much smaller than that of Tempus, and thus he became much more successful as a watchmaker.

The biological and evolutionary meaning of homologues was discussed, e.g., by Riedl (1978), Donoghue (1989), Donoghue and Sanderson (1994), Wagner (1989a,b, 1995, 1996), Müller and Wagner (1991), and Wagner and Altenberg (1996). Stratified complexity with homologues at different levels is also present in flowers, as will be shown in the following sections.

AGE OF FLORAL ELEMENTS

A. Phylogeny of Flowering Plants

The flowering plants in the narrow sense (angiosperms) are a group of the seed plants (spermatophytes). Extant angiosperms are seen as a relatively young diversification, the "crown group" of an older clade, the "stem group" angiophytes, without well-established fossils and without surviving branches other than the angiosperms (Fig. 1) (Doyle and Donoghue, 1993; for definition of "stem group" and "crown group," see also Jefferies, 1979). The flowering plants in the wide sense, the anthophytes, comprise, in addition to the angiophytes, a few other spermatophytes, such as Bennettitales and Gnetales. It is uncertain, however, whether the anthophytes are monophyletic (Goremykin *et al.,* 1996; Chaw *et al.,* 1997). The angiosperms consist of some small relic basal clades and the two main clades monocots and eudicots (APG, 1998). The eudicots are the largest of these main clades of the angiosperms, and within the eudicots the asterids are the largest and in some way the biologically most elaborate clade.

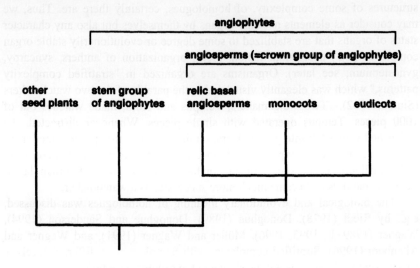

FIGURE 1   Phylogeny of angiosperms (adapted from Doyle and Donoghue, 1993)

Structural elements of flowers have various evolutionary ages. They can be grouped loosely in three categories: (1) Some elements are older than the angiophytes, since they also occur in other seed plants. (2) Others are characteristic for all extant angiosperms but do not occur in other seed plants, they must have arisen in the stem group of the angiophytes. (3) Still others are present in large parts of the angiosperms but are lacking in the basal angiosperms, and thus are younger than the early angiosperms.

B. Preangiophyte Floral Elements

1. Ovules and Seeds

Ovules and seeds characterize all seed plants (spermatophytes) (Bateman and DiMichele, 1994; Kenrick and Crane, 1997). Ovules consist of a nucellus and one or two integuments, with the nucellus being the megasporangium where meiosis takes place. One or more megaspores develop into a haploid female gametophyte, which eventually produces one or more egg cells. In part of the spermatophytes, including the angiosperms, the sperm cells are transported to the egg cell by a pollen tube, which grows through the upper part of the nucellus. It is guided to the nucellus through a narrow canal, the micropyle, which is formed by the tip of the integument. The integument is an integral part of the ovule throughout the spermatophytes. In basal angiosperms there are two integuments (see later). However, the inner commonly still forms the micropyle, while the outer one is shorter and in later development forms a substantial part of the seed coat. Only in asterids (a large group of highly advanced angiosperms) with the change of ovule structure from crassinucellar and bitegmic into tenuinucellar and unitegmic (see later), does the mature egg cell come to lie immediately at the apex of the nucellus, and the pollen tube reaches it through the micropyle without growing through several cell layers of the nucellar apex.

The inner integument of angiosperm ovules is probably homologous to the single integument of other spermatophytes (see, e.g., Endress, 1996, for Gnetales). Studies in molecular developmental genetics of angiosperms ovules are most promising for further elucidation of the homology of parts (e.g., Schneitz *et al.*, 1998).

2. Microsporangia

The male counterpart of ovules are the microsporangia. They occur in all spermatophytes and even in some nonspermatophytes (all heterosporous plants). Microsporangia are more simple than ovules. A tissue of meiocytes is surrounded by a tapetum layer, which has secretory functions during pollen differentiation, and a mechanical layer, which helps opening the mature sporangium.

3. Outer Male – Inner Female Organs

The close association of outer male and inner female structures in flowers or flower-like complexes is an innovation that may have arisen already before angiosperms existed, if the concept of anthophytes is correct (the concept has recently been challenged, see before). This positional relationship is fixed in angiosperm flowers. A single exception is known with a reversal of stamen and carpel position: *Lacandonia* (Triuridaceae) (Márquez-Guzmán *et al.*, 1989). A loose association of peripheral male and central female units is also present in Gnetales, although the "elements" involved may not be homologous in Gnetales and in angiosperms. Nevertheless, it is not impossible that a general homology of the positional relationship of outer male and inner female units could be present, irrespective of the more detailed homology of those organs (Endress, 1996; Frohlich and Meyerowitz, 1997).

C. Angiosperm Floral Elements

1. Carpels, Angiospermy, and Postgenital Fusion

The most prominent key innovation in angiosperm phylogeny was the advent of carpels. One to numerous ovules are enclosed in a covering part and secluded from the outside by secretion or by postgenital fusion of a part of the inner carpellary surface (Endress and Igersheim, 1997). In uniovulate carpels the ovule is usually positioned in the median plane on the adaxial side. In carpels with more than one ovule the ovules are usually positioned on both carpel flanks (or margins).

Secretion or postgenital fusion not only secludes the ovules from the outside, but also forms the site of pollen tube growth and guidance from the stigma to the ovary. Seclusion as such of originally exposed parts is not a novelty in angiosperms. It also occurs by thickening growth in the integument of Gnetales and other gymnosperms and of cone scales of some conifers or by folding of cupular walls.

However, the real innovation in angiosperms is postgenital fusion with tight adhesion of epidermal cells of two contiguous surfaces, which seems to be restricted to angiosperms. Perhaps the evolutionary advent of postgenital fusion and of carpels went hand in hand. The complete and early postgenital fusion of carpel flanks also led to the novelty of inner morphological surfaces, which is otherwise restricted to animals where it arises by other mechanisms.

A circumstantial observation that should be critically and comparatively studied is that in basal angiosperms the carpel seems to close before the ovule is initiated so that the ovule is not visible from the outside in the intact carpel in any stage of development, whereas in a few more advanced groups, carpel closure is delayed, as compared to the ovules, and the young ovule(s) is (are) visible before closure. This is most extreme in some caryophyllids, e.g., in Chenopodiaceae (Hakki, 1971) or Plumbaginaceae (De Laet et al., 1995). There also seems to be a correlation between this precocity of ovule appearance and basal ovule position in these caryophyllids. This differential behavior was partly responsible for the once hotly debated stachyospory (ovules borne on an axis) vs. phyllospory (ovules borne on a leaf) within angiosperms concomitant with an alleged diphyletic origin of angiosperms (e.g., Lam, 1950).

The morphological nature of carpels in a phylogenetic context is still uncertain. A carpel may be a "megasporophyll," i.e. a leaf-like structure bearing ovules, or it may be a composite organ consisting of an ovule-bearing part that is surrounded by a bract or by a newly formed organ.

2. Thecal Organization in Stamens

Angiosperm stamens have anthers with four microsporangia (pollen sacs), organized into two thecae basically in all clades (Endress and Stumpf, 1990). Thecal organization of stamens does not occur in other seed plants. The most similar structures in other seed plants are disporangiate (e.g., in conifers or in *Gnetum*). Thus, in angiosperms, instead of a simple sporangium on each side of the anther, there are two sporangia that form a theca. At maturity the theca opens by a longitudinal slit between the two sporangia and by disintegration of the septum between the two sporangia. It is supposed that the thecal structure provides a more efficient apparatus for pollen presentation than a single sporangium (Hufford and Endress, 1989). The evolutionary origin of thecal organization is unclear.

There are very rare cases where stamens do not have a thecal organization (Endress and Stumpf, 1990) or where the ovules are reduced so much that the female meiosis takes place directly in the base of the ovary (Fagerlind, 1945). From the systematic distribution of these exceptional features within the angiosperms (see, e.g., Chase *et al.,* 1993; Soltis *et al.,* 1997) it can be concluded that they are not basal in angiosperms. Interestingly, both these cases of morphological evolutionary "dedifferentiation" are present in some parasitic flowering plants.

D. Floral Elements That Occur in Only Part of the Angiosperms

1. The Advent of Petals and the Four Organ Categories

A perianth with sepals and petals is characteristic for the eudicots, while in the basal relic angiosperm groups and monocots this distinction is problematic in most cases. The distinction between sepals and petals has also been shown by the ABC model of developmental genetics for the two distantly related plants of the eudicots, *Arabidopsis* (rosids) and *Antirrhinum* (asterids) (Coen and Meyerowitz, 1991; Theissen and Saedler, 1995). However, there are many groups among the eudicots where it is difficult to define petals and sepals. On the other hand, there are a few basal angiosperms, which apparently have sepals and petals (Nymphaeales, *Saruma* of Aristolochiaceae). In the basal eudicots (e.g., Ranunculaceae, Hamamelidaceae) the presence of sepals and petals is labile (Hiepko, 1965; Endress, 1994).

Stamens and carpels are clearly defined by their male and female functions, respectively. Stamens produce microsporangia, and carpels produce megasporangia. The microsporangia in turn produce pollen and pollen tubes, each with two sperm cells, and the megasporangia produce the embryo sac with the egg cell. Sepals and petals are less clearly defined. Although sepals commonly have a protective function for the young flowers and structural differentations adapted to this function, whereas petals have attractive function at anthesis and structural differentiations adapted to this function, this is not exclusively so. Many flowers have two whorls of nonreproductive floral organs. Then the outer whorl could be identified as sepals and the inner as petals, irrespective of function. This is definition by position, not by quality. However, flowers with only one whorl or with more than two whorls also occur. Then the identification of sepals and petals may become difficult. However, if viewed in the phylogenetic context such cases may become more clear. Santalaceae are an example with one whorl of organs outside the stamens, which are like sepals in texture, position in bud and protective function. However, the sister family Olacaceae has an additional, outer, whorl of small organs, which indicates that the sepal-like organs of Santalaceae may have evolved from petals.

The question remains whether position (if fixed), quality, or in some way both of them should be decisive for identification. The ABC model of flower developmental genetics mentioned earlier opened a window to better understand homeotic changes of organ properties in the four floral whorls of *Arabidopsis* and *Antirrhinum*. In this context it would also be interesting to know more about the potential of an evolutionary transfer of properties between whorls (for discussion, see Endress, 1994; Baum, 1998; Irish and Kramer, 1998; Albert *et al.*, 1998).

2. Syncarpy

In basal angiosperms the carpels of a flower are free from each other or there is only one carpel in a flower. However, in the majority of the angiosperms as a whole, the carpels of each flower are congenitally united. This is true for the majority of eudicots and of monocots. In many cases carpel union goes up to the top. In extreme cases, union is so integrated that the number of elements, of carpels, is no longer clear, e.g., in Primulaceae. Here it can be deduced only from the total floral symmetry that originally five carpels made up the gynoecium. The major advantage of syncarpy over apocarpy is the potential of centralized pollen tube selection by a compitum (common pollen tube transmitting tract of all carpels) and enhanced diversity of dispersal units (see discussion in Endress, 1982). Only exceptionally has syncarpy reverted to apocarpy, and in such cases a compitum is retained by postgenital fusion of the carpel tips (Endress *et al.*, 1983).

3. Sympetaly and Floral Tubes

Sympetaly, the congenital union of petals, is a conspicuous key innovation in flowers at the level of the asterids. It occurs here and there in a number of other groups, also in monocots. However, in the asterids, it is a real breakthrough. The asterids comprise about one-fourth of all angiosperm species. Except for the more basal asterids, only on very few occasions has sympetaly reversed to choripetaly (free petals).

Sympetaly comes about by a combination of two developmental processes: (1) by meristem fusion of the petal primordia and, in addition, (2) by an intercalary meristem that produces a corolla tube by intercalary elongation.

Sympetaly provides an enormous potential for the diversification of flowers. The architecture of the flowers becomes more stable. Large flowers can be made more easily. All the largest flowers on earth have united perianth parts. In the asterids, *Brugmansia* (Solanaceae) and *Fagraea* (Gentianaceae) have species with flowers half a meter in length. The largest flowers in the world are in species of *Rafflesia* and *Aristolochia*; they are not representatives of the asterids, but they are also characterized by flowers with united perianth parts.

The evolutionary flexibility of floral tube length and shape provides an excellent means for isolation mechanisms in pollination. The easy and rapid evolution of different pollination modes is greatly enhanced by simple differences in length and width of the floral tube formed by the united petals, which are easily achieved by subtle differential growth of the intercalary meristems. Thus many large families with sympetalous flowers have a similarly wide spectrum of pollination biological forms, e.g., Polemoniaceae (Grant and Grant, 1965), Gentianaceae, Solanaceae (Vogel, 1990), Acanthaceae, Gesneriaceae, and Bignoniaceae (Endress, 1994).

Comparative developmental studies of a wide range of asterids showed two kinds of sympetaly: either the petals appeared as distinct primordia and the fused part appeared only later ("late sympetaly") or the fused part appeared first as a ring meristem on which the individual petals appeared later ("early sympetaly") (Erbar, 1991; Erbar and Leins, 1996a). These authors showed that these two kinds of sympetaly, to some extent, coincide with two major groups of asterids that appear in molecular systematic works. The asterid I group in Chase *et al.* (1993) is largely characterized by "late sympetaly" and the asterid II group by "early sympetaly" (see also Olmstead *et al.,* 1993).

In addition to the question of the systematic distribution of these traits, it is worthwhile to ask about correlations with other floral traits. It seems that there is a tendency of "early sympetaly" to be associated with inferior ovaries and with reduced calyces. Both lines of inquiry, the systematic and the structural, should be integrated into an evolutionary framework. Thus the question is whether the predominance of "early sympetaly" in asterids II is a consequence of the predominance of inferior ovaries in this group. This hypothesis should be studied by investigating the sympetaly pattern in those rarer cases of reverse ovary position in both asterids I and II (e.g., Rubiaceae, some Gesneriaceae in asterids I, or Pittosporaceae, Aquifoliaceae, in asterids II; for Pittosporaceae, see Erbar and Leins, 1996b). Some critical groups should also be restudied from this point of view.

In some basal clades of asterids (the asterid III group in Chase *et al.,* 1993) sympetaly is in general less constant than in the higher asterids. It either occurs only in parts of a family (e.g., Loasaceae, Pittosporaceae, Cornaceae) or there are apparent reversals to choripetaly (e.g., some Ericaceae s.l.), as seen from various cladistic analyses, molecular and nonmolecular (Anderberg, 1993, 1994; Judd and Kron, 1993; Kron and Chase, 1993; Kron, 1996, 1997). It may also be weakly expressed or be difficult to recognize in mature stages (e.g., Araliaceae, Pittosporaceae; Erbar and Leins, 1988, 1996b). It would seem that in these latter cases meristem fusion of the petal primordia already occurs, but the potential to work with intercalary meristems to form floral tubes is not yet established or elaborated. It may be hypothesized that sympetaly is not genetically deeply rooted in the floral organization of the lower asterids so that it could easily disappear or reappear. In the higher asterids it is genetically more stabilized so that it is more difficult for choripetaly to return.

4. Floral Spurs

Floral spurs are hollow outgrowths of a laminar organ, most often from petals, sometimes from sepals or from both petals and sepals. They occur in sympetalous and in choripetalous flowers. Usually they contain nectar, which is produced in the spur itself or near the spur and then lead into the spur by capillary forces. Floral spurs are in some way analogous to floral tubes in sympetalous flowers. However, flowers with spurs are much less common than flowers with tubes throughout the entire angiosperms. This may be due to the fact that floral tubes can fluctuate in length and width, whereas floral spurs commonly fluctuate only in length. Flowers with tubes show a wide spectrum of pollination syndromes, from small insects to large animals, such as bats and large birds, whereas flowers with spurs are more restricted to insects (and hummingbirds). This may be due to the fact that floral tubes end in the floral base, which has a robust architecture, while floral spurs have no reinforced end. Nevertheless, some groups with spurred flowers were highly successful.

While flowers with tubes characterize almost the entire asterids, flowers with spurs commonly characterize single genera, rarely an entire family. The best studied genus in an evolutionary context is *Aquilegia* (Ranunculaceae) (Hodges and Arnold, 1995; Hodges, 1997a,b). *Aquilegia* is unusual in having a spur in each petal and the flowers being polysymmetric. Most other spurred flowers are monosymmetric with only one spur. Vochysiaceae is a family, Antirrhineae a tribe of the Scrophulariaceae (Sutton, 1988) with spurs throughout. Tropaeolaceae, Balsaminaceae, Valerianaceae, and Orchidaceae are other families with spurred flowers in many taxa. *Pelargonium* (Geraniaceae) has an "internal" spur in the peduncle and is highly diverse in pollination biology (Vogel, 1954; Manning and Goldblatt, 1996; Struck and Van der Walt, 1996). *Diascia* (Scrophulariaceae) is a special case because instead of nectar it produces oil in the spurs, which is collected by specialized bees *(Rediviva)*; since the oil is collected not with the proboscis but with the two forelegs, these flowers have evolved two spurs side by side (Vogel, 1974).

Bird pollination seems to be absent in basal angiosperms and basal eudicots

(Endress, 1990b). Hummingbird-pollinated species of *Aquilegia* and *Delphinium* (both Ranunculaceae) of the basal eudicots are the only exceptions, obviously due to their invention of spurs.

5. Tenuinucellar, Unitegmic Ovules

Ovules occur in all seed plants. In angiosperms there are two predominant forms: (1) crassinucellar (with a thick nucellus that forms more than one cell layer around the meiocyte) and bitegmic (with two integuments) and (2) tenuinucellar (with a thin nucellus that forms only one cell layer around the meiocyte, which disintegrates at maturity) and unitegmic (with a single integument) (Philipson, 1977). The former type is basal in angiosperms. The second type characterizes a more advanced, large, especially successful clade of the eudicots, the asterids. Therefore, the second type may be seen as a key innovation within the angiosperms, although there is no new structure involved at the morphological level. Histologically, they are associated with an "endothelium" (a distinct layer of secretory tissue at the integument surface adjacent to the nucellus), which is lacking in the basal type; the formation of embryo sac haustoria or endosperm haustoria is also often associated with the second type. An evolutionary advantage of tenuinucellar, unitegmic ovules may be their potential to form larger numbers of ovules in an ovary because of the smaller size of the ovules (Endress, 1994). It results in an enlarged diversity in dispersal biology.

There are also scattered taxa with smaller ovules in basal angiosperms (Saururaceae, Rafflesiaceae, Lactoridaceae) (Igersheim and Endress, 1998). Histological differentiations analogous to an endothelium in asterids are also present in *Lactoris* and *Rafflesia* but different in detail from that in asterids.

In basal asterids and their sister groups in rosids the ovule structure is transitional and the tenuinucellar, unitegmic type is not stabilized. This may be seen in Malpighiales (APG, 1998) (ovules still bitegmic, but fluctuating between crassinucellar and tenuinucellar), in Ericales (already tenuinucellar, but fluctuating between bitegmic and unitegmic), or in Garryales (crassinucellar but unitegmic).

CHANGING EMPHASIS OF ELEMENTS DURING EVOLUTION OF ANGIOSPERMS

In basal angiosperms floral organs are differentiated and function as individual entities. In higher advanced groups with more elaborate flowers, different organs may be synorganized into complex structures, which may be seen as novel elements (Endress, 1990a). Synorganization may encompass (1) all organs of one kind, (2) part of the organs of one kind, or (3) organs of different kinds, into one unified structure. Widespread cases of the first are syncarpy (congenital union of all carpels), sympetaly (congenital union of all petals), synandry (congenital union of all stamens), or synanthery (postgenital union of all anthers). Examples of the second are the lips in Zingiberaceae and Costaceae, which consist of two or five congenitally united staminodes, respectively, or the androecial lip of Lecythidaceae, which consists of the lower stamens or staminodes of the flower (Endress, 1994). Examples for the third synorganization are the lips in Orchidaceae, which may consist of a petal and three staminodes that are congenitally fused (for different interpretation, see, e.g., Endress, 1995). In many asterids and in some monocots petals and stamens are congenitally fused. The most elaborate synorganizations are the gynostemium in orchids (congenital fusion of stamens and carpels) and the gynostegium in asclepiads (postgenital fusion of stamens and carpels). In both cases the new structure produces complicated devices for pollen transport (pollinaria), which originate from parts of both stamens and carpels (see Endress, 1994).

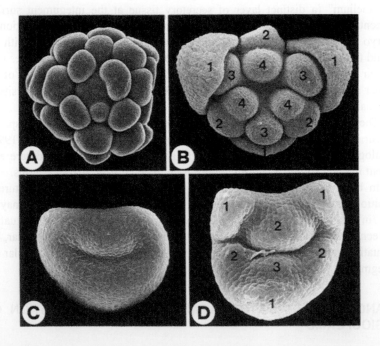

FIGURE 2 Young developmental stages of flowers of successively more integrated organization. A. *Hypserpa laurina* (Menispermaceae): Floral organs not in whorls, not fixed in number (x 100). B. *Veratrum album* (Melanthiaceae): Floral organs in whorls, fixed in number; numbers of whorls numbered (x 130). C and D. *Oncididum ornithorrhynchum* (Orchidaceae): Floral organs in whorls, fixed in number, organs of the outer whorls all initially fused forming a ring meristem (C); later organs of different categories synorganized (D); numbers of whorls numbered (both x 200).

In general, one can say that at a basal, biologically "primitive" level of angiosperms, flowers play with organ **number and arrangement**. At a higher level, the more the organs become synorganized and the more number and arrangement of elements become fixed, the flowers instead play with an enhanced **flexibility in shape** and other properties of the elements themselves and with **new structures** produced by the elements and superimposed on the elements (e.g., corona in Apocynaceae or lip elaborations in orchids) (Fig. 2). In basal angiosperms floral organization is relatively "open," in higher advanced groups it is generally more "closed." Thus floral evolution went from a relatively "open" to a more "closed" organization.

In an "open" floral organization the number (and position) of floral organs is not fixed (although the sequence sepals-petals-stamens-carpels is fixed, see before). It may vary within an individual, depending on the position in the ramification system or depending on the vigor of the individual plant or of the branch on which the flower forms.

Distylium and *Distyliopsis* (Hamamelidaceae) are examples of a position-dependent floral structure of an "open" organization (Endress, 1978). The flowers lack a perianth. They are bisexual or male with a variable number of stamens. The bisexual flowers have two carpels. The inflorescence is a panicle, in which each axis is terminated by a flower. The most complete flowers, bisexual and with the highest number of stamens, are those that terminate a main inflorescence axis. The shorter the axis, the more reduced is its flower: at first stamen number decreases, then the gynoecium is reduced and disappears. The most reduced flowers consist of a single stamen. *Nigella* (Ranunculaceae) and *Papaver* (Papaveraceae) are examples of a vigor-dependent floral structure. Both are annuals. When grown under dry or poor conditions the plants are dwarfish and they produce a single small flower with a reduced number of organs (for *Nigella*, see Troll, 1969, p. 440; for *Papaver*, see Murbeck, 1912, p. 36). All these taxa are members of the basal eudicots. Papaveraceae is especially interesting because it contains an "open" and a "closed" subfamily. Papaveroideae have simple flowers with a relatively "open" organization and high variability of organ numbers (see earlier discussion), while Fumarioideae have highly elaborated flowers with a "closed" organization and constant organ numbers.

In a "closed" floral organization the number (and position) of floral organs is fixed (and only changes in teratological cases) within a species or also at higher levels. In these cases variation is only in the size but not in the number of organs.

Solanum hirtum in the asterids is an example of a position-dependent floral structure of a "closed" organization (Diggle, 1995). All flowers have the same number of organs but some flowers with a distinct position in the branching system have the gynoecium reduced, resulting in functionally male flowers.

CONCLUSIONS

In contrast to the evolutionary advent of early floral structures, such as of flowers themselves and of carpels, which evolved in preangiophytes and in the stem group of angiophytes, the advent of some younger key innovations discussed here can be evaluated in more detail. The main reason is that those groups that would be crucial for the evolutionary origin of flowers and of carpels are extinct, whereas for younger innovations there is a diversity of living representatives of crucial groups available, which allow better reconstruction of the processes. The progressively more detailed molecular systematic research of large clades of the angiosperms provides the basis for pursuing this question.

Preliminary studies, e.g., for the advent of sympetaly or of the tenuinucellate, unitegmic type of ovules, or of monosymmetry in Lamiales (Endress, 1998), suggest that key innovations do not arise in full-fledged form but start with trials. They begin unnoticed as minor changes and become prominent and rooted in the organization much later when they are established in the genetic structure of the group, enhanced by their evolutionary success. In the beginning, reversals to the ground state are still easy and common, but later this becomes progressively more restricted. The details of the evolutionary success have to be studied by such different fields as population genetics, developmental genetics, ecology, and systematics.

Thus the advent of key innovations seems to be in two steps: (1) first occurrence of a new feature and (2) consolidation in the genetic apparatus to provide stability to the new, favorable feature, which manifests itself now as an innovation.

It should be emphasized again that the evolution of these key innovations in flowers has not been critically studied as yet, and now with so many new techniques available, it will be fascinating to tackle them in a more diverse way.

ACKNOWLEDGMENTS

I thank Günter P. Wagner and an anonymous reviewer for critically reviewing the text. This study profited from a grant (Nr. 3100-040327.94) by the Swiss National Foundation.

LITERATURE CITED

Albert, V.A., Gustafsson, M.H.G., and Di Laurenzio, L. (1998). Ontogenetic systematics, molecular developmental genetics, and the angiosperm petal. *In* "Molecular Systematics of Plants" (D. E. Soltis, P. S. Soltis, and J. J. Doyle, eds.), 2nd Ed.) pp. 349-374. Kluwer Academic Publishers, Boston.

Anderberg, A. A. (1993). Cladistic interrelationships and major clades of the Ericales *Plant Syst. Evol.* **184**:207-231.

Anderberg, A. A. (1994). Phylogeny of the Empetraceae, with special emphasis on character evolution in the genus Empetrum. *Syst. Bot.* **19**:35-46.

APG (The Angiosperm Phylogeny Group). (1998). An ordinal classification for the families of flowering plants. *Ann. Mo. Bot. Gard.* **85**:531-553.

Bateman, R. M., and DiMichele, W. A. (1994). Heterospory: the most iterative key innovation in the evolutionary history of the plant kingdom. *Biol. Rev.* **69**:345-417.

Baum, D. A. (1998). The evolution of plant development. *Curr. Opin. Plant Biol.* **1**:79-86.

Baum, D. A., and Larson, A. (1991). Adaptation reviewed: a phylogenetic methodology for studying character macroevolution. *Syst. Zool.* **40**:1-18.

Berenbaum, M. R., Favret, C., and Schuler, M. A. (1996). On defining "key innovations" in an adaptive radiation: Cytochrome P450S and Papilionidae *Am. Nat. Suppl.* **148**:S139-S155.

Burger, W. C. (1977). The Piperales and the monocots. Alternative hypotheses for the origin of the monocotyledonous flowers. *Bot. Rev.* **43**:345-393.

Chase, M. W., Soltis, D. E., Olmstead, R. G., Morgan, D., Les, D. H., Mishler, B. D., Duvall, M. R., Price, R. A., Hills, H. G., Qiu, Y.-L., Kron, K. A., Rettig, J. H., Conti, E., Palmer, J. D., Manhart, J. R., Sytsma, K. J., Michaels, H. J., Kress, W. J., Karol, K. G., Clark, W. D., Hedrén, M., Gaut, B. S., Jansen, R. K., Kim, K.-J., Wimpee, C. F., Smith, J. F., Furnier, G. R., Strauss, S. H., Xiang, Q.-Y., Plunkett, G. M., Soltis, P. S., Swensen, S. M., Williams, S. E., Gadek, P. A., Quinn, C. J., Eguiarte, L. E., Golenberg, E., Learn, G. H., Jr., Graham, S. W., Barrett, S. C. H., Dayanandan, S., and Albert, V.A. (1993). Phylogenetics of seed plants: an analysis of nucleotide sequences from the plastid gene *rbcL. Ann. Mo. Bot. Gard.* **80**:528-580.

Chaw, S.-M., Zharkikh, A., Sung, H-M., Lau, T-C., and Li, W.-H. (1997). Molecular phylogeny of extant gymnosperms and seed plant evolution: analysis of nuclear 18S rRNA sequences. *Mol. Biol. Evol.* **14**:56-68.

Cheverud, J. M. (1990). The evolution of morphological variation patterns. *In* "Evolutionary Innovations" (M. H. Nitecki, ed.), pp.134-145. University of Chicago Press, Chicago.

Coen, E.S., and Meyerowitz, E. M. (1991). The war of the whorls: genetic interactions controlling flower development. *Nature* **353**:31-37.

Cracraft, J. (1990). The origin of evolutionary novelties: pattern and process at different hierarchical levels. *In* "Evolutionary Innovations" (M. H. Nitecki, ed.), p. 304. University of Chicago Press, Chicago.

Crane, P. R., and Kenrick, P. (1997). Diverted development of reproductive organs: a source of morphological innovation in land plants. *Plant Syst. Evol.* **206**:161-174.

De Laet, J., Clinckemaillie, D., Jansen, S., and Smets, E. (1995). Floral ontogeny in the Plumbaginaceae. *J. Plant Res.* **108**:289-304.

Diggle, P. K. (1995). Architectural effects and the interpretation of patterns of fruit and seed development. *Ann. Rev. Ecol. Syst.* **26**:531-552.

Donoghue, M. J. (1989). Phylogenies and the analysis of evolutionary sequences, with examples from seed plants. *Evolution* **43**:1137-1156.

Donoghue, M. J., and Sanderson, M. J. (1994). Complexity and homology in plants. *In* " Homology: The Hierarchical Basis of Comparative Biology" (B. K. Hall, ed.), pp. 393-421. Academic Press, San Diego.

Doyle, J. A., and Donoghue, M. J. (1993). Phylogenies and angiosperm diversification. *Paleobiology* **19**:141-167.

Endress, P. K. (1978). Blütenontogenese, Blütenabgrenzung und systematische Stellung der perianthlosen Hamamelidoideae. *Bot. Jahrb. Syst.* **100**:249-317.

Endress, P. K. (1982). Syncarpy and alternative modes of escaping disadvantages of apocarpy in primitive angiosperms. *Taxon* **31**:48-52.

Endress, P. K. (1990a). Patterns of floral construction in ontogeny and phylogeny. *Biol. J. Linn. Soc.* **39**:153-175.

Endress, P. K. (1990b). Evolution of reproductive structures and functions in primitive angiosperms (Magnoliidae). *Mem. N. Y. Bot. Gard.* **55**:5-34.

Endress, P. K. (1994). "Diversity and Evolutionary Biology of Tropical Flowers." Cambridge University Press, Cambridge.

Endress, P. K. (1995). Major evolutionary traits of monocot flowers. In "Monocotyledons: Systematics and Evolution" (P. J. Rudall, P. J. Cribb, D. F. Cutler, and C. J. Humphries eds.), pp. 43-79. Royal Botanic Gardens, Kew.

Endress, P. K. (1996). Structure and function of female and bisexual organ complexes in Gnetales. Int. J. Plant Sci. 157:S113-S125.

Endress, P. K. (1998). Antirrhinum and Asteridae – evolutionary changes of floral symmetry. Soc. Exp. Biol. Symp. Ser. 51:133-140.

Endress, P. K., and Igersheim, A. (1997). Patterns of angiospermy in basal angiosperms.Am. J. Bot. Suppl. 84(6):190.

Endress, P. K., and Stumpf, S. (1990). Non-tetrasporangiate stamens in the angiosperms: structure, systematic distribution and evolutionary aspects.Bot. Jahrb. Syst. 112:193-240.

Endress, P. K., Jenny, M., and Fallen, M.E. (1983). Convergent elaboration of apocarpous gynoecia in higher advanced dicotyledons (Sapindales, Malvales, Gentianales).Nord. J. Bot. 3:293-300.

Erbar, C. (1991). Sympetaly - a systematic character? Bot. J. Syst. 112:417-451.

Erbar, C., and Leins, P. (1988). Blütenentwicklungsgeschichtliche Studien an Aralia und Hedera (Araliaceae). Flora 180:391-406.

Erbar, C., and Leins, P. (1996a). Distribution of the character states "early" and "late" sympetaly within the "Sympetalae Tetracyclicae" and presumably related groups.Bot. Acta 109:427-440.

Erbar, C., and Leins, P. (1996b). An analysis of the early floral development ofPittosporum tobira (Thunb.) Aiton and some remarks on the systematic position of the family Pittosporaceae. Feddes Repert. 106:463-473.

Fagerlind, F. (1945). Blüte und Blütenstand der GattungBalanophora. Bot. Not. 1945:330-350.

Friis, E. M., and Endress, P. K. (1990). Origin and evolution of angiosperm flowers.Adv. Bot. Res. 17:99-162.

Frohlich, M. W., and Meyerowitz, E. M. (1997). The search for flower homeotic gene homologs in basal angiosperms and Gnetales: a potential new source of data on the evolutionary origin of flowers. Int. J. Plant Sci. 158:S131-S142.

Goremykin, V., Bobrova, V., Pahnke, J., Tritsky, A., Antonov, A., and Martin, W. (1996). Noncoding sequences from the slowly evolving chloroplast inverted repeat in addition torbcL data do not support gnetalean affinities of angiosperms Mol. Biol. Evol. 13:383-396.

Grant, V., and Grant, K. A. (1965). "Flower Pollination in thePhlox Family." Columbia University Press, New York.

Hakki, M. I. (1971). Blütenmorphologische und embryologische Untersuchungen anChenopodium capitatum und Chenopodium foliosum sowie weiteren Chenopodiaceae. Bot. Jahrb. Syst. 92:178-330.

Heard, S. B., and Hauser, D. L. (1995). Key evolutionary innovations and their ecological mechanisms. Hist. Biol. 10:151-173.

Hiepko, P. (1965). Vergleichend-morphologische und entwicklungsgeschichtliche Untersuchungen über das Perianth bei den Polycarpicae.Bot. Jahrb. Syst. 84:359-508.

Hodges, S. A. (1997a). Rapid radiation due to a key innovation in columbines (Ranunculaceae: Aquilegia). In "Molecular Evolution and Adaptive Radiations" (T. Givnish and K. Sytsma, eds.), pp. 391-405. Cambridge University Press, Cambridge.

Hodges, S. A. (1997b). Floral nectar spurs and diversification.Int. J. Plant Sci. 158:S81-S88.

Hodges, S. A., and Arnold, M. L. (1995). Spurring plant diversification: are floral nectarspurs a key innovation? Proc. R. Soc. Lond. B 262:343-348.

Hufford, L. D., and Endress, P. K. (1989). The diversity of anther structures and dehiscence patterns among Hamamelididae. Bot. J. Linn. Soc. 99:301-346.

Hunter, J. P. (1998). Key innovations and the ecology of macroevolution. Trends Ecol. Evol. 13:31-36.

Igersheim, A., and Endress, P. K. (1998). Gynoecium diversity and systematics in paleoherbs.Bot. J.

Linn. Soc. **127**:289-370.

Irish, V. F., and Kramer, E. M. (1998). Genetic and molecular analysis of angiosperm flower development. *Adv. Bot. Res.* **28**:199-230.

Jefferies, R. P. S. (1979). The origin of chordates – a methodological essay. *In* "The Origin of Major Invertebrate Groups" (M. R. House, ed.), pp. 443-477. Academic Press, London.

Judd, W. S., and Kron, K. A. (1993). Circumscription of Ericaceae (Ericales) as determined by preliminary cladistic analyses based on morphological, anatomical, and embryological features. *Brittonia* **45**:99-114.

Kenrick, P., and Crane, P. R. (1997). "The Origin and Early Diversification of Land Plants: A Cladistic Study." Smithsonian Institution Press, Washington, D.C.

Kron, K. A. (1996). Phylogenetic relationships of Empetraceae, Epacridaceae, Ericaceae, Monotropaceae, and Pyrolaceae: evidence from nuclear ribosomal 18S sequence data. *Ann. Bot.* **77**:293-303.

Kron, K. A. (1997). Phylogenetic relationships of Rhododendroideae (Ericaceae). *Am. J. Bot.* **84**:973-980.

Kron, K. A., and Chase, M. W. (1993). Systematics of the Ericaceae, Empetraceae, Epacridaceae and related taxa based upon *rbcL* sequence data. *Ann. Mo. Bot. Gard.* **80**:735-741.

Lam, H. J. (1950). Stachyospory and phyllospory as factors in the natural system of the Cormophyta. *Sven. Bot.Tidskr.* **44**:517-534.

Lauder, G. V., and Liem, K. F. (1989). The role of historical factors in the evolution of complex organismal functions. *In* "Complex Organismal Functions: Integration and evolution in Vertebrates" (D. B. Wake, and G. Roth, eds.). Wiley, London.

Leroy, J.-F. (1993). "Origine et évolution des plantes à fleurs: les Nymphéas et le génie de la nature." Masson, Paris.

Manning, J. C., and Goldblatt, P. (1996). The *Prosoeca peringueyi* (Diptera: Nemestrininidae) pollination guild in Southern Africa: long-tongued flies and their tubular flowers. *Ann. Mo. Bot. Gard.* **83**:67-86.

Márquez-Guzmán, J., Engleman, E. M., Martínez-Mena, A., Martínez, E., and Ramos, C. (1989). Anatomía reproductiva de *Lacandonia schismatica* (Lacandoniaceae). *Ann. Mo.Bot. Gard.* **76**:124-127.

Masters, J. C., Rayner, R. J., and Hunter, J. P. (1998). Key innovations? *Trends Ecol. Evol.* **13**:281-282.

McKinney, F. K. (1988). Multidisciplinary perspectives on evolutionary innovations. *Trends Ecol. Evol.* **3**:220-222.

Meeuse, A. D. J. (1992). "Angiosperm Evolution, No Abominable Mystery." Eburon, Delft.

Müller, G. B. (1990). Developmental mechanisms at the origin of morphological novelty: A side-effect hypothesis. *In* "Evolutionary Innovations" (M.H. Nitecki, ed.), pp. 99-130. University of Chicago Press, Chicago.

Müller, G. B., and Wagner, G. P. (1991). Novelty in evolution. *Ann. Rev. Ecol. Syst.* **22**:229-256.

Murbeck, S. (1912). Untersuchungen über den Blütenbau der Papaveraceen. Kungl. Svenska Vetenskapsakademiens Handlingar **50** (1):1-168.

Nitecki, M. H. (1990). The plurality of evolutionary innovations. *In* "Evolutionary Innovations" (M.H. Nitecki, ed.), pp. 3-18. University of Chicago Press, Chicago.

Olmstead, R. G., Bremer, B., Scott, K. M., and Palmer, J. D. (1993). A parsimony analysis of the Asteridae sensu lato based on *rbcL* sequences. *Ann. Mo. Bot. Gard.* **80**:700-722.

Philipson, W. R. (1977). Ovular morphology and the classification of dicotyledons. *Plant Syst. Evol. Suppl.* **1**:123-140.

Raff, R. (1996). "The Shape of Life: Genes, Development and the Evolution of Animal Form." University of Chicago Press, Chicago.

Riedl, R. (1978). "Order in Living Organisms." Wiley, Chichester.

Rohde, K. (1996). Robust phylogenies and adaptive radiations: a critical examination of methods used to identify key innovations. *Am.Nat.* **148**:481-500.

Sanderson, M. J., and Donoghue, M. J. (1994). Shifts in diversification rate with the origin of angiosperms. *Science* 264:1590-1593.

Schneitz, K., Baker, S. C., Preuss, D., Gasser, C. S., and Redweik, A. (1998). Pattern formation and growth during floral organogenesis: *HUELLENLOS* and *AINTEGUMENTA* control the formation of the proximal region of the ovule primordium in *Arabidopsis thaliana*. *Development* 125:2555-2563.

Seilacher, A. (1990). The sand-dollar syndrome: a polyphyletic constructional breakthrough. *In* "Evolutionary Innovations" (M. H. Nitecki, ed.), pp. 232-252. University of Chicago Press, Chicago.

Simon, H. A. (1962). The architecture of complexity.*Proc. Am. Philos. Soc.* 106:467-482.

Smith, J. M., and Szathmáry, E. (1995). "The Major Transitions in Evolution." Oxford University Press, Oxford.

Soltis, D. E., Soltis, P. S., Nickrent, D. L., Johnson, L. A., Hahn, W. J., Hoot, S. B., Sweere, J. A., Kuzoff, R. K., Kron, K. A., Chase, M. W., Swensen, S. M., Zimmer, E. A., Chaw, S-M., Gillespie, L. J., Kress, W. J., and Sytsma, K. J. (1997). Angiosperm phylogeny inferred from 18S ribosomal DNA sequences.*Ann. Mo. Bot. Gard.* 84:1-49.

Struck, M., and Van der Walt, J. J. A. (1996). Floral structure and pollination in *Pelargonium. In* "The Biodiversity of African Plants" (L. J. G. van der Maesen *et al.,* eds.), pp. 631-638. Kluwer, Amsterdam.

Sutton, D. A. (1988). "A Revision of the Tribe Antirrhineae." British Museum (Natural History), London/Oxford University Press, Oxford.

Szathmáry, E., and Smith, J. M. (1995). The major evolutionary transitions.*Nature* 374:227-232.

Theissen, G., and Saedler, H. (1995). MADS-box genes in plant ontogeny and phylogeny: Haeckel's 'biogenetic law' revisited. *Curr. Opin. Genet. Dev.* 5:628-639.

Thomson, K. S. (1992). Macroevolution: the morphological problem *Am. Zool.* 32:106-112.

Troll, W. (1969). "Die Infloreszenzen," Vol. 2, 1. Fischer, Stuttgart.

Vogel, S. (1954). Blütenbiologische Typen als Elemente der Sippengliederung, dargestellt anhand der Flora Südafrikas. *Bot.Stud.* (Fischer, Jena.) 1:1-338.

Vogel, S. (1974). Ölblumen und ölsammelnde Bienen. Tropische und Subtropische Pflanzenwelt. (Steiner, Wiesbaden) 7:283-547

Vogel, S. (1990). Radiación adaptiva del síndrome floral en las familias neotropicales.*Bol. Acad. Nac. Cienc.* (Cordoba, Argentina) 59:5-30.

Wagner, G. P. (1989a). The origin of morphological characters and the biological basis of homology. *Evolution* 43:1157-1171.

Wagner, G. P. (1989b). The biological homology concept.*Ann. Rev. Ecol. Syst.* 20:51-69.

Wagner, G. P. (1995). The biological role of homologues: a building block hypothesis. *Neues Jahrb. Geol. Palaeontol. Abh.* 195:279-288.

Wagner, G. P. (1996). Homologues, natural kinds and the evolution of modularity. Am. Zool.36:36-43.

Wagner, G. P., and Altenberg, L. (1996). Complex adaptations and the evolution of evolvability. *Evolution* 50:967-976.

22

ORIGIN OF BUTTERFLY WING PATTERNS

H. FRED NIJHOUT

Evolution, Ecology and Organismal Biology Group, Duke University, Durham, NC 27708

INTRODUCTION

Butterfly color patterns are very different from those of leopards and zebras. The color patterns of leopards and zebras are made up of spots and stripes that are placed either randomly or evenly and whose number and position differ from individual to individual. These coat patterns have the same characteristics of randomness and individual variability as the ridge patterns of human fingerprints. In butterflies, by contrast, the same spot or stripe occurs in exactly the same location in all individuals of a species. More importantly, a given spot or stripe can be traced from species to species within a genus and often from genus to genus within a family. The elements that make up the wing pattern of butterflies are an anatomical system that is as organized and diverse as the vertebrate skeleton and the body segmentation and tagmatization of arthropods. It is a system in which there is homology and in which problems of developmental and evolutionary origin, adaptation, and diversification can be analyzed.

A butterfly color pattern element is an individuated character in the way that a bone is an individuated character but a leopard spot is not. A color pattern element is, however, a character of a peculiar sort. A pattern element is not an object that has a substance and that can be isolated like one can dissect out a bone or an imaginal disk; rather, a pattern element is the product of a localized event of pigment synthesis that results from a spatially patterned activation of an enzymatic pathway. Why then is the resultant patch of pigment a character? Why is a spot on a butterfly wing an individuated character that can be given a name and whose identity can be traced across phylogenetic space, whereas a leopard spot is not? The answer to these questions bear on the issue of how characters originate in development and evolution, and how characters become individuated. In the following sections I will first outline the general structure and properties of the elements of butterfly wing patterns and then discuss what we have learned about their developmental and evolutionary origins. We will see that the principles that underly color pattern development, evolution, and individuation are general ones that apply also to more conventional morphological characters.

THE STRUCTURE OF WING PATTERNS

The main organizing principle of butterfly color patterns is the symmetry system. In its simplest form, a symmetry system consists of a pair of pigment bands that run roughly parallel to each other from the anterior to the posterior margin of the wing and approximately normal to the longitudinal veins of the wing. It is called a symmetry system because the pigment distribution in each band of the pair mirrors that of the other (Nijhout, 1991). The field between the bands of each symmetry system typically differs in pigmentation (usually darker) from the general background of the wing. Three such systems of bands make up the basic wing pattern of butterflies: the basal symmetry system, the central symmetry system, and the border symmetry system. In addition there may be one or two narrow bands that run closely parallel to the distal wing margin, the submarginal bands (Fig. 1A).

Along the midline of the border and central symmetry systems there is often a distinctive set of pigmented marks. Those within the border symmetry system are often highly elaborated into eyespots and are called the border ocelli (Fig. 1A). Along the midline of the central symmetry system there is typically a single large mark called the discal spot, a name that derives from the fact that this spot always occurs at the apex of the so-called discal cell.[1] In some moths the discal spot is elaborated into a large eyespot.

---

[1] A "cell" in this case is an area bounded by wing veins. The spatial pattern of veins on insect wings is relatively constant for all species and genera within a taxonomic family, although it differs greatly between families. All butterflies therefore have the same number of wing cells, although the shape of these cells varies with the overall shape of the wing. In about half the species the discal cell is closed off by a set of crossveins, and in those cases the discal spot coincides precisely with the position of these crossveins.

In most cases each pigment band is interrupted wherever it crosses a wing vein so that it looks like a linear series of short segments. The alignment of such a series is typically imperfect, so that the whole band has the appearance of a geologic fault zone with each segment of the band having slipped proximally or distally along a wing vein. By breaking up the bands of the symmetry systems into series of short segments, the wing veins provide the second organizing principle for butterfly wing patterns. The wing veins behave as boundaries that compartmentalize the wing surface with respect to color pattern formation. This compartmentalization breaks the three symmetry systems up into parallel series of isolated and, as we shall see, semi-independent pattern elements. The result of this interaction between symmetry systems and wing veins is illustrated in Fig. 1B. This figure emphasizes the dislocation of the pigment bands and presents the so-called nymphalid groundplan: the set of pattern elements out of which butterflies build their wing patterns (Nijhout, 1991). The groundplan thus consists of eight parallel series of pattern elements: one for each band of each of the three symmetry systems, a row of border ocelli, and a set of submarginal bands. Each wing cell has a representative of each of these pattern elements so the overall groundplan can be viewed as a serial repetition of the same set of pattern elements.

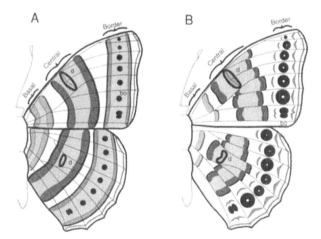

FIGURE 1 The nymphalid groundplan is made up of paired pigment bands, forming a set of three symmetry systems, the border symmetry system, central symmetry system, and basal symmetry system. The discal spot (d) and the border ocelli (bo) are found along the midlines of the central and border symmetry systems, respectively. (A) The vertical structure of the groundplan and its constitutive symmetry systems, while (B) emphasizes the role of wing veins which break up the bands of the symmetry systems into semi-independent pattern elements. One example of how real patterns are derived from this groundplan by selective expression, displacement, and morphological modification of these elements is illustrated in Fig. 9.

PATTERN ELEMENTS AS CHARACTERS

Color pattern evolution and diversification occur by the selective modification of the pattern elements illustrated in Fig. 1B. During evolution, each pattern element can change independently from the other elements in each of its modalities, of color, shape, size, and position (see Fig. 9 for an example). The degree of independence of the pattern element is most easily demonstrated by examining their correlated variation among individuals. There is always a small amount of individual variation in the size, shape, and position of pattern elements, that comes about through individual variation in genotypes or environment experienced during development. By an appropriate experimental and breeding design it is possible to partition the relative contribution of genetic and environmental variation to variation in the phenotype (Paulsen, 1994; Falconer and Mackay, 1996; Lynch and Walsh, 1998). If two pattern elements covary in size or position from individual to individual, it could indicate that the two share developmental determinants for the processes that determine size or position. If two pattern elements do not covary (that is, if the second varies in the opposite direction from the first as often as it varies in the same direction, or if the first element varies and the second does not), then the two either share no variable developmental determinants that affect the characteristic in question or the variation in one direction imposed by one set of determinants is balanced exactly by variation in the opposite direction imposed by another set of determinants. There are some interesting correlation patterns among the elements of the nymphalid groundplan in *Precis coenia*. The *size* of pattern elements that belong to a homologous series (i.e., that are members of the same symmetry system band) covary significantly whereas the sizes of nonhomologous elements are uncorrelated. The *position* of pattern elements within the same wing cells are moderately correlated (even if they belong to different symmetry systems), but the positions of elements in different wing cells are not (Paulsen and Nijhout, 1993; Paulsen, 1994). Moreover, the correlated variation among pattern elements diminishes with distance between them (elements physically farther apart on the wing have less correlated variation than elements closer together) as well as with the degree to which they have diverged from each other morphologically (homologous elements that differ in properties such as pigmentation or complexity covary less than homologous elements that do not differ in these properties).

Finally, studies on the genetics of wing pattern diversity in *Heliconius* have revealed a hierarchy of genetic regulation of pattern element characteristics. Some genes affect the color, position, or size of single pattern elements, while others affect the properties of an entire rank of serial homologues. Several genes are known that affect the pattern on one wing surface only (i.e., either fore wing or hind wing), while others affect the pattern in homologous regions on both fore and hind wing. Finally, some genes affect the properties of many unrelated pattern elements at once (Nijhout *et al.*, 1990; Nijhout, 1991). There appear to exist genetic mechanisms that control the properties of pattern elements individually, in homologous series, or in ensembles over large areas of the wing.

With independent control of the various properties (presence/absence, size, shape, color, position) of the individual members of the eight series of pattern elements, the number of possible permutations that can be used to build up the overall wing pattern is enormous. One of the consequences of this great permutational complexity is that no two species of butterflies have exactly the same overall pattern, although within a species the color pattern is so characteristic and constant that it can be used for species identification. Insofar as pattern elements and their properties are genetically and developmentally independent they are compartmented and free to respond independently to natural selection.

The function of color patterns is visual communication. Accordingly, butterfly wing patterns have undergone extensive adaptive evolution for such functions as sexual signaling, aposematism, camouflage, and mimicry. In building up the wing pattern, some pattern elements are eliminated while others are brought into mutual alignment and their color and shape are adjusted to achieve a particular overall visual effect. The extraordinary diversity of patterns, and the fact that extremely accurate Batesian, Muellerian, and dead leaf mimicry can evolve, suggests the operation of a highly versatile developmental mechanism that places and shapes the pattern elements in a precise yet flexible way. The evolution of such a flexible developmental mechanism and the way in which it produced a highly evolvable color pattern will be the subject of the remainder of this chapter.

EVOLUTION OF A PATTERNING MECHANISM

The Origin of Symmetry Systems

In order to understand the evolution of the highly compartmented nymphalid groundplan, we need to first understand the nature and origin of symmetry systems. Symmetry systems are probably the most widespread of all color patterns, not only in butterflies and moths, but also in fish and mammals. The general nature of symmetry systems in moths and other animals was first recognized by Henke (1933). He noted that the bands of a symmetry system sometimes merge together and fuse to form a U-shaped pattern (Fig. 2A,B) or a variety of closed patterns (Fig. 2C,D). A symmetry system appears to be nothing more than a closed loop figure, or a circle, that has been truncated. An eyespot is therefore a symmetry system, as is any pattern that has a differentiation of color or structure that runs from center to periphery. Symmetry systems can therefore be radial, as in the case of eyespots (or Fig. 2C), or bilateral, when there is extensive truncation (as in Fig. 2A).

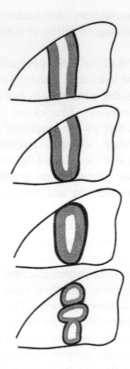

FIGURE 2 Various simple modifications of symmetry systems found in moths which give clues to the nature of symmetry systems. A. Normal symmetry bands. B. Symmetry bands fused together at one end. C. Symmetry bands fused at both ends to form a circle. D. Multiply closed pattern. For real examples of such patterns, see Henke (1933) and Nijhout (1991).

The development and evolution of symmetry systems can perhaps be best understood by first looking at some simple color patterns of vertebrates. Zebra stripes, for instance, are symmetry systems, although this is not always easy to detect. Some races of Burchell's zebra (*Equus burchelli* variety *antiquorum* from Namibia) have gray so-called shadow stripes in the white areas between their black stripes and some individuals have yet narrower and paler gray stripes between the shadow stripes and the black stripes (Fig. 3A). I'll call these the primary and secondary shadow stripes, respectively. Shadow stripes are expressed only where the main black stripes are very far apart, so it looks like the patterning system is attempting to fit additional stripes in where the separation between existing stripes becomes large. A more spectacular intercalation of stripes can be seen in lionfish (various species of *Pterois*; Fig. 3B), where species differ in the number of levels of intercalation.

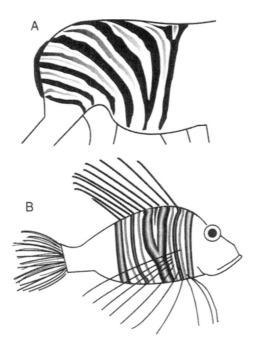

FIGURE 3   A. Striping pattern on a Burchell's zebra from Namibia (*Equus burchelli* var *antiquorum*) with intercalated shadow stripes. B. Color pattern of the lionfish, *Pterois volitans*, illustrating several levels of intercalated stripes.

Intercalation of elements is fairly common in pattern-forming systems that are growing or that occur on curved surfaces where the spacing between existing elements must necessarily expand.   Brain corals are a good example. Here the spacing of polyps gradually increases as the coral head grows and the surface of the colony expands; new polyps then form where the distance between existing polyps exceeds some value.   An overall even spacing is maintained by intercalation when the gap between existing structures exceeds a critical value.   Pattern intercalation of the sort we see in brain corals and zebra stripes actually tells us something important about the mechanism of pattern formation.   In order to understand what is going on we need to recognize that there are two alternative ways in which the space between two diverging stripes can be filled: by branching of one of the stripes or by intercalation of a stripe unconnected to either of the flanking stripes.   The difference between these two patterns lies in the timing of the processes of pattern formation relative to growth.   Understanding the mechanism by which repeated spatial patterns arise during development helps explain why this is so.   It is well established that in development and physiology the formation of evenly spaced repeated patterns requires a mechanism that involves lateral inhibition (Bard, 1977, 1981; Meinhardt, 1982; Murray, 1989; Oster, 1988; Held, 1992). In lateral inhibition the presence of a given structure inhibits the formation of similar structures in its immediate vicinity. This inhibition diminishes with distance, so the next structure can only form where the inhibition falls below a particular threshold. Whether a new structure will actually form outside the area of inhibition depends on the presence of a mechanism that produces the conditions for the initiation of that structure. Typically this would be the same mechanism that gave rise to the initial structure, and if this initiation mechanism is constitutive and spatially widespread then the conditions exist for the generation of a relatively evenly spaced repeated pattern.

Depending on the nature of the inhibitory mechanism and the way it interacts with the initiation mechanism, the pattern that is formed can be either evenly spaced points (such as bristles of insects, or the spots on cheetahs) or evenly spaced stripes (as the stripes of zebras), with the spacing between the elements being determined by the strength and range of the inhibition. Suppose then that we are considering two neighboring stripes and suppose that the field on which they occur is expanding. Then there are three possible outcomes, depending on the stage of pattern formation: (1) if pattern formation is fully complete, then the stripes and interstripe region simply become broader; (2) if pattern formation is still ongoing and the stripes are not yet fixed, then the pattern reorganizes itself so that three stripes can be formed instead of two, and if growth is more extensive in one region of the field than in another so that the stripes diverge in a wedge-shaped pattern, then one stripe will produce a branch that occupies the widening interband region; and (3) if the first set of stripes is fully determined but the interstripe region is still capable of pattern formation, then a new stripe can form halfway between the existing stripes, but unconnected to them. The second mechanism is believed to be responsible for stripe branching in zebras, and the third mechanism for the formation of shadow stripes (Bard, 1977, 1981). Because shadow stripes are determined *after* the main stripes, and not simultaneously with them, it is possible for different factors to influence the developmental details of each. This can account for the fact that the two types of stripes differ in pigmentation, with the shadow stripes being gray rather than black.

In the lionfish *Pterois volitans*, the primary stripes and shadow stripes differ not only in pigmentation but also in structure (Fig. 3B). Not only are the shadow stripes paler than the main stripes, but the main stripes have a graded pigment distribution, darker near the periphery and lighter near the center of the stripe. This gradation is not seen in the secondary shadow stripes, although it is occasionally evident in the primary shadow stripes, wherever these are particularly broad. This graded pigment distribution makes each stripe *self-symmetrical* with regard to color pattern.

We have now arrived at the point where we can conceptualize two kinds of symmetry systems. A pair of stripes and their intervening shadow stripe is a symmetry system, and each band is itself also a symmetry system. We have already seen that in the first case the symmetry arises from the lateral inhibition that existing stripes exert on the formation of new stripes. For this to work, the existing stripes need to be developmentally fixed, or determined, before the shadow stripe is induced. In the second case, the self-symmetry of a stripe tells us something about the shape of the gradient that induces pigment synthesis. This gradient must also be self-symmetrical, which means that a graph of its crossection must be hump-shaped. The two edges of a stripe differentiate in the same way because they respond to the same gradient value, which is different than the value at the central core of the stripe. For the purposes of the present argument I'll assume that the value at the margins is lower than at the core, but it is also possible that the difference is one of timing if, for instance, determination proceeds like a grass fire from a line-shaped central core outward. In the latter case, the dark outer margins represent the active fire, while the paler core represents the burnt grass left behind; lower and higher values of the gradient can be thought of shorter or longer times since ignition. As we'll see later, the grass fire model cannot easily explain the next stages in the development of a symmetry system, and a simple self-symmetrical spatial gradient is a more useful conceptual model for understanding what happens next.

The symmetry systems of the nymphalid groundplan appear to have evolved from simple self-symmetrical pattern elements (Nijhout, 1994). It is possible to trace the steps by which this evolution occurred by examining the symmetry systems and their precursors in various groups of moths. Moths in the suborders most closely related to the butterflies have color patterns based on symmetry systems, but moths in the more basal suborders of the Lepidoptera do not. In the latter, patterns are typically made up of fairly irregular spots or blotch-like figures, placed in species-specific but irregular patterns on the wing surface. Often these figures are monochrome (like zebra stripes), but when they are not, their pigmentation exhibits self-symmetry (like lionfish stripes), so that they look like irregularly rounded concentric figures or rings (Figs. 2B and 4). In some species there is much intraspecific variation in the size and shape of these pattern elements. In individuals where two neighboring elements are enlarged, they simply fuse smoothly at their point of contact to form a single closed figure (Figs. 4 and 5). The fact that two self-symmetrical patterns can fuse smoothly to form a single closed pattern indicates that the two are produced by identical developmental processes. When these composite figures enlarge so that they extend to the anterior and posterior wing margins, the pattern that results resembles a pair of irregular bands and constitutes what we recognize as a symmetry system. Symmetry systems are constructed by the fusion of the outer pigment rings of a row of adjoining circular elements (Henke, 1933; Nijhout, 1991).

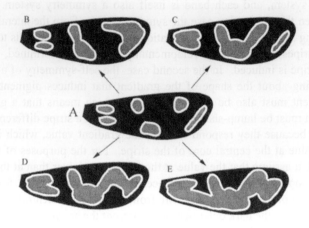

FIGURE 4 Fusion among rounded pattern elements in moths of the genus *Zygaena* (Lepidoptera: Zygaenidae). These various patterns of fusion are found as interspecific diversity as well as intraspecific variation and indicate that the rounded elements of the pattern are produced by identical processes (after Nijhout, 1994).

In the course of pattern evolution in the moths, the arrangement of the elements that produced symmetry systems became increasingly regular. They came to be arranged in several parallel series, running from the anterior to the posterior margins of the wing. Some species of moths have as many as six such parallel symmetry systems, presumably derived from six parallel rows of centers of origin. In many moths, particularly in the Arctiidae and Geometridae, the bands of adjacent symmetry system can fuse (Fig. 5), indicating again that the different systems are produced by identical developmental processes (Nijhout, 1994).

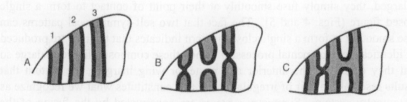

FIGURE 5 Fusion among bands of symmetry systems in moths of the family Arctiidae. Bands belonging to the same symmetry system can fuse, as can bands belonging to adjacent symmetry systems. Such fusions indicate that identical processes produce each of the symmetry systems (for photos of specimens illustrating these kinds of fusion patterns, see Nijhout, 1991, 1994).

The butterflies typically have three symmetry systems, although in many species the basal symmetry system is missing or is represented only by its distal-most band. The bands of each symmetry system usually have a distinctive pigmentation and morphology. The central elements of the border symmetry system (called the border ocelli; Fig. 1B) are almost always distinct and highly differentiated, in contrast to the situation in moths where these elements are usually absent or only faintly expressed. There are no cases in which the bands of adjacent symmetry systems fuse so as to make the central fields of those systems contiguous, as we see in certain moths (Fig. 5). It appears then that as the butterflies diverged from the moths, each of their symmetry systems acquired the ability to express a distinctive pigmentation and morphology (Fig. 6) and they became sufficiently differentiated from each other that fusion between adjacent systems is no longer possible. The most complex evolution has occurred in the border symmetry system. Not only have the central elements of this system undergone complex elaboration and differentiation (Nijhout, 1991), but even the two bands of the symmetry pair have diverged from each other in pigmentation and morphology so that in many species it is difficult to see that the two are sister bands (or homologous members) of the same symmetry system.

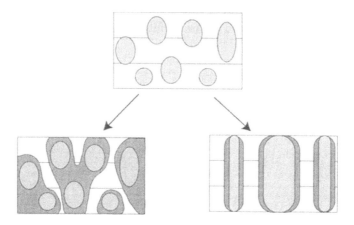

FIGURE 6   Origin of symmetry systems form random spotting patterns (A). Primitive moths have patterns that originate by fusion of irregularly arranged centers of origin (B, see also Fig. 4). In the ancestors of the butterflies these centers became arranged in three regular rows, producing three symmetry systems (after Nijhout, 1994).

A Hierarchy of Symmetry Systems

We have seen that there are two ways of making sets of stripes that look like symmetry systems: (1) by the intercalation of stripes between existing ones (Figs. 3 and 7A) and (2) by the expansion and truncation of radially symmetrical patterns (Figs. 2 and 5). In the first case, exemplified earlier by zebra and *Pterois* color patterns, it is necessary that the primary elements (the main pigment stripes) first be irreversibly determined so that the process that produces the intercalated element (the shadow stripe) can no longer alter the two flanking primary elements. In this case symmetry is imposed by the temporal sequence in which bands are determined. In the second case, symmetry is a preexisting condition that is the consequence of the concentric organization of a pattern element. We can think of the concentric rings of such a pattern as corresponding to isoclines or contours of the determination gradient. Spot-like patterns would be produced by humped or circular mound-like gradients and stripes or bands by elongated mounds or ridge-shaped gradients.

An added complication to this system is that in some butterflies and moths the bands of a symmetry system are themselves composed of a symmetrical arrangement of pigments (Nijhout, 1991). In order for a band to be self-symmetrical in pigmentation, a band cannot simply be a threshold or contour on a gradient but must be a ridge of some sort. The simplest way to obtain this would be if the band is initially formed at a threshold on a gradient but then that site becomes the source or high point of a new gradient (Fig. 7B), very much in the way the sequence of gradients in early embryonic development of *Drosophila* is set up (Lawrence, 1992). A symmetry system thus begins as a single self-symmetrical band and each of its edges then becomes the organizing center of a new symmetry system. The edge of a band, initially merely an aspect of a structure (the band), can thus acquire an identity independent from the structure of which it was a part.

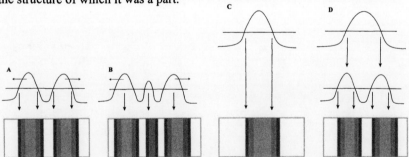

FIGURE 7 Mechanisms for the formation and multiplication of symmetry systems. Pigment bands are formed at several threshold levels of a symmetrical gradient for pigment determination, here depicted as a roughly bell-shaped curve. A and B. Intercalation of a new symmetry system (or stripe) when two existing symmetry systems (or stripes) move apart. C and D. Each threshold becomes the origin of a new symmetrical gradient for pigment determination.

Individuation in Symmetrical Structures

The hierarchy of symmetry systems outlined earlier poses an interesting problem in character evolution. One can ask the question: are the members of a pair of symmetrical structures different characters? Two appendages of a symmetrical pair or the petals of a radially symmetrical flower, for instance, are certainly different structures; they have independent developmental origins, and if the development of one is perturbed the other is unaffected. However, insofar as they are perfect mirror images of each other it is difficult to see them as different characters because exactly the same information went into their manufacture and exactly the same information can be extracted from their analysis. Symmetrical structures may have developmental independence, but they have no individuality, if by individuality we mean the possession of characteristics that distinguish one from the other. In bilateral symmetry (but not in radial symmetry) the two are at least mirror images of each other, so they are not perfectly identical and are at least recognizable as left or right instantiations of the character, but that does not qualify them as different characters. Only if the two for some reason diverged in their morphology, like the cutter and crusher claws of a lobster, do they become different characters, and then only in those features in which they diverged. The causes of divergence can be manifold. A simple change in absolute size, if it is accompanied by allometric changes, can lead to the appearance of many differences between two structures that, had they been the same size, would have been absolutely identical in all details. What is needed in such a case is some developmental event that influences the two members of the pair in a qualitatively or quantitatively different way.

If members of a symmetrical pair develop in a spatially asymmetrical environment then during its ontogeny each will experience unique interactions not shared with its sister structure. Differences in physical interactions and differences in local patterns of gene expression will cause each member to develop unique characteristics. This is the case during the development of symmetry systems, since they develop on a wing with distinct proximo-distal and antero-posterior differentiation. This is why the proximal and distal bands of a symmetry system can differ in shape, pigmentation, and distance from their common center. Individuation of a pigment band requires that it be subject to some developmental influences that are not shared with other bands, so that it develops unique characteristics by which it can be distinguished from all other pigment bands.

Presumably any symmetrical structure can be the locus for the multiplication and subsequent differentiation of parts. Differentiation of an initially homogeneous structure, whether a zygote, an insect segment, or a symmetry system, into many different parts must necessarily occur by a mechanism that somehow generates differences in different regions of that structure. Those differences can be imposed from the outside by a spatially heterogeneous environment or they can be generated from within by processes that break up the initial homogeneity. Recent findings about the control of embryonic determination show that this happens primarily by means of concentration gradients of gene products that control the differential expression of genes. Different thresholds then give rise to expression of new genes in some areas and not in others. The new gene products then diffuse and produce new gradients, and a succession of such gradient and threshold events gradually subdivides a field into ever smaller and more specialized regions, each of which can become recognized as a different part or character.

In order for such a process of successive subdivision of a developmental field to produce discrete structures there must exist boundaries of some sort so that the effects of region-specific gradients remain localized. If communication is by diffusion of transcription factors, for instance, it would be difficult to constrain their effect and it would be impossible to produce a sharply defined bounded part, character, or pattern element. Some mechanism for establishing boundaries is therefore necessary and several quite different mechanisms appear to exist. Boundary sharpening of gradients can be achieved by positive feedback and lateral inhibition processes of the general kind that are believed to be involved in the formation of zebra stripes and insect segments (Bard, 1977, 1981; Meinhardt, 1982). In the case of symmetry systems, it is the large physical distance between the two bands of a system that provides an effective barrier. One band can be subject to developmental influences that simply decay below some threshold before they reach the region where the other band develops. This is in some sense a trivial mechanism and it would probably not be useful for producing closely spaced or densely packed features, unless the decay can vary sharply in space, or thresholds are very sharply defined (and cells can differentiate between minute differences in a gradient), or unless those structures are not determined simultaneously. However, when different features are determined sequentially, so that one is fixed before the process that determines the next one begins, it should be possible to develop tightly packed features without sharp boundaries on the spread of information. Finally, it is possible to have physical barriers that block communication between different parts of a developing field. In such a case the field would be effectively compartmentalized into developmentally independent units and the boundaries of the compartment would at least in part define the boundaries of structures and parts that subsequently develop. Butterfly wings become compartmentalized this way, as we will see next.

Evolution of Serial Homology

In butterfly wing patterns the symmetry system bands do not run continuously across the wing but are interrupted and dislocated at the wing veins so that the band is broken up into a series of short pattern elements (Fig. 1B). The wing veins appear to act as a barrier to communication between different parts of the wing epithelium and they effectively compartmentalize the wing with regard to pattern formation. Evidence for this compartmentalization comes from several sources: pattern elements are often abruptly truncated precisely at the wing veins; experimental perturbation by damaging the wing epithelium can severely distort the pattern that develops but this effect remains restricted to the wing compartments in which the damage is localized; genetic studies have revealed that there are genes that can alter the pattern within a single compartment without affecting the pattern elsewhere on the wing (Nijhout *et al.*, 1990; Nijhout, 1991); finally, species-specific patterns are often characterized by specific modifications of the pattern within a single compartment (Nijhout, 1991). Thus observational, experimental, and genetic evidence indicates that veins somehow constrain pattern-forming processes to areas within their boundaries. We have preliminary evidence from electron microscopy that gap junctions, which are abundant in the epithelial cells of the wing, are actually absent in epithelial cells near the wing veins. These junctions are required for cytoplasmic communication between cells, and in their absence, diffusible signals cannot get from one cell to another. Gap junction-less regions of epithelium at the wing veins would thus act as effective barriers to the diffusion of morphogenetic signals, and communication across such a barrier would be difficult.

Symmetry systems evolved long before they became compartmentalized by the wing veins. In the vast majority of moths, for instance, symmetry system bands run smoothly across the wing without interruption or dislocation. The origin of compartmentalization of the pattern can be understood by comparative studies of patterns in the moths and their sister group, the caddisflies (Trichoptera). Although caddisflies do not have symmetry systems, many have a color pattern made up of irregular random bands, called ripple patterns, and these bands are always interrupted and truncated at the wing veins. Many moths have similar ripple patterns and wherever these are found they are always truncated at the wing veins, even in species that have symmetry system bands that are uninterrupted by wing veins (Nijhout, 1994). From perturbation studies we know that during ontogeny, ripple patterns are determined before the elements of the nymphalid groundplan. This finding implies that in moths the wing is compartmentalized for color pattern formation at the time ripple patterns are determined but not some time later when the symmetry systems are determined. The simplest mechanism to account for this is that gap junctions are absent in the epithelium around the wing veins during the early stages of color pattern determination, and that gap junctions form in these areas during the later stages of pattern determination.

The ability to form compartments at the wing veins is evidently very old, predating the origin of the Lepidoptera, and was thus present in the lineage of moths within which the butterflies evolved.[2] What happened in the course of butterfly evolution is that the compartmentalization persisted during a longer period of ontogeny so that it came to overlap the time at which the symmetry systems were determined. Compartmentalization of the wing for symmetry systems could thus be due to a simple heterochronic shift in the timing of gap junction inactivation. The consequence of this compartmentalization, however, has proven to be completely out of proportion to the simplicity of the mechanism, as it enabled the production of an unprecedented and unrivalled diversification of patterns.

Because there is no communication between compartments, it is difficult for processes in one compartment to influence those in adjacent ones. Only factors that that are shared by two regions of the wing before compartments are formed, or that are transmitted through the extracellular medium and can thus avoid compartment boundaries, can have common causal effects on the events in two compartments. In either event, uncoupling between compartments assures that independent instances of pattern formation take place in each. If these instances use exactly the same developmental information, then one would expect identical patterns to be produced in each compartment. Therefore, the net result of compartmentalization is the serial repetition of pattern elements in each compartment, with each series corresponding to one of the original symmetry system bands.

Individuation of Serial Homologues

The compartments do not, however, contain exactly the same developmental information. They differ in size and shape and those differences can have profound effects on diffusion-dependent developmental events. The size and shape of a developmental field affects the shape of diffusion gradients that develop within it and this alters the position and shape of thresholds. So simple and unavoidable differences in the initial conditions of compartment formation should affect pattern development in each compartment differently. These differences can cause misalignment of the replicates that develop in each compartment, so that a single smoothly continuous band is now broken up into unconnected and dislocated elements. Differences in initial and boundary conditions can also cause the replicate elements in each compartment to differ in shape and size, even though they all share the same genetic determinants.

In each compartment there will therefore be systematic differences in the pattern that develops. Subsequent evolutionary events can then gradually reduce or magnify these differences. In order for genetic changes to alter the pattern in one compartment and not in another there has to be a nonlinear association between genetic variation and pattern variation so that a small

---

[2] Exactly which family of moths forms the sister group to the butterflies is still unresolved and a matter of much speculation and controversy.

genetic change can have a big effect in one location on the wing and a small effect elsewhere. Such nonlinearities are actually part of the very nature of a diffusion-threshold process (Nijhout and Paulsen, 1997; Klingenberg and Nijhout, 1998) and are characteristic of many other types of developmental and pattern-forming processes as well (Murray, 1989). Indeed, genetical studies of *Heliconius* have shown that genes have evolved that affect the pattern in one compartment only (Nijhout *et al.*, 1990; Nijhout, 1991). It is, therefore, possible for the pattern elements in each compartment to acquire characteristics that are not shared by their homologues in other compartments, and thus to acquire a modicum of genetic and developmental independence from its serial homologues. The degree to which mutation and selection can alter one pattern element without affecting other pattern elements within its compartment, or other pattern elements in its homologous series, depends on the number of shared developmental determinants. These shared determinants are expressed as genetic correlations among pattern elements (Paulsen and Nijhout, 1993; Paulsen, 1994; and see section "Pattern Elements as Characters").

Each pattern element shares a few relatively weak genetic correlations for size, shape, and position with its serial homologues and with its neighbors in a compartment, but is otherwise remarkably free of such correlations (Paulsen, 1994). This finding reveals that the normal variation in size, shape, and position of different pattern elements is *not* dominated by the effects of the many developmental factors the two have in common, but by factors that are unique to each element of the pattern. By compartmentalization and subsequent evolution, each pattern element has acquired unique morphological properties and unique patterns of variation, and through this uniqueness it has acquired an identity that sets it apart from all the other elements of the color pattern.

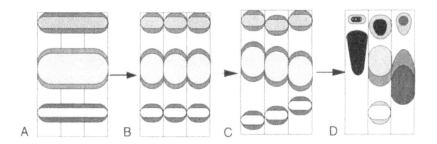

FIGURE 8 In butterflies, the wing pattern is compartmentalized by the wing veins so that the position, pigmentation, and shape of the bands in a compartment can be modified independently of those in an adjoining compartment (after Nijhout, 1994).

E PLURIBUS UNUM AND EX UNO PLURIA

The hypothesis presented earlier for the origin and individuation of pattern elements emerges from studies of comparative morphology, genetics, and development, and proposes the following scenario for the evolution of this evolvable system. The origin of individuated pattern elements probably began with a simple system of circular patterns, all formally identical to one another, like the stripes of a zebra or the spots of a leopard. The self-symmetrical nature of these original patterns led to their fusion into parallel systems of paired bands (Fig. 6). By interactions with the asymmetrical developmental environment of the wing, these simple bands each acquired different characteristics, so that distinctive systems of patterns evolved along the proximo-distal axis of the wing. Subsequent compartmentalization of the wing surface allowed the development of each segment of a band to be uncoupled from that of its neighbors. In each compartment the developmental environment then diverged sufficiently so that each pattern element was able to acquire a unique morphology (Fig. 8) and independent genetic variation. The consequence of the developmental uncoupling of pattern elements was a system of morphological parts virtually unencumbered by mutual constraints on their variation and evolution. Each element could be independently modified in form, color, and position, and the ensemble is able to build up a virtually unlimited diversity of color patterns (e.g., Fig. 9).

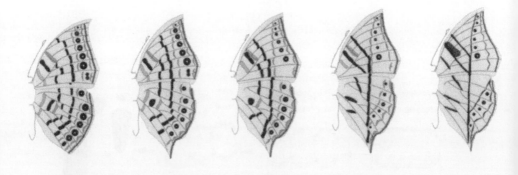

FIGURE 9   Derivation of the dead-leaf mimicking wing pattern of *Kallima inachus* from the elements of the nymphalid groundplan (Fig. 1B). The sequence of patterns shown is not intended to represent an evolutionary transformation series but is given to illustrate how each element of the pattern is modified and displaced to produce the overall image of a veined leaf.

ACKNOWLEDGMENTS

I would like to thank Louise Roth, Dan McShea, and Chris Klingenberg for critical comments on the manuscript and Monique Nijhout for drawing Fig. 3. This work was supported in part by grants from the National Science Foundation.

LITERATURE CITED

Bard, J. B. L. (1977). A unity underlying the different zebra striping patterns. *J. Zool.* **183**:527-539.

Bard, J. B. L. (1981). A model for generating aspects of zebra and other mammalian coat patterns. *J. Theor. Biol.* **93**:363-385.

Falconer, D. S., and Mackay, T. F. C. (1996). "Introduction to Quantitative Genetics," 4th ed. Longman Sci. and Techn., UK.

Held, L. I. (1992). "Models for Embryonic Periodicity." Karger, Basel.

Henke, K. (1933). Untersuchungen an *Philosamia cynthia* Drury zur Entwicklungsphysiologie des Zeichnungsmusters auf dem Schmetterlingsflügel. *Wilhelm Roux' Arch.* **128**:15-107.

Klingenberg, P. C., and Nijhout, H. F. (1999). Genetics of fluctuating asymmetry: a developmental model of developmental instability. *Evolution* **53**:358-375.

Lawrence, P. A. (1992). "The Making of a Fly." Blackwell, London.

Lynch, M., and Walsh, B. (1997). "Genetics and Analysis of Quantitative Traits." Sinauer, Sunderland, MA.

Meinhardt, H. (1982). "Models of Biological Pattern Formation." Academic Press, New York.

Murray, J. D. (1989). "Mathematical Biology." Springer-Verlag, New York.

Nijhout, H. F. (1991). "The Development and Evolution of Butterfly Wing Patterns." Smithsonian Inst. Press, Washington, DC.

Nijhout, H. F. (1994). Symmetry systems and compartments in lepidopteran wings: the evolution of a patterning mechanism. *Development* (Suppl.) 225-233.

Nijhout, H. F., and Paulsen, S. M. (1997). Developmental models and polygenic characters. *Am. Nat.* **149**:394-405.

Nijhout, H. F., Wray, G., and Gilbert, L.E. (1990). An analsis of the phenotypic effects of certain color pattern genes in *Heliconius* (Lepidoptera: Nymphalidae). *Biol. J. Linn. Soc.* **40**:357-372.

Oster, G. F. (1988). Lateral inhibition models of developmental processes. *Math. Biosci.* **90**:265-286.

Paulsen, S. M. (1994). Quantitative genetics of butterfly wing color patterns. *Dev. Genet.* **15**:79-91.

Paulsen, S. M., and Nijhout, H. F. (1993). Phenotypic correlation structure among the elements of the color pattern of *Precis coenia* (Lepidoptera: Nymphalidae). *Evolution* **47**:593-618.

23

PERSPECTIVES ON THE EVOLUTIONARY ORIGIN OF TETRAPOD LIMBS

JAVIER CAPDEVILA AND JUAN CARLOS IZPISÚA BELMONTE

The Salk Institute for Biological Studies, La Jolla, CA 92037

INTRODUCTION

The basic structure of the vertebrate limb is remarkably conserved among amniote tetrapods. It consists of a proximal part (stylopod) with a single skeletal element, a medial part (zeugopod) with two elements, and a distal part (autopod) composed of carpus or tarsus and radiating digits. This morphological conservation appears to be the consequence of the utilization of similar molecular mechanisms to control growth and patterning in the limb of most tetrapods (reviewed by Johnson and Tabin, 1997; Ng *et al.*, 1999). The vertebrate limb bud is probably the structure for which more molecular data are available in terms of understanding patterning mechanisms in developing embryos. Classical embryological studies in avian and amphibian embryos and gene knockouts in mouse have contributed greatly to our understanding of how positional information is established and how growth and patterning are integrated in the limb.

In the last two decades, the study of limb development has benefited from a new synthesis of developmental and evolutionary biology based on the reinterpretation of the fossil record and on new discoveries from developmental genetics (Gilbert *et al.*, 1996; Raff, 1996; Gerhart and Kirschner, 1997; Wilkins, 1998). The analysis of the fossil record (especially of Late Devonian tetrapods discovered in recent years) has helped to establish the evolutionary history of the vertebrate limb, and developmental biologists have attempted to explain this evolutionary history in terms of changes in the activity of genes involved in

The Character Concept in Evolutionary Biology

embryonic development. Traditionally, two types of developmental changes have been postulated to explain the origin of macroevolutionary novelties (Raff and Kaufman, 1983; Gilbert *et al.*, 1996; Raff, 1996): heterochrony (alteration in the timing of some developmental event) and allometry (differential growth of parts). The mechanisms that could explain rapid morphological changes in evolution have been discussed at length by several authors (see for instance Gould, 1977; Raff; 1996). The challenge for developmental biologists is to provide detailed and plausible molecular explanations for the developmental changes that are proposed to lead to macroevolutionary novelties.

In this scenario, what does developmental biology tell us about the origin and evolution of the vertebrate limb? First, that the molecular processes that control growth and patterning in the vertebrate limb also operate in the development of other appendages (and body wall outgrowths in general) in many organisms (reviewed by Shubin *et al.*, 1997). There seems to be a very specific molecular formulation to make an outgrowth that sticks out of the main body axis. That molecular formulation, composed of several genetic networks, constitutes a "developmental module" that is used time and again during embryonic development to control growth and patterning of many body wall outgrowths, including limbs. Second, that changes in the spatial or temporal pattern of expression of key developmental genes, like the *Hox* genes, are associated with the appearance of morphological innovations such as digits, or with a change in the number and shape of digits. We are reaching the point where developmental biologists should be able to provide a plausible mechanistic explanation for the sequence of genetic events that culminated in the adoption of the basic tetrapod limb plan.

In this review, we focus on some aspects related to the evolution of vertebrate limbs from a molecular perspective. We first discuss what is known about appendages in general and limb induction and limb positioning in particular. We then analyze the role of some key genes involved in outgrowth and patterning, and we finally discuss the genetic control of digit formation and the developmental constraints that influence the evolution of vertebrate limb morphology.

ORIGIN OF APPENDAGES

Animal appendages are outgrowths of the body wall that are adapted for specialized functions such as feeding and locomotion. The tetrapod limb is just one type of appendage, a particular example of body wall outgrowth that played a key role in the appearance of terrestrial vertebrates in the late Devonian. In recent years, it has also become clear that, even after more than 500 million years of independent evolution, a very similar genetic regulatory machinery operates in both vertebrate and arthropod appendages (reviewed by Shubin *et al.*, 1997). The prevailing interpretation of this fact is that an ancient genetic network, operating in a common ancestor, controlled the formation of sensory, feeding, or locomotor outgrowths that were already patterned along the anteroposterior, dorsoventral, and proximodistal axes (Panganiban *et al.*, 1997). Thus, the genetic machinery required to build a tetrapod limb is considered to be a modern version of an ancient genetic mechanism that controlled body wall outgrowths in many primitive organisms.

One of the best examples to support this interpretation comes from the study of the homeodomain-containing transcription factor Distalless (Dll/Dlx), which seems to play a major role in the mechanism controlling appendage formation. The *Dll* gene was originally cloned in *Drosophila*, where it was shown to be expressed at the distal tip of growing limbs (Cohen *et al.*, 1989; Panganiban *et al.*, 1994, 1995). Clonal analysis demonstrated that *Dll* activity is required for appendage outgrowth in *Drosophila* (Cohen and Jurgens, 1989). Since its discovery in the fly, *Dll/Dlx* genes have been cloned in many different organisms. A comparison of the expression of *Dll/Dlx* in several metazoan phyla has revealed that *Dll/Dlx* is associated with outgrowths from the body wall in many different animals. For example, *Dll/Dlx* genes are expressed in the tubefeet and spines of sea urchins, in Onychophoran lobopodia and antennae, in ampullae and siphons of ascidians, in the parapodia of annelids, and in vertebrate limb buds (Panganiban *et al.*, 1997).

What is known about the evolutionary history of the *Dll/Dlx* genes? In vertebrates, *Dll/Dlx* genes are distributed in pairs that are located on the same chromosomes (Rossi *et al.*, 1994; Simeone *et al.*, 1994; Stock *et al.*, 1996). Phylogenetic analyses of *Dll/Dlx* gene sequences suggest that a primitive tandem duplication resulted in a linked pair of *Dll/Dlx* genes, presumably after the divergence of arthropods and chordates, but prior to the divergence of tunicates and vertebrates. That pair of *Dll/Dlx* genes was later duplicated in the chromosomal events that also originated the four clusters of *Hox* genes present in bony fish and tetrapods (Stock *et al.*, 1996). Some regulatory elements that direct *Dll/Dlx* expression in specific organs or structures (including the vertebrate limb bud) have been identified (Morasso *et al.*, 1995). Also, linked *Dll/Dlx* genes show overlapping patterns of expression during embryonic development, which suggests that they share cis-acting sequences (Ellies *et al.*, 1997). Cross-regulatory interactions between *Dll/Dlx* genes may be responsible for their overlapping patterns of expression in several organisms (Zerucha *et al.*, 1997).

So far, a direct role for *Dll/Dlx* in the outgrowth of vertebrate appendages has not been demonstrated, although the pattern of expression of several *Dll/Dlx* genes in the vertebrate limb bud suggests their involvement in apical ectodermal ridge (AER) activity, cartilage differentiation, and other functions (Dollé *et al.*, 1992; Bulfone *et al.*, 1993; Ferrari *et al.*, 1995; Panganiban *et al.*, 1997; Ferrari *et al.*, 1999).

How do *Dll/Dlx* genes actually work? The exact mechanism is still unknown, but it could involve local control of cell division. Campbell and

Tomlinson (1998) have also recently proposed a role for *Dll/Dlx* in the control of adhesive properties of cells. *Dll* null clones in leg imaginal disks in *Drosophila* appear to sort out from wild-type surrounding cells, which seems to indicate that *Dll*-expressing cells share specific adhesive properties. The authors speculate that this function of controlling cell adhesion could be related to an ancient role in the formation of outgrowths from the body wall in different organisms. It should be pointed out, however, that *Dll/Dlx* genes are not associated with all types of outgrowths. For example, in several crustaceans, *Dll/Dlx* expression is not associated with the appearance of limb branches (Williams, 1998). This suggests that not all outgrowths are patterned by the simple iteration of the *Dll/Dlx*-dependent mechanism that patterns the unbranched leg of *Drosophila*.

Panganiban and collaborators speculate that, most likely, the ancestral role of *Dll/Dlx* genes was unrelated to appendages or body wall outgrowths. For example, *Dll/Dlx* genes are expressed in arthropods in the CNS and PNS, and in vertebrates in the CNS, which suggests that *Dll/Dlx* function arose in the CNS before being coopted to direct outgrowths from the body wall. This observation can be generalized to state that the genetic network that controls appendage growth and patterning was most likely already present before appendages and outgrowths appeared (Panganiban *et al.*, 1997).

LIMB INDUCTION AND OUTGROWTH

Basal chordates (Amphioxus) do not have appendages. Tetrapods have two sets of paired appendages, derived from the pectoral and pelvic fins of gnathostomes (reviewed by Coates, 1994). The first paired appendages appeared in some jawless fish, and gnathostomes already had pectoral and pelvic paired fins (Carroll, 1988). We do not know how fins appeared in these primitive fish. Our knowledge about how limbs are induced at specific locations along the main body axis derives almost exclusively from experiments conducted with avian embryos. Limbs originate in the flanks of the early embryo, where groups of cells in the lateral plate mesoderm develop into small buds of mesenchymal cells encased in an ectodermal jacket (Searls and Janners, 1971). In many tetrapods, as development proceeds, mesodermal signals induce the distal part of the ectoderm to form a thickening called apical ectodermal ridge (AER), which runs along the anteroposterior axis of the limb bud, separating the dorsal aspect of the limb from the ventral aspect (Saunders, 1948; Todt and Fallon, 1984). The integrity of the AER is essential to keep the limb cells proliferating and the limb bud growing. When the AER is surgically removed, proliferation of the limb bud cells is affected and the resulting limb is truncated (Saunders, 1948). The AER interacts with the zone of polarizing activity (ZPA), a group of mesenchymal cells in the posterior margin of the limb bud that control anteroposterior patterning in the limb (Saunders and Gasseling, 1968). The polarizing activity of the ZPA is mediated by *sonic hedgehog* (*shh*; Riddle *et al.*, 1993; López-Martínez *et al.*, 1995). The *shh* gene encodes a secreted factor homologous to the product of the *Drosophila* segment polarity gene *hedgehog* (*hh*), involved in many patterning processes in the embryo and imaginal discs (Lee *et al.*, 1992; Mohler and Vani, 1992; Tabata *et al.*, 1992; Tashiro *et al.*, 1993; reviewed by Ingham, 1998). The mesenchymal cells in the distal part of the limb bud constitute the progress zone (PZ), which is kept in a proliferative state by the AER (Summerbell *et al.*, 1973). Cells in the PZ give rise to most of the skeletal elements of the limb. As the limb grows, mesenchymal cells leave the PZ moving proximally and acquiring positional information to give rise to the skeletal elements (Summerbell *et al.*, 1973), which develop in a proximodistal direction that follows a specific sequence of prechondrogenic condensations and bifurcations. This sequence is highly conserved among tetrapods, as revealed by comparative morphogenetic analyses (Shubin and Alberch, 1986).

This general scheme of limb development does not apply to all vertebrates. For example, in osteichthyan fish like the zebrafish, *Danio rerio*, the apical fin bud ectoderm does not form an AER, but it rather transforms into a protruding fold that encloses the dermal rays and terminates proliferation of the mesenchyme of the fin bud (Geraudie, 1978). The result is a proximodistal subdivision of the mesenchyme that forms four radials, and several peripheral foci form other distal radials. Recently, several mutations in zebrafish that affect fin formation have been identified (van Eeden *et al.*, 1996), but their exact roles in fin bud patterning have not been examined in detail.

The AER is clearly a major organizing structure in the limb bud of higher tetrapods, but not all tetrapod limbs have an AER. Richardson and collaborators speculate that the AER was present in the common ancestor of anurans and amniotes, and that it has been lost in several species which are direct developers, including several species of frog. In these frogs there is a thickened apical ectoderm, but no AER (Richardson *et al.*, 1998). In slow worms and other reptilians, the AER appears but later on it degenerates and the adult is limbless. The molecular basis of this phenomenon is not well known, but Raynaud and collaborators have demonstrated that treatment with basic fibroblast growth factor (FGF) protein is able to maintain proliferation of slow worm limb bud cells (Raynaud and Kan, 1992; Raynaud *et al.*, 1995). FGF and chick AER are also able to rescue leg bud outgrowth in python embryos (Cohn and Tickle, 1999). FGF proteins (which are expressed in the AER in vertebrate limb buds; reviewed by Martin, 1998) have been previously shown to be able to substitute for the AER in maintaining the proliferation of PZ cells and outgrowth of the limb (Niswander *et al.*, 1993; Fallon *et al.*, 1994; Mahmood *et al.*, 1995). It would be interesting to analyze the expression of endogenous *Fgfs* genes and other genes involved in limb induction and outgrowth in slow worm embryos and other limbless reptiles (see an example of this approach in Cohn and Tickle, 1999). This type of comparative analysis should provide useful information about the factors required to induce limbs and to support their growth in the embryonic flank.

How are limbs positioned along the main body axis in the first place? The mechanism of limb induction is still a matter of controversy, but some molecular players have already been identified (reviewed by Martin, 1998). The *Fgf-8* gene is expressed transiently and dynamically in the intermediate mesoderm at the forelimb and hindlimb levels before limbs are induced, and *Fgf-8* activity can maintain cells in a proliferative state at the flank positions that correspond to the limb fields (Crossley *et al.*, 1996; Vogel *et al.*, 1996). FGF-8 protein (and other proteins of the same family) seems to be able to direct initiation and normal development of a limb bud from the embryonic flank (Cohn *et al.*, 1995; Ohuchi *et al.*, 1995; Crossley *et al.*, 1996; Vogel *et al.*, 1996). Interestingly, another member of the same gene family, FGF-10, could be involved in mediating the effect of FGF-8 on the lateral plate mesoderm cells (Ohuchi *et al.*, 1997; Xu *et al.*, 1998). The *Fgf-10* gene has been shown to be induced by FGF-8, and FGF-10 is able to induce ectodermal cells to form an AER in a more direct way than FGF-8 (Ohuchi *et al.*, 1997). Targeted mutation of the *Fgf-10* gene in mice results in the absence of limbs (Min *et al.*, 1998; Sekine *et al.*, 1999).

Although ectopic FGF proteins can induce the development of an ectopic limb in the flank of the embryo, factors other than FGF proteins could also be involved in limb induction. For example, the Hensen's node has limb-inducing activity (Dealy, 1997), and insulin-like growth factor-I (IGF-I) and insulin are produced by medial tissues next to the prospective limb-forming regions of the lateral plate, and they are able to induce limb bud-like structures *in vitro* (Dealy and Kosher, 1996). Thus, although enormous progress has been made in the last few years in our molecular understanding of limb induction, the exact molecular nature of the limb inducer remains elusive.

Irrespective of the nature of the endogenous limb inducer, its function clearly depends on its adequate spatiotemporal pattern of expression in axial structures prior to limb induction. *Hox* genes, which encode homeodomain transcription factors initially identified in *Drosophila*, play a key role in this process. It is generally accepted that a combinatorial expression of *Hox* genes in the embryonic trunk and lateral plate mesoderm (Cohn *et al.*, 1997) determines the level at which the limbs are going to develop, and perhaps the initial anteroposterior polarity also. An interesting result that supports the involvement of a combinatorial *Hox* code in positioning the vertebrate limbs is that mice lacking the *Hoxb-5* gene have the shoulder girdle slightly shifted (Rancourt *et al.*, 1995), which is consistent with a role for *Hoxb-5* in establishing positional cues in the embryonic axis. According to this view, limbs would be induced in the embryonic flank at specific positions that contain certain combinations of *Hox* gene expression. Once the limb bud has been induced, several *Hox* genes also seem to be important in delimiting the region where the ZPA (and shh expression) is going to be located. Thus, *Hoxb-8* was proposed to be required for the initiation of *shh* expression in the posterior mesenchyme of the limb bud (Lu *et al.*, 1997a; Stratford *et al.*, 1997), although it would not be required for *shh* maintenance (Charité *et al.*, 1994). Besides, ectopic *Hoxb-8* in the anterior margin of the mouse limb bud is able to induce ectopic *shh*, which results in pattern duplications (Charité *et al.*, 1994). However, the fact that *Hoxb-8*-deficient mice have normal limbs (Van den Akker *et al.*, 1999) clearly indicates that *Hoxb-8* is not necessary for *shh* expression, which could be controlled by a combination of several *Hox* genes. Another *Hox* gene, *Hoxd-12*, has also been shown to be involved in the control of *shh* in the limb (Knezevic *et al.*, 1997; Hérault *et al.*, 1998; Mackem and Knezevic, 1999).

Hox genes provide spatial cues in a variety of embryonic structures in vertebrates (reviewed by Krumlauf, 1994; Burke *et al.*, 1995; Cohn and Tickle, 1999), and the study of *Hox* gene regulation is an active field of research (reviewed by Duboule, 1998). In the embryonic axis, *Hox* gene expression is controlled by several factors. Retinoic acid (RA) seems to be a key regulator of *Hox* genes (reviewed by Marshall *et al.*, 1996). Interestingly, RA is involved in controlling *Hox* gene expression in the lateral plate mesoderm at the time at which the limb fields are determined. Inhibition of RA activity in the embryonic flank prevents initiation of *shh* expression in the limb, most likely by down-regulating *Hoxb-8* (Lu *et al.*, 1997). Also, it has been demonstrated that the *Hoxb-8* gene has regulatory elements that bind Cdx proteins (Charité *et al.*, 1998). These proteins are homologues of *Drosophila* Caudal, a protein involved in anteroposterior patterning in the embryo. Other *Hox* genes are also regulated by Cdx proteins (Shashikant *et al.*, 1995; Subramanian *et al.*, 1995). Charité and collaborators (1998) have proposed an ancestral role for Cdx/Caudal proteins in specifying anteroposterior axial patterning in a variety of organisms through the control of *Hox* gene expression boundaries. The TGF-ß factor Gdf-11 also appears to act upstream of *Hox* genes, since *Gdf-11*-deficient mice show a posterior displacement of the hindlimbs that correlates with alterations in *Hox* gene expression in the trunk (McPherron *et al.,* 1999).

It is important to point out that both *Hox* and *Fgf* genes are older than limbs. Some authors have speculated that a small number of ancestral *Hox* genes were initially involved in the specification of polarity in the digestive tract and they were later on recruited in the patterning of other axial structures and of body wall outgrowths, including limbs (van der Hoeven *et al.*, 1996). The establishment and maintenance of the spatial and temporal distribution of *Hox* gene expression play a key role in determining axial pattern in many organisms. It is becoming clear that specific differences in *Hox* transcription patterns can be correlated with anatomical modifications in different species, which clearly suggests that evolutionary variation of *Hox* cis-regulatory elements has played an important role in the origin and evolution of body plans (reviewed by Gellon and McGinnis, 1998). In the case of *Fgfs*, several members of the superfamily have been shown to be required during gastrulation in the mouse embryo (reviewed by Rossant *et al.*, 1997), which suggests that their ancestral role was related to very basic mechanisms of axial patterning. *Fgfs* were probably recruited later for derived functions such as the induction of the limb buds. Coulier and collaborators (1997) have proposed a model of evolution of the *Fgf* superfamily through phases of gene duplications, one of which may have coincided with the emergence of vertebrates.

ORIGIN OF DIGITS

The key event in the evolutionary transition from fins to limbs was the acquisition of digits, about 360 million years ago. Until very recently, the prevailing theory stated that digits (both fingers and toes) were a morphological novelty that appeared in osteolepiform fish (such as *Eusthenopteron foordi*) as an adaptation to the need of supporting the weight of the animal during terrestrial locomotion. However, recent analyses of fossils of one of the most primitive known tetrapods, the amphibian *Acanthostega gunnari* (Coates, 1996), and of an aquatic rhizodontid sarcopterygian fish with finger-like structures in the pectoral fin (Daeschler and Shubin, 1998) have suggested that digits evolved in water as an event unrelated to the need of terrestrial locomotion. The osteolepiform-tetrapod relationship has also been recently revised by Ahlberg and Johanson (1998) in a way that suggests that we may reasonably expect that the discovery of fossils of more primitive forms will provide more details about the entire lineage that leads to tetrapods (Janvier, 1998; Johanson and Ahlberg, 1998).

On the matter of the origin of digits, two main views have dominated the debate for more than a century: one that considers digits as being unique to tetrapods (Holmgren, 1933), and another that considers radials of fins (of sarcopterygian fish) homologous to digits (Gregory and Raven, 1941). Indeed, both tetrapod digits and fin radials seem to branch both anteriorly and posteriorly from the metapterygial axis (the main axis of cartilaginous condensation in the limb), but only if this axis is considered to be a straight line. In 1986, Shubin and Alberch expanded Holmgren's initial observation that the digital arches in tetrapods are discontinuous with the rest of the limb (Holmgren, 1952) and, based on comparative morphological analyses of prechondrogenic limb patterns, they proposed a redefinition of the metapterygial axis as bending anteriorly through the digital arch. According to this view, tetrapod digits would now lie now postaxially to the metapterygial axis (Shubin and Alberch, 1986; see Fig. 1). In 1991, Coates noticed a correspondence between the pattern of expression of *Hoxd* genes in the limb bud and the "bent" metapterygial axis of the tetrapod limb (Coates, 1991), as redefined by Shubin and Alberch (1986). Since then, it has become clear that the proper development of digits in the vertebrate limb requires the activity of *Hoxa* and *Hoxd* gene complexes (Tabin, 1992; Dollé *et al.*, 1993; reviewed by Rijli and Chambon, 1997; Kondo *et al.*, 1998).

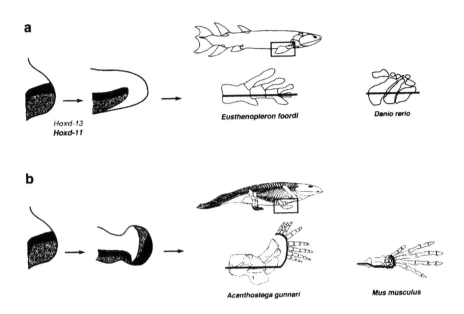

FIGURE 1 According to some interpretations, the origin of digits correlates with a reversal in the polarity of *Hoxd* expression. Sarcopterygian fin buds, like those of *Eusthenopleron foordi* (a), presumably had nested expression of *Hoxd-11, -13* in the posterior margin of the bud. Expression persists in older buds, fading in the most posterior edge. This gives rise to an adult fin with a straight metapterygial axis (in black), similar to the one observed in the zebrafish (*Danio rerio*). In contrast, tetrapod limb buds like those of *Achantostega gunnari* or the mouse (*Mus musculus*) (b) have a later phase of *Hoxd* expression where nested expression of *Hoxd-11, -13* expands anteriorly and distally and the anterior limits of the genes are reversed with respect to the early phase. This "bending" of the metapterygial axis and reversed *Hoxd* polarity correlates with bending of the appendicular axis and the appearance of digits. The illustrations were adapted from Coates and Clack (1990), Ahlberg and Milner (1994), Ahlberg and Johanson (1998), Coates (1995), and Sordino *et al.* (1995).

The prevailing view at this time is that the appearance of digits seems to be associated with the adoption of novel spatial and temporal patterns of expression of *Hox* genes in the vertebrate limb bud (Sordino *et al.*, 1995; Nelson *et al.*, 1996; Sordino and Duboule, 1996; reviewed by Shubin *et al.*, 1997). In an early stage during limb bud development, *Hoxd-11, -13* genes are expressed in a nested sequence in the posterior edge of the limb bud in tetrapods, and of the fin bud in fish like the teleost *Danio rerio* and (presumably) sarcopterygian fish like the osteolepiform *Eusthenopleron* (Fig. 1). As the limbs elongate, expression of *Hoxd-11, -13* maintains the same pattern in fin buds, but in tetrapods (including, presumably, *Acanthostega gunnari*), *Hoxd-11, -13* expand more anteriorly in the distal part of the limb bud, which is the region that eventually gives rise to the digital arch, from which the digits form. Thus, a novel late phase of *Hoxd-11, -13* expression in the distal part of the limb bud seems to be closely associated with the appearance of digits, perhaps through the induction of new patterns of cell proliferation (Zákány *et al.*, 1997). Shubin and collaborators (1997) have proposed that the reversal of the relative anterior boundaries of the 5' most *Hoxd* genes (that were posteriorly restricted at an early stage) is closely correlated with the origin of digits.

Clearly, comparative studies in several organisms are necessary to get a complete picture of the origin of digits. For example, it would be useful to investigate *Hox* gene expression during fin bud development in lungfish, which are much closer to tetrapods than teleost fish like *Danio rerio*. Also, although in most tetrapods skeletal elements appear in a posterior to anterior sequence in the zeugopod and autopod, in urodele amphibians the sequence goes from anterior to posterior (Shubin and Alberch, 1986). Recent studies have revealed that in the axolotl limb there is no anterior distal expansion of the posterior domain of *Hoxd-11* expression, contrary to what is observed in mice and chicks. This result suggests that the anterior expansion of expression of *Hoxd* genes in higher vertebrates is probably linked to the formation of a handplate (axolotls do not form an expanded paddle-like handplate prior to digit differentiation), but is not necessary for digit differentiation, at least in limbs in which the sequence of digit formation goes from anterior to posterior (Torok *et al.*, 1998).

How are the different phases of *Hoxd* expression in the limb bud controlled at the genetic level? A complex set of enhancers within the regulatory regions of each *Hoxd* gene is used during the early phase (Beckers *et al.*, 1996; van der Hoeven *et al.*, 1996), characterized by nested domains of expression in the posterior margin of the limb bud (Nelson *et al.*, 1996). In contrast, the late phase of *Hoxd* expression, characterized by antero-distal expansion and polarity reversal of *Hoxd* expression (Nelson *et al.*, 1996), is regulated by a single enhancer (Gérard *et al.*, 1993; van der Hoeven *et al.*, 1996). The fact that up to five posterior *Hoxd* complex genes act through a shared regulatory element suggests that digits are evolutionary novelties that originated through the simultaneous recruiting of several *Hoxd* genes to be expressed in the developing limb in an antero-distal domain (the autopod). This scenario would imply that a relatively small number of evolutionary steps would be sufficient to cause a change in the expression of multiple genes. It is now the time for developmental biologists to suggest a plausible mechanistical account of the molecular changes that explain the appearance of digits.

Regarding the number of digits, most tetrapods have a pentadactyl digit formula, but polydactyly seems to be common in early limbed tetrapods. For example, *Tulerpeton curtum* had six digits, and *Acanthostega gunnari* had eight. Some authors speculate that paddle-like autopods, with multiple short digits, could have been advantageous to these primitive aquatic tetrapods. Locomotion in a terrestrial environment, conversely, would favor autopods with fewer, longer digits (Zákány *et al.*, 1997). Zákány and collaborators have recently proposed a model for a dose-dependent regulation of number and size of digits by *Hox* genes. According to this model (Fig. 2), which is based on the analysis of mouse mutants, digit size and number are quantitatively specified by the dose of Hox proteins, and not by a qualitative *Hox* code (Zákány *et al.*, 1997; Zákány and Duboule, 1999). Thus, a progressive reduction in the dose of *Hox* gene products causes a succession of phenotypes where ectrodactyly (reduction in digit size) is observed first, then polydactyly (extra digits), oligodactyly (loss of digits), and finally adactyly (absence of digits). Noting that the succession goes through a step of polydactyly, the authors speculate that *Hoxa* genes were predominantly active in the distal appendage of polydactylous short-digited ancestral tetrapods (such as *Acanthostega*), and that *Hoxd* genes were recruited later in evolution, which resulted in a reduction in digit number and an increase in digit length (Sordino *et al.*, 1996; Zákány *et al.*, 1997). This would provide an account of specific genetic changes responsible for the morphological transitions observed during vertebrate limb development.

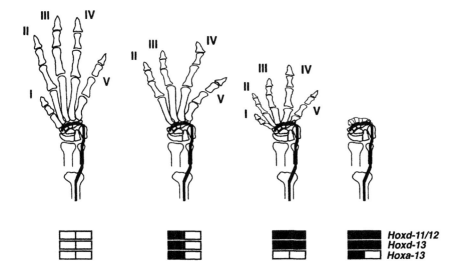

FIGURE 2   In the mouse, *Hoxd* and *Hoxa* genes control the size and number of digits in a dose-dependent fashion. A dark square indicates a loss-of-function allele of the corresponding gene, and a grey square indicates wild-type allele. Progressive reduction in the dose of *Hox* gene products causes first ectrodactyly, then olygodactyly, and finally adactyly, but going through a step of polydactyly where up to seven digit vestiges can be identified in the severely affected limbs (similar to the last limb in the figure). The figure shows mouse forelimb skeletons. Adapted from Zákány *et al.* (1997).

This model notwithstanding, *Hoxa* and *Hoxd* complexes seem to have different roles in digit patterning. For example, it does not seem possible to cause polydactyly without affecting *Hoxd* genes, and oligodactyly is only observed when perturbing *Hoxa* genes. Also, the proposed model would not apply to anamniotes, since they seem to have reached pentadactyly independently (Coates, 1994).

The dose of Hox proteins is not the only factor that controls digit number in vertebrates. Experiments in chick embryos have shown that the number of digits also depends on the extension of the AER. Consistent with this, experimental manipulations that extend the AER result in extra digits. Also, local alterations in the pattern of programmed cell death in the limb may lead to the appearance of extra digits, presumably because in that situation cells that are committed to die can now form cartilage (Katagiri *et al.*, 1998). Several mutant mice display ectopic expression of *shh* or *Indian hedgehog* (*Ihh*) in the anterior margin of the limb bud, which results in digit duplications (Chan *et al.*, 1995; Masuya *et al.*, 1995, 1997; Büscher *et al.*, 1997; Qu *et al.*, 1997, 1998; Takahashi *et al.*, 1998; Yang *et al.*, 1998). Some other mutant mice have extra digits that do not seem to correlate with alterations in the pattern of expression of any member of the *hh* gene family in the limb (Dudley *et al.*, 1995; Fawcett *et al.*, 1995; Luo *et al.*, 1995; Hofmann *et al.*, 1996; Rodríguez *et al.*, 1996; Dunn *et al.*, 1997; Katagiri *et al.*, 1998; ten Berge *et al.*, 1998). Clearly, a variety of local alterations in the mesenchyme and the AER of the developing limb bud can result in the appearance of extra digits, which suggests a multigenic control of digit patterning during vertebrate limb development.

What about the genetic mechanisms that determine the differences between the digits? In vertebrates, most likely there are no specific genes for specific fingers. In other organisms there is some evidence that specific pattern elements in the appendages are closely associated with restricted patterns of expression of certain genes. For example, in *Drosophila melanogaster*, the expression of the genes *knirps*, *knirps-related*, and *spalt* in the wing imaginal disc has been shown to be closely linked to the development of wing vein number two (Sturtevant *et al.*, 1997; Biehs *et al.*, 1998; Lunde *et al.*, 1998), and expression of the *collier* gene is related to the development of vein three and the space between veins three and four (Vervoort *et al.*, 1999; Mohler *et al.*, 2000).

We still do not know what makes the differences between the digits. Recently, however, Bone Morphogenetic Proteins (BMPs) have been proposed to play a key role during digit determination in the chick (Drossopoulou *et al.*, 2000), although the mechanism by which they do is still unclear. Of course, the problem of digit identity is just part of the bigger problem of describing the molecular mechanism that patterns the limb along the anteroposterior axis. In this process, the protein encoded by the gene *shh* is a key player.

THE ROLE OF *SONIC HEDGEHOG*

Shortly after the limb bud is induced, *shh* expression is detected in the posterior margin of the limb bud, colocalizing with the ZPA (Echelard *et al.*, 1993; Krauss *et al.*, 1993; Riddle *et al.*, 1993). In 1993, Riddle and collaborators demonstrated that ectopic application of *shh* to the anterior margin of the limb bud was able to induce mirror-image duplications of the digits, thus mimicking ZPA activity. Since then, other *hh* genes have been cloned in many different organisms, and they have been shown to be involved in many developmental decisions in several organs and tissues. Several components of its transduction pathway have already been identified (reviewed by Ingham, 1998 and Johnson and Scott, 1998).

Shh is required neither for the initiation of limb development nor for the establishment of initial anteroposterior polarity of the limb, since even in the complete absence of Shh protein there is some anteroposterior polarity in the limb (Noramly *et al.*, 1996; Ros *et al.*, 1996). However, *shh* is absolutely required for the maintenance of limb bud growth and patterning. *Shh* acts in a regulatory loop with FGF proteins expressed in the AER to maintain cell growth and proliferation in the mesenchyme, and to maintain the integrity of the AER (Niswander *et al.*, 1994; Laufer *et al.*, 1994), through the control of the BMP antagonist Gremlin (reviewed by Vogt and Duboule, 1999). Mice that are null for *shh* (Chiang *et al.*, 1996) have limbs, but they are reduced and the skeletal pattern is altered, displaying severe distal truncations that include absence of digits.

Although *shh* is able to mimic the ZPA activity, it seems unlikely that Shh itself gives positional information to all the cells in the limb bud. Shh protein does not seem to diffuse a long distance *in vivo* (Martí *et al.*, 1995), and a membrane-tethered form of Shh is still able to elicit a dose-dependent patterning response, which suggests that (at least) part of the organizing activity of Shh is mediated by secondary signals (Yang *et al.*, 1997) such as BMPs and others. A new *Drosophila* gene, named *tout-velu*, has been shown to be required for the diffusion of Hh proteins (Bellaiche *et al.*, 1998). The gene encodes an integral membrane protein that belongs to the *EXT* gene family, which is involved in the human multiple exostoses syndrome that affects bone morphogenesis. Since one of the vertebrate Hh proteins, Ihh, is involved in bone morphogenesis, the authors propose that the exostoses syndrome could be caused by the abnormal diffusion of Hh proteins. The mechanism by which EXT proteins regulate Hh diffusion has been reviewed by Perrimon and Bernfield (2000).

Shh expression has been detected in the posterior margin of the limb or fin bud in all types of vertebrate limbs and fins examined so far (Echelard *et al.*, 1993; Krauss *et al.*, 1993; Riddle *et al.*, 1993; Sordino *et al.*, 1995; Zardoya *et al.*, 1996; Endo *et al.*, 1997; Imokawa and Yoshizato, 1997; Stark *et al.*, 1998).

One exception is a recently described newt homologue of *Xenopus banded hedgehog* (Ekker *et al.*, 1995), called *N-bhh*, which is uniformly expressed both in the early limb bud of the newt embryo and in the mesenchymal blastemal cells from the initial stages of regeneration. Since *N-bhh* has not been detected in developing limbs of higher vertebrates, it has been speculated that its expression in developing and regenerating newt limbs may somehow be related to the regenerative capability of urodeles (Stark *et al.*, 1998). In the vertebrate species that can regenerate limbs or fins, *shh* expression has always been observed to be closely associated with the regenerating areas (Endo *et al.*, 1997; Imokawa and Yoshizato, 1997; Stark *et al.*, 1998). In general, tetrapod limb buds present a distalized pattern of *shh* expression when compared to fin buds. For example, in the zebrafish pelvic fin *shh* is confined to the posterior margin, never reaching the most distal mesenchymal cells (Sordino *et al.*, 1995). Thus, distalization of *shh* expression seems to correlate well with the fin to limb transition. It has been shown in chick and mouse limbs that several *Hoxd* genes are controlled (at least partially) by *shh*, so it could be argued that, most likely, the distalization of *shh* expression has played a major role in the appearance of novel spatial and temporal patterns of expression of *Hoxd* genes in the vertebrate limb bud.

As pointed out by Coates (1995), it is reasonable to assume that a basic network of signaling factors involved in the control of growth and patterning of the limb, most likely including *shh*, *Fgfs*, *Wnts*, and others (all of them interacting with *Hox* genes), was already present before the evolutionary split between ray- and lobe-finned bony fishes. The common ancestor of both groups presumably had two sets of paired fins containing dermal rays, and the pelvic and pectoral fins were probably dissimilar. These and other morphological features suggest the existence of elaborated mechanisms that control the timing of transition from ectodermal ridge to fin fold, and also the existence of mechanisms to impose asymmetry upon the endoskeletal pattern and the pattern of cellular proliferation in the fin bud. Expression studies in the fin buds of extant ray-finned fishes might help test this assumption.

EVOLVABILITY OF THE VERTEBRATE LIMB

Why is the basic plan of the tetrapod limb so remarkably conserved? The prevailing view is that the appearance of the basic plan provided the best adaptation to the necessities of feeding and locomotion in an aquatic environment during Ordovician and late Devonian. Successful evolutionary novelties (like the appearance of fingers and toes) were selected and fixed, in some way "locked-up." Thus, although parts of the limb show more variability than others (the autopod being the more variable in number and shape of digits), the basic architecture has been maintained, which has led many evolutionary biologists to speculate on the nature of the "developmental constraints" that have limited the evolution of the tetrapod limb. Development constrains evolution by limiting the universe of possible phenotypes that may result from selection (Alberch, 1982).

In order to understand why a successful structure such as the tetrapod limb becomes locked-up, that is, fixed during evolution, one needs to argue that the genetic mechanisms that build the limb are not exclusive. A common theme in molecular evolution is the reiterated use of successful genes or even complete genetic algorithms ("modules" or "syntagmata") in the making of what appear to be very different structures in the same developing organism (see for instance García Bellido, 1994; Gilbert *et al.*, 1996). For example, many genes involved in limb development (such as *shh* and *Hox* genes) are also involved in the development of the nervous system or the urogenital system. The capability of the *shh* gene or a particular *Hox* gene to mutate in a way that has some chance to be positively selected is severely restricted by the need to keep the integrity of a genetic module that is required for the development of several other organs or structures. Thus, the fact that key regulatory genes are required for many different functions in the organism imposes a constraint in the "evolvability" of organs or structures that are linked by the use of common genetic modules. This evolutionary constraint is an important factor to be considered when studying how evolving developing systems are controlled by genetic networks (Duboule and Wilkins, 1998; Hérault and Duboule, 1998; Kirschner and Gerhardt, 1998).

Of course, this constraint does not entirely prevent tetrapod limbs from evolving. One example is the independent evolution of forelimbs and hindlimbs. According to fossil evidence, forelimbs and hindlimbs appeared at the same time, as did fingers and toes (Coates, 1994). Thus, forelimbs and hindlimbs are considered to be "serially homologous" structures, which implies that, in some way, the same molecular formulation has been used to build two pairs of very similar structures at different locations along the body flank (Shubin *et al.*, 1997). What makes forelimbs different from hindlimbs must be some kind of differential gene expression or regulation. We are just beginning to understand the molecular mechanisms that could have driven the independent evolution of pectoral and pelvic appendages. In recent years several genes have been described that are expressed exclusively in forelimbs or hindlimbs in mouse, chick, and other organisms. For example, *Tbx-5* is expressed exclusively in the forelimb, and *Tbx-4* in the hindlimb (Bollag *et al.*, 1994; Simon *et al.*, 1997; Gibson-Brown *et al.*, 1998; Isaac *et al.*, 1998; Logan *et al.*, 1998; Ohuchi *et al.*, 1998). Both genes are members of the family of T-box transcription factors. *Pitx-1*, a member of the Otx-related subclass of paired-type homeodomain proteins (Lamonerie *et al.*, 1996; Szeto *et al.*, 1996), is expressed exclusively in the hindlimb (Shang *et al.*, 1997; Logan *et al.*, 1998). *Hoxc-4* and *Hoxc-5* are restricted to the forelimb (Nelson *et al.*, 1996). Experiments involving loss of gene function in mice and ectopic expression in chick embryos have recently demonstrated the role of *Tbx-5* as a forelimb determinant (Rodriguez-Esteban *et al.*, 1999; Takeuchi *et al.*, 1999) and of *Tbx-4* and *Pitx-1* as hindlimb determinants (Lanctot *et al.*, 1999; Logan and Tabin, 1999; Rodriguez-Esteban *et al.*, 1999; Szeto *et al.*, 1999; Takeuchi *et al.*, 1999; reviewed by Graham and McGonnell, 1999). The role of these genes as "selectors" of limb identity has been reviewed by Weatherbee and Carroll (1999).

FUTURE EXPERIMENTAL APPROACHES AND CONCLUSION

The remarkable genetic similarity between vertebrate and arthropod appendages has made possible the use of homology cloning to isolate and characterize many genes involved in the control of vertebrate limb growth and patterning. This "homology approach" to study the mechanisms of vertebrate limb development has been, by far, the most successful one in the last two decades. We can anticipate that this approach will still be useful in the future, although the focus will probably shift slightly toward the understanding of the mechanisms that determine specific differences in development. We know that similar genes make similar structures in very different organisms, and we marvel at the conservation of very complex genetic mechanisms, but what makes a fly wing so different from a chicken wing if the genes involved are so similar? Clearly, a deeper understanding of the spatial and temporal regulation of key developmental genes and their targets is required in order to even begin scratching the surface of this problem.

We should count on technical improvements that will expand the number of species available for comparative studies. For example, gene transfer is a powerful tool to analyze gene activity during embryonic development but, so far, efficient methods have only been developed for a few species. Gene targeting in mouse and transgenic approaches in mouse, chick, and *Drosophila* have provided invaluable information about appendage development. In the mouse, new techniques that use ultrasound microscopy for guiding injections of cells or viruses into early stage mouse embryos *in utero* (Liu *et al.*, 1998) should make possible gain-of-function studies in mutant embryos. If we want to study limb development in species other than the "traditional" model systems, we need methods to introduce foreign DNA into a variety of embryos, tissues, and organs in vertebrates and invertebrates. In recent years, several techniques to deliver DNA into developing embryos of several species have been described. Preliminary results of gene transfer have been reported in insect embryos (Leopold *et al.*, 1996; Lewis *et al.*, 1999; Oppenheimer *et al.*, 1999), medaka fish (Lu *et al.*, 1997b), newt limbs (Burns *et al.*, 1994; Pecorino *et al.*, 1994), and others. Future improvement in the efficiency and convenience of these methods would undoubtedly help us to analyze many aspects of limb development in different species.

In theory, to obtain a complete understanding of the genetic events that led to the evolutionary changes of the vertebrate limb would require a massive effort of comparison of regulatory sequences that control the spatial and temporal expression of key developmental genes. The availability of the entire sequences of several genomes in the near future should provide a great opportunity to compare regulatory sequences of genes involved in key developmental decisions in the limb. But what genes should we focus on? Strong candidates for this approach are genes that belong to the *Hox, hh, Wnt, Fgf,* and *BMP* superfamilies. All of these genes are involved in many developmental decisions in vertebrate embryos, but we should not limit our efforts to the analysis of only a few candidates. New molecular players have been added to the list of factors likely to control growth and patterning in the vertebrate limb, so that the Hox code might not be the only "molecular code" operating in the limb. For example, Eph receptor tyrosine kinases and their ligands are involved in the guidance of motor axons and neural crest cells through the anterior half of the somites (Krull *et al.,* 1997), and they are also required for somite development (Durbin *et al.,* 1998). In the vertebrate limb bud, there is a reciprocal compartmentalization of Eph receptors and ligands that suggests that these molecules could be involved in the formation of spatial boundaries that may help to organize the pattern of the vertebrate limb (Flenniken *et al.,* 1996; Gale *et al.,* 1996; Patel *et al.,* 1996).

A method that has already been used to identify regulatory sequences of genes involved in embryonic development in mouse is the enhancer-trap system. Mouse embryonic stem cell lines are generated that carry *β-galactosidase* trap constructs integrated in their genome. The analysis of the pattern of expression of *β-galactosidase* at different stages of embryonic development makes possible the identification of lines that express the reporter in interesting patterns, driven by regulatory sequences located around the integration site. The cloning of sequences flanking the insertion site is relatively easy (reviewed by Korn *et al.,* 1992; Voss *et al.,* 1998). Expression of the same or other reporters in transgenic lines can also be driven by putative regulatory elements already isolated from the vicinity of interesting genes. This kind of transgenic promoter analysis is very important in identifying single regulatory elements that direct expression of key genes in specific areas of the embryo. A related technique is the GAL4-UAS system, which has been used extensively in *Drosophila melanogaster* (Brand and Perrimon, 1993) and, lately, has also been applied to the mouse (originally described by Ornitz *et al.,* 1991; McMahon, 1998). In this system, a target strain carries a transgene controlled by yeast UAS regulatory sequences, which respond only to the yeast transcriptional activator GAL4, while a transactivator strain expresses an active GAL4 gene driven by any selected promoter. When both lines are crossed, mouse embryos that carry both lines express the transgene under the control of the selected promoter. This system can be used to express any transgene in a controlled way in any organ or tissue for which specific promoters are available, or it can be modified to be used as an enhancer-trap method. The development of new systems of conditional genome alteration on mice (reviewed by Lobe and Nagy, 1998) is a very active field of research which promises to be particularly useful for the study of limb development in vertebrates.

Interestingly, important regulatory elements have been shown to be remarkably conserved in different species. For example, a short regulatory element from a *Xenopus Dll/Dlx* gene is able to confer appropriate expression on a *ß-galactosidase* reporter gene in ectodermal structures in the mouse (Morasso *et al.*, 1995). Also, regulatory elements that control AER-specific gene expression seem to be highly conserved between mouse and human (Liu *et al.*, 1994). Finally, an interesting project would be to analyze the regulatory sequences of genes specifically expressed either in forelimbs or hindlimbs but not both (*Tbx-4*, *Tbx-5*, *Pitx-1*, and others) in order to unveil the genetic mechanisms that determine the differences between forelimbs and hindlimbs. It would be interesting to see if similar exclusive patterns of expression are also observed in other vertebrates.

What can be expected from the detailed analysis of the key molecular players in limb development? In the best scenario, developmental biologists should be able to explain how specific changes in gene expression lead to the appearance of limbs and their subsequent modifications during evolution. However, there are also important methodological questions to be asked: is that massive effort of analysis and comparison of regulatory sequences feasible and worthwhile? The analysis of the data obtained from the massive genome sequencing projects still in progress should help answer these questions. The completed sequences of key genomes could reveal, taken together, a remarkable pattern of conservation of regulatory elements that would make further analysis much faster than expected before. There is hope for this possibility, since it has been shown that the genomes of several species (see for example Schofield *et al.*, 1997) show remarkable conservation of synteny (that is, similar genes and even key regulatory sequences are located in similar relative chromosomic positions in several related species).

The new synthesis that we call evolutionary developmental biology (Gilbert *et al.*, 1996; Raff, 1996; Gerhart and Kirschner, 1997; Wilkins, 1998) has already had a tremendous impact in the way biologists think about evolution, but there is an ongoing debate about the scope and methods of this relatively new discipline. Several authors have pointed out that evolutionary developmental biology seems too focused on developmental mechanisms, without providing convincing explanations of how the molecular changes responsible for evolutionary changes actually occur and spread through populations (see for example Gilbert *et al.*, 1996; Wilkins, 1998). These authors claim that a "population genetics of regulatory genes" is necessary in order to complement the molecular hypotheses that, so far, only provide partial answers to the question of how evolutionary novelties actually appear. Developmental biology usually provides hypotheses about what kind of genetic changes could be responsible for a certain evolutionary novelty, but it usually fails at providing a detailed account of how the changes actually occurred and were selected (see

Palopoli and Patel, 1996; Wilkins, 1998). We should expect a shift toward more integrated approaches to evolutionary developmental biology so that problems such as the origin, evolution, and morphogenetic roles of genetic networks are studied in the context of their actual relevance in populations.

In the emerging field of evolutionary developmental biology, the problem of the origin and evolution of the tetrapod limb has attracted considerable attention. The abundance of fossil specimens, especially from the late Devonian, has allowed paleontologists to formulate detailed hypotheses on the sequence of character acquisition in early tetrapods, including detailed predictions on limb evolution. The view that digits appeared in Devonian fishes as an adaptation to an aquatic environment is now generally accepted, after the analysis of recently discovered transitional forms, and there is reasonable hope that new fossil discoveries will help us reconstruct the entire lineage that led to the tetrapods. As mentioned before, developmental biologists have proposed molecular hypotheses to account for the appearance of digits, which are evolutionary novelties. However, for our molecular predictions to be of explanatory value, we should aim to make a detailed reconstruction of the genetic changes that led to the appearance of appendages in the first place and, later in evolutionary history, to the fin-limb transition, a landmark event in the appearance of vertebrates adapted to new terrestrial, aquatic, and aerial environments. In conclusion, the goal is to achieve a detailed account of how the genetic changes appeared and how and when these changes were selected and fixed in specific phyla to eventually give rise to the basic plan of the tetrapod limb.

ACKNOWLEDGMENTS

J.C. was supported by a Hoffman Foundation Fellowship. Work in the laboratory is supported by grants from N.I.H and the G. Harold and Leila Y. Mathers Charitable Foundation. J.C.I.B. is a Pew Scholar.

LITERATURE CITED

Ahlberg, P. E., and Milner, A. R. (1994). The origin and early diversification of tetrapods. *Nature* **368:**507-513.

Ahlberg, P. E., and Johanson, Z. (1998). Osteolepiforms and the ancestry of tetrapods. *Nature* **395:**792-793.

Alberch, P. (1982). Developmental constraints in evolutionary processes. *In* "Evolution and Development: Report of the Dahlem Workshop on Evolution and Development" (J. T. Bonner, ed.), pp. 313-332. Springer-Verlag, New York.

Beckers, J., Gérard, M., and Duboule, D. (1996). Transgenic analysis of a potential Hoxd-11 limb regulatory element present in tetrapods and fish. *Dev. Biol.* **180:**543-553.

Bellaiche, Y., The, I., and Perrimon, N. (1998). Tout-velu is a Drosophila homologue of the putative tumour suppressor EXT-1 and is needed for Hh diffusion. *Nature* **394:**85-88.

Biehs, B., Sturtevant, M. A., and Bier, E. (1998). Boundaries in the Drosophila wing imaginal disc organize vein-specific genetic programs. *Development* **125:**4245-4257.

Bollag, R., Siegfried, Z., Cebra-Thomas, J. A., Garvey, N., Davison, E. M., and Silver, L. M. (1994). An ancient family of embryonically expressed mouse genes sharing a conserved protein motif with the T locus. *Nat. Genet.* **7**:383-389.

Brand, A. H., and Perrimon, N. (1993). Targeted gene expression as a means of altering cell fates and generating dominant phenotypes. *Development* **118**:401-415.

Bulfone, A., Kim, H. J., Puelles, L., Porteus, M. H., Grippo, J. F., and Rubenstein, J. L. (1993). The mouse Dlx-2 (Tes-1) gene is expressed in spatially restricted domains of the forebrain, face and limbs in midgestation mouse embryos. *Mech. Dev.* **40**:129-140.

Burke, A. C., Nelson, C. E., Morgan, B. A., and Tabin, C. (1995). Hox genes and the evolution of vertebrate axial morphology. *Development* **121**:333-346.

Burns, J. C., Matsubara, T., Lozinski, G., Yee, J. K., Friedmann, T., Washabaugh, C. H., and Tsonis, P. A. (1994). Pantropic retroviral vector-mediated gene transfer, integration, and expression in cultured newt limb cells. *Dev. Biol.* **165**:285-289.

Büscher, D., Bosse, B., Heymer, J., and Rüther, U. (1997). Evidence for genetic control of Sonic hedgehog by Gli3 in mouse limb development. *Mech. Dev.* **62**:175-182.

Campbell, G., and Tomlinson, A. (1998). The roles of the homeobox genes aristaless and Distal-less in patterning the legs and wings of Drosophila. *Development* **125**:4483-4493.

Carroll, R. L. (1988). "Vertebrate Paleontology." Freeman, San Francisco.

Chan, D. C., Laufer, E., Tabin, C., and Leder, P. (1995). Polydactylous limbs in Strong's Luxoid mice result from ectopic polarizing activity. *Development* **121**:1971-1978.

Charité, J., Graaff, W. D., Shen, S., and Deschamps, J. (1994). Ectopic expression of Hoxb-8 causes duplication of the ZPA in the forelimb and homeotic transformation of axial structures. *Cell* **78**:589-601.

Charité, J., de Graaff, W., Consten, D., Reijnen, M. J., Korving, J., and Deschamps, J. (1998). Transducing positional information to the Hox genes: critical interaction of cdx gene products with position-sensitive regulatory elements. *Development* **125**:4349-4358.

Chiang, C., Litingtung, Y., Lee, E., Young, K. E., Corden, J. L., Westphal, H., and Beachy, P. A. (1996). Cyclopia and defective axial patterning in mice lacking sonic hedgehog gene function. *Nature* **383**:407-413.

Coates, M., and Clack, J. (1990). Polydactyl in the earliest tetrapod limbs. *Nature* **347**:66-69.

Coates, M. I. (1991). New palaeontological contributions to limb ontogeny and phylogeny. *In* "Developmental Patterning of the Vertebrate Limb" (J. R. Hinchliffe, J. Hurle, and D. M. Summerbell, eds.), pp. 328-338. Plenum, New York.

Coates, M. I. (1994). The origin of vertebrate limbs. *Development*, **Suppl.** 169-180.

Coates, M. I. (1995). Fish fins or tetrapod limbs -a simple twist of fate? *Curr. Biol.* **5**:844-848.

Coates, M. I. (1996). The Devonian tetrapod *Acanthostega gunnari* Jarvik: postcranial anatomy, basal tetrapod interrelationships and patterns of skeletal evolution. *Trans. R. Soc. Edinb. Earth Sci.* **87**:363-421.

Cohen, S. M., and Jurgens, G. (1989). Proximal-distal pattern formation in Drosophila: cell autonomous requirement for Distal-less gene activity in limb development. *EMBO J.* **8**:2045-2055.

Cohen, S. M., Bronner, G., Kuttner, F., Jurgens, G., and Jackle, H. (1989). Distal-less encodes a homoeodomain protein required for limb development in Drosophila. *Nature* **338**:432-434.

Cohn, M. J., and Tickle, C. (1999). Developmental basis of limblessness and axial patterning in snakes. *Nature* **399**:474-479.

Cohn, M. J., Izpisúa Belmonte, J.-C., Abud, H., Heath, J. K., and Tickle, C. (1995). Fibroblast growth factors induce additional limb development from the flank of chick embryos. *Cell* **80**:739-746.

Cohn, M. J., Patel, K., Krumlauf, R., Wilkinson, D. G., Clarke, J. D., and Tickle, C. (1997). Hox9 genes and vertebrate limb specification. *Nature* **387**:97-101.

Coulier, F., Pontarotti, P., Roubin, R., Hartung, H., Goldfarb, M., and Birnbaum, D. (1997). Of worms and men: an evolutionary perspective on the fibroblast growth factor (FGF) and FGF receptor families. *J. Mol. Evol.* **44:**43-56.

Crossley, P. H., Minowada, G., MacArthur, C. A., and Martin, G. R. (1996). Roles for FGF8 in the induction, initiation, and maintenance of chick limb development. *Cell* **84:**127-136.

Daeschler, E. B., and Shubin, N. (1998). Fish with fingers?. *Nature* **391:**133.

Dealy, C. N., and Kosher, R. A. (1996). IGF-I and insulin in the acquisition of limb-forming ability by the embryonic lateral plate. *Dev. Biol.* **177:**291-299.

Dealy, C. N. (1997). Hensen's node provides an endogenous limb-forming signal. *Dev. Biol.* **188:**216-223.

Dollé, P., Price, M., and Duboule, D. (1992). Expression of the murine Dlx-1 homeobox gene during facial, ocular and limb development. *Differentiation* **49:**93-99.

Dollé, P., Dierich, A., LeMeur, M., Schimmang, T., Schuhbaur, B., Chambon, P., and Duboule, D. (1993). Disruption of the Hoxd-13 gene induces localized heterochrony leading to mice with neotenic limbs. *Cell* **75:**431-441.

Drossopoulou, G., Lewis, K. E., Sanz-Ezquerro, J. J., Nikbakht, N., McMahon, A. P., Hofmann, C., and Tickle, C. (2000). A model for anteroposterior patterning of the vertebrate limb based on sequential long- and short-range Shh signaling and Bmp signaling. *Development* **127:**1337-1348.

Duboule, D. (1998). Vertebrate hox gene regulation: clustering and/or colinearity? *Curr. Opin. Genet. Dev.* **8:**514-518.

Duboule, D., and Wilkins, A. S. (1998). The evolution of "bricolage." *TIG* **14:**54-59.

Dudley, A.T., Lyons, K. M., and Robertson, E. J. (1995). A requirement for bone morphogenetic protein-7 during development of the mammalian kidney and eye. *Genes Dev.* **9:**2795-2807.

Dunn, N. R., Winnier, G. E., Hargett, L. K., Schrick, J. J., Fogo, A. B., and Hogan, B. L. (1997). Haploinsufficient phenotypes in Bmp4 heterozygous null mice and modification by mutations in Gli3 and Alx4. *Dev. Biol.* **188:**235-247.

Durbin, L., Brennan, C., Shiomi, K., Cooke, J., Barrios, A., Shanmugalingam, S., Guthrie, B., Lindberg, R., and Holder, N. (1998). Eph signaling is required for segmentation and differentiation of the somites. *Genes Dev.* **12:**3096-3109.

Echelard, Y., Epstein, D. J., St-Jacques, B., Shen, L., Mohler, J., McMahon, J. A., and McMahon, A.P. (1993). Sonic hedgehog, a member of a family of putative signaling molecules, is implicated in the regulation of CNS polarity. *Cell* **75:**1417-1430.

Ekker, S. C., McGrew, L. L., Lai, C. J., Lee, J. J., von Kessler, D. P., Moon, R. T., and Beachy, P. A. (1995). Distinct expression and shared activities of members of the hedgehog gene family of *Xenopus laevis*. *Development* **121:**2337-2347.

Ellies, D. L., Stock, D. W., Hatch, G., Giroux, G., Weiss, K. M., and Ekker, M. (1997). Relationship between the genomic organization and the overlapping embryonic expression patterns of the zebrafish dlx genes. *Genomics* **45:**580-590.

Endo, T., Yokoyama, H., Tamura, K., and Ide, H. (1997). Shh expression in developing and regenerating limb buds of *Xenopus laevis*. *Dev. Dyn.* **209:**227-232.

Fallon, J. F., López, A., Ros, M. A., Savage, M. P., Olwin, B. B., and Simandl, B. K. (1994). FGF-2: apical ectodermal ridge growth signal for chick limb development. *Science* **264:**104-107.

Fawcett, D., Pasceri, P., Fraser, R., Colbert, M., Rossant, J., and Giguere, V. (1995). Postaxial polydactyly in forelimbs of CRABP-II mutant mice. *Development* **121:**671-679.

Ferrari, D., Sumoy, L., Gannon, J., Sun, H., Brown, A. M., Upholt, W. B., and Kosher, R. A. (1995). The expression pattern of the Distal-less homeobox-containing gene Dlx-5 in the developing chick limb bud suggests its involvement in apical ectodermal ridge activity, pattern formation, and cartilage differentiation. *Mech. Dev.* **52:**257-264.

Ferrari, D., Harrington, A., Dealy, C. N., and Kosher, R. A. (1999). Dlx-5 in limb initiation in the chick embryo. *Dev. Dyn.* **216:**10-15.

Flenniken, A. M., Gale, N. W., Yancopoulos, G. D., and Wilkinson, D. G. (1996). Distinct and overlapping expression patterns of ligands for Eph-related receptor tyrosine kinases during mouse embryogenesis. *Dev. Biol.* **179**:382-401.

Gale, N. W., Holland, S. J., Valenzuela, D. M., Flenniken, A., Pan, L., Ryan, T. E., Henkemeyer, M., Strebhardt, K., Hirai, H., Wilkinson, D. G., Pawson, T., Davis, S., and Yancopoulos, G. D. (1996). Eph receptors and ligands comprise two major specificity subclasses and are reciprocally compartmentalized during embryogenesis. *Neuron* **17**:9-19.

García-Bellido, A. (1994). How organisms are put together. *Eur. Rev.* **2**:15-21.

Gellon, G., and McGinnis, W. (1998). Shaping animal body plans in development and evolution by modulation of Hox expression patterns. *Bioessays* **20**:116-125.

Gérard, M., Duboule, D., and Zákány, J. (1993). Cooperation of regulatory elements involved in the activation of the Hoxd-11 gene. *C. R. Acad. Sci. III* **316**:985-994.

Geraudie, J. (1978). The fine structure of the early pelvic fin bud of the trouts *Salmo gairdneri* and *S. trutta fario. Acta Zool.* **59**:85-96.

Gerhart, J., and Kirschner, M. (1997). "Cells, Embryos and Evolution." Blackwell Scientific, Oxford.

Gibson-Brown, J. J., Agulnik, S. I., Silver, L. M., Niswander, L., and Papaioannou, V. E. (1998). Involvement of T-box genes Tbx2-Tbx5 in vertebrate limb specification and development. *Development* **125**:2499-2509.

Gilbert, S. F., Opitz, J. M., and Raff, R. A. (1996). Resynthesizing evolutionary and developmental biology. *Dev. Biol.* **173**:357-372.

Gould, S. J. (1977). "Ontogeny and Phylogeny." Harvard University Press, Cambridge.

Graham, A., and McGonnell, I. (1999). Limb development: farewell to arms. *Curr. Biol.* **9**:R368-370.

Gregory, W. K., and Raven, H. C. (1941). Studies on the origin and early evolution of paired fins and limbs. *Ann. N. Y. Acad. Sci.* **42**:273-360.

Hérault, Y., and Duboule, D. (1998). Comment se construisent les doigts? *Recherche* **305**.

Hérault, Y., Beckers, J., Kondo, T., Fraudeau, N., and Duboule, D. (1998). Genetic analysis of a Hoxd-12 regulatory element reveals global versus local modes of controls in the HoxD complex. *Development* **125**:1669-1677.

Hofmann, C., Luo, G., Balling, R., and Karsenty, G. (1996). Analysis of limb patterning in BMP-7-deficient mice. *Dev. Genet.* **19**:43-50.

Holmgren, N. (1933). On the origin of the tetrapod limb. *Acta Zool.* **14**:185-295.

Holmgren, N. (1952). An embryological analysis of the mammalian carpus and its bearing upon the question of the origin of the tetrapod limb. *Acta Zool.* (Stockh.) **33**:1-115.

Imokawa, Y., and Yoshizato, K. (1997). Expression of Sonic hedgehog gene in regenerating newt limb blastemas recapitulates that in developing limb buds. *PNAS USA* **94**:9159-9164.

Ingham, P. W. (1998). Transducing Hedgehog: the story so far. *EMBO J.* **17**:3505-3511.

Isaac, A., Rodríguez-Esteban, C., Ryan, A., Altabef, M., Tsukui, T., Patel, K., Tickle, C., and Izpisúa Belmonte, J. C. (1998). Tbx genes and limb identity in chick embryo development. *Development* **125**:1867-1875.

Janvier, P. (1998). Forerunners of four legs. *Nature* **395**:748-749.

Johanson, Z., and Ahlberg, P. E. (1998). A complete primitive rhizodont from Australia. *Nature* **394**:569-573.

Johnson, R., and Tabin, C. (1997). Molecular models for vertebrate limb development. *Cell* **90**:979-990.

Johnson, R. L., and Scott, M. P. (1998). New players and puzzles in the Hedgehog signaling pathway. *Curr. Opin. Genet. Dev.* **8**:450-456.

Katagiri, T., Boorla, S., Frendo, J. L., Hogan, B. L., and Karsenty, G. (1998). Skeletal abnormalities in doubly heterozygous Bmp4 and Bmp7 mice. *Dev. Genet.* **22**:340-348.

Kirschner, M., and Gerhart, J. (1998). Evolvability. *PNAS USA* **95**:8420-8427.

Knezevic, V., De Santo, R., Schughart, K., Huffstadt, U., Chiang, C., Mahon, K. A., and Mackem, S. (1997). Hoxd-12 differentially affects preaxial and postaxial chondrogenic branches in the limb and regulates Sonic hedgehog in a positive feedback loop. *Development* 124:4523-4536.

Kondo, T, Hérault ,Y., Zákány. J., and Duboule, D. (1998). Genetic control of murine limb morphogenesis: relationships with human syndromes and evolutionary relevance. *Mol. Cell. Endocrinol.* 140:3-8.

Korn, R., Schoor, M., Neuhaus, H., Henseling, U., Soininen, R., Zachgo, J., and Gossler, A. (1992). Enhancer trap integrations in mouse embryonic stem cells give rise to staining patterns in chimaeric embryos with a high frequency and detect endogenous genes. *Mech. Dev.* 39:95-109.

Krauss, S., Concordet, J. P., and Ingham, P. W. (1993). A functionally conserved homolog of the Drosophila segment polarity gene hh is expressed in tissues with polarizing activity in zebrafish embryos. *Cell* 75:431-1444.

Krull, C. E., Lansford, R., Gale, N. W., Collazo, A., Marcelle, C., Yancopoulos, G. D., Fraser, S. E., and Bronner-Fraser, M. (1997). Interactions of Eph-related receptors and ligands confer rostrocaudal pattern to trunk neural crest migration. *Curr. Biol.* 7:571-580.

Krumlauf, R. (1994). Hox genes in vertebrate development. *Cell* 78:191-201.

Lamonerie, T., Tremblay, J. J., Lanctot, C., Therrien, M., Gauthier, Y., and Drouin, J. (1996). Ptx1, a bicoid -related homeo box transcription factor involved in transcription of the pro-opiomelanocortin gene. *Genes Dev.* 10:1284-1295.

Lanctot, C., Moreau, A., Chamberland, M., Tremblay, M. L., and Drouin, J. (1999). Hindlimb patterning and mandible development require the Ptx1 gene. *Development* 126:1805-1810.

Laufer, E., Nelson, C. E., Johnson, R. L., Morgan, B. A., and Tabin, C. (1994). Sonic hedgehog and Fgf-4 act through a signaling cascade and feedback loop to integrate growth and patterning of the developing limb bud. *Cell* 79:993-1003.

Lee, J. J., von Kessler, D. P., Parks, S., and Beachy, P. A. (1992). Secretion and localized transcription suggest a role in positional signaling for products of the segmentation gene hedgehog. *Cell* 71:33-50.

Leopold, R. A., Hughes, K. J., and DeVault, J. D. (1996). Using electroporation and a slot cuvette to deliver plasmid DNA to insect embryos. *Genet. Anal.* 12:197-200.

Lewis, D. L., DeCamillis, M. A., Brunetti, C. R., Halder, G., Kassner, V. A., Selegue, J. E., Higgs, S., and Carroll,S. B. (1999). Ectopic gene expression and homeotic transformations in arthropods using recombinant sindbis viruses. *Curr. Biol.* 9:1279-1287.

Liu, Y. H., Ma, L., Wu, L.Y., Luo, W., Kundu, R., Sangiorgi, F., Snead, M. L., and Maxson, R. (1994). Regulation of the Msx2 homeobox gene during mouse embryogenesis: a transgene with 439 bp of 5' flanking sequence is expressed exclusively in the apical ectodermal ridge of the developing limb. *Mech. Dev.* 8:187-197.

Liu, A., Joyner, A. L., and Turnbull, D. H. (1998). Alteration of limb and brain patterning in early mouse embryos by ultrasound-guided injection of Shh-expressing cells. *Mech. Dev.* 75:107-115.

Lobe, C. G., and Nagy, A. (1998). Conditional genome alteration in mice. *Bioessays* 20:200-208.

Logan, M., Simon, H.-G., and Tabin, C. (1998). Differential regulation of T-box and homeobox transcription factors suggests roles in controlling chick limb-type identity. *Development* 125:2825-2835.

Logan, M., and Tabin, C. J. (1999). Role of Pitx1 upstream of Tbx4 in specification of hindlimb identity. *Science* 283:1736-1739.

López-Martínez, A., Chang, D. T., Chiang, C., Porter, J. A., Ros, M.A., Simandl, B. K., Beachy, P. A., and Fallon, J. F. (1995). Limb-patterning activity and restricted posterior localization of the amino-terminal product of hedgehog cleavage. *Curr. Biol.* 5:791-796.

Lu, H.-C., Revelli, J.-P., Goering, L., Thaller, C., and Eichele, G. (1997a). Retinoid signaling is required for the establishment of a ZPA and for the expression of Hoxb-8, a mediator of ZPA formation. *Development* 124:1643-1651.

Lu, J. K., Burns, J. C., and Chen, T. T. (1997b). Pantropic retroviral vector integration, expression, and germline transmission in medaka *(Oryzias latipes)*. *Mol. Mar. Biol. Biotechnol.* **6:**289-295.

Lunde, K., Biehs, B., Nauber, U., and Bier, E. (1998). The knirps and knirps-related genes organize development of the second wing vein in Drosophila. *Development* **125:**4145-4154.

Luo, G., Hofmann, C., Bronckers, A. L., Sohocki, M., Bradley, A., and Karsenty, G. (1995). BMP-7 is an inducer of nephrogenesis, and is also required for eye development and skeletal patterning. *Genes Dev.* **9:**2808-2820.

Mackem, S., and Knezevic, V. (1999). Do 5'Hoxd genes play a role in initiating or maintaining A-P polarizing signals in the limb? *Cell Tissue Res.* **296:**27-31.

Mahmood, R., Bresnick, J., Hornbruch, A., Mahony, C., Morton, N., Colquhoun, K., Martin, P., Lumsden, A., Dickson, C., and Mason, I. (1995). A role for FGF-8 in the initiation and maintenance of vertebrate limb bud outgrowth. *Curr. Biol.* **5:**797-806.

Marshall, H., Morrison, A., Studer, M., Popperl, H., and Krumlauf, R. (1996). Retinoids and Hox genes. *FASEB J.* **10:**969-798.

Martí, E., Takada, R., Bumcrot, D. A., Sasaki, H., and McMahon, A. P. (1995). Distribution of Sonic hedgehog peptides in the developing chick and mouse embryo. *Development* **121:**2537-2547.

Martin, G. R. (1998). The roles of FGFs in the early development of vertebrate limbs. *Genes Dev.* **12:**1571-1586.

Masuya, H., Sagai, T., Wakana, S., Moriwaki, K., and Shiroishi, T. (1995). A duplicated zone of polarizing activity in polydactylous mouse mutants. *Genes Dev.* **9:**1645-1653.

Masuya, H., Sagai, T., Moriwaki, K., and Shiroishi, T. (1997). Multigenic control of the localization of the zone of polarizing activity in limb morphogenesis in the mouse. *Dev. Biol.* **182:**42-51.

McMahon, A. P. (1998). "Inducible Gene Activation and Inactivation in Utero." Keystone Symposia on Molecular and Cellular Biology, April 3-8. Steamboat Springs, Colorado.

McPherron, A. C., Lawler, A. M., and Lee, S. J. (1999). Regulation of anterior/posterior patterning of the axial skeleton by growth/differentiation factor 11. *Nat. Genet.* **22:**260-264.

Min, H., Danilenko, D. M., Scully, S. A., Bolon, B., Ring, B. D., Tarpley, J. E., DeRose, M., and Simonet, W. S. (1998). Fgf-10 is required for both limb and lung development and exhibits striking functional similarity to Drosophila branchless. *Genes Dev.* **12:**3156-3161.

Mohler, J., and Vani, K. (1992). Molecular organization and embryonic expression of the hedgehog gene involved in cell-cell communication in segmental patterning of Drosophila. *Development* **115:**957-971.

Mohler, J., Seecoomar, M., Agarwal, S., Bier, E., and Hsai, J. (2000). Activation of knot (kn) specifies the 3-4 intervein region in the Drosophila wing. *Development* **127:**55-63.

Morasso, M. I., Mahon, K. A., and Sargent, T. D. (1995). A Xenopus distal-less gene in transgenic mice: conserved regulation in distal limb epidermis and other sites of epithelial-mesenchymal interaction. *PNAS USA* **92:**3968-3972.

Nelson, C. E., Morgan, B. A., Burke, A. C., Laufer, E., DiMambro, E., Murtaugh, C., Gonzales, E., Tessarollo, L., Parada, L. F., and Tabin, C. (1996). Analysis of Hox gene expression in the chick limb bud. *Development* **122:**1449-1466.

Ng, J. K., Tamura, K., Büscher, D., and Izpisúa Belmonte, J. C. (1999). Molecular and cellular basis of pattern formation during vertebrate limb development. *Curr. Top. Dev. Biol.* **41:**37-66.

Niswander, L., Tickle, C., Vogel, A., Booth, I., and Martin, G. R. (1993). FGF-4 replaces the apical ectodermal ridge and directs outgrowth and patterning of the limb. *Cell* **75:**579-587.

Niswander, L., Jeffrey, S., Martin, G. R., and Tickle, C. (1994). A positive feedback loop coordinates growth and patterning in the vertebrate limb. *Nature* **371:**609-612.

Noramly, S., Pisenti, J., Abbott, U., and Morgan, B. (1996). Gene expression in the limbless mutant: polarized gene expression in the absence of Shh and an AER. *Dev. Biol.* **179:**339-346.

Ohuchi, H., Nakagawa, T., Yamauchi, M., Ohata, T., Yoshioka, H., Kuwana, T., Mima, T., Mikawa, T., Nohno, T., and Noji, S. (1995). An additional limb can be induced from the flank of the chick embryo by FGF4. *Biochem. Biophys. Res. Commun.* **209:**809-816.

Ohuchi, H., Nakagawa, T., Yamamoto, A., Araga, A., Ohata, T., Ishimaru, Y., Yoshioka, H., Kuwana, T., Nohno, T., Yamasaki, M., Itoh, N., and Noji, S. (1997). The mesenchymal factor, FGF10, initiates and maintains the outgrowth of the chick limb bud through interaction with FGF8, an apical ectodermal factor. *Development* **124**:2235-2244.

Ohuchi, H., Takeuchi, J., Yoshioka, H., Ishimaru, Y., Ogura, K., Takahashi, N., Ogura, T., and Noji, S. (1998). Correlation of wing-leg identity in ectopic FGF-induced chimeric limbs with the differential expression of chick Tbx5 and Tbx4. *Development* **125**: 51-60.

Oppenheimer, D. I., MacNicol, A. M., and Patel, N. H. (1999). Functional conservation of the wingless-engrailed interaction as shown by a widely applicable baculovirus misexpression system. *Curr. Biol.* **9**:1288-1296.

Ornitz, D. M., Moreadith, R. W., and Leder, P. (1991). Binary system for regulating transgene expression in mice: targeting int-2 gene expression with yeast GAL4/UAS control elements. *PNAS USA* **88**:698-702.

Palopoli, M. F., and Patel, N. H. (1996). Neo-Darwinian developmental evolution: can we bridge the gap between pattern and process? *Curr. Opin. Genet. Dev.* **6**:502-508.

Panganiban, G., Nagy, L., and Carroll, S. B. (1994). The role of the Distal-less gene in the development and evolution of insect limbs. *Curr. Biol.* **4**:671-675.

Panganiban, G., Sebring, A., Nagy, L., and Carroll, S. (1995). The development of crustacean limbs and the evolution of arthropods. *Science* **270**:1363-1366.

Panganiban, G., Irvine, S. M., Lowe, C., Roehl, H., Corley, L. S., Sherbon, B., Grenier, J. K., Fallon, J. F., Kimble, J., Walker, M., Wray, G. A., Swalla, B. J., Martindale, M. Q., and Carroll, S. B. (1997). The origin and evolution of animal appendages. *PNAS* **94**:5162-5166.

Patel, K., Nittenberg, R., D'Souza, D., Irving, C., Burt, D., Wilkinson, D. G., and Tickle, C. (1996). Expression and regulation of Cek-8, a cell to cell signalling receptor in developing chick limb buds. *Development* **122**:1147-1155.

Pecorino, L. T., Lo, D. C., and Brockes, J. P. (1994). Isoform-specific induction of a retinoid-responsive antigen after biolistic transfection of chimaeric retinoic acid/thyroid hormone receptors into a regenerating limb. *Development* **120**:325-333.

Perrimon, N., and Bernfield, M. (2000). Specificities of heparan sulphate proteoglycans in developmental processes. *Nature* **404**:725-728.

Qu, S., Niswender, K. D., Ji, Q., van der Meer, R., Keeney, D., Magnuson, M. A., and Wisdom, R. (1997). Polydactyly and ectopic ZPA formation in Alx-4 mutant mice. *Development* **124**:3999-4008.

Qu, S., Tucker, S. C., Ehrlich, J. S., Levorse, J. M., Flaherty, L. A., Wisdom, R., and Vogt, T. F. (1998). Mutations in mouse Aristaless-like4 cause Strong's luxoid polydactyly. *Development* **125**:2711-2721.

Raff, R. A., and Kaufman, T. C. (1983). "Embryos, Genes, and Evolution: The Developmental-Genetic Basis of Evolutionary Change." Macmillan, New York (reprinted in 1991 by Indiana University Press, Bloomington).

Raff, R. A. (1996). "The Shape of Life: Genes, Development, and the Evolution of Animal Form." University of Chicago Press, Chicago.

Rancourt, D. E., Tsuzuki, T., and Capecchi, M. R. (1995). Genetic interaction between hoxb-5 and hoxb-6 is revealed by nonallelic noncomplementation. *Genes Dev.* **9**:108-22.

Raynaud, A., and Kan, P. (1992). DNA synthesis decline involved in the developmental arrest of the limb buds in the embryos of the slow worm, *Anguis fragilis* (L.). *Int. J. Dev. Biol.* **36**:303-310.

Raynaud, A., Kan, P., Bouche, G., and Duprat, A. M. (1995). Fibroblast growth factors (FGF-2) and delayed involution of the posterior limbs of the slow-worm embryo. *C. R. Acad. Sci. III* **318**:573-578.

Richardson, M. K., Carl, T. F., Hanken, J., Elinson, R. P., Cope, C., and Bagley, P. (1998). Limb development and evolution: a frog embryo with no apical ectodermal ridge (AER). *J. Anat.* **192**:379-390.

Riddle, R. D., Johnson, R. L., Laufer, E., and Tabin, C. (1993). Sonic hedgehog mediates the polarizing activity of the ZPA. *Cell* **75**:1401-1416.

Rijli, F.M., and Chambon, P. (1997). Genetic interactions of Hox genes in limb development: learning from compound mutants. *Curr. Opin. Genet. Dev.* **7**:481-487.

Rodríguez, C., Kos, R., Macías, D., Abbott, U. K., and Izpisúa Belmonte, J.C. (1996). Shh, HoxD, Bmp-2, and Fgf-4 gene expression during development of the polydactylous talpid2, diplopodia1, and diplopodia4 mutant chick limb buds. *Dev. Genet.* **19**:26-32.

Rodriguez-Esteban, C., Tsukui, T., Yonei, S., Magallon, J., Tamura, K., and Izpisúa Belmonte, J.C. (1999). The T-box genes Tbx4 and Tbx5 regulate limb outgrowth and identity. *Nature* **398**:814-818.

Ros, M. A., López-Martínez, A., Simandl, B. K., Rodríguez, C., Izpisúa Belmonte, J. C., Dahn, R., and Fallon, J. F. (1996). The limb field mesoderm determines initial limb bud anteroposterior asymmetry and budding independent of sonic hedgehog or apical ectodermal gene expressions. *Development* **122**:2319-2330.

Rossant, J., Ciruna, B., and Partanen, J. (1997). FGF signaling in mouse gastrulation and anteroposterior patterning. *Cold Spring Harb. Symp. Quant. Biol.* **62**:127-133.

Rossi, E., Faiella, A., Zeviani, M., Labeit ,S., Floridia, G., Brunelli, S., Cammarata, M., Boncinelli, E., and Zuffardi, O. (1994). Order of six loci at 2q24-q31 and orientation of the HOXD locus. *Genomics* **24**:34-40.

Saunders, J. W. J. (1948). The proximodistal sequence of the origin of the parts of the chick wing and the role of the ectoderm. *J. Exp. Zool.* **108**:363-404.

Saunders, J. W., Jr., and Gasseling, M. T. (1968). Ectoderm-mesenchymal interaction in the origins of wing symmetry. *In* "Epithelial-Mesenchymal Interactions" (R. Fleischmajer and R. E. Billingham, eds.), pp. 78-97. Williams and Wilkins, Baltimore.

Schofield, J. P., Elgar, G., Greystrong, J., Lye, G., Deadman, R., Micklem, G., King, A., Brenner, S., and Vaudin, M. (1997). Regions of human chromosome 2 (2q32-q35) and mouse chromosome 1 show synteny with the pufferfish genome *(Fugu rubripes). Genomics* **45**:158-167.

Searls, R. L., and Janners, M. Y. (1971). The initiation of limb bud outgrowth in the embryonic chick. *Dev. Biol.* **24**:198-213.

Sekine, K., Ohuchi, H., Fujiwara, M., Yamasaki, M., Yoshizawa, T., Sato, T., Yagishita, N., Matsui, D., Koga, Y., Itoh, N., and Kato, S. (1999). Fgf10 is essential for limb and lung formation. *Nat. Genet.* **21**:138-141.

Shang, J., Luo, Y., and Clayton, D. A. (1997). Backfoot is a novel homeobox gene expressed in the mesenchyme of developing hind limb. *Dev. Dyn.* **209**:242-253.

Shashikant, C. S., Bieberich, C. J., Belting, H.-G., Wang, J. C. H., Borbély, M. A., and Ruddle, F. H. (1995). Regulation of *Hoxc-8* during mouse embryonic development: identification and characterization of critical elements involved in early neural tube expression. *Development* **121**:4339-4347.

Shubin, N. H., and Alberch, P. (1986). A morphogenetic approach to the origin and basic organization of the tetrapod limb. *In* "Evolutionary Biology" (M. K. Hecht, B. Wallace, and Prance, eds.), Vol. 20, pp. 319-387. Plenum, London.

Shubin, N., Tabin, C., and Carroll, S. (1997). Fossils, genes and the evolution of animal limbs. *Nature* **388**:639-648.

Simeone, A., Acampora, D., Pannese, M., D'Esposito, M., Stornaiuolo, A., Gulisano, M., Mallamaci, A., Kastury, K., Druck, T., Huebner, K., and Boncinelli, E. (1994). Cloning and characterization of two members of the vertebrate Dlx gene family. *PNAS USA* **91**:2250-2254.

Simon, H.-G., Kittappa, R., Khan, P. A., Tsilfidis, C., Liversage, R. A., and Oppenheimer, S. (1997). A novel family of T-box genes in urodele amphibian limb development and regeneration: candidate genes involved in vertebrate forelimb/hindlimb patterning. *Development* **124**:1355-1366.

Sordino, P., van der Hoeven, F., and Duboule, D. (1995). Hox gene expression in teleost fins and the origin of vertebrate digits. *Nature* **375**:678-81.

Sordino, P., and Duboule, D. (1996). A molecular approach to the evolution of vertebrate paired appendages. *Trends Ecol. Evol.* **11**:114-119.

Sordino, P., Duboule, D., and Kondo, T. (1996). Zebrafish Hoxa and Evx-2 genes: cloning, developmental expression and implications for the functional evolution of posterior Hox genes. *Mech. Dev.* **59**:165-175.

Stark, D. R., Gates, P. B., Brockes, J. P., and Ferretti, P. (1998). Hedgehog family member is expressed throughout regenerating and developing limbs. *Dev. Dyn.* **212**:352-363.

Stock, D. W., Ellies, D. L., Zhao, Z., Ekker, M., Ruddle, F. H., and Weiss, K. M. (1996). The evolution of the vertebrate Dlx gene family. *PNAS USA* **93**:10858-10863.

Stratford, T. H., Kostakopoulou, K., and Maden, M. (1997). Hoxb-8 has a role in establishing early anterior-posterior polarity in chick forelimb but not hindlimb. *Development* **124**:4225-4234.

Sturtevant, M. A., Biehs, B., Marin, E., and Bier, E. (1997). The spalt gene links the A/P compartment boundary to a linear adult structure in the Drosophila wing. *Development* **124**:21-32.

Subramanian, V., Meyer, B. I., and Gruss, P. (1995). Disruption of the murine homeobox gene Cdx1 affects axial skeletal identities by altering the mesodermal expression domains of Hox genes. *Cell* **83**:641-653.

Summerbell, D., Lewis, J. H., and Wolpert, L. (1973). Positional information in chick limb morphogenesis. *Nature* **244**:492-496.

Szeto, D. P., Rodriguez-Esteban, C., Ryan, A. K., O'Connell, S. M., Liu, F., Kioussi, C., Gleiberman, A. S., Izpisúa Belmonte, J. C., and Rosenfeld, M. G. (1999). Role of the Bicoid-related homeodomain factor Pitx1 in specifying hindlimb morphogenesis and pituitary development. *Genes Dev.* **13**:484-494.

Szeto, D. P., Ryan, A. K., O'Connell, S. M., and Rosenfeld, M. G. (1996). P-OTX: A PIT-1-interacting homeodomain factor expressed during anterior pituitary gland development. *PNAS USA* **93**:7706-7710.

Tabata, T., Eaton, S., and Kornberg, T.B. (1992). The Drosophila hedgehog gene is expressed specifically in posterior compartment cells and is a target of engrailed regulation. *Genes Dev.* **6**:2635-2645.

Tabin, C. J. (1992). Why we have (only) five fingers per hand: hox genes and the evolution of paired limbs. *Development* **116**:289-296.

Takahashi, M., Tamura, K., Büscher, D., Masuya, H., Yonei-Tamura, S., Matsumoto, K., Naitoh-Matsuo, M., Takeuchi, J., Ogura, K., Shiroishi, T., Ogura, T., and Izpisúa Belmonte, J. C. (1998). The role of Alx-4 in the establishment of anteroposterior polarity during vertebrate limb development. *Development* **25**:417-4425.

Takeuchi, J. K., Koshiba-Takeuchi, K., Matsumoto, K., Vogel-Hopker, A., Naitoh Matsuo, M., Ogura, K., Takahashi, N., Yasuda, K., and Ogura, T. (1999). Tbx5 and Tbx4 genes determine the wing/leg identity of limb buds. *Nature* **398**:810-814.

Tashiro, S., Michiue, T., Higashijima, S., Zenno, S., Ishimaru, S., Takahashi, F., Orihara, M., Kojima, T., and Saigo, K. (1993). Structure and expression of hedgehog, a Drosophila segment-polarity gene required for cell-cell communication. *Gene* **124**:183-189.

ten Berge, D., Brouwer, A., Korving, J., Martin, J. F., and Meijlink, F. (1998). Prx1 and Prx2 in skeletogenesis: roles in the craniofacial region, inner ear and limbs. *Development* **125**:3831-3842.

Todt, W. L., and Fallon, J. F. (1984). Development of the apical ectodermal ridge in the chick wing bud. *J. Embryol. Exp. Morphol.* **80**:21-41.

Torok, M. A., Gardiner, D. M., Shubin, N. H., and Bryant, S. V. (1998). Expression of HoxD genes in developing and regenerating axolotl limbs. *Dev. Biol.* **200**:225-233.

van den Akker, E., Reijnen, M., Korving, J., Brouwer, A., Meijlink, F., and Deschamps, J. (1999). Targeted inactivation of Hoxb8 affects survival of a spinal ganglion and causes aberrant limb reflexes. *Mech. Dev.* **89**:103-114.

van der Hoeven, F., Zákány, J., and Duboule, D. (1996). Gene transpositions in the HoxD complex reveal a hierarchy of regulatory controls. *Cell* **85**:1025-1035.

van Eeden, F. J. M., Granato, M., Schach, U., Brand, M., Furutani-Seiki, M., Haffter, P., Hammerschmidt, M., Heisenberg, C.-P., Jiang, Y.-J., Kane, D. A., Kelsh, R. N., Mullins, M. C., Odenthal, J., Warga, R. M., and Nusslein-Volhard, C. (1996). Genetic analysis of fin formation in the zebrafish, Danio rerio. *Development* **123**:255-262.

Vervoort, M., Crozatier, M., Valle, D., and Vincent, A. (1999). The COE transcription factor Collier is a mediator of short-range Hedgehot-induced patterning of the Drosophila wing. *Curr. Biol.* **9**:632-639.

Vogel, A., Rodríguez, C., and Izpisúa Belmonte, J.-C. (1996). Involvement of FGF-8 in initiation, outgrowth and patterning of the vertebrate limb. *Development* **122**:1737-1750.

Vogt, T. F., and Duboule, D. (1999). Antagonists go out on a limb. *Cell* **99**:563-566.

Voss, A.K., Thomas, T., and Gruss, P. (1998). Efficiency assessment of the gene trap approach. *Dev. Dyn.* **212**:171-180.

Weatherbee, S. D., and Carroll, S. B. (1999). Selector genes and limb identity in arthropods and vertebrates. *Cell* **97**:283-286.

Wilkins, A. S. (1998). Evolutionary developmental biology: where is it going?. *Bioessays* **20**:783-784.

Williams, T. A. (1998). Distalless expression in crustaceans and the patterning of branched limbs. *Dev. Genes Evol.* **207**:427-434.

Xu, X., Weinstein, M., Li, C., Naski, M., Cohen, R. I., Ornitz, D. M., Leder, P., and Deng, C. (1998). Fibroblast growth factor receptor 2 (FGFR2)-mediated reciprocal regulation loop between FGF8 and FGF10 is essential for limb induction. *Development* **125**:753-765.

Yang, Y., Drossopoulou, G., Chuang, P.-T., Duprez, D., Martí, E., Bumcrot, D., Vargesson, N., Clarke, J., Niswander, L., McMahon, A., and Tickle, C. (1997). Relationship between dose, distance and time in Sonic Hedgehog-mediated regulation of anteroposterior polarity in the chick limb. *Development* **124**:4393-4404.

Yang, Y., Guillot, P., Boyd, Y., Lyon, M. F., and McMahon, A. P. (1998). Evidence that preaxial polydactyly in the Doublefoot mutant is due to ectopic Indian Hedgehog signaling. *Development* **125**:3123-3132.

Zákány, J., and Duboule, D. (1999). Hox genes in digit development and evolution. *Cell Tissue Res.* **296**:19-25.

Zákány, J., Fromental-Ramain, C., Warot, X., and Duboule, D. (1997). Regulation of number and size of digits by posterior Hox genes: a dose dependent mechanism with potential evolutionary implications. *PNAS* **94**:13695-13700.

Zardoya, R., Abouheif, E., and Meyer, A. (1996). Evolutionary analyses of hedgehog and Hoxd-10 genes in fish species closely related to the zebrafish. *PNAS USA* **93**:13036-13041.

Zerucha, T., Muller, J. P., Chartrand, N., and Ekker, M. (1997). Cross-interactions between two members of the Dlx family of homeobox-containing genes during zebrafish development. *Biochem. Cell. Biol.* **75**:613-622.

EPIGENETIC MECHANISMS OF CHARACTER ORIGINATION

STUART A. NEWMAN[1] AND GERD B. MÜLLER[2]

[1]*Department of Cell Biology and Anatomy, New York Medical College, Valhalla, NY 10595*
[2]*Department of Anatomy, University of Vienna, A-1090 Vienna, Austria*
[2]*Konrad Lorenz Institute for Evolution and Cognition Research, A-3422 Altenberg, Austria*

INTRODUCTION

Evolutionary biology is currently the scene of debates around such topics as the tempo and mode of phenotypic evolution, the degree to which genetic change can result from selectively neutral mechanisms, and the universality of adaptation in accounting for complex traits. However, in all the contending views the notion that an organism's morphological phenotype is determined by its genotype is taken for granted. This tenet is also essentially undisputed in developmental biology, which today is commonly characterized as the study of "genetic programs" for the generation of body plan and organ form. In this chapter we explore the validity of this widely held notion and suggest that an alternative way of looking at the causal relationship between genes and form can resolve some of the debates in evolutionary theory, as well as apparent

The Character Concept in Evolutionary Biology

paradoxes that have arisen with recent findings of extensive functional redundancy in developmental systems. In particular, we propose that the correlation of an organism's form with its genotype, rather than being a *defining condition* of morphological evolution, is a *highly derived property*. This implies that other causal determinants of biological morphogenesis have been active over the course of evolution and that a theory of morphological evolution based on neo-Darwinian mechanisms alone must remain incomplete.

We set out from the observation that many organisms, particularly among the bacteria, protists, and fungi, but also among higher animals such as arthropods, molluscs, and vertebrates (as well as many plants), exhibit phenotypic polymorphisms and morphological plasticity. Radically different forms occur in different settings or different phases of the life cycle. These distinct forms could represent independent adaptations, each realized by a separately evolved genetic subroutine. Alternatively, rather than being adaptive, morphological plasticity could reflect the influence of external physicochemical parameters on any material system. If the latter is the case in at least some instances in contemporary organisms, it is plausible that in earlier multicellular forms this externally conditioned kind of morphological determination was even more prevalent. This is because ancient organisms undoubtedly exhibited less genetic redundancy, and metabolic integration and homeostasis, than modern organisms, and were thus more subject to external molding forces. Thus it is proposed that morphological variation in response to the environment is a primitive, physically based property. This property is characteristic of all "soft matter" (deGennes, 1992), and "excitable media" (Mikhailov, 1990; Winfree, 1994) (see later), and would have been an inevitable feature of the viscoelastic cell aggregates that constituted the first multicellular organisms.

Examining the morphological plasticity of some modern organisms can provide insight into the flexible, environment-dependent relationship between genotype and form that still prevails in most of the living world. *Candida albicans*, for instance, a frequent fungal pathogen in humans, is able to switch among forms ranging from single budding cells, to thread-like hyphae, to strings of yeast-like cells plus long septated filaments, known as pseudohyphae (Braun and Johnson, 1997; Ishii, Yamamoto *et al.*, 1997). These and other considerations have led to the suggestion that *C. albicans* has no "default" morphology (Magee, 1997). Even in vertebrates the environment can play a decisive role in morphological development. For example, incubation temperature determines sex in reptiles in a species-dependent fashion –high temperatures produce males in lizards and crocodiles, but females in chelonians (Deeming and Ferguson, 1988). In mice the number of vertebrae can depend on the uterine environment: fertilized eggs of a strain with 5 lumbar vertebrae preferentially develop into embryos with 6 vertebrae when transferred into the uteri of a 6 vertebrae strain (McLaren and Michie, 1958). Animals that undergo metamorphosis, such as echinoderms, tunicates, arthropods, and amphibians, also exhibit multiple morphological phenotypes, and metamorphosis can be influenced by environmental change as well as intrinsic timing mechanisms (Gilbert *et al.*, 1996).

The pervasiveness of plasticity and polymorphism suggests that the correspondence of a genotype to one morphological phenotype, as is frequently the case in higher animals, should be considered exceptional–a highly derived condition in which an "overdetermining" genetic circuitry ensures that changes of extrinsic or intrinsic variables have less impact on the morphological outcome. If modern organisms are "Mendelian," in the sense that genotype and phenotype are inherited in close correlation, and that morphological change is most typically dependent on genetic change, then the polymorphic primitive metazoans we postulate would have constituted a "pre-Mendelian world" of living organisms, whose genotypes and morphological phenotypes were connected in only a loose fashion.

In this exploratory period of organismal evolution the mapping of genotype to morphological phenotype would have been one to many, rather than one to one. The prototypes of modern forms, however phenotypically distinct, were probably totally or partially interconvertible at the generative level. Only later, with the evolution of genetic redundancies (Tautz, 1992; Picket and Meeks-Wagner, 1995; Wagner, 1996; Cooke, *et al.*, 1997; Nowak, *et al.*, 1997; Wilkins, 1997) and other mechanisms supporting reliability of developmental outcome (Rutherford and Lindquist, 1998), a closer linkage between genetic change and phenotypic change was established, with evolution under selective criteria favoring the maintenance of morphological phenotype in the face of environmental or metabolic variability. Organisms thus characterized by a closer mapping of genotype to phenotype marked the transition from the pre-Mendelian to the Mendelian world.

This scenario of different phases in morphological evolution raises the possibility that the origination of organismal forms and characters and their adaptive fine tuning are based on different mechanisms. Moreover, it points to an important conceptual gap in current evolutionary theory. Neo-Darwinism, in its present form, deals competently and successfully with the variation and adaptation of characters, but sidesteps the problem of their causal origin. Thus the emergence and organization of discrete morphological units still remains an open problem, recognized under the terms of "novelty" or "innovation" (Müller, 1990; Müller and Wagner, 1991).

The essence of the concept we will develop in the following pages is that epigenetic mechanisms, rather than genetic change, are the major sources of morphological innovation in evolution. We do not use the term "epigenetic" to refer to DNA-related mechanisms of inheritance, such as methylation and chromatin assembly [see Jablonka and Lamb (1995) for a review]. The epigenetic mechanisms that we consider are *conditional, nonprogrammed determinants of individual development*, of which the most important are (i) interactions of cell metabolism with the physicochemical environment within and external to the organism, (ii) interactions of tissue masses with the physical environment on the basis of physical laws inherent to condensed materials, and (iii) interactions among tissues themselves, according to an evolving set of rules. We suggest that different epigenetic processes have prevailed at different stages of morphological evolution, and that the forms and characters assumed by metazoan organisms originated in large part by the action of such processes.

A number of earlier authors have discussed the role of epigenetic factors in evolution. Some of these works have argued for the importance of developmental constraints in influencing the direction of phenotypic change (Alberch, 1982; Maynard Smith *et al.*, 1985; Stearns, 1986) or emphasized environmental effects on development (Johnston and Gottlieb, 1990). Other authors have pointed to the intrinsic dynamical structure of developmental systems in accounting for a nonrandom variation of traits (Ho and Saunders, 1979; Kauffman, 1993; Goodwin, 1994). Our concept goes beyond these suggestions in postulating that the processes by which morphological characters are determined are different at different phases of evolution, with genetic integration taking on a more prominent role after a character is established. In particular, our view involves the recognition that forms and characters produced by epigenetic factors can serve as templates for the accumulation of overdetermining genetic mechanisms. As a result, the action of the originating epigenetic factors may be obscured or even superseded in modern developmental systems.

The relationship between genes and biological form is not simple, and the standard notion of the "genetic program" is increasingly seen as problematic (Oyama, 1985; Nijhout, 1990; Müller and Wagner, 1991; Bolker and Raff, 1996; Neumann-Held, 1998). We propose a revised interpretation of that relationship: with regard to the origin of morphology, we take the physical nature of living organisms to be their most salient property. This implies that epigenetic processes, which are contingent and conditional, are the motive forces in the evolution of biological form. As evolution proceeds, genetic change that favors maintenance of morphological phenotype in the face of environmental or metabolic variability coopts the morphological outcomes of epigenetic processes, resulting in the heritable association of particular forms with particular genealogical lineages.

We note that the notion of "evolvability"—the inherent potential of certain lineages to change during the course of evolution—is interpreted in an entirely different fashion in light of the ideas presented here than it has been in other recent discussions (Gerhart and Kirschner, 1997; Kirschner and Gerhart, 1998). For us evolvability represents the continued efficacy of epigenetic processes in a lineage—some of them quite ancient, and some of more recent origin—and as such is tied to the primitive morphogenetic plasticity hypothesized earlier. Genetic evolution, particularly of the cooptative kind, will tend to suppress such evolvability and buffer the development of form. This contrasts with the view that evolvability is a product of advanced evolution, achieved by the emergence of new genetic mechanisms that favor plasticity.

EPIGENESIS IN A "PRE-MENDELIAN" WORLD: THE PHYSICS OF TISSUE MASSES AND THE ORIGIN OF BODY PLANS

Multicellular organisms first arose more than 600 million years ago (Conway Morris, 1993). By approximately 540 million years ago, at the end of the "Cambrian explosion," virtually all the "bauplans" or body types seen in modern organisms already existed (Whittington, 1985; Conway Morris, 1989; Briggs *et al.*, 1992). The original multicellular forms were established with cells that were metabolically and structurally sophisticated – the first eukaryotic cells appeared at least a billion years earlier (Knoll, 1992). Although many, if not most, of the genes present in modern multicellular organisms were already in place, encoding corresponding proteins with well-defined roles in unicellular structure and function, these genes and proteins had not been selected for the construction of multicellular characters.

The most ancient multicellular forms must have been simple cell aggregates that arose by adhesion of originally free-living cells, or by the failure of the same to separate after mitosis (Bonner, 1998). The precise chemical or physical nature of the adhesive interaction would have been unimportant, as long as it served to keep the organism's cells from dispersing. Indeed, the advent of a cell-cell adhesion mechanism early in the history of multicellular life, although dependent of the preexistence of particular gene products, need not have required additional gene sequence change. For example, some modern cell surface proteins, such as the cadherins, mediate cell attachment only in the presence of calcium ions (Takeichi, 1991). Protein chemists are well aware that many proteins that perform no adhesive function at all exhibit different degrees of "stickiness" under different ionic conditions. It is thus plausible that a protein on the surface of an ancient unicellular eukaryote could have acquired a new function—adhesion—by virtue of a simple change in the ionic content of the organism's aqueous environment, leading to simple multicellular forms by fiat (Kazmierczak and Degens, 1986).

While the appearance of primitive multicellular forms in the fossil record may have thus been a relatively straightforward matter (Bonner, 1998), not dependent on the evolution of any complex developmental machinery, the "heritability" of the multicellular state would have depended either on the persistence of the new external conditions or on the evolution of adhesion proteins that were less dependent on context. The earliest multicellular organisms, however, were unlikely to have generated their complex forms using the baroque, hierarchical, molecular machinery that guides morphogenesis in modern organisms (Nüsslein-Vollhard, 1996). Rather, the existence of a simple mechanism of adhesion, whereby cells could remain attached to one another after they divided, would have been sufficient to establish multicellularity.

Compartmentalization, Tissue Multilayering, and Segmentation

Once one or several adhesive mechanisms were in place, other more complex morphological consequences could have inevitably followed, simply by virtue of variations in cell adhesivity brought about by random processes like metabolic noise and by the way in which the relevant physical laws act on such heterogeneous cell aggregates. Cells with different amounts of adhesion molecules on their surfaces, for example, tend to sort out into islands of more cohesive cells within lakes composed of their less cohesive neighbors. Eventually, by random cell movement, the islands coalesce and an interface is established, across which cells will not intermix (Steinberg and Takeichi, 1994; Steinberg, 1998). What is observed is similar to what happens when two immiscible liquids, such as oil and water, are poured into the same container. An important feature of this mechanism is that the final morphological outcome is independent of the initial conditions – in effect it is "goal directed." Thus, when several differentially adhesive cell populations arise within the same tissue mass, multilayered structures can form automatically, comprising distinct "compartments" (Crick and Lawrence, 1975; Garcia-Bellido *et al.,* 1976) (Fig. 1A). Indeed, two of the five major types of gastrulation seen in modern metazoans, *epiboly and involution* (and possibly a third, *delamination)* (Fig. 1C), could have originated as simple consequences of differential adhesion (Newman, 1994).

Thus, somewhat counterintuitively, lax regulation of the abundance of adhesion proteins, in conjunction with thermodynamic processes, can lead rather directly to novel, multilayered organismal forms. Furthermore, if variations in metabolic or biosynthetic activity, rather than being purely random across the tissue mass, affected cell-cell adhesion in a temporally or spatially periodic fashion, then *compartmentalization*—the establishment of boundaries of immiscibility—takes the form of *segmentation* (Newman, 1993) (Fig. 1D). Moreover, the generation of periodicities is all but inevitable in the complex, "excitable media" represented by even the simplest aggregates of cells.

Excitable media are materials that actively respond to their environment, mechanically, chemically, or electrically. Nonliving examples have been well studied (Gerhardt *et al.,* 1990; Mikhailov, 1990; Starmer *et al.,* 1993; Winfree, 1994). Aggregates of living cells, embodying metabolic and genetic networks responsive to the external environment and containing positive and negative feedback loops and diffusible components, must have tended spontaneously to develop chemical oscillations (Goldbeter, 1995) and spatial periodicities (Turing, 1952; Boissonade *et al.,* 1994). From such biochemical periodicities it is only a few steps to segmental tissue organization (Palmeirim *et al.,* 1997; Pourquié, 1998), which is therefore likely to have arisen numerous times in the history of life (Newman, 1993).

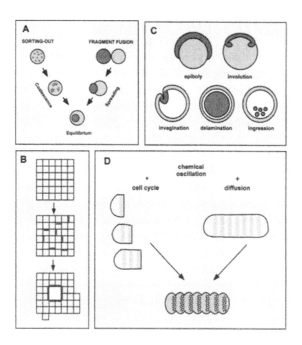

FIGURE 1   Generic processes in tissue morphogenesis. **A.** Schematic representation of the behavior of intermixed cells and corresponding tissue fragments in the case where the two cell populations are differentially adhesive. The cell mixture will sort out as the more adhesive cells establish more stable bonds with one another than with cells of the other population. Random motion leads to the formation of cohesive islands of these cells, and these will ultimately coalesce into a separate tissue phase, or compartment. The equilibrium configuration of the cell mixture is identical to that which would be formed by fusion and spreading of fragments of tissue consisting of the same differentially adhesive cell populations. **B.** Schematic view of formation of a lumen or internal cavity by differential adhesion in an epithelioid tissue consisting of polarized cells. In the original state (*top*) the cells are uniformly adhesive and make contacts around their entire peripheries. Upon expression of an antiadhesive protein in a polarized fashion in a random subpopulation of cells (*center*), and random movement of the cells throughout the mass, bonds between adhesive surfaces are energetically favored over those between adhesive and nonadhesive surfaces, resulting in lumen formation (*bottom*). **C.** Schematic cross-sectional views of the five main types of gastrulation. In each case a new population of cells differentiates from a solid or hollow embryo and assumes a position which would be attained by a similarly situated differentially adhesive population. **D.** Schematic representation of two modes of tissue segmentation that can arise when the tissue's cells contain a biochemical circuit that generates a chemical oscillation or "molecular clock," and the oscillating species directly or indirectly regulates the strength or specificity of cell adhesivity. In the mechanism shown on the left, the periodic change in cell adhesivity occurs in a growth zone in which the cell cycle has a different period from the regulatory oscillator; as a result, bands of tissue are sequentially generated with alternating cohesive properties. In the mechanism shown on the right, one or more of the biochemical species can diffuse, leading to a set of standing waves of concentration of the regulatory molecule by a reaction-diffusion mechanism. This leads to the simultaneous formation of bands of tissue with alternating cohesive properties. See Newman (1993) for additional details. (A, with changes, from Steinberg, 1998; B after Newman and Tomasek, 1996.)

CELL POLARITY AND LUMEN FORMATION

The first multicellular organisms plausibly were composed of cells with a uniform, or random, distribution of adhesive molecules on their surfaces. Many modern cell types, in contrast, are *polarized*, capable of allocating different molecular species to their apical and basolateral regions (Rodriguez-Boulan and Nelson, 1993). The targeting of adhesive molecules, or antiadhesive molecules, to specific regions of the cell surface has dramatic consequences. A tissue mass consisting of motile epithelioid cells that are nonadhesive over portions of their surfaces would readily develop cavities or lumens. If such spaces were to come to adjoin one another, as a result of random cell movement, they would readily fuse (Fig. 1B). Lumen formation may therefore have originated as a simple consequence of differential adhesion in cells that express adhesive properties in a polarized fashion. The formation of lumens in masses of mammary carcinoma cells by the forced, polar expression of the *met* oncogene (Tsarfaty *et al.,* 1992) is a model for this morphological innovation in a contemporary system.

Significantly, the first morphologically complex multicellular organisms, represented by the Vendian fossil deposits dating from as early as 700 million years ago, appear to have been flat, often segmented, but apparently solid-bodied creatures (Seilacher, 1992; Conway Morris, 1993). Among modern phyla the coelenterates, such as hydra, are forms with a single lumen; echinoderms (e.g., starfish) and vertebrates have both a digestive tube and a surrounding body cavity. It is thought to have taken up to 100 million years after the appearance of the Vendian fauna for organisms to develop distinct body cavities, although recent evidence suggests that this may have occurred more rapidly (Seilacher *et al.,* 1998). Once these triploblastic forms arose, all the modern body plans burst onto the scene in short order.

It is interesting to consider the possibility that the advent of polarized cells may have provided the physical basis for the rapid profusion of body types during the Cambrian explosion. Depending on the taxon, cell polarity could have arisen before or after the evolutionary event that led to multicellularity. In either case, lumen or cavity formation would have been an inevitable physical consequence of the conjunction of these two properties. Two of the major types of gastrulation—*invagination* and *ingression* (Fig. 1C)—whatever the mechanisms of their realization in modern animal phyla, could have originated in ancient organisms by the actions of differential adhesion in establishing multiple tissue layers in conjuction with lumens in cell aggregates (Newman, 1994).

The combined effects of the various physical properties that were generic to the earliest multicellular aggregates considered as chemically excitable, viscoelastic soft matter, may thus have ensured the production of a profusion of multilayered, hollow, segmented forms – a pre-Mendelian world of fully or partially interconvertible prototypes for the genetically routinized body plans to come. While not every physically attainable multicellular form would necessarily prosper, many strikingly different kinds would. Moreover, the surviving morphotypes define, in a real sense, their own ecological niches rather than representing merely adaptations to pre-existing ones. A novel implication of this interpretation of the burst of forms during the early history of metazoan life is that the disparate organismal forms would have been achieved with no requirement for competition or differential fitness. Since function would follow form, rather than the other way around, the pre-Mendelian world would thus also have been, in this sense alone, a "pre-Darwinian" one.

EPIGENESIS IN A MENDELIAN WORLD: SOURCE OF INNOVATION AND HOMOLOGY

Once major body plans were established, selection for biochemical integration, which promoted physiological homeostasis and developmental reliability, stabilized the relationship between genotype and ecological setting referred to as fitness or adaptedness. This increasingly unique matching between genotype and phenotype led ultimately to Mendelian heritability. Morphological innovation leading to diversification at the subphylum level was to follow. While the standard picture holds that this was the virtually exclusive result of incremental selection of minor, random, phenotypic variants, we suggest that epigenesis was also an important driving force in these later events.

As a consequence of compartmentalization, organisms came to contain differentiated subpopulations of cells with the potential to perform specialized functions. The biochemically divergent tissues formed from such cells provided components of one another's environment, and as the forms produced began to depend on their interactions, *embryonic induction* came into existence. The *conditionality* of tissue interactions, along with residual generic morphogenetic properties, guaranteed that the resulting systems retained a significant degree of "play." Variations in the natural developmental environment, like experimental perturbation (Hall, 1984; Müller, 1989), can divert even highly evolved systems into alternative pathways, with physical factors continuing to play an influential role. Even in the developmental systems represented by contemporary metazoa by no means are all components strictly determined by the genome (see below). Rather, such systems are characterized by an interplay between epigenetic and genetic control, which generates reliable phenotypic outcomes. As a consequence of the continued conditional nature of evolved development, evolutionary modifications that affect one part of a system can have strong effects on other parts, leading to "unexpected" phenotypic innovations.

The Continuity of Physical Influences

Earlier we indicated how the generic physical properties of tissues may have strongly influenced the array of forms generated in early organismic evolution. Although the role of these physical processes in the formation of body plans must have receded as more developmental interactions and the associated biochemical inertia set in, physical principles, and biomechanical factors in particular, remained active in secondary developmental fields and had important consequences for the further evolvability of phenotypic design.

Evolved morphogenesis is largely a matter of molding clusters of dividing cells into physical shapes. Layers, sheaths, tubes, rods, spheres, etc. are formed by aggregates of cells, mobilizing a wide range of biomechanical forces that result from the different properties of different cell types and their extracellular products (Fig. 2A-C). Once these macro shapes have formed, their macro properties in turn become important parameters for further development, not only creating geometric templates and barriers, but also controlling gene activity. These higher level physical factors become a part of the developmental program that is not explicitly specified in any inherited code of information. Their existence, however, determines what may result from a developmental system, both in a constraining and a generative manner.

As an illustration we consider the vertebrate limb. The origin and evolution of limbs are largely consequences of evolving an internal skeleton. Skeletogenesis is based on the generic capacity of mesenchymal cells to adhere and condense, and produce cartilage matrix. During limb development this sequence of events is constrained by the spatial confinements of the limb bud and is modulated genetically through differential cell adhesion (Yokouchi *et al.*, 1995; Newman, 1996). In developing mesenchyme the presence of diffusible, positively autoregulatory effectors of extracellular matrix production (such as transforming growth factor-beta), along with diffusible inhibitory factors, can lead to spatial periodicities in the conditions required for chondrogenesis (Newman and Frisch, 1979; Newman, 1988; Leonard *et al.*, 1991; Newman, 1996; Miura and Shiota, 2000a, b). Spatial self-organization of the limb bud mesenchyme thus leads to a basic pattern of repeating skeletal elements. The evolution of the vertebrate limb can be viewed as the history of molecular and genetic modulation of developmental mechanisms (Shubin *et al.*, 1997), which are fundamentally generic and physical. Later, additional physical factors become important as muscle contractions and embryonic movements begin to influence bone and joint formation (Drachman and Sokoloff, 1966; Persson, 1983; Amprino, 1985; Hall, 1986), muscle and tendon differentiation (Scott *et al.*, 1987; Giori *et al.*, 1993), and consequently innervation (Dahm and Landmesser, 1991), blood vessel patterns, etc. This means that the generic properties of limb tissues contribute not merely to skeletogenesis but eventually influence "downstream" development. Physical factors thus continue to be of decisive importance even in contemporary ontogenies.

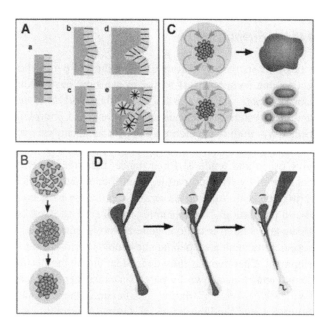

FIGURE 2 Epigenetic mechanisms of tissue morphogenesis and organogenesis. **A.** Schematic representation of major modes of epithelial morphogenesis resulting from extrinsic alteration of cell parameters. In *a*, a pattern formation mechanism (e.g., a reaction-diffusion system) is activated in a flat epithelial sheet, possibly mediated by a subjacent mesenchymal layer, and marks a subset of cells to undergo alteration of one or more "potential functions" (e.g., adhesive strength, cytoskeletal tension.). In *b-e*, resulting epithelial morphologies are indicated. A placode, *b*, will form if the lateral regions of the epithelial cells become more adhesive than the apical and basal regions. An evagination, *b*, as in a developing intestinal villus, or an invagination, *d*, as in a developing hair or feather (Chuong and Widelitz, 1998) will form if the change in cell potential gives rise to a "bending moment" (Newman, 1998) that destabilizes the flat configuration. Progressive cycles of patterning and invagination will give rise to a branched tubular structure, *e*, as in salivary gland morphogenesis (Kashimata and Gresik, 1996). **B.** Schematic representation of mesenchymal condensation, as occurs during skeletal morphogenesis and many other developmental processes. Such condensations can be initiated by local patches of elevated production of extracellular matrix (ECM) molecules, and consolidated by cell-cell adhesion. **C.** Morphogenesis of connective tissue elements, such as cartilage rods and nodules, occurs by the regulation of the pattern of mesenchymal condensation formation. One way that this can occur is by the interplay of a positively autoregulatory diffusible activator of ECM production, such as TGF-beta (curved arrows), with a diffusible inhibitor of its activity (straight arrows). In the absence of the inhibitor (*top*) resulting cartilage forms as an amorphous mass; in its presence, patterns of well-spaced nodules and rods can form as centers of activation become surrounded by domains of inhibition. **D.** Origin of the "fibular crest" in archosauran hindlimbs by mechanical regulation of mesenchymal morphogenesis. Progressive evolutionary reduction of the fibula increases the mechanical load on the connective tissue between the tibia and the fibula, exerted by the pulling action of the iliofibularis muscle. A stress-dependent cartilage forms in response and becomes later incorporated into the ossifying tibia to form a prominent crest, a homologue shared by theropod dinosaurs and carinate birds. This tight fixation of the proximal fibula permits its further distal reduction in avian limbs. (B and C after Newman, 1996; D adapted from Müller and Streicher, 1989).

Innovation at the Phenotypic Level

As a result of the increasingly homeostatic nature of evolved development, changes affecting one component of a system can now have strong effects on associated components. The existence of thresholds in developmental processes, and the systemic consequences of modified morphogenesis, can create unexpected by-products, which may appear as phenotypic innovations at the subphylum level (Müller, 1990; Müller and Wagner, 1991).

Connective tissue and tendons, for instance, have the capacity to react to biomechanical stimuli by forming cartilage and bone. Such skeletal elements, known as cartilaginous or ossified sesamoids, can arise as a consequence of changes of bone proportions, for example: the altered stresses on embryonic connective tissue and tendon insertions generate novel sesamoids. In the avian hindlimb, four such movement-dependent sesamoids form during the course of normal development. They remain fully dependent on the biomechanical stimuli of embryonic movement, as shown by paralysis experiments which inhibit their formation (Wu, 1996). Later, during ossification, these skeletal elements become incorporated into the long bones of the limb. Evolutionary changes of bone proportion (Streicher and Müller, 1992) will generate similar changes in embryonic biomechanics, and the resulting skeletal elements appear as novel characters of avian bones, such as the supratendinal bridge, the cnemial process, or the fibular crest of the tibiotarsus (Müller and Streicher, 1989) (Fig. 2D). They all represent significant changes of bone morphology, yet none of these characters will have arisen as a direct result of a mutation or "new genes" for that specific character. Rather they arise as a side effect of mutations affecting other characters, such as the size or the growth rate of the tibia. The novelties that result from the consequently altered biomechanical conditions will then become incorporated into the bauplan of the limb. In addition, the general stress dependency of the skeletogenic system further modulates the external shape and inner architecture of bones during postnatal activity (Carter and Orr, 1992).

In the highly advanced developmental systems of recent vertebrates such epigenetic side effects probably account only for relatively minor character innovations. However, it is plausible that the entire endoskeleton of vertebrates arose in a similar fashion. *In vitro* and *in vivo* studies demonstrate that cells, and connective tissue cells in particular, arrange themselves along stress fields (Harris *et al.*, 1980; Bard, 1990). Moreover, cartilage matrix secretion is an autonomous property of mesenchymal cells, highly dependent on cell number and density (Cottrill *et al.*, 1987) and compression (Vogel and Koob, 1989; Robbins *et al.*, 1997). Therefore it is likely that any mesenchymal tissue mass above a certain threshold size, such as the embryonic body axis or lateral outgrowths from it, may have automatically begun to generate dense cores of matrix-secreting cell arrays along the stress fields generated by passive and active movement, thus stabilizing large mesenchymal cell aggregates and allowing the further increase of body size. The dynamic interaction between diffusible cytokines promoting and inhibiting the expansion of these aggregates [which themselves are induced by mechanical loading (Klein-Nulend *et al.*, 1995)] will have readily led to their periodic arrangement (Newman and Frisch, 1979; Newman, 1984, 1996), giving rise to the vertebrae, ribs, digits, etc., of the modern endoskeleton.

While the evolution of biochemical circuitry and developmental control mechanisms would have subsequently fixed new traits that arose in ways such as described above, the "strength" of such fixation can be variable. The susceptibility of movement-dependent sesamoids to paralysis, for example, differs significantly (Wu, 1996). Thus it appears that the generation and the fixation of novelty are quite distinct processes, governed by different mechanisms (Müller and Wagner, 1991).

From Homoplasy to Homology

We have argued that epigenesis is a primary factor directing morphological evolution, even in evolved developmental systems. In particular, we have suggested that structural innovations were probably largely epigenetic in their origin. Although the population-level establishment of any morphological innovation will depend on the ecological conditions under which its carriers live (Liem, 1990; Galis and Drucker, 1996), innovations initially originate as "pure" consequences of ubiquitous material and developmental propensities. Therefore generic processes can lead to similar forms in unrelated organismal lineages, manifested as the characteristic "homoplasies" of morphological evolution (Wake, 1991; Sanderson and Hufford, 1996; Moore and Willmer, 1997). However, another characteristic of advanced morphological evolution was to prove crucial, namely the establishment of heritable anatomical units. This principle of organismal design, commonly referred to as "homology," is central for any conceptual understanding of morphological evolution (Hall, 1994). The question thus arises of a causal relationship between homoplasy and homology.

We propose the following scenario: As "Mendelian" organisms with increased matching between genotype and phenotype began to emerge, development originally based on generic physical tissue properties was stabilized, and specific outcomes of morphogenetic processes became templates for the organization of newly evolving and integrated biochemical circuitry. This led to developmental individualization and modular building units (Wagner, 1989, 1995). However as these modules became functionally integrated and fixed at the bauplan level of a lineage, they in turn assumed a specific "constructional" identity at that level, becoming the elements of macroscopic design referred to as "homologues". This morphological identity eventually transcends all processes involved in the ontogeny of an individual homologue, be they genetic, cellular, biochemical, or physical, since these can change over the course of evolution (Wagner, 1989; Wray and Raff, 1991; Hall, 1994; Bolker and Raff, 1996). Thus the stabilized macro patterns (homologues) became more decisive for the further path of morphological evolution than the generic conditions on which they were initially based.

This means that although homologues may first arise by the same epigenetic processes that produce homoplasies, they eventually become independent of their underlying molecular, epigenetic, and generic constituents and increasingly play an organizational role in morphological evolution. They take on a "life of their own" and are thus inherited as structural units of morphological organization, not tied to any particular generative process. Homoplasies reflect the origin of morphological innovation in the generic material properties of tissues – they are an echo of the pre-Mendelian world. Homologues, in contrast, act as formal "attractors" of design, around which more design is added (Müller and Newman, 1999).

GENES AND FORM: A REINTERPRETATION

We propose that a synthetic, causal understanding of both development and evolution of morphology can be achieved by relinquishing a gene-centered view of these processes. This is not to say that programmed gene expression plays an unimportant role during embryogenesis or that random genetic change is not a major factor of evolution. However, we argue, in agreement with some earlier writers (Ho and Saunders, 1979; Oyama, 1985; Seilacher, 1991; Goodwin, 1994), that these factors are not explanatory of morphology in either of these settings. What replaces gene sequence variation and gene expression as morphological determinants in our framework are epigenetic processes: initially the physics of condensed, excitable media represented by primitive cell aggregates, and later conditional responses of tissues to each other, as well as to external forces. These determinants are considered to have set out the original morphological templates during the evolution of bodies and organs and to have remained, to varying extents, effective causal factors in all modern multicellular organisms.

We emphasize that the indirect relationship of genes to form, which we postulate for tissue morphogenesis, is analogous to what is generally accepted to constitute this relationship in the most fundamental role of genes: protein synthesis. Here genes also influence the realization of form without being its determinants. The three-dimensional, folded structure of a protein—its biologically functional morphology—is defined by interactions of the polypeptide chain within itself and with its external environment. The typical functional form of a protein is identical to that decreed by the thermodynamics of spontaneous processes. Correspondingly, the universe of protein secondary structures and folded motifs in existing organisms is limited to a relatively small number of forms (perhaps 1000) out of an astronomically large number of potential random compact structures (Chothia, 1992; Li *et al.*, 1996). Although spontaneous folding does not always take place in the cytoplasmic environment of the modern cell, which frequently employs energy-dependent chaperoning processes instead (Beissinger and Buchner, 1998), evolution has clearly used the spontaneously achieved morphologies as templates for the accumulation of sophisticated reinforcing mechanisms.

Just as an understanding of the set of preferred protein motifs and the morphologies of particular proteins depends on an appreciation of the originating role of physical mechanisms, we contend that an understanding of the forms assumed by metazoan organisms requires knowledge of the generative epigenetic processes that originally (in evolutionary history) produced those forms. Morphological development in ancient or modern metazoans was and is dependent on genetically specified biochemical constituents acting in the context of dynamic material systems with characteristic generic properties. As long as these generic properties dominated, the genes were merely suppliers of building blocks and catalysts, with little direct influence on the architectural outcome. However, genetic evolution is highly suited for enhancing the reliability of generation of "generically templated" forms. Standard modalities of gene evolution, such as promoter duplication and diversification (Goto *et al.*, 1989; Small *et al.*, 1991), metabolic integration, and functional redundancy (Wilkins, 1997), can add parallel routes to the same end point (Newman, 1994). Eventually, some of the parallel routes may come to predominate and constitute what has the appearance of a "genetic program," although (strangely, from the point of view of the computer metaphor) the physical outcome of the program's execution actually preexists the programmed "instructions." Thus physical morphogenesis would become secondarily captured and routinized by genetic circuitry, possibly involving mechanisms such as genetic assimilation (Waddington, 1961).

The "unit character" notion was considered problematic and was dismissed early on by such pioneers as Johannsen and Morgan (see for example Dunn, 1965). A number of more recent writers have continued to point out the inadequacy of a view of genes that takes them individually to correspond to particular complex traits, or even collectively to constitute programs for the production of such traits (Neumann-Held, 1998; Nijhout, 1990; Oyama, 1985) (although this has not prevented some latter day commentators from writing about genes for stripes, tails, or fingers). As application of more precise methods in developmental and evolutionary biology provides increasing evidence that there is no necessary relation between genetic and morphological change (Atchley *et al.,* 1988; Meyer *et al.*, 1990; Sturmbauer and Meyer, 1992; and Bruna *et al.*, 1996, the need for a new synthesis along these lines becomes more compelling. What has been missing from previous discussions is any positive account of what (if not genes *per se*) causes metazoan organisms to take on the forms they do in the course of development and evolution. We propose that epigenetic processes play this positive role, either (to use Aristotle's famous distinction) as "efficient causes" early in evolution, and to a more limited extent in contemporary organisms, or as "formal causes" (i.e., templates) as evolution progresses and new ways are invented to achieve the same morphological ends. Indeed, evolution can be considered an engine for turning efficient causes into formal causes.

CONCLUSIONS

The formal framework of neo-Darwinian theory requires morphological characters to be given and, therefore, does not constitute a theory of how they arise. We have proposed a concept for the origination of morphological characters and their cooptation by the genome. We summarize our position in the following points:

1.   The origin of form and characters is based on epigenetic principles acting both in the pre-Mendelian and in the Mendelian world.

• The earliest epigenetic mechanisms to influence biological form were the physics of chemically active condensed materials, which include primitive cell masses, resulting in a delimited, and essentially exhaustive, array of body plans and organ forms: segmented, hollow, mutlilayered, and branched structures.

• As a consequence of the biochemical and genetic integration of interactions, development increasingly takes place in a Mendelian arena in which genotype and morphological phenotype become more closely matched. Development also becomes susceptible to Darwinian modifiers, leading to the exploration of the residual morphogenetic "play" remaining in multicellular systems. In particular, physical properties and threshold effects of the developmental systems under modification generate morphogenetic by-products that become the kernels of morphological innovations, which elaborate on a smaller scale the major morphological themes of the earlier phase.

2.   The epigenetic concept addresses a number of open problems in evolutionary theory, such as the origin of body plans, morphological innovation, and homology..

• If epigenesis can account for the origins of bauplans and morphological innovation, competition among marginally different forms for adaptive advantage is not a sine-qua-non for morphological change. Darwinian adaptation-driven evolution can therefore be considered to be a limiting case of the epigenetic model. Selection, in this view, functions to release and consolidate inherent developmental potential, rather than guiding morphological evolution directly.

• Homology, the principle of morphological organization, is a consequence of the interplay between generic, morphogenetic templates and evolving, stabilizing biochemical circuitry. Fixed at the bauplan level, their molecular and developmental bases free to drift, homologues persevere and become attractors of morphological design.

3. *The epigenetic concept entails a new interpretation of the relationship between genes and biological form.*
   • The relationship between genotype and phenotype in the earliest metazoan organisms is hypothesized to have been different from that in modern organisms. The present relationship between genes and form is a derived condition, a product of evolution rather than its precondition.
   • Evolvability, in general, represents the carryover of epigentic determination from an earlier epoch of even greater morphogenetic plasticity, rather than the evolution of sophisticated genetic mechanisms selected to undermine rigid genetic determination.
   • Genetic change is required for evolution to progress, but with respect to morphology it mainly plays a consolidating role, rather than an innovating one. Physically-determined morphogenesis becomes secondarily captured and routinized by genetic circuitry which thus serves to channel and reinforce epigenetic propensities.

LITERATURE CITED

Alberch, P. (1982). Developmental constraints in evolutionary processes. *In* "Evolution and Development" (J. T. Bonner, ed.), pp. 313-332. Springer-Verlag, Berlin,

Amprino, R. (1985). The influence of stress and strain in the early development of shaft bones. *Anat. Embryol.* **172**:49-60.

Atchley, W. R., Newman, S. *et al.* (1988). Genetic divergence in mandible form in relation to molecular divergence in inbred mouse strains. *Genetics* **120**:239-253.

Bard, J. B. L. (1990). Traction and the formation of mesenchymal condensations in vivo. *Bioessays* **12**:389-395.

Beissinger, M., and Buchner, J. (1998). How chaperones fold proteins. *Biol. Chem.* **379**:245-259.

Boissonade, J., Dulos, E. *et al.* (1994). Turing patterns: from myth to reality. *In* "Chemical Waves and Patterns" (R. Kapral and K. Showalter, eds.), pp. 221-268. Kluwer, Boston.

Bolker, J. A., and Raff, R. A. (1996). Developmental genetics and traditional homology. *Bioessays* **18**(6): 489-494.

Bonner, J. T. (1998). The origins of multicellularity. *Integr. Biol.* **1**:27-36.

Braun, B. R., and Johnson, A. D. (1997). Control of filament formation in *Candida albicans* by the transcriptional repressor TUP1. *Science* **277**:105-109.

Briggs, D. E. G., Fortey, R. A. *et al.* (1992). Morphological disparity in the cambrian. *Science* **256**:1670-1673.

Carter, D. R., and Orr, T. E. (1992). Skeletal development and bone functional adaptation. *J. Bone. Miner. Res.* **7**(Suppl 2):389-395.

Chothia, C. (1992). Proteins. One thousand families for the molecular biologist. *Nature* **357**:543-544.

Chuong, C.-M., and Widelitz, R. B. (1998). Feather formation: a model of the formation of epithelial appendages. *In* "Molecular Basis of Epithelial Appendage Morphogenesis" (C.-M. Chuong, ed.), pp. 57-74. R.G. Landes, Austin, TX.

Conway Morris, S. (1989). Burgess Shale faunas and the Cambrian explosion. *Science* **246**:339-346.

Conway Morris, S. (1993). The fossil record and the early evolution of the Metazoa. *Nature* **361**:219-225.

Cooke, J., Nowak, M. A. *et al.* (1997). Evolutionary origins and maintenance of redundant gene expression during metazoan development. *Trends Genet.* **13**:360-364.

Cottrill, C. P., Archer, C. W. *et al.* (1987). Cell sorting and chondrogenic aggregate formation in micromass culture. *Dev. Biol.* **122**:503-515.

Crick, F. H. C., and Lawrence, P. A. (1975). Compartments and polyclones in insect development. *Science* **189**:340-347.

Dahm, L. M., and Landmesser, L. T. (1991). The regulation of synaptogenesis during normal development and following activity blockade. *J. Neurosci.* **11**:238-255.

Deeming, D. C., and Ferguson, M. W. (1988). Environmental regulation of sex determination in reptiles. *Philos. Trans. R. Soc. Lond. B Biol. Sci.* **322**:19-39.

deGennes, P. G. (1992). Soft matter. *Science* **256**:495-497.

Drachman, D. B., and Sokoloff, L. (1966). The role of movement in embryonic joint development. *Dev. Biol.* **14**:401-420.

Galis, F., and Drucker, E. G. (1996). Pharyngeal biting mechanics in centrarchid and cichlid fishes: insights into a key evolutionary innovation. *J. Evol. Biol.* **9**:641-670.

Garcia-Bellido, A., Ripoll, P. *et al.* (1976). Developmental compartmentalization in the dorsal mesothoracic disc of Drosophila. *Dev. Biol.* **48**:132-147.

Gerhardt, M., Schuster H., *et al.* (1990). A cellular automation model of excitable media including curvature and dispersion. *Science* **247**:1563-1566.

Gerhart, J., and Kirschner, M. (1997). "Cells, Embryos, and Evolution." Blackwell, Malden, MA.

Gilbert, L. I., Tata, J. R. *et al.*, eds. (1996). "Metamorphosis: Postembryonic Reprogramming of Gene Expression in Amphibian and Insect Cells." Academic Press, San Diego.

Giori, N. J., Beaupre, G. S. *et al.* (1993). Cellular shape and pressure may mediate mechanical control of tissue composition in tendons. *J. Orthop. Res.* **11**:581-592.

Goldbeter, A. (1995). "Biochemical Oscillations and Cellular Rhythms: The Molecular Bases of Periodic and Chaotic Behaviour." Cambridge University Press, Cambridge.

Goodwin, B. C. (1994). "How The Leopard Changed Its Spots." Weidenfeld and Nicolson. London.

Goto, T., MacDonald, P. *et al.* (1989). Early and late periodic patterns of *even skipped* expression are controlled by distinct regulatory elements that respond to different spatial cues. *Cell* **57**:413-422.

Hall, B. K. (1984). Developmental mechanisms underlying the formation of atavisms. *Biol. Rev.* **59**:89-124.

Hall, B. K. (1986). The role of movement and tissue interactions in the development and growth of bone and secondary cartilage in the clavicle of the embryonic chick. *J. Embryol. Exp. Morph.* **93**:133-152.

Hall, B. K., ed. (1994). "Homology." Academic Press, San Diego.

Harris, A. K., Stopak, D. *et al.* (1980). Fibroblast traction as a mechanism for collagen morphogenesis. *Nature* **290**:249-251.

Ho, M. W., and Saunders, P. T. (1979). Beyond neo-Darwinism - An epigenetic approach to evolution. *J. Theor. Biol.* **78**:573-591.

Ishii, N., Yamamoto, M. *et al.* (1997). A DNA-binding protein from *Candida albicans* that binds to the RPG box of *Saccharomyces cerevisiae* and the telomeric repeat sequence of *C. albicans*. *Microbiology* **143**:417-427.

Jablonka, E., and Lamb, M. J. (1995). "Epigenetic Inheritance and Evolution." Oxford University Press, Oxford.

Johnston, T. D., and Gottlieb, G. (1990). Neophenogenesis: a developmental theory of phenotypic evolution. *J. Theor. Biol.* **147**:471-495.

Kashimata, M., and Gresik, E. W. (1996). Contemporary approaches to the study of salivary gland morphogenesis. *Eur. J. Morphol.* **34**:143-147.

Kauffman, S. A. (1993). "The Origins of Order." Oxford University Press, New York.

Kazmierczak, J., and Degens, E. T. (1986). Calcium and the early eukaryotes. *Mitt. Geol.- Palaeont. Inst. Univ. Hamburg* **61**:1-20.

Kirschner, M. and Gerhart, J. (1998). Evolvability. *Proc. Natl. Acad. Sci. U S A* **95**:8420-7.

Klein-Nulend, J., Roelofsen, J. *et al.* (1995). Mechanical loading stimulates the release of transforming growth factor-beta activity by cultured mouse calvariae and periosteal cells. *J. Cell. Physiol.* **163**:115-119.

Knoll, A. H. (1992). The early evolution of eukaryotes: a geological perspective. *Science* **256**:622-627.

Leonard, C. M., Fuld, H. M. *et al.* (1991). Role of transforming growth factor-beta in chondrogenic pattern formation in the embryonic limb: stimulation of mesenchymal condensation and fibronectin gene expression by exogenous TGF-beta and evidence for endogenous TGF-beta-like activity. *Dev.Biol.* **145**:99-109.

Li, H., Helling, R. *et al.* (1996). Emergence of preferred structures in a simple model of protein folding. *Science* **273**:666-669.

Liem, K. F. (1990). Key evolutionary innovations, differential diversity, and symecomorphosis. *In* "Evolutionary Innovations" (M. H. Nitecki, ed.), pp.147-170. University of Chicago Press, Chicago.

Magee, P. T. (1997). Which came first, the hypha or the yeast? *Science* **277**:52-53.

Maynard Smith, J., Burian, R. *et al.* (1985). Developmental constraints and evolution. *Q. Rev. Biol.* **60**(3):265-287.

McLaren, A., and Michie, D. (1958). An effect of the uterine environment upon skeletal morphology in the mouse. *Nature* **181**:1147-1148.

Meyer, A., Kocher, T. D. *et al.* (1990). Monophyletic origin of Lake Victoria cichlid fishes suggested by mitochondrial DNA sequences. *Nature* **347**:550-553.

Mikhailov, A. S. (1990). "Foundations of Synergetics I." Springer-Verlag, Berlin.

Miura, T., and Shiota, K. (2000a). Extracellular matrix environment influences chondrogenic pattern formation in limb bud micromass culture: Experimental verification of theoretical models. *Anat. Rec.* **258**:100-107.

Miura, T. and Shiota, K. (2000b). TGFbeta2 acts as an "activator" molecule in reaction-diffusion model and is involved in cell sorting phenomenon in mouse limb micromass culture. *Dev. Dyn.* **217**:241-9.

Moore, J., and Willmer, P. (1997). Convergent evolution in invertebrates. *Biol. Rev. Camb. Philos. Soc.* **72**:1-60.

Müller, G. B. (1989). Ancestral patterns in bird limb development: a new look at Hampé's experiment. *J. Evol. Biol.* **2**:31-47.

Müller, G. B. (1990). Developmental mechanisms at the origin of morphological novelty: a side-effect hypothesis. *In* "Evolutionary Innovations" (M. H. Nitecki, ed.), pp. 99-130. The University of Chicago Press, Chicago.

Müller, G. B., and Newman, S. A. (1999). Generation, integration, autonomy: three steps in the evolution of homology. *In* "Homology" (Novartis Foundation Symposium 222) pp. 65-79. Wiley, Chichester.

Müller, G. B., and Streicher, J. (1989). Ontogeny of the syndesmosis tibiofibularis and the evolution of the bird hindlimb: a caenogenetic feature triggers phenotypic novelty. *Anat. Embryol.* **179**:327-339.

Müller, G. B., and Wagner, G. P. (1991). Novelty in evolution: restructuring the concept. *Annu. Rev. Ecol. Syst.* **22**:229-256.

Neumann-Held, E. M. (1998). The gene is dead – Long live the gene! Conceptualizing genes the constructionist way. *In* "Sociobiology and Bioeconomics: The Theory of Evolution in Biological and Economic Theory" (P. Koslowsky, ed.), pp. 105 - 137. Springer, Berlin.

Newman, S. A. (1984). Vertebrate bones and violin tones: music and the making of limbs. *Sciences (NY)* **24**:38-43.

Newman, S. A. (1988). Lineage and pattern in the developing vertebrate limb. *Trends Genet.* **4**:329-332.

Newman, S. A. (1993). Is segmentation generic? *Bioessays* **15**:277-283.

Newman, S. A. (1994). Generic physical mechanisms of tissue morphogenesis: a common basis for development and evolution. *J. Evol. Biol.* **7**:467-488.

Newman, S. A. (1996). Sticky fingers: Hox genes and cell adhesion in vertebrate limb development. *Bioessays* **18**:171-174.

Newman, S. A. (1998). Epithelial morpohogenesis: a physico-evolutionary interpretation. *In* "Molecular Basis of Epithelial Appendage Morphogenesis" (C. M. Chuong, ed.), pp. 341-358. R. G. Landes, Austin, TX.

Newman, S. A., and Frisch, H. L. (1979). Dynamics of skeletal pattern formation in developing chick limb. *Science* **205**:662-668.

Nijhout, H. F. (1990). Metaphors and the roles of genes in development. *Bioessays* **12**:441-446.

Nowak, M. A., Boerlijst, M. C. *et al.* (1997). Evolution of genetic redundancy. *Nature* **388**:167-171.

Nüsslein-Vollhard, C. (1996). Gradients that organize embryo development. *Sci. Am.* **275**:54-55; 58-61.

Oyama, S. (1985). "The Ontogeny of Information." Cambridge University Press, Cambridge.

Palmeirim, I., Henrique, D. *et al.* (1997). Avian hairy gene expression identifies a molecular clock linked to vertebrate segmentation and somitogenesis. *Cell* **91**:639-648.

Persson, M. (1983). The role of movements in the development of sutural and diarthrodial joints tested by long-term paralysis of chick embryos. *J. Anat.* **137**:591-599.

Picket, F. B., and Meeks-Wagner, D. R. (1995). Seeing double: appreciating genetic redundancy. *Plant Cell* **7**:1347-1356.

Pourquié, O. (1998). Clocks regulating developmental processes. *Curr. Opin. Neurobiol.* **8**: 665-670.

Robbins, J. R., Evanko, S. P. *et al.* (1997). Mechanical loading and TGF-beta regulate proteoglycan synthesis in tendon. *Arch. Biochem. Biophys.* **342**:203-211.

Rodriguez-Boulan, E., and Nelson, W. J. (1993). "Epithelial and Neuronal Cell Polarity." Company of Biologists, Cambridge.

Rutherford, S. L., and Lindquist, S. (1998). Hsp90 as a capacitor for morphological evolution. *Nature* **396**:336-342.

Sanderson, M. J., and Hufford, L., eds. (1996). "Homoplasy: The Recurrence of Similarity in Evolution." Academic Press, San Diego.

Scott, J. E., Haigh, M. *et al.* (1987). The effect of muscle paralysis on the radial growth of collagen fibrils in developing tendon. *Clin. Sci.* **72**:359-363.

Seilacher, A. (1991). Self-organizing mechanisms in morphogenesis and evolution. *In* "Constructional Morphology and Evolution" (N. Schmidt-Kittler and K. Vogel, eds.), pp. 251-271. Springer, Berlin.

Seilacher, A. (1992). Vendobionta and Psammocorallia - lost constructions of precambrian evolution. *J. Geol. Soc. (Lond.)* **149**:607-613.

Seilacher, A., Bose, P. K. *et al.* (1998). Triploblastic animals more than 1 billion years ago: trace fossil evidence from india [In Process Citation]. *Science* **282**:80-83.

Shubin, N., Tabin, C. *et al.* (1997). Fossils, genes, and the evolution of animal limbs. *Nature* **388**:639-648.

Small, S., Kraut, R. *et al.* (1991). Transcriptional regulation of a pair-rule stripe in *Drosophila*. *Genes. Dev.* **5**:827-839.

Starmer, C. F., Biktashev, V. N. *et al.* (1993). Vulnerability in an excitable medium: analytic and numerical studies of initiating unidirectional propagation. *Biophys. J.* **65**:1775-1787.

Stearns, S. C. (1986). Natural selection and fitness, adaptation and constraint. *In* "Patterns and Processes in the History of Life" (D. M. Raup and D. Jablonski, eds.), pp. 23-44. Dahlem Konferenzen, Springer-Verlag, Berlin.

Steinberg, M. S. (1998). Goal-directedness in embryonic development. *Integr.Biol.* **1**:49-59.

Steinberg, M. S., and Takeichi, M. (1994). Experimental specification of cell sorting, tissue spreading, and specific spatial patterning by quantitative differences in cadherin expressions. *Proc. Natl. Acad. Sci.* **91:**206-209.

Streicher, J., and Müller, G. B. (1992). Natural and experimental reduction of the avian fibula: Developmental thresholds and evolutionary constraint. *J. Morphol.* **214:**269-285.

Sturmbauer, C., and Meyer, A. (1992). Genetic divergence, speciation, and morphological stasis in a lineage of African cichlid fishes. *Nature* **358:**578-581.

Takeichi, M. (1991). Cadherin cell adhesion receptors as a morphogenetic regulator. *Science* **251:**1451-1455.

Tautz, D. (1992). Redundancies, development and the flow of information. *Bioessays* **14:**263-266.

Tsarfaty, I., Resau, J. H. *et al.* (1992). The met proto-oncogene receptor and lumen formation. *Science* **257:**1258-1261.

Turing, A. (1952). The chemical basis of morphogenesis. *Philos. Trans. R. Soc. Lond. B* **237:** 37-72.

Vogel, K. G., and Koob, T. J. (1989). Structural specialization in tendons under compression. *Int. Rev. Cytol.* **115:**267-293.

Waddington, C. H. (1961). Genetic assimilation. *Adv. Genet.* **10:**257-293.

Wagner, A. (1996). Genetic redundancy caused by gene duplications and its evolution in networks of transcriptional regulators. *Biol. Cybern.* **74:**557-567.

Wagner, G. P. (1989). The biological homology concept. *Annu. Rev. Ecol. Syst.* **20:**51-69.

Wagner, G. P. (1995). The biological role of homologues: a building block hypothesis. *N. JB. Geol. Paläont. Abh.* **195:**279-288.

Wake, D. B. (1991). Homoplasy: the result of natural selection or evidence of design limitations? *Am. Nat.* **138:**543-567.

Whittington, H. B. (1985). "The Burgess Shale." Yale University Press, New Haven, CT.

Wilkins, A. S. (1997). Canalization: a molecular genetic perspective. *Bioessays* **19:**257-262.

Winfree, A. T. (1994). Persistent tangled vortex rings in generic excitable media. *Nature* **371:**233-236.

Wray, G. A. and Raff, R. A. (1991). The evolution of developmental strategy in marine invertebrates. *TREE* **6**(2):45-50.

Wu, K. C. (1996). Entwicklung, Stimulation und Paralyse der embryonalen Motorik. *Wien. Klin. Wochenschr.* **108:**303-305.

Yokouchi, Y., Nakazato, S. *et al.* (1995). Misexpression of *Hoxa-13* induces cartilage homeotic transformation and changes cell adhesiveness in chick limb buds. *Genes Dev.* **9:**2509-2522.

25

KEY INNOVATIONS AND RADIATIONS

FRIETSON GALIS

Institute for Evolutionary and Ecological Sciences, University of Leiden, 2300RA Leiden, The Netherlands

KEY INNOVATIONS, DIVERSITY, AND RATE OF EVOLUTIONARY CHANGE

Evolution does not proceed as a gradually continuous process; the rate of evolutionary change varies over time (e.g., Simpson, 1944; Fitch and Ayala, 1995). Accelerations of evolutionary changes are the result of important environmental, behavioral, structural, genetical, or physiological changes. When such important changes are not environmental but concern traits of an organism, they are usually called key innovations, especially when they are seen as triggers of diversification.

The Character Concept in Evolutionary Biology

The concept of a key innovation has been defined in rather different ways since it was used by Miller (1949) as adjustments (key adjustments) in the morphological and physiological mechanism which are essential to the origin of new major groups. In all definitions the basic idea is that some evolutionary changes are more important than others (see Hunter, 1998). The importance can be with respect to the appearance of higher taxa, the proliferation of species, and/or the generation of new morphologies (Vermeij, 1995; Hunter 1998). Most definitions include the aspect of key innovations as triggers to diversification (e.g., Miller, 1949; Levinton, 1988; Heard and Hauser, 1995; Bond and Opell, 1998).

The early literature saw key innovations as innovations that allowed the invasion of new adaptive zones (e.g., Miller, 1949; Mayr, 1960; see also Simpson, 1944). The occupation of a new adaptive zone subsequently allowed the origin and diversification of higher taxa. These ideas on the central importance of the invasion of new adaptive zones led to the still popular notion that behavioral change usually precedes morphological change. This logically follows from the presumed fact that the occupation of a new adaptive zone will usually be caused by a behavioral change (Ewer, 1960: "a mutation causing slight webbing in a non-swimmer will not cause accumulation of genes making for the habit of swimming"; Simpson and Roe, 1958; Mayr, 1960).

The rise of cladistics and the comparative method (e.g., Harvey and Pagel, 1991; Brooks and McLennon, 1991; Martins, 1996) has had a major impact on the study of key innovations as causative factors for evolutionary sucess. Cladistics allows the recognition of monophyletic groups. Knowledge of monophyletic groups makes it possible to compare the species richness of clades and, thanks to this, species richness is now an important measure of evolutionary succes.

The shift in emphasis from causative factors that allow the occupation of new adaptive zones to ones that allow species diversification has led to a search for key innovations that increase the ecological potential in a wider sense than just allowing the occupation of a new adaptive zone (Liem, 1973; Jernvall et al., 1996; Schaefer and Lauder, 1986; Hodges and Arnold, 1995; Berenbaum et al., 1996; Galis and Drucker, 1996; Bond and Opel, 1998). Innovations that markedly increase the ecological potential can be behavioural and morphological as well as physiological. In snakes, both a venomous bite (e.g., in colubrids, elapids, and viperids) and constriction (e.g., in colubrids and boids) are employed to catch large prey items (Greene, 1978, 1997). Both these traits can be seen as innovations with a far-reaching influence on the evolution of these clades. Similarly, protection of the young can be achieved by behavioral traits (e.g., the later discussed mouth-brooding in some cichlid fish families) and by morphological and physiological traits (e.g., internal gestation in mammals). The notion that behavioral key innovations precede morphological ones is, therefore, not always true. The mutual influences of behavior and structures (and other traits) on each other make it useless to try to predict in general whether behavioral innovations precede morphological ones or vice versa.

Several authors have proposed that key innovations must be tested by comparing the speciosity of taxa with and without a certain key innovation (Lauder, 1981; Lauder and Liem, 1989; Jensen, 1990; Heard and Hauser, 1995). The same authors recognize that major difficulties are involved with the testing, most importantly a lack of suitable taxa for comparison as key innovations are often unique characters. These serious difficulties concerning the testability have made the concept of key innovations controversial (e.g., Heard and Hauser, 1995; Hunter, 1998), but this has not led to a reduction in attempts to recognize key innovations (e.g., Hodges and Arnold, 1995; Jernvall et al., 1996; Galis and Drucker, 1996; Berenbaum et al., 1996; Bond and Opel, 1998; Simmons et al., 1998; Rey et al., 1999). Evidently, there is still a need to identify character changes that have a disproportionate effect on evolution.

An even more serious problem than the possible uniqueness of key innovations is that in most frequently used definitions the key innovation is causally linked to diversification. This makes these definitions untestable. The definition of Heard and Hauser (1995) is, for instance, "an evolutionary change in individual trait(s) that is causally linked to an increased diversification rate in the resulting clade (for which it is a synapomorphy)." Innovations can never do more than allow diversification (Levinton, 1988; Liem, 1990) because diversification depends not only on innovations, but also on ecological circumstances and the genetic makeup of organisms. These other factors make it impossible to test whether an innovation is a key innovation on the basis of the presence of the innovation in speciose taxa and the absence in less speciose taxa. It is, for example, possible that one group possessing a key innovation has radiated and another group possessing the same key innovation has not. An example may be found in the fish family Embiotocidae, which is not speciose, but the members possess the same innovation in their pharyngeal jaw apparatus which is supposed to be causally linked to the speciosity of the Pomacentridae, Labridae, and Cichlidae (Galis and Drucker, 1996). There is circularity in the definition when characters that are supposed to have been a cause for diversification can only be found in diverse groups.

Some authors have defined key innovations such that the innovation is causally related to the development of a major new body plan, which opens up a new character space (Van Valen, 1971; Baum and Larson, 1991; Rosenzweig and McCord, 1991). These mechanistic definitions do not define a key innovation as an innovation that necessarily increases the potential for diversification. Instead, these definitions focus on a key innovation as changing the possibilities and impossibilities of an organism, resulting in an improved level of functioning. This is of course a valid way of defining a key innovation. However, when analyzing the evolutionary success of taxa it is useful to explicitly include the increased potential for diversification. Therefore, I propose the following definition, which is based on these mechanistic definitions, but includes the aspect of key innovations as triggers of diversification: "A key innovation is an innovation which opens up a new character space (or breaks constraints) that potentially allows the occupation of more niches." The inclusion of more niches and not just new niches is, thus, important as otherwise innovations are included that cause drastic changes, but do not necessarily allow diversification. The constraints that are broken and in this way open up a new character space may be functional, constructional, developmental, and genetic (see section "Factors Inhibiting the Evolution of Structural Key Innovations").

Testing. This definition of a key innovation allows us to test whether characters are key innovations without the need for historical testing. To test whether an innovation is a key innovation one has to measure the potential change in ecological performance that is the result of the possession of the character. For structural innovations this requires a functional morphological analysis. Designing quantitative criteria to test may be more straightforward for structural (and physiological) innovations than for behavioral ones, as e.g., degrees of freedom of a construction can be quantified (Koolstra *et al.*, 1988; Galis, 1992; Muller, 1993; van Gennip and Berkhoudt, 1992). See the section 'Structural Changes and Key Innovations" for a discussion of the relationship between degrees of freedom and diversification. Behavioral key innovations are usually complex and variable and it is therefore more difficult to quantify their effect.

Although quantitative criteria may sometimes be difficult to find, qualitative criteria to test for the opening up of a new character space and for an increase in potential niche width will usually be feasible for structural, behavioral and physiological innovations. It is to be expected that expansion of potential niche width will most often be caused by either an increase in the potential range of prey items or an increase in the potential number of habitats. This can be measured if not in a quantitative sense, at least in a qualitative sense. For example, in the behavioral key innovation of constriction in snakes (Greene, 1978) it can be convincingly argued that the upper size limit of prey items caught by biting prey with the jaws is considerably lower than that by coiling around the prey with a long body. Another clear example provides the shell around the egg of amniotes. The shell allows eggs to be deposited on land and therefore inevitibly increases the number of potential habitats that can be occupied, without any precise quantitative estimation of the niche expansion. The decision how much an innovation should increase the potential niche width to be called a key innovation is of course arbitrary and will vary dependent on the author. Although strictly historical testing is not possible, the presence of an innovation in species-rich groups gives support to the notion that it is a key innovation. The presence of an innovation in a species-poor group is, however, not sufficient to conclude that the innovation is not a key innovation.

Although behavioral key innovations are also discussed, the main focus of this chapter is on structural key innovations.

KEY INNOVATIONS AND THE OPENING UP OF A NEW CHARACTER SPACE

Key innovations drastically change the possibilities of an organism, i.e., they open up a new character space. As a consequence there is the option that, depending on the ecological background, the selection pressures on *many* characters of the organism suddenly change. When the selection pressures on many characters change, a cascade of changes is expected to follow the acquisition of a key innovation. I shall illustrate this with the following example of a key behavioral innovation, mouth brooding in lake-dwelling African cichlids (Fig. 1).

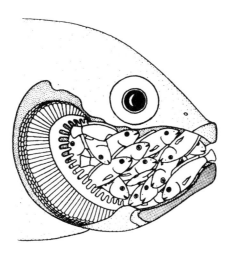

FIGURE 1 A brooding cichlid with head partly cut away to reveal the brood of well-developed young within its mouth. From Fryer and Iles (1972).

The initial functional significance of mouth brooding was almost certainly protection of the young against predation and other environmental hazards. The decreased predation pressure is, however, not the only change in selection pressure that is the result of this novel behavior. The shielding of the young in the mouth and not in a nest, as substrate brooding cichlid do, allows the fish to live in areas where substrate brooders cannot live, e.g., over muddy substrates and in open water (an extensive part of the lake). This must have led to large-scale changes in selection pressures on the adult. I shall focus on the muddy substrates as in muddy areas the oxygen content and the transparency of the water are lower. This has consequences among others for the predation pressure exerted by visual predators. Furthermore, the coloration that makes males distinguishable from females and from males of other species is concealed (Seehausen et al., 1997) and the visibility of prey items is reduced. Therefore, selection pressures on the eyes, lateral line system, brains, gills, coloration (natural and sexual selection), and feeding apparatus will all change. In addition, selection pressures on the ovaries will change as optimal clutch size changes. Size and shape changes of structures will then further lead to changed selection pressures on the surrounding structures. For example, when the size of the gills changes, this will affect either the streamline and/or it will affect the size of surrounding structures (Chapman and Galis, in prep.).

Similarly, key structural and physiological innovations lead to changed selection pressures on many, if not all, traits of the organism. Consider for example the acquisition of oral jaws in jawed fishes. The possession of jaws has far-reaching consequences for the ease with which a wide range of prey items can be caught. This has not only led to evolutionary changes in the feeding apparatus, but the increased pressure on prey items supposedly selected for higher swimming speed in prey, better vision, and in general led to an arms race between predators and prey influencing a multitude of traits (Vermeij, 1994).

STRUCTURAL CHANGES AND KEY INNOVATIONS

Some body plans can be modified and diversified more easily than others. This is an important factor in explaining why some taxa are more speciose than others. A structural key innovation leads to a body plan that can be more easily diversified than the original body plan and, therefore, can lead to an increase in niche width.

Vermeij (1974) has proposed that an increase in potential versatility of a Bauplan comes from an increase in the number of independent parameters or structures. An increased number of independent parameters leads to a construction with a higher degree of freedom and thus to a higher number of possible mechanical solutions for functional problems. With an increase in the number of solutions for functional problems comes the possibility of greater mechanical efficiency and effectiveness of resource exploitation (Vermeij, 1974). He hypothesized that more versatile taxa or body plans tend to replace less potentially versatile taxa in the course of time (see also Schaeffer and Rosen, 1961). Lauder (1981) and Lauder and Liem (1989) argued further that taxa exhibiting a large number of structural decouplings are expected to be more versatile and hence more speciose than taxa with fewer decouplings. An example of a high degree of freedom of a construction associated with high speciosity is provided by the bony fishes that are both very diverse and speciose compared to other vertebrate classes   (bony fish species make up 96% of all fish species; Nelson, 1994) and are characterized by a particularly large number of loosely connected bony elements in their heads (Fig. 2).

Innovations that lead to an increase in the degrees of freedom are (1) duplications, (2), decouplings, (3) increased complexity of a structure, and (4) new structures.

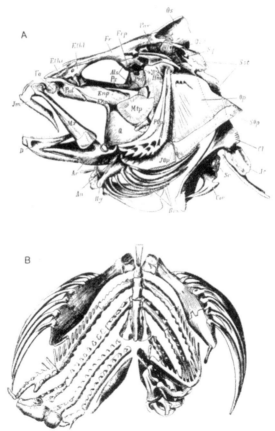

FIGURE 2 (A) Bony elements of the head of a perch (*Perca fluviatilis*). The bony elements of the branchial arches and hyoid apparatus are not visible and are shown in (B). From Claus (1880).

Another type of structural innovation is the loss of a function by a structure that subsequently becomes available for a new function or for increased specialization of an already existing function (Roth and Wake, 1989; see also Schwenk, 1993). This type of innovation does not lead to an increase in the number of independent elements, and nonetheless it is easy to find examples of key innovations that do belong to this type. A striking example is presented by the evolution of wings in insects. Kukalova-Peck (1983, 1992) has proposed a transformation scheme that supports an old idea that the wings of insects have evolved from gills that were attached to the legs of their ancestors. Part of the leg became incorporated in the thorax and the wing articulates with this part. The selective forces that explain the initial part of the transformation are still unclear, but there is both molecular and structural evidence for the transformation from gills to wings (Kukalova-Peck, 1983, 1992; Averof and Akam, 1995; Averof and Cohen, 1997; Dickinson et al., 1997). It is clear that the acquisition of wings greatly expands the number of possible niches that can be occupied and it is, without doubt, one of the best examples of a key structural innovation. However, in this case an increase in the number of independent elements can also be demonstrated because the freedom of movement of the wings of insects must have involved the evolution of an elaborate articulation between the wings and thorax.

This example supports the idea that for a structural key innovation a loss of function is important, but not sufficient and that it needs to be accompanied by a change that increases the number of degrees of freedom. The increase in degrees of freedom can also emerge from an extra structure that takes over the lost function.

Here, I present examples of the different types of structural key innovations that can increase the number of degrees of freedom. Usually, examples where the analysis has been carried out in detail are rare. Therefore, I have only selected those examples that, even without a detailed analysis, seem beyond a doubt to be key innovations.

1. Structural Duplications

Duplications of structural elements followed by specialization for different functions are very important in evolutionary history (e.g., Lauder 1981; Bonner, 1988; Müller and Wagner, 1991). A beautiful example of a key innovation is formed by the vertebral column. This structure with repeated elements has been of outstanding importance in the evolution of the large variety of body plans in vertebrates (e.g., Slijper 1946; Radinsky 1987).

The vertebral column is involved in many functions, among others it protects the spinal cord and other structures (e.g., heart and lungs in many vertebrates), provides support, transmits the force of limbs in terrestrial tetrapods, is the site of insertion of numerous muscles and tendons (e.g., Slijper 1946; Radinsky, 1987), allows the mobility of the head, and provides a rudder in birds (Gatesy and Dial, 1996). In addition, a very large number of muscles and ligaments can be attached to protrusions of the vertebrae (see for example, the large neural spines for the support of the head in the giraffe; Fig. 3. An important aspect is the ease with which evolutionary changes in the number of vertebrae have apparently happened repeatedly to accommodate size changes in different parts of the body, e.g., the long necks in ichthyosaurs which contained up to 200 vertebrae (Starck, 1979), the generally long bodies of snakes (435 vertebrae in the python; Starck, 1979), and the reduction of the tail in birds and humans. The only exception is the powerful constraint on cervical vertebral number in mammals (Galis, 1999), which has prevented giraffes and camels from developing more than 7 vertebrae in their long necks (Fig. 3).

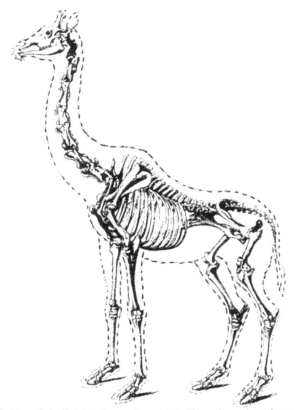

Figure 3   Skeleton of giraffe (*Giraffa camelopardalis*). The neck consists of seven large cervical vertebrae. Note the large neural (dorsal) spines on the rostral thoracic vertebrae. Ligaments and muscles that support the head and neck are attached to these spines. From Owen (1866).

An evolutionary change which nicely illustrates the evolutionary flexibility provided by the vertebral column is the transition from water to land that was made by the early tetrapods. On land the weight of the body needs to be supported and this support is provided by the fins which were transformed into limbs. The force of the limbs needs to be transmitted to the body and to achieve this a connection is necessary between the limbs and the vertebral column. The pectoral girdle was already well developed in fishes, but the pelvic girdle was very small and initially the force transmission between vertebral column and pelvic girdle occurred via a rib (Radinsky, 1987). Later the pelvic girdle of tetrapods increased in size and the transmission of force no longer required a rib. The vertebral column with its many variable elements has undoubtedly provided vertebrates with a body plan that can easily be modified and diversified and allows the occupation of many more feeding niches and habitats.

Other examples of duplications that were followed by differentiation of the duplicated elements are the pharyngeal (gill) arches in fishes. The pharyngeal arches have been differentiated into many new structures in fishes and other vertebrates. The most well-known example is the modification of a pair of pharyngeal arches into jaws in jawed fishes, in itself a key innovation. Although the morphological evidence for this transition is compelling (Mallatt, 1996, 1997), the transition is contested (Janvier, 1996). The genetic data are not conclusive (reviewed in Köntges and Lumsden, 1996; Kuratani et al., 1997), but the Otx gene expression patterns support the traditional idea of the homology of the jaws of jawed fishes and the first gill arches in lampreys (Tomsa and Langeland, 1999). The evolution of the hyoid and pharyngeal jaws from other pharyngeal arches is, however, well supported and shows the evolutionary importance and potential of this set of duplicated elements (e.g., Mallatt, 1996; Janvier, 1996).

Other examples include teeth in vertebrates (see increases in complexity, Fig. 4) and feathers in birds. A beautiful example is provided by the duplication of sensory organs followed by specialization for different types of sensory perception. Staaden and Römer (1992) documented the evolutionary transition from stretch to hearing organs in ancient grasshoppers (see also Meier and Reichert, 1995; Hodos and Butler, 1997). Undoubtedly the increase in types of sensory perception will potentially allow organisms to occupy more niches.

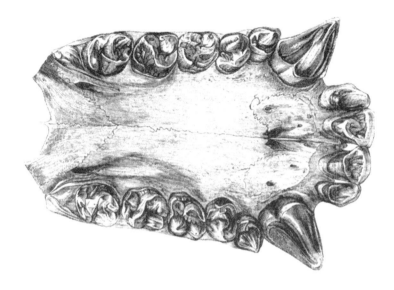

FIGURE 4 Dentition of the upper jaw of a male gorilla (*Gorilla gorilla*). Teeth provide a good example of duplicated elements that have become specialized for different functions. The specialization involved increases in complexity of shape. From Owen (1866).

2. Decouplings

Decouplings are particularly common evolutionary changes and they can have a major effect on evolution (Lauder and Liem, 1989; Schaefer and Lauder, 1986; Vermij, 1974; Galis and Drucker, 1996).

Soft tissues. Possibly the most common type of decoupling in soft tissues is an invagination of a tissue and the subsequent development into a structure which is separate from the original structure. The digestive tract has produced numerous organs this way during evolution (e.g., stomach, thymus, see Fig. 5). An example which certainly fits the definition of a key innovation as it breaks an important constraint which allows the occupation of a large number of niches is the evolution of lungs in bony fishes from a pouch in the digestive tract (e.g., Johansen, 1970; Graham, 1997). In the Sarcopterygians the possession of lungs enabled the transition from water to land, which led to an immense radiation of tetrapods. Not all taxa which possess lungs or other air-breathing organs are speciose (Graham, 1997) and, thus, this provides a clear example of how a key innovation can be present in many taxa, without necessarily leading to radiation.

FIGURE 5 Digestive tract of a salmon (*Salmo salar.*) In soft tissues invaginations are a commoi type of decoupling.

A second common type of decoupling in soft tissues is the subdivision o muscles into components with different lines of action, ultimately leading t(separate muscles (Winterbottom, 1974; Friel and Wainwright, 1998) (although ii this paper called duplications). Many of the individual muscles in fishes ar(assumed to have evolved in this way from the hypaxial and epaxial bod) musculature (Winterbottom, 1974).

Hard tissues. Structural decouplings of hard tissues usually involve th(development and modification of articulations between bony elements or shift: of muscles and ligaments such that connections between bony elements ar(changed. A convincing and well-known example is the increased mobility of th(mouth by the decoupling of an upper jaw bone from the cheekbones ii neopterygian fishes (including among others Amia, salmonids, and all teleosts which was followed in the succesful teleosts by the decoupling of yet anothe upper part of the upper jaw Schaeffer and Rosen, 1961). These two decouplings have led to a mobile mouth with greatly improved suction and biting capacity, allowing teleosts to feed on a host of different prey items.

A second example of simple structural decouplings with a major effect is also found in fishes. In cichlid fishes two decouplings are found in the pharyngeal jaw apparatus which provide them not only with a stronger bite (presumably the inital selective advantage), but, more importantly, also with a much greater freedom of movements (Fig. 6; Galis and Drucker, 1996; for the freedom of movements of the cichlid pharyngeal jaw apparatus, Galis, 1992, 1993; see also Liem, 1974). A biomechanical analysis shows that at first the upper and lower pharyngeal jaws must have become decoupled, followed by a decoupling of the fourth gill arch and the lower pharyngeal jaw (modified fifth gill arches). A comparison of pomacentrids and cichlids supports this order of decouplings (Galis and Snelderwaard, 1997). The effect of the two decouplings is a pharyngeal jaw apparatus that is not only versatile but can also easily be modified and diversified evolutionarily.

3. Increased Structural Complexity

A key innovation that leads to more structural complexity is the formation of a hypocone (cusp) on molars in mammals (Hunter *et al.*, 1996). This hypocone, which is supposed to have independently evolved 20 times, can be modified into many different shapes (cusps and ridges, Fig. 4) and this allows differentiation of feeding habits (see also Jenkins *et al.*, 1997). Changes of tooth shape are supposed to have been particularly important for the diversification of herbivorous habits in mammals (Jernvall *et al.*, 1996).

Another key innovation of this type is the folding of the columella in pulmonates (Gittenberger, 1996, in prep.). The functional significance of this change in shape is an enlarged surface area for attachment of the important muscle that retracts the head and foot inside the shell (Fretter and Graham, 1994), and in addition when the folds reach a certain size they partially close the aperture. These folds thus allow improved protection against predators and dessication and, therefore, lead to an increase in potential habitats that can be occupied. In the speciose Clausiliidae this protection of the aperture by folds has been perfected by a decoupling of a fold which has produced a movable "door" (clausilium), which can completely close the aperture and thus provide better protection (Gittenberger, in prep.). At the same time the movable "door" allows a larger aperture for the rostral part of the snail's body to move out of the shell during activity (Gittenberger, 1996).

It is to be expected that decouplings will usually be preceded by an increase in structural complexity. In duplicated structures increases in complexity will presumably follow the event of duplication. Increases in complexity can thus be key innovations by themselves, but they are also commonly found in combination with other types of changes.

4. New Structures

Changes in development can lead to the formation of novelties that cannot be traced back to a character in the ancestor (Müller and Wagner, 1991). The most convincing examples of key innovations among such new structures are probably the coverings that are formed around organisms protecting them against the environment (e.g., against cold, drought, predators). The hard shell around eggs certainly allows amniotes to occupy many more habitats. Similar examples are the development of a shell in molluscs and brachiopods and an armor in insects and many other organisms. Scales, feathers, and hairs can also be mentioned in this respect. These examples all increase the independence of organisms with respect to their environment and, in this way, allow the occupation of many more niches.

FACTORS INHIBITING THE EVOLUTION OF STRUCTURAL KEY INNOVATIONS

Despite their drastic consequences, structural key innovations can be tiny structural changes. This especially holds for decouplings and increases in structural complexity. It is an interesting question why such tiny changes with often large functional advantages often only occur after long periods of stasis of the ancestral state.

Wagner and Schwenk (2000) argue that the functional integration of a (functional) unit will prevent selection for internal changes that disrupt the integration (see also Schwenk, this volume). They call a functional unit that resists evolutionary change by self-stabilization an evolutionary stable configuration (ESC, analogous to evolutionary stable strategies). They hypothesize that the initial event that allows a break up of the integration is likely to be an external change because the functional integration strongly protects against internal changes that disrupt the integration. A structural change in a different functional unit provides one sort of external change. One can plausibly also include behavioral changes as external events, i.e., ones that change the functional demands of the structure itself. Drastic changes in the physical environment can be included as well for the same reason.

There are additional reasons to expect constraints on change of an integrated character complex. It is likely that the selective advantages of functional integration will lead to an increase in the developmental integration. An increase in developmental integration (ultimately leading to a developmental unit, see also Wagner, 1996) will further increase the difficulty of breaking up functional integration. The increased developmental integration will, thus, impose a developmental constraint on the emergence of key innovations.

Another important reason for constraint on change is that functional units are usually involved in more than one function, and selection for improvement of one function may interfere with the efficiency of the other function (Roth and Wake, 1989). An example of a structure with more than one function is the gill apparatus, which in many teleost fish has a function in both respiration and food processing.

Constructional constraints form yet another category. A complex of characters which does not form a functional unit may resist change because it forms part of several functional units, each with their own functional demands (see Barel, 1984, for a similar reasoning). A good example is provided by the neck in mammals. The neck is a complex structure which consists mainly of cervical vertebrae, the spinal cord, and axons leaving and entering the spinal cord, ligaments, muscles (mainly for support and mobility of the head), the esophagus, and the trachea. An important contribution to the complexity comes from the many axons that leave and enter the spinal cord, passing the cervical vertebrae mostly on their way to the forelegs (brachial plexus). Variations in the structures of the neck easily interfere with the position of other structures. A structural abnormality that infrequently occurs in mammals is the development of a rib on the seventh cervical vertebra, a cervical rib. In humans this abnormality leads in about half of the cases to severe problems in the arms (thoracic outlet syndrome; Roos, 1996; Makhoul and Machleder, 1992). The supernumerary rib presses on the nerves and blood vessels that lead to the arms. The symptoms may lead to severe degeneration of the arms and operation on the supernumarary rib is a common treatment of the aberration. The same syndrome can also be caused by abnormalities in some muscles and tendons of the neck which also press on the nerves and blood vessels. This constructional constraint is hypothesized to be one of the factors that select against changes in cervical vertebrae in mammals (Galis, 1999).

It is likely that the selective advantages of structural integration will lead to increased developmental integration, similar to what is expected in the case of functional integration. Indeed, in mice the developmental control of the shape of cervical vertebrae is much stronger than that of more caudal vertebrae (Sofaer, 1978, 1983).

Finally, genetic constraints can also constrain the evolution of key structural innovations. The number of mutational events necessary to produce an innovation will vary and the probability of occurrence will be inversely related to the number. In addition, pleiotropic effects may be inhibiting as well. The evolutionary constraint on the number of cervical vertebrae in mammals again provides an example. The activity of *Hox* genes during the early development of the vertebrae (somitogenesis) appears to be coupled to the activity of *Hox* genes in cell proliferation as changes in cervical vertebral number are associated with an increased susceptibility to neonatal cancer, and a coupling of these abnormalities is found in transgenic mice in which *Hox* gene activity has been manipulated (Galis, 1999).

Summarizing the Factors That Constrain the Evolution of Key Structural Innovations

1. Functional constraints caused by:
    a. functional integration within a functional unit
    b. combination of functional demands within one functional unit
2. Constructional constraints caused by spatial relations within a structural unit
3. Developmental constraints caused by:
    a. developmental integration of a functional unit (developmental unit)
    b. developmental integration of a structural unit (developmental unit)
4. Genetic constraints caused by:
    a. the probability of mutational events
    b. pleiotropic effects

Key innovations, thus, break functional, constructional, developmental, and genetic constraints.

FACTORS PROMOTING THE EVOLUTION OF STRUCTURAL KEY INNOVATIONS

Relaxation of Stabilizing Selection

The disappearance of stabilizing selection facilitates changes that break up functional, structural, and developmental integration. Stabilizing selection does not give a chance to many variations that are present in a population to get established, but as soon as this regime is relaxed the chances for establishment are increased. Stabilizing selection can disappear for a variety of reasons, e.g., because of the occupation of a new niche, the disappearance of competitors or predators, and the appearance of new prey types.

Developmental Mechanisms of Plasticity That Allow the Incorporation of Structural Novelties

During normal development, conditions are often variable and the organism responds to this variability with a variety of phenotypically plastic responses (e.g., Stearns, 1989; Schlichting and Pigliucci, 1998). Phenotypic plasticity and other mechanisms of plasticity provide a flexible response to environmental changes and to structural changes in other parts of the body. These mechanisms of plasticity will facilitate the incorporation of key and other evolutionary innovations, including structural ones.

Developmental mechanisms that provide plasticity include excess structural capacity, decoupling of developmental pathways, variable developmental pathways, and phenotypic plasticity. These mechanisms are discussed in more detail in Galis (1996).

KEY INNOVATIONS AND SPECIATION

Diversification of taxa involves speciation events and these speciation events can occur allopatrically or sympatrically. Allopatric speciation occurs in the absence of gene flow and sympatric speciation in the presence of gene flow. The two processes of speciation are very different and it is likely that the role of key innovations as a trigger of diversity differs as a consequence.

In the sympatric scenario, polymorphic populations are kept divided by assortative mating habits (Bush, 1994). Disruptive selection then leads to divergence of the polymorphic populations. Sexual selection which acts on the mating system can be disruptive (Lande, 1982; Wu, 1985; Turner and Burrows, 1995; Payne and Krakauer, 1997). Natural selection on other traits (e.g., body size) with pleiotropic effects on mate preference can also have a disruptive effect (Thoday and Gibson, 1959; Maynard Smith, 1966; Rosenzweig, 1978; Kondrashov and Mina, 1986; Johnson *et al.*, 1996). A large number of speciation events and mating barriers produced by sexual selection is not enough to maintain species diversity. Species diversity is determined by the balance of the numbers of species that originate and become extinct. When, after speciation events, the new species are indistinguishable ecologically, species will be lost in a process akin to random drift (Wright, 1931). However, the species will never be exactly similar ecologically, in which case the extinction process will be considerably more rapid, except when the differences lead to niche differentiation (MacArthur and Levins, 1967; Meszéna and Metz, in press). Thus, disruptive natural selection that avoids extinction because of limiting similarity is necessarily involved in diversification.

Diversification of the body plan occurs in both sympatrically and allopatrically evolved radiations. An important difference is the nature of the selection processes, disruptive followed by directional selection in sympatric speciation and directional in allopatric speciation. Evolution in a speciose allopatric clade requires a body plan that can be diversified, but the modifications can occur slowly since there is no competition between diverging incipient species. Evolution of a speciose sympatric clade is only possible when *rapid* disruptive selection can repeatedly occur within the evolving clade. Rapid disruptive selection poses high demands on the relative ease with which diversification of a body plan can occur. It is, thus, to be expected that several structural key innovations will have been involved in the history of such a radiation, each of which increased the number of degrees of freedom of the body plan.

The cichlid species flock of Lake Victoria in Africa provides one of the most spectacular examples of speciation and diversification (e.g., Fryer and Iles, 1972; Lowe-McConnel, 1987). Diversification was in this case facilitated by the availability of an empty habitat (relaxation of stabilizing selection), subsequent to the formation of the lake after the most recent ice age (Johnson, 1996). Speciation has apparently mainly been driven by sexual selection for strikingly coloured males (Seehausen *et al.*, 1997; Seehausen and van Alphen, 1997), although allopatric speciation will certainly have also played a role given the size of the lake and the diversity of habitats. Cichlid fishes have indeed acquired several structural key innovations, at least two more structural key innovations relative to their presumed ancestors (see section "Decouplings and radiations," Fig. 6). These key innovations have led to a flexible and versatile pharyngeal jaw apparatus (Galis and Drucker, 1996).

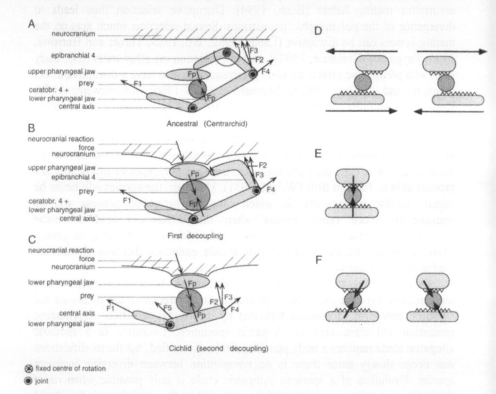

FIGURE 6 Schematic representations of biting with pharyngeal jaws in (A) presumed cantrarchid-like ancestors of cichlids (coupled movement of the upper and lower pharyngeal jaw caused by F1 and F2; rotation of epibranchial 4, part of the fourth gill arch, causes the upper pharyngeal jaw to go down), (B) fishes with the hypothesized intermediary state after the decoupling of the upper and lower pharyngeal jaw (there is no rotation and, thus, the upper pharyngeal jaw is not pushed down but pushed up against the neurocranium; the fourth gill arch and the lower pharyngeal jaw are together pulled up by F1 and F2), (C) cichlids after the decoupling of the lower pharyngeal jaw and the fourth gill arch (ligamentous and muscular connections have disappeared between the lower pharyngeal jaw and the fourth gill arch and the muscles producing F1 and F2 are inserting directly on the lower pharyngeal jaw instead of on epibranchial 4). Possibilities of movement of the pharyngeal jaw apparatus of a centrarchid-like ancestor of cichlids (D *or* E) and of a cichlid (D *and* E *and* F). Note the much increased versatility of the pharyngeal jaw apparatus after the two decouplings. Based on models of Galis (1992) and Galis and Drucker (1996).

There are two reasons why the flexible and versatile pharyngeal jaw apparatus of cichlids promotes evolutionary diversification (Galis and Metz, 1998). First, it provides behavioral plasticity; second, it provides evolvability. Although cichlids usually act as specialists, occupying particular feeding niches, they can eat very diverse food items when necessary, albeit with lower efficiency. This is probably relevant right from the start of the speciation process, because if competition forces a polymorphic population toward diversification, this type of phenotypic plasticity immediately allows rapid shifts. The second reason why the versatile pharyngeal jaw apparatus is important for evolutionary diversification is that quite small evolutionary changes in morphology and behavior allow cichlids to specialize on different food items. The striking diversity of feeding niches that characterizes cichlids of Lake Victoria suggests that niche differentiation occurred by rapid specialization for different feeding niches. This implication is strengthened by the observation that sibling species are always pigeonholed by small differences in feeding behavior (Hoogerhoud et al., 1983; Bouton et al., 1997).

In the case of the haplochromine cichlids, two spectacular radiations have independently occurred, in Lake Victoria and in Lake Malawi. In rivers, no substantial radiation of haplochromine cichlids has taken place, which is probably because of the much lower structural diversity of a fast-flowing river. The lower structural diversity has both consequences for potential niche diversification and for drift in mate recognition systems. Tilapiine cichlids possess the same key structural innovations as the haplochromines and two of the three big genera (*Oreochromis and Sarotherodon*) also possess the same key behavioral innovation (mouth brooding), but they have not radiated (Trewavas, 1983). As tilapiines also occur in the same lakes where haplochromines have radiated it is not immediately apparent which ecological factors could be responsible for this striking difference. Sexual dichromatism, which presumably has played an important role in the diversification of haplochromines (Seehausen et al., 1997; Seehausen and van Alphen, 1997), is also common in *Oreochromis* (Trewavas, 1983). Possibly, genetic factors have constrained diversification of tilapiines.

The low diversity of haplochromines in rivers and of tilapiines in lakes provides a good example of how diversity is the result of many interacting factors of which key innovations are but a few.

KEY INNOVATIONS AS CHARACTERS IN EVOLUTIONARY STUDIES

Key innovations potentially allow diversification. When analyses show that the possession of certain key innovations is causally linked to diversification, this gives information about the order of evolutionary changes. This implies that key innovations should turn up at low nodes in phylogenetic reconstructions. The analysis of key innovations and the diversifying changes that have followed the key innovations, therefore, provide an independent test to phylogenetic reconstructions. The possession of a key innovation could be of even more use to phylogenetic reconstructions when methods used for phylogenetic reconstruction would weigh characters more explicitly. It is plausible to assume that the possession of a key innovation that has been followed by a cascade of diversifying changes is of more indicative value than one of the diverse characters that have evolved after this key innovation.

In this chapter special emphasis is given to structural key innovations; however, behavioral and physiological key innovations are as important and sometimes have very similar effects. For example, mouth breeding in cichlids has a similar effect as the possession of a uterus in mammals. Another example, migrating in winter and seed caching are behavioral key innovations that allow animals to use habitats that are unfavorable in winter, which is comparable to the physiological key innovation of hibernation.

It is more difficult to categorize behavioral key innovations as they are usually complex, specific, and variable. It would be useful when the analysis of behavior would allow a similar structural characterization as is now possible in morphology. It is now only possible to categorize them with respect to their effect on niche width. Physiological key innovations have not received much attention at all, with some exceptions, e.g., endothermy. Especially in insects it seems that physiological changes (e.g., changes in sensitivity for chemical cues in relation to radiation over different host plants) play a crucial role in speciation (Bush, 1994). Further study of physiological and behavioral key innovations is, therefore, eagerly awaited to improve evaluation of the role of these characters in evolution.

Most of the discussed structural key innovations are relatively simple structural changes. Suites of changes can of course also lead to important increases of adaptation, and the resulting complex characters do sometimes fit the definition of a key innovation (e.g., eyes, syrinx). However, for an understanding of evolutionary change it is more useful to analyze the individual changes that have led to a complex character and to identify the changes that are the key steps in the development of these complex characters than to try to generalize on the basis of unique complex structures (see also Lauder and Liem, 1989).

Most current research on diversification focuses on the role of sexual selection (e.g., Mitra et al., 1996; Barraclough et al., 1998; Price et al., 1998) and key innovations receive little attention in comparison. Hopefully this chapter is part of a renewed interest in the study of these most interesting of character changes.

ACKNOWLEDGMENTS

I have greatly benefitted from discussions with Hans Metz, Jan Sevenster, Ole Seehausen, and Jacques van Alphen. I thank Russ Lande, Hans Metz, Yuri Robbers, Jan Sevenster, Hans Slabbekoorn, Günter Wagner, Jacques van Alphen, and Elisabeth van Ast-Gray for helpful comments on the manuscript, Edi Gittenberger for information on gastropods, and Marten Brittijn and Adri 't Hooft for help with the figures.

LITERATURE CITED

Averof, M., and Akam, M. (1995). Insect-crustacean relationships: insights from comparative developmental and molecular studies. *Phil. Trans. R. Soc. Lond. B* **347**:293-303.

Averof, M., and Cohen, S. M. (1997). Evolutionary origin of insect wings from ancestral gills. *Nature* **385**:627-630.

Barel, C. D. N., (1984). Form-relations in the context of constructional morphology: the eye and syspensorium of lacustrine cichlidae (Pisces, Teleostei). *Neth. J. Zool.* **34**:439-502.

Barraclough, T. G., Vogler, A. P., and Harvey, P. H. (1998). Revealing the factors that promote speciation. *Phil. Trans. R. Soc. Lond. B.* **353**:241-249.

Baum, D. A., and Larson, A. (1991). Adaptation reviewed: a phylogenetic methodology for studying character macroevolution. *Syst. Zool.* **40**:1-18.

Berenbaum, M. R., Favret, C., and Schuler, M. A. (1996). On defining "key innovations" in an adaptive radiation: cytochrome P450s and Papilionidae. *Am. Nat.* **148**:S139-S155.

Bond, J. E., and Opell, B. D. (1998). Testing adaptive radiation and key innovation hypotheses in spiders. *Evolution* **52**:403-414.

Bonner, J. T. (1988). "The Evolution of Complexity by Means of Natural Selection." Princeton University Press, Princeton.

Bouton, N., Seehausen, O., and van Alphen, J. J. M. (1997). Resource partitioning among rock-dwelling haplochromines (Pisces: Cichlidae) from Lake Victoria. *Ecol. Freshwater Fish* **6**:225-240.

Brooks, D. R., and McLennan, D. A. (1991). "Phylogeny, Ecology, and Behavior." University of Chicago Press, Chicago.

Bush, G. (1994). Sympatric speciation in animals: new wine in old bottles. *Trends Ecol. Evol.* **9**:285-288.

Claus, C. (1880). "Lehrbuch der Zoologie." Elwert'sche Verlagsbuchhandlung, Marburg.

Dickinson, M. H., Hannaford, S., and Palka, J. (1997). The evolution of insect wings and their sensory apparatus. *Brain Behav. Evol.* **50**:13-24.

Ewer, R. F. (1960). Natural selection and neoteny. *Acta Biotheor.* **13**:161-184.

Fitch, W. M., and Ayala, F. J. (1995). "Tempo and Mode in Evolution: Genetics and Paleontology 50 years after Simpson." National Academy Press, Washington, DC.

Fretter, V., and Graham, A. (1994). "British Prosobranch Molluscs. Their Functional Anatomy and Ecology." The Ray Society, London.

Friel, J. P., and Wainwright, P. C. (1998). Evolution of motor patterns in tetraodontiform fishes: does muscle duplication lead to functional diversification? *Brain Behav. Evol.* **52**:159-170.

Fryer, G., and Iles, T. D. (1972). "The Cichlid Fishes of the Great Lakes of Africa: Their Biology and Evolution." Oliver and Boyd, London.

Galis, F. (1992). A model for biting in the pharyngeal jaws of a cichlid fish: *Haplochromis piceatus*. *J. Theor. Biol.* **155**:343-368.

Galis, F. (1993). Interactions between the pharyngeal jaw apparatus, feeding behaviour and ontogeny in the cichlid fish, *Haplochromis piceatus*. A study of morphological constraints in evolutionary ecology. *J. Exp. Zool.* **267**:137-154.

Galis, F. (1996). The application of functional morphology to evolutionary studies. *Trends Ecol. Evol.* **11**:124-129.

Galis, F., and Drucker, E. G. (1996). Pharyngeal biting mechanisms in centrarchid and cichlid fishes: insights into a key evolutionary innovation. *J. Evol. Biol.* **9**:641-670.

Galis, F., and Metz, J. A. J. (1998). Why are there so many cichlid species? *Trend. Ecol. Evol.* **13**:1-2.

Galis, F. (1999). Why do almost all mammals have seven cervical vertebrae? Developmental constraints, Hox genes, and cancer. *J. Exp. Zool. (Mod. Dev. Biol.)* **285**:19-26.

Galis, F., and Snelderwaard, P. (1997). A novel biting mechanism in damselfishes (Pomacentridae): the pushing up of the lower pharyngeal jaw by the pectoral girdle. *Neth. J. Zool.* **47**:405-410.

Gatesy, S. M., and Dial, K. P. (1996). From frond to fan: Archaeopteryx and the evolution of short-tailed birds. *Evolution* **50**:3027-2048.

Gennip, E. M. S. J. van, and Berkhoudt, H. (1992). Skull mechanics in the pigeon, *Columb livia*, a three-dimentional kinematic model. *J. Morph.* **213**:197-224.

Gittenberger E. (1996). Aperture in gastropod shells. *Neth. J. Zool.* **46**:191-205.

Graham, J. B. (1997). "Air-Breathing Fishes. Evolution, Diversity, and Adaptation." Academic Press, San Diego.

Greene, H. W. (1997). "Snakes. The Evolution of Mystery in Nature." University of California Press, Berkeley.

Greene, H. W., and Burghardt, G. M. (1978). Behavior and phylogeny: constriction in ancient and modern snakes. *Science* **200**:74-77.

Harvey, P. H., and Pagel, M. D. (1991). "The Comparative Method in Evolutionary Biology." Oxford University Press, Oxford.

Heard, S. B., and Hauser, D. L. (1995). Key evolutionary radiations and their ecological mechanisms. *Hist. Biol.* **10**:151-173.

Hodges, S. A., and Arnold, M. L. (1995). Spurring plant diversification: are floral nectar spurs a key innovation? *Proc. R. Soc. London Ser. B.* **262**:343-348.

Hodos, W., and Butler, A. B. (1997). Evolution of sensory pathways in vertebrates. *Brain Behav. Evol.* **50**:189-197.

Hoogerhoud, R. J. C., Witte, F., and Barel, C. D. N. (1983). The ecological differentiation of two closely resembling *Haplochromis* species from Lake Victoria (*H. iris* and *H. hiatus*: Pisces, Cichlidae). *Neth. J. Zool.* **33**:283-305.

Hunter, J. (1998). Key innovations and the ecology of macroevolution. *Trends Ecol. Evol.* **13**:31-36.

Hunter, J., Jernvall, J. P., and Forteliu, M. (1996). The hypocone as a key innovation in mammalian evolution. *Proc. Natl. Acad. Sci. U.S.A.* **92**:10718-10722.

Janvier, P. (1996). "Early Vertebrates." Clarendon Press, Oxford.

Jernvall, J., Hunter, J. P., and Fortelius, M. (1996). Molar tooth diversity, disparity, and ecology in cenozoic ungulate radiations. *Science* **274**:1489-1492.

Jenkins F. A., Gatesy, S. M., Shubin, N. H., and Amaral, W. W. (1997). Haramiyids and triassic mammalian evolution. *Nature* **385**:715-718.

Johansen, K. (1970). Air breathing in fishes. *In* "Fish Physiology" (W.S. Hoar and D.J. Randall, eds.), Vol. IV, pp. 361-411. Academic Press, London.

Johnson, P. A., Hoppenstaedt, F. C., Smith, J. J., and Bush, G. L. (1996). Conditions for sympatric speciation: a diploid model incorporating habitat fidelity and non-habitat assortative mating. *Evol. Ecol.* **10**:187-205.

Johnson, T. C., Scholz, C. A., Talbot, M. R., Kelts, K., Ricketts, R. D., Ngobi, G., Beuning, K., Ssemmanda, I., and McGill, J. W. (1996). Late pleistocene desiccation of lake Victoria and rapid evolution of cichlid fishes. *Science* **273**:1091-1093.

Kondrashov, A. S., and Mina, M. V. (1986). Sympatric speciation: when is it possible? *Biol. J. Linn. Soc.* **27**:201-223.

Köntges, G., and Lumsden, A. (1996). Rhombencephalic neural crest segmentation is preserved throughout craniofacial ontogeny. *Development* **122**:3229-3242.

Kukalová-Peck, J. (1983). Origin of the insect wing and wing articulation from the arthropodan leg. *Can. J. Zool.* **61**:1618-1669.

Kukalová-Peck, J. (1992). The "uniramia" do not exist: the ground plan of the Pterygota as revealed by Permian Diaphanopterodea from Russia (Insecta: Paleodictyopteroidea). *Can. J. Zool.* **70**:236-255.

Kuratani, S., Matsuo, I., and Aizawa, S. (1997). Developmental patterning and evolution of the mammalian viscerocranium: genetic insights into comparative morphology. *Devel. Dyn.* **209**:139-155.

Lande, R. (1982). Rapid origin of sexual isolation and character divergence in a cline. *Evolution* **36**:1-12.

Lauder, G. V. (1981). Form and function: structural analysis in evolutionary morphology. *Paleobiology* **7**:430-442.

Lauder, G. V., and Liem, K. F. (1989). The role of historical factors in the evolution of complex organismal functions. *In* "Complex Organismal Functions: Integration and Evolution in Vertebrates" (D. B. Wake and G. Roth, eds.), pp. 63-278. Dahlem Conference, John Wiley and Sons Ltd., New York.

Levinton, J. S. (1979). A theory of diversity equilibrium and morphological evolution *Science* **204**:335-336.

Liem, K. F. (1974). Evolutionary strategies and morphological innovations: cichlid pharyngeal jaws. *Syst. Zool.* **22**:425-441.

Liem, K. F. (1990). Key evolutionary innovations, differential diversity, and Symecomorphosis. *In* "Evolutionary Innovations" (M. H. Nitecki, ed.), pp. 147-170. University of Chicago Press, Chicago.

Lowe-McConnell, R. H. (1987). "Ecological Studies in Tropical Fish Communities." Cambridge University Press, Cambridge.

Martins, E. P. (1996). " Phylogenies and the Comparative Method in Animal Behavior." Oxford University Press, Oxford.

Maisey, J. G. (1996). "Discovering Fossil Fishes." Henry Holt and Co., New York.

Makhoul, R. G., and Machleder, H. I. (1992). Developmental anomalies at the thoracic outlet: an analysis of 200 consecutive cases. *J. Vasc. Surg.* **16**:534-545.

Mallatt, J. (1996). Ventilation and the origin of jawed vertebrates: a new mouth. *Zool. J. Linn. Soc.* **117**:329-404.

Mallatt, J. (1997). Crossing a major morphological boundary: the origin of jaws in vertebrates. *Zoology* **100**:128-140.

Maynard Smith, J. (1966). Sympatric speciation. *Am. Nat.* **100**:637-650.

Mayr, E. (1960). The emergence of evolutionary novelties. *In* "Evolution after Darwin," Vol. I, pp. 349-380. Univ. Chicago Press, Chicago.

MacArthur, R. H., and Levins, R. (1997). The limiting similarity, convergence and divergence of coexisting species. *Amer. Nat.* **101**:377-385.

Meier, T., and Reichert, H. (1995). Developmental mechanisms, homology and evolution of the insect peripheral nervous system. *In* "The Nervous Systems of Invertebrates: An Evolutionary and Comparative Approach" (O. Breidbach and W. Kutsch, eds.), pp. 249-271. Birkhaüser Verlag, Basel.

Meszéna, G., and Metz, J. A. J. (1999). The role of effective environmental dimensionality. *In* "Adaptive Dynamics in Context" (U. Dieckmann and J. A. J. Metz, eds.), Cambridge Univ. Press, Cambridge.

Miller, A. H. (1949). Some ecologic and morphologic considerations in the evolution of higher taxonomic categories. *In* "Ornithologie als biologische Wissenschaft" (E. Mayr and E. Schüz, eds.), pp. 84-88. Carl Winter.

Mitra, S., Landel, H. S., and Pruett-Jones, S. (1996). Species richness covaries with mating system in birds. *Auk* **113**:544-551.

Muller, M. (1993). The angles of femoral and tibial axes with respect to the cruciate ligament four-bar system in the knee joint. *J. Theor. Biol.* **161**:221-230.

Müller, G. B., and Wagner, G. P. (1991). Novelty in evolution: restructuring the concept. *Annu. Rev. Ecol. Syst.* **22**:229-256.

Nelson, J. S. (1994). "Fishes of the World," 3rd ed. Wiley, New York.

Owen, R. (1866). "On the Anatomy of Vertebrates." Vols. I, II, III. Longmans, Green, and Co., London.

Payne, R. J. H., and Krakauer, D. C. (1997). Sexual selection, space, and speciation. *Evolution* **51**:1-9.

Price, T. (1998). Sexual selection and natural selection in bird speciation. *Phil. Trans. R. Soc. Lond.* B **353**:251-260.

Radinsky, L. B. (1987). "The Evolution of Vertebrate Design." University of Chicago Press, Chicago.

Rey, D., Cuany, A., Pautou, M. P., and Meyran, J. C. (1999). Differential sensitivity of mosquito taxa to vegetable tannins. *J. Chem. Ecol.* **25**:537-548.

Roe, A., and Simpson, G. G. (1958). "Behaviour and Evolution." Yale University Press, New Haven.

Roos, D. B. (1996) Historical perspectives and anatomic considerations. *Sem. Thoracic Cardiovasc. Surg.* **8**:183-189.

Rosenzweig, M. L. (1987). Competitive speciation. *Biol. J. Linn. Soc.* **10**:275-289.

Rosenzweig, M. L., and McCord, R. D. (1991). Incumbent replacement: evidence for long-term evolutionary progress. *Paleobiology* **17**:202-213.

Roth, G., and Wake, D. B. (1989). Conservatism and innovation in the evolution of feeding in vertebrates. *In* "Complex Organismal Function: Integration and Evolution in Vertebrates" (D. B. Wake, and G. Roth, eds.), pp. 7-21. John Wiley and Sons Ltd.

Schaeffer, B., and Rosen, D. E. (1961). Major adaptive levels in the evolution of the actinopterygian feeding mechanism. *Am. Zool.* **1**:187-204.

Schaefer, S. A., and Lauder, G. V. (1986). Testing historical hypotheses of morphological change: biomechanical decoupling in loricarioid catfishes. *Evolution* **50**:1661-1675.

Schwenk, K. (1993). The evolution of chemoreception in squamate reptiles: a phylogenetic approach. *Brain Behav. Evol.* **41**:124-137.

Schlichting, C. D., and Pigliucci, M. (1998). "Phenotypic Evolution. A Reaction Norm Perspective." Sinauer Associates, Sunderland.

Seehausen, O., van Alphen, J. J. M., and Witte, F. (1997). Cichlid fish diversity threatened by eutrophication that curbs sexual selection. *Science* **277**:1808-1811.

Seehausen, O., and van Alphen, J. J. M. (1997). The effect of male coloration on female mate choice in closely related Lake Victoria cichlids (*H. nyererei* complex), *Behav. Ecol. Sociobiol.* **42**:1-8.

Simmons, N. B., and Geisler, J. H. (1998). Phylogenetic relationships of *Icaronycteris*, *Archaeonycteris*, *Hassianycteris*, and *Palaeochiropteryx* to extant bat lineages, with comments on the evolution of echolocation and foraging strategies in *Microchiroptera*. *Bull. Amer. Mus. Nat. Hist.* **235**:4-182.

Simpson, G. G. (1944). "Tempo and Mode in Evolution." Columbia University Press, New York.

Slijper, E. J. (1946). Comparative biologic-anatomical investigations on the vertebral column and spinal musculature of mammals. *Kon. Ned. Akad. Wet., Verh.* (Section 2, DL 42, No.5), 1-128.

Sofaer, J. A. (1978). Morphogenetic influences and patterns of developmental stability in the mouse vertebral column. *In* "Development, Function and Evolution of teeth" (P. M. Buckler and K. A. Joysey, eds.), pp. 215-227. Academic Press, London.

Sofaer, J. A. (1983). Developmental stability in the mouse vertebral column. *J. Anat.* **140**:131-141.

Staaden, M.J., and Römer, H. (1992). Evolutionary transition from stretch to hearing organs in ancient grasshoppers. *Nature* **394**:773-776.

Starck, D. (1979). "Vergeleichende Anatomie der Wirbeltiere." Springer Verlag, Berlin.

Stearns, S. C. (1989). The evolutionary significance of phenotypic plasticity. *Bioscience* **39**:436-445.

Thoday, J. M., and Gibson, J. B. (1962). Isolation by disruptive selection. *Nature* **193**:1164-1166.

Tomsa, J. M., and Langeland, J. A. (1999). Otx expression during Lamprey embryogenesis provides insights into the evolution of the vertebrate head and jay. *Dev. Biol.* **207**:26-37.

Trewavas, E. (1983). "Tilapiine Fishes of the Genera *Sarotherodon, Oreochromis* and *Danakilia.*" British Museum, London.

Turner, G. F., and Burrows, M. T. (1995). A model of sympatric speciation by sexual selection. *Proc. R. Soc. Lond. B* **260**:287-292.

Van Valen, L. M. (1971). Adaptive zones and the orders of mammals. *Evolution* **25**:420-428.

Vermeij, G. J. (1974). Adaptation, versatility, and evolution. *Syst. Zool.* **22**:466-477.

Vermeij, G. J. (1994). The evolutionary interaction among species: selection, escalation, and coevolution. *Annu. Rev. Ecol. Syst.* **25**:219-236.

Vermeij, G.J. (1995). Economics, volcanoes, and Phanerozoic revolutions. *Paleobiology* **21**:125-152.

Wagner, G. P. (1996). Homologues, natural kinds and the evolution of modularity. *Am. Zool.* **36**:36-43.

Wagner G. P., and Schwenk, K. (2000). Evolutionary stable configurations: Functional integration and the evolution of phenotypic stability. *Evol. Biol.* **31**:155-217

Winterbottom, R. (1974). A descriptive synonymy of the striated muscles of the teleostei. *Proc. Acad. Nat. Sci. Phil.* **125**:225-317.

Wright, S. (1931). Evolution in Mendelian populations. *Genetics* **16**:97-159.

Wu, C. (1985). A stochastic simulation study on speciation by sexual selection. *Evolution* **39**:66-82.

INDEX

Abdominal bristle number, 393
Absolute constraints, 117
Acanthostega gunnari, 538, 540–541
Acting factors, 51
Adaptive system
  anatomical coordination, 27
  character identification, 153–154
  complex system, 154
  developmental programs, 27
  organismic property evolution, 143
  plasticity, 381
Adductor mandibulae, tetradontiform fish
  burst duration, 296–297
  motor patterns, 289–293
Adult characters, 371
AER, *see* Apical ectodermal ridge
African butterfly, *see Bicyclus*
AG, *see* Angular process
Alignment, DNA sequence, 304–307
Allelic analysis, 151
Allelic association tests, 402–403
Allometry, QTL role, 428–429

Amino acid sequence, in deep phylogeny, 103
Amphibian organizer, *Xenopus,* 442–443
Analogous characters, 22–23
Anatomical coordination
  in adaptation, 27
  primacy of function, 26
Anatomy, bryozoans, 276–278
Angiosperms
  angiospermy, 498
  carpels, 498
  floral element evolution, 503–505
  floral spurs, 502–503
  floral tubes, 500–502
  organ categories, 499–500
  ovules, 503
  petals, 499–500
  phylogeny, 495–496
  postgenital fusion, 498
  stamen thecal organization, 499
  sympetaly, 500–502
  syncarpy, 500
Angiospermy, 498

Angular process, QTL detection, 420–421
Animals
  appendage origin, 532
  generation, 61, 65–67
  life ordering, 58–59
Antirrhinum, organ categories, 499–500
A–P circuitry
  arthropod body wall extensions, 472–473
  limb positioning, 467–468
Apical ectodermal ridge
  in chick digits, 542
  ZPA interaction, 534–535
Appendages
  development, 467
  origin, 532–534
  patterning, 460–461
Approximations, object parts, 271
Aquilegia, 502–503
Arabidopsis, 120, 499–500
Arabidopsis thaliana, 373
Aristotelian method, 18, 25–26, 58–60
Arthropods
  body wall extensions, 472–473
  homology comparisons, 311–313
  limbs
    axial differentiation, 481–483
    boundary delineation, 473–476
    branching homology problems, 476–477
    development mechanisms, 456–460
    Drosophila model, 466
    identity, 456–459
    integration, 481–483
    modularity, 456–459, 481–483
    morphological variation, 472
  modular morphology, 459–460
  segmentation, 478–481
Association tests, 403–406
Atomism
  Aristotle's theory, 60
  Buffon, Georges, 67
  and Darwinism, 68–69
  nature and organism conception, 71–72
  supporters, 64
Attribute character, 215
Axial differentiation, arthropod limbs, 481–483
Axial patterning, arthropod segmentation,
    478–481
Axial skeleton, as mechanical unit, 182–183

Bacon, Francis, 30
Balistes capriscus, 289
Base-to-base homology, 307–313
Basic plan, tetrapod limb, 544

Baupläne
  diversity, 167
  and functional units, 192
  higher-order ESCs, 189
  key innovations, 586
  organism congealing, 191
  phenotypic stasis, 167
  in structuralist approach, 167–168
Behavior
  measurement, 286–287
  pufferfish inflation, 293–294
Bicyclus
  divergent eyespot patterns, 358–359
  eyespot morphological organization, 348–349
  eyespot morphological pattern, 344–347
  eyespot probing, 349–350
  eyespot shape, 355–356
  eyespot size and color, 351
  mutagenesis, 358–359
  unit character, 357
Bigeye, 355
Biochemical circuitry, 571
Bioklise, 47
Biological characters
  description, 363–364
  evolution, 216
  usages, 215
Biological classification
  and chemical classification, 202
  and Darwinian theory, 204
  as developmental concept, 209–214
  and natural selection, 205
  organisms, 205
  from taxonomy, 201–202
Biological complexity, 111
Biological criteria, for characters, 22
Biological fields, 32, 58, 437
Biological objects
  higher level ontological primacy, 144–146
  as inheritance unit, 151
  organism role, 146–149
Biological organization, from process homol-
    ogy, 439
BMP, *see* Bone morphogenetic proteins
Body plans, *see also* Baupläne
  diversification, 597
  and epigenesis, 563–564
  key innovations, 583–584, 586–588
Body wall
  animal appendages, 532
  extensions in arthropods, 472–473
Bone morphogenetic proteins
  in chick digit determination, 542
  process homology, 448

Bonnet, Charles, 63–64
Botanical systematics, 44–45
Brain
  mapping by Vogts, 42–43
  pathological variations, 50
Bristle characters, *Drosophila simulans,* 396
Broca's area, in neo-cortex, 40
Bryozoans
  anatomy, 276–278
  parts, 278–279
Buckeye butterfly, *see Precis coenia*
Buffon, Georges, 65–67
Building block hypothesis, 101–102
Bumble bees, geographic variation, 48–50
Burst duration, tetradontiform fish, 292, 296–297
Burst onset, tetradontiform fish, 292
Butterfly
  Bicyclus, 344–347, 358–359
  Bicyclus anynana, 348–351, 355–359
  Precis coenia, 347, 349–350
  symmetry systems, 521–522
  wing patterns, 359, 512–513, 524–526

Caenorhabditis elegans, 131–132
Candida albicans, 560
Candidate genes, 398–400
Carpels, angiosperms, 498
Cartilaginous sesamoids, 570
CAS, *see* Complex adaptive system
β-Catenin, *Xenopus,* 442
Causal homeostasis characters, 9–10
Cautery, in eyespot experiments, 353
Cell differentiation, from developmental genetics, 436
Cell ontological primacy, 144–145
Cell polarity, 566–567
Central symmetry systems, butterfly wing, 512
Centrarchid motor patterns, 298
Character-attributes, 264
Character avoidance, 24
Character complexes
  genetic basis, 429–430
  individuation, 430–431
  locus effects, 430
Character decomposition
  consequences, 226
  with neo-Darwinism, 165–166
  organisms, 153–158
  in population evolution, 228–229
  selection context-dependent units, 149–153
Character delineation, 82
Character designation, 30–31

Character divergence, 390
Character–environment relationship
  genetic correlation, 367–368
  multivariate dimensions, 368–371
Character expression, 373
Character fitness, 373
Character homology, 310
Character identification, 153–154
Character independence, 158
Character interaction, 26–27
Character-parts, 264
Character polarity, *see also* Polarity
  from commonality principle, 325
  dependencies, 320
  direct *vs.* indirect methods, 330–332
  by ontogenetic method, 323–324
  ontogenetic *vs.* phylogenetic methods, 332–333
  from outgroup comparison, 322–323
  from paleontology, 322, 325
  from pattern cladistics, 322
  by stratigraphic method, 325
Character production, data compression, 30
Character ranking system, 22
Character replication
  copy relationships, 83–84
  copy templates, 85–86
  hierarchy, 91–95
  indirectness, 87–91
  mechanisms, 104
  multiple copies, 84
  parental copy, 85
  previously produced copies, 84
  process, 82–83
  tree reasoning, 95–104
Character space, and key innovation, 584–586
Character state
  generality, 329–330
  via ontogeny, 330
Character systems, *see also specific characters*
  causal homeostasis, 9–10
  definition, 2
  from descriptors, 224–226
  evolutionary existence, 10
  in evolutionary theory, 9
  modularity, 227–228
  pseudodefinition, 3
  working concept, 168
Character testing, key innovation, 584
Character variables
  as biological character type, 215
  part comparison, 264
Character variation, in population, 390
Character weighting, *a priori,* 21–22

Cheekbones, jaw bone decoupling, 592–593
Chemical classification
  biological classification comparison, 202
  for predictions, 201
Chemosensory structure, *Drosophila* bristles,
    391
Chilomycterus schoepfi, 289
Chromosomes, *Drosophila,* 394
Cichlids
  pharyngeal jaw, 599
  radiations, 599
  speciation and diversification, 597–598
Cinctipora elegans, 279
Cladistic analysis
  in character designation, 30–31
  in character polarity, 322
  and homology, 2
  independence, 21
  for key innovation, 582
  polarity methods, 333–335
  quantitative characters, 23–24, 25
  rooting methods, 333–335
Cladograms
  DNA sequence alignment, 305–307
  optimization, 308–310
Classification
  character reference, 15–16
  functional individuation, 208–209
  genealogy, 203–209
  synthetic approach, 26
CN, *see* Condyloid process
Columella folding, 593
comet, 355–356
Commonality principle
  for character polarity, 325
  in rooting, 327–328
Comparative analysis
  appendage development, 467
  for key innovation, 582
Comparative biology
  characterization, 32
  organism, 58
Comparative patterning, 472
Compartmentalization
  butterfly evolution, 526
  and epigenesis, 564
  serial homologue individuation, 526–527
Complex adaptive system, 154
Complexity, in structures, 248–249, 593
Complex systems, properties, 4
Composition, part differences, 271–274
Compound eye genetics, 5
Computers, for character counting, 19
Condensations, mouse mandible, 415

Condyloid process, QTL detection, 420–421
Constraints
  detection, 120–122
  in life history, 116–117
  from selection gradients, 124
  structural key innovation, 594–595
  in structures, 249–250
Constructional constraints, structural key inno-
    vation, 595
Copies, in character replication
  multiple copies, 84
  parental copies, 85
  previously produced copies, 84
  relationship, 83–84
  from repetition, 89
  templates, 85–86
Copying, *see* Replication
Coronoid process, QTL detection, 420–421
Correlated characters, Darwin's weighting, 23
Correlation
  organic integration, 20
  parts in systematics, 20
Coughing, in tetradontiform fish, 293–294
Coupling, limbs and segments, 480–481
CR, *see* Coronoid process
Cyclicity, 247–248
cyclops, 355–356

Danio rerio, 535
Darwinism
  and atomism, 68–69
  and biological classification, 204
  character criteria, 22
  for classification, 203
  pros and cons, 69–71
Data
  compression in character production, 30
  developmental data, 213
  gene expression homology, 446–447
  molecular data, 325–326
Dead-end replicators, 89–90
de Beer, Gavin, 435–436
Decomposition
  characters
    consequences, 226
    with neo-Darwinism, 165–166
    organisms, 153–158
    in population evolution, 228–229
    selection context-dependent units,
      149–153
  descriptors, 221–224
  parts, 269
  in structures, 250–252

Decoupling
  hard tissue, 592–593
  phylogenetic decoupling, 92
  soft tissue, 591–592
Deep homology, 447–449
Delta, 403
Descriptors
  for character system, 224–226
  decomposition, 221–224
  frequency space, 219
  G–P map, 231
  types, 217–219
Development
  Aristotle's model, 60
  arthopod limb mechanisms, 456–460
  control in evolution, 571
Developmental biology, homology role, 437
Developmental constraints, structural key inno-
    vation, 595
Developmental correlation of parts, 63–64
Developmental data, in taxonomy, 213
Developmental genetics, 435–436
Developmental integration, arthropod limbs,
    481–483
Developmental plasticity, 371–372, 596
Developmental process, as characters, 443
Developmental programs, in adaptation, 27
Diderot, Denis, 64
Differential epistasis
  in pleiotropic patterns, 414
  and QTLs, 425, 428–429
Differentiation
  arthropod limb, 481–483
  cell differentiation, 436
Digits, origin, 538–542
Dispositional properties, 8
Distal-less, 349–350
Distyliopsis, 505
Distylium, 505
Dll/Dlx, 533–534
DNA replication, 87
DNA sequences, 303–307
Drosophila
  appendage development, 467
  appendage patterning, 460–461
  developmental mechanisms, 459–460
  eye development, 193
  genetic analysis, 437–438
  homonomy, 482–483
  later limb patterning, 469
  as limb development model, 466
  limb primordia, 462–465
  limbs, 467–468, 475–476
  as model system, 49

  morphological characters, 351
  process homology, 439, 441
  QTL to gene, 398–399
  Rel pathway, 442
  segmentation, 479–480
  wing imaginal disc, 474–475
  Wnt pathways, 440
Drosophila melanogaster
  sensory bristle number, 390–398
  starvation resistance, 125
Drosophila simulans, 396
Drought response, 375–376
D–V circuitry, 467–468
Dynamic homology, 304–307, 310–313
Dynamic programming, for cladograms, 308

Ecological guild, from character, 14
Electromyograms, tetradontiform fish, 289
Electrons, in chemical classification, 201
Embryo
  Aristotle's theory, 59–60
  homology studies, 437–438
  induction, 442
  mouse enhancer-trap system, 547–548
  preexistence, 63
Embryology, Harvey, *vs.* Malpighi, 61–62
Embryonic characters, reliability, 23
Eminooecia carsonae, 278–279
Empirical correlation, 21
Endogenous factors, in Vogt's theory, 51
Enhancer-trap system, 547–548
Environment
  character relationship, 367–371
  context from character, 14
  effect on phenotype, 364–365
  genotype interaction, 374, 393, 425
  heterogeneity analysis, 372
  heterogeneity relationships, 373
  interenvironmental homology, 376–378
  phenotypic responses, 374–375
  QTL effects, 397
  variation, 375
Epigenesis
  and cell polarity, 566–567
  definition, 60
  functional correlation of parts, 62–63
  and generation, 72–73
  genes and form, 572–574
  Harvey's theory, 61–62
  homoplasy and homology, 571–572
  and lumen formation, 566–567
  in Mendelian world, 567
  phenotypic innovation, 570–571

Epigenesis *(continued)*
  physical properties, 567–568
  in pre-Mendelian world, 563–564
Epistasis
  for allele system, 151
  differential epistasis, 414, 425, 428–429
Erigeron annuus, 373
ESC, *see* Evolutionarily stable configurations
Essential characters, 25–26
Eunomie, 48
Eusthenopteron foordi, 538
Evolution
  adaptive evolution, 143
  angiosperm floral elements, 503–505
  biochemical circuitry, 571
  biological characters, 216
  butterfly wing color, 359, 514
  developmental control, 571
  Dll/Dlx genes, 533
  genetic architecture, 413
  G–P map, 230–235
  life history, 115
  modularity, 227, 231–232
  morphogenesis, 568
  morphological evolution, 561
  patterning mechanism, 515–524
  population, 228–229
  from process homology, 436–437, 450
  pufferfish inflation behavior, 293–294
  serial homology, 524–526
  structural key innovation, 594–596
  theory for dynamical systems, 233–234
  vertebrate limb, 568
Evolutionarily stable configurations
  component part stabilization, 190–191
  definition, 173–174
  as functional units, 183–189
Evolutionary attributes, 190–191
Evolutionary biology, homology role, 437
Evolutionary characters
  alteration, 128
  alternative solutions, 130–131
  and functional architecture, 110–114
  genomic approach, 131–132
  identification, 123–124
  key innovations as, 599–600
  phenomics approach, 134
  QTL mapping, 132–134
Evolutionary explanatory models, 28
Evolutionary potential, populations, 127
Evolutionary theory
  character existence, 10
  characterization, 32
  character roles, 9

genes as units, 31–32
  and systematics, 32
Evolution operators
  compatibility, 234
  definition, 228
Excitable media, 564
Exogenous factors, in Vogt's theory, 51
Ex ovo omnia, 61
Expressivity, and brain pathological variations, 50
Eye
  compound eye genetics, 5
  as functional unit, 193–194
Eyespots
  Bicyclus, divergent patterns, 358–359
  genetic variation, 350–357
  morphological pattern, 344–347

Factor analysis, evolutionary characters, 123–124
Factor descriptors, for organism, 219–222
Faithful copying, 92–93
Faithful replication through history, 92–93
Feature, as biological character type, 215
Ferehault de Ráumur, René-Antoine, 65
Fertilization, tree representation, 99
FGF, *see* Fibroblast growth factor
Fibroblast growth factor, 535–536
Fin radials, 538
Fish
  adaptive plasticity, 381
  Balistes capriscus, 289
  Chilomycterus schoepfi, 289
  Danio rerio, 535
  Monacanthus hispidus, 289
  neopterygian fish, 592–593
  osteolepiform fish, 538
  pharyngeal arches, 590
  Pterois volitans, 518
  pufferfish, 293–294
  tetradontiform fish, 289–294, 296–297
Fitness function
  definition, 155
  Pi-additivity, 159–160
Fitness structure, definition, 234
Fixed-state optimization, 312
Floral elements
  angiosperms
    angiospermy, 498
    carpels, 498
    evolution, 503–505
    floral spurs, 502–503
    floral tubes, 500–502

organ categories, 499–500
ovules, 503
petals, 499–500
phylogeny, 495–496
postgenital fusion, 498
stamen thecal organization, 499
sympetaly, 500–502
syncarpy, 500
preangiophytes
microsporangia, 497
outer male–inner female organs, 497
ovules, 496–497
seeds, 496–497
Floral organs, 494–495
Floral spurs, 502–503
Floral tubes, 500–502
Flowering plant phylogeny, 495–496
Form, epigenetic principles, 572–574
Formative factors, in Vogt's theory, 51
Formative process, in Vogt's theory, 51
Fossils, in rooting, 335
Fragment homology, 307–313
Frequency distribution
definition, 155–157
Pi-LE, 160–161
Frequency space, descriptor association, 219
Frequency vectors, decomposition, 222–224
Fringe protein, deep homology, 448–449
Function
in homology, 444–445
as organismal trait, 444
in process homology, 439
Functional architecture, and evolutionary characters, 110–114
Functional constraints, structural key innovation, 594–595
Functional context, from character, 14
Functional correlation of parts, 62–64
Functional individualization, 39–40
Functional individuation, 208–209
Functional integration, 147–148
Functionality, parts, 267–268
Functional kinds, in generalizations, 207–208
Functional units
as characters, 192–194
concepts and early work, 168–174
ESCs, 183–189
evolutionary attributes, 190–191
mechanical units, 179–183
structural units, 177–179
types, 174–176
Fundamental variation, in Darwinism, 69

β-Galactosidase, 547–548
Gassendi, Pierre, 61
Gastrulation character, 93
GEI, *see* Genotype–environment interaction
Genealogy, as classification, 203–209
Gene homology
expression data, 446–447
level placement, 446
Generality
character state distribution, 329–330
interpretation, 324
Generation
animal generation, 61, 65–67
degeneration, 67
epigenetic conceptualization, 72–73
progeneration, 85
spontaneous generation, 66–67
Genes
additive effects, 150
assembling, 105
bigeye, 355
comet, 355–356
cyclops, 355–356
Delta, 403
disruptions in mouse, 104
distal-less, 349–350
Dll/Dlx, 533–534
epigenetic principles, 572–574
in evolutionary theory, 31–32
Hox, 536–537, 539–542
melanine, 353
in mortality rates, 132
PAX6, 438
Pax-6, 5
from QTL, 398–406
scabrous, 403
sonic hedgehog, 543–544
spotty, 353, 355, 356
structural homology, 437–438
tree representation, 100–101
Wnt pathways, 440
Genetic algorithms, 28–29
Genetic analysis
Drosophila, 437–438
mammalian mandible, 418–420
Genetic architecture
Drosphila bristle number, 392
evolution, 413
mammalian mandible, 414–416
morphological characters, 389–390
Genetic assimilation
phenotypic plasticity role, 379
signature, 381
Genetic constraints, structural key innovation, 595

Genetic correlation
  and pleiotropic patterns, 412–413
  reaction norms, 367–368
Genetics
  for character complexes, 429–430
  compound eye, 5
  developmental genetics, 435–436
  digit difference mechanism, 542
  Drosophila model system, 49
  Heliconius wing pattern, 514–515
  in morphological integration, 429
  in morphological variation, 4–5
  quantitative, *see* Quantitative genetics
  Vogt's approach, 45–46, 48–49, 51–52
Genetic sampling, *Drosphila,* 392–393
Genetic variation
  additive, 149–150
  eyespot pattern, 350–357
  for quantitative traits, 390
  reaction norm, 366–367
Gene transfer, 546
Genomics, 131–132
Genotype
  effect on phenotype, 364–365
  in Vogt's theory, 51
Genotype–environment interaction, 374, 393, 425
Genotype–phenotype map, 230–235
Geographic variation, Vogt's theory, 44, 48–50
Germ-line replicators, 89–90
G matrix
  for constraints, 120–122
  for evolutionary characters, 123–124
G–P map, *see* Genotype–phenotype map
Gray triggerfish, *see Balistes capriscus*
Groups
  character designation, 17–18
  character distribution, 19
  classification, 18
  ρ Groups *vs.* parts, 264–265
  F Groups *vs.* parts, 264–265
Gryllus wing morphs, 129

Hartsoeker, Nicolaas, 63
Harvey, William, 61–63
Heliconius, 514–515
Heterogeneity, as material process, 1–2
Heterophylly, 377
Hierarchy of parts, 265–267
Histoire des Animaux, 65–66
Homologous characters, 22–23, 92
Homology
  base-to-base homology, 307–313

character homology, 310
concept and cladistics, 2
deep homology limitations, 447–449
definition, 377
dynamic homology, 304–307, 310–313
in embryos, 437–438
and epigenesis, 571–572
fragment homology, 307–313
function problem, 444–445
gene homology, 446–447
interenvironmental homology, 376–378
intraorganismal homology, 376
morphological homology, 446
parts, 260
process, *see* Process homology
serial homology, 524–526
serially reiterated structures, 476–477
as synapomorphy, 330
trait phenotypic plasticity, 376
in vertebrate limb development studies, 546
Homology: An Unsolved Problem, 435
Homonomy, 481–482
Homoplasy, 31, 571–572
Homo sapiens
  classification, 212
  and Human Genome Project, 213
  taxonomy, 209–210
Hordeum vulgare, 373
Hox
  in digit development, 539–542
  limb induction, 536–537
Human Genome Project, 213
Hydra viridis, 65–67
Hypocone, 593

Identification, in RA, 252–254
Identity, arthopod limbs, 456–459
Idiotype, in Vogt's theory, 51
Iguania, lingual feeding system, 187–188
Incisor alveolus, QTL detection, 420–421
Individuals
  character reference, 15–16
  parts comparison, 264
Individuation
  character complexes, 430–431
  functional individuation, 208–209
  objects, 207
  serial homologues, 526–527
  symmetrical structures, 523–524
Inflation behavior, pufferfish, 293–294
Information-theoretic framework
  complexity, 248–249

cyclicity, 247–248
decomposition losses, 250–251
in RA, 238–241
Inheritance
    biological object role, 151
    genetic variance objects, 149–150
    Vogt's theory, 48–49
Innovation
    key, *see* Key innovations
    phenotypic innovation, 570–571
Insect leg, deep homology, 448–449
Institution building, by Vogts, 46
Integration
    definition, 262
    developmental integration, 481–483
    functional integration, 147–148
    morphological integration, 172, 428, 429
    organic integration, 20
    plasticity integration, 370
    structural integration, 147–148
Intercalation, in pattern-forming systems, 517
Interenvironmental homology, 376–378
Internal selection
    in component part stabilization, 190–191
    in ESC characterization, 186
Interspecific variation, 182
Intraorganismal homology, 376
Intrapopulation variability, 24
Introgression
    candidate gene alleles, 399–400
    in QTL combination tests, 403–406
Invagination, 591
Invariant characters, 228–230
Isolation, in part integration, 262

Jaw bone, decoupling in fish, 592–593
Juvenile hormone, 129

Key innovations
    and body plan development, 583–584
    character space opening, 584–586
    character testing, 584
    cladistics, 582
    comparative methodology, 582
    definition, 582
    diversification, 583
    as evolutionary characters, 599–600
    hard tissue decoupling, 592–593
    new structures, 594

soft tissue decoupling, 591–592
    and speciation, 597–599
    structural
        changes, 586–588
        complexity, 593
        duplications, 588–590
        inhibiting factors, 594–596
        promoting factors, 596
    testing, 583
Kinematic patterns, 288

Language ability, 40
Latent variables, 247–248
Lattice of relations, 241–242
Lattice of structures, 243–246
Laws, in biological classification, 209
Leg, deep homology, 448–449
Leibniz, Gottfried Wilhelm, 63
L'Homme Machine, 65
Life, from replicating molecules, 89
Life cycle, organism, 114–115
Life history
    adaptive evolution, 143
    organism, 114–116
    quantitative genetics approach, 119
    trade-offs, 116
    traits, 126–130
    Y model, 113–114
Limb primordia, *Drosophila*
    developmental control, 464–465
    early and late development, 462
    segment position, 462–464
Limbs
    arthropods
        axial differentiation, 481–483
        boundary delineation, 473–476
        branching, 476–477
        development mechanisms, 456–460
        diversification, 478–481
        identity, 456–459
        integration, 481–483
        modularity, 456–459, 481–483
        morphological variation, 472
    development conservation, 468–471
    Drosophila
        branching, 469
        as development model, 466
        positioning regulation, 467–468
    induction and outgrowth, 534–538
    Sonic hedgehog role, 543–544
    vertebrate, 544–546, 568
Lingual feeding system, Iguania, 187–188

Linkage disequilibrium
  for allele system, 151
  in genetic correlation, 412
Linkage equilibrium
  in character independence, 158
  definition, 155
  in population, 157
  in quantitative genetics, 157
  Pi-LE, 160–161
Linnean classification, 18
Lionfish, *see Pterois volitans*
Locus, on character complexes, 430
Log-linear model, and RA information-
    theoretic, 238
Lumen, and epigenesis, 566–567

Macroevolution, plasticity role, 379–381
MALIGN, in homology comparison, 312
Malpighi, Marcello, 61–62
Mammals
  mandible, 178–179, 414–420, 426–428
  teeth, 185
Mandible
  mammalian
    as functional unit, 178–179
    genetic architecture, 414–416
    measurements, 418–420
    morphology, QTL role, 426–428
    population source, 416–418
  tetradontiform fish, 289–293, 296–297
Maps
  GP map, 230–235
  QTLs, 422, 428
Masseteric region, QTL detection, 420–421
Materialism, triumphs, 65–67
Material processes
  heterogeneity, 1–2
  during ontogeny and phylogeny, 1–2
Mathematical models
  in biology, 142–143
  testing, 5
Mating habits, polymorphic population, 597
Maxilla, 178–179

Mechanical units, 179–183
Mechanosensory structure, *Drosophila* bristles,
    391
melanine, 353
Mendelev periodic table, 199–202
Mendelian world, epigenesis, 567–572
Metapterygial axis, digit and fin branching, 538

Microevolutionary processes, 24–25
Microsporangia, preangiophytes, 497
MO, *see* Molar alveolus
Models
  development, Aristotle's theory, 59–60
  Drosophila bristles, 390–391
  evolutionary characters, 111–112
  evolutionary explanatory models, 28
  for genetics, 49
  limb development, 466
  log-linear model, 238
  mathematical models, 5, 142–143
  optimality, 126–130
  state models, 247–248
  Y model, 113–114, 121
Modularity
  arthropod limbs, 456–459, 481–483
  character systems, 227–228
Modular morphology, arthropods, 459–460
Modular operators, and invariant characters,
    228–230
Modules
  homologous pathways, 445
  part comparison, 264–265
Molar alveolus, QTL detection, 420–421
Molars, hypocone formation, 593
Molecular data, in midpoint network rooting,
    325–326
Molecules, replicating, life orgination, 89
Monacanthus hispidus, 289
Moreau de Maupertuis, Pierre Louis, 64
Morphogenesis
  arthropod limb, 477
  characters, 91–92
  from developmental genetics, 435–436
  evolution, 568
Morphological characters
  Drosophila, 351
  in taxonomy, 389–390
Morphological evolution, 561
Morphological homology, 446
Morphological integration
  genetic basis, 429
  mammalian mandible, 428
  statistical approach, 172
Morphological organization, eyespot, 347–350
Morphological pattern, eyespot, 344–347
Morphological plasticity, modern organisms,
    560–561
Morphological variation
  arthropod limbs, 472
  via genetics, 4–5

Morphology
  arthropods, 459–460
  flowers, 494–495
  functional morphology, 169–170
  mandible, QTL role, 426–429
  mouse sphenoid, 104
Mortality
  gene effects, 132
  organism, 114–116
Moth
  pattern evolution, 520
  symmetry system hierarchy, 522
Motor patterns
  centrarchids, 298
  definition, 287
  kinematic pattern comparison, 288
  tetradontiform fish, 289–294, 296
Mouse
  embryo enhancer-trap system, 547–548
  gene disruptions, 104
  as mandible source, 416–418
  transformed mouse, 129–130
MS, *see* Masseteric region
Multicellular organisms, 563
Multivariate interval mapping, 428
Muscles
  subdivision, 592
  tetradontiform fish, 294
Mutagenesis, in *Bicyclus anynana,* 358–359
Mutagen screening, *Caenorhabditis elegans,*
  131–132

Natural descriptor, 218
Natural kinds
  analysis implications, 7–8
  concept challenges, 6
  conceptualization, 7
  definition, 6
  stability, 8
Natural selection
  in biological classification, 205
  inheritance objects, 149
  intrapopulation variability recognition,
    24
  variant choice, 206
Nature, atomistic conception, 71–72
Naturphilosophen, 30
Needham, John Turberville, 66
Neo-cortex
  areas, 40
  characterization, 38

Neo-Darwinism
  for character decomposition, 165–166
  phenotypic plasticity, 364
Neopterygian fish, 592–593
Network rooting, 325–326
Neutron, in chemical classification, 201
Nigella, element evolution, 505
Nucleotides
  in deep phylogeny, 103
  tree representation, 99, 101–102
Nymphalid groundplan, 515, 519

Objects
  biological objects, 144–149, 151
  individuation, 207
  part approximations, 271
  part levels, 271
  as parts strategy, 270
Observation, and character, 57–58
Offspring, phenotype, selection effects,
    118–119
Ontogeny
  in character polarity, 323–324, 332–333
  in character states, 330
  material organization heterogeneity, 1–2
  in rooting, 327–328
Ontological primacy
  higher level biological objects, 144–146
  population, 145
Optimality model
  life history testing, 127–130
  life history traits, 126–127
Optimization-alignment
  DNA sequences, 305–307
  from POY, 312
Organic integration, for character correlation, 20
Organisms
  atomistic conception, 71–72
  in biological object theory, 146–149
  character decomposition, 153–158
  character division, 27–28
  classification, 205
  in comparative biology, 58
  detailed description, 19
  and environmental variation, 375
  factor descriptors, 219–222
  identification, 14–15
  intraorganismal homology, 376
  life cycle, 114–115
  life history theory, 114–116
  morphological plasticity, 560–561

Organisms *(continued)*
multicellular organisms, 563
natural subsystems, 3
partitioning into characters, 26–27
parts, 260
product descriptors, 219–222
property adaptive evolution, 143
reproductive success, 92–93
structural homologues, 28
suborganismal objects, 144
Organs
angiosperm categories, 499–500
preangiophytes, outer male–inner female, 497
Ornithogalum montanum, 370
Ossified sesamoids, 570
Osteolepiform fish, 538
Outgroup comparison
in character polarity, 322–323, 330
in rooting, 326
Outgrowth
limbs, 534–538
vertebrate appendages, 533
Ovules
angiosperms, 503
preangiophytes, 496–497

Paleontology
in character polarity, 322, 325
in rooting, 327–328
Pangenesis, in Darwinism, 68–69
Papaver, element evolution, 505
Paracrine factors, 439
Parietal foramen, 186
Partitioning, organism into characters, 26–27
Parts
associated problems, 265
character comparison, 264
character-parts, 264
in *Cinctipora elegans,* 279
classification, 281–282
compositional differences, 271–274
decomposition, 269
developmental correlation, 63–64
duplicate parts, 274–275
in *Eminooecia carsonae,* 278–279
functional correlation, 62–64
functionality, 267–268
ρ group comparison, 264–265
F group comparison, 264–265
hierarchies, 265–267
homology, 260
individual comparison, 264
integration and isolation, 262

module comparison, 264–265
object part strategy, 270–271
shape differences, 274
spatial relationships, 271–274
technical definition, 262–263
temporal scale and range, 268–269
Pathology
brain variations, 50
in Vogt's theory, 42
Paths, 247–248
Patterning evolution
symmetrical structure individuation, 523–524
symmetry systems, 515–522
PAX6, 438
Pax-6, 5
Pelargonium, 502
Penetrance, and brain pathological variations, 50
Periodic table, 199–202
Petals, angiosperms, 499–500
Pharyngeal arches, key innovation, 590
Pharyngeal jaw, cichlids, 599
Phase transitions, in symmetry breaking, 234
PHAST, in homology comparison, 312
Phenetics, characters as artifacts, 29
Phenomics, for evolutionary character problem, 134
Phenotypes
conceptualization, 166–168
genotype and environmental effects, 364–365
gross, characters, 103–104
modularization, 429
multivariate, *Ornithogalum montanum,* 370
offspring, select effects, 118–119
in Vogt's theory, 51
Phenotypic characters, phylogenetic decoupling, 92
Phenotypic innovation, epigenesis, 570–571
Phenotypic novelty, definition, 378–379
Phenotypic plasticity
concept development, 364–365
development interaction, 371–372
in homology definition, 377
in macroevolution, 381
neodarwinist view, 364
role, 379
in trait homology, 376
Phenotypic response, and environment, 374–375
Phyletic history, from character, 14
Phylogenetic continuity, homologous characters, 92
Phylogenetic decoupling, phenotypic characters, 92
Phylogenetic reconstitution, procedure, 95

Phylogenetic stability, in functional units, 192
Phylogenetic systematics, part homology, 260
Phylogeny
    with amino acid sequence, 103
    in character polarity, 320, 332–333
    DNA sequence alignment, 304–307
    DNA sequences, 303
    flowering plants, 495–496
    material organization heterogeneity, 1–2
    with nucleotide, 103
    pattern inferrence, 31
    PHAST, 312
    tetradontiform fish, 294
Pi-additivity
    in character independence, 158
    definition, 155
    fitness function, 157, 159–160
Pigment band, butterfly wing, 513
Pi-LE, *see* Linkage equilibrium
Planehead filefish, *see Monacanthus hispidus*
Plant life ordering, 58–59
Plasticity
    adaptive plasticity, 381
    correlations, 368–371
    developmental interaction, 371–372, 596
    integration, 370
    in macroevolution, 379–380
    morphological plasticity, 560–561
    phenotypic, *see* Phenotypic plasticity
    reaction norm, 366–367
Pleiotropy
    differential epistasis role, 414
    genetic correlation, 412–413
    locus on character complexes, 430
    and QTL detection, 420–424
Polarity, *see also* Character polarity
    cell polarity, 566–567
    in cladistic analysis, 333–335
    and rooting, 326–328
Polymorphic population, mating habit division,
        597
Polymorphisms, wings, 129
Population
    character system, 224–226
    character variation, 390
    evolution, 228–229
    evolutionary potential, 127
    frequency vector, 223–224
    genetic correlation, 413
    intrapopulation variability, 24
    local fitness optimum, 120
    for mandibles, 416–418
    nucleotide entrance, 102
    ontological primacy, 145

Pi-additivity and Pi-LE, 157
    polymorphic population, 597
    tree representation, 96
Population-level characters, 96
Population-lineages, variants, 103
Postgenital fusion, angiosperms, 498
POY, in optimization-alignment, 312
Preangiophytes
    microsporangia, 497
    outer male–inner female organs, 497
    ovules, 496–497
    seeds, 496–497
Precis coenia, 347, 349–350
Preformation
    Buffon, Georges, 67
    mechanical conceptualization, 62
Pre-Mendelian world, epigenesis, 563–564
Prey, effect on burrfish, 292
Primitive descriptor, 217–218
Principle of continuity, 58–60
Process homology
    BMP pathway example, 448
    concept extension, 438–439
    criteria, 443–444
    in evolution, 436–437
    examples, 439–441
    general principles, 441–442
    level issues, 445–446
    merits, 449–451
Product descriptors, for organism, 219–222
Progeneration, in parental copying, 85
Proper descriptor, 218
Proteins
    BMP, 448, 542
    fringe protein, 448–449
Proton, in chemical classification, 201
Psychiatric diseases
    classification, 43–44
    Drosophila model system, 49
    Vogt classification, 47
Pterois volitans, 518
Pufferfish, 293–294

QTL, *see* Quantitative trait locus
Quantitative characters
    in biological analysis, 25
    cladistics, 24–25
Quantitative complementation, 400–401,
        403–406
Quantitative constraints, 117
Quantitative genetics
    for character problem, 117–119
    in constraint detection, 120–122

Quantitative genetics *(continued)*
 evolutionary character identification, 123–124
 gene additive effects, 150–151
 Pi-additivity and Pi-LE, 157
 selection experiments, 125
 selection gradients, 124
 trade-offs, 124
Quantitative trait locus
 Drosophila bristle number, 392–398
 to gene, 398–406
 in mandibular morphology, 426–429
 mapping, 132–134
 pleiotropic patterns, 420–424
Quasi-independence, in adaptationist theory,
 153

RA, *see* Reconstructability analysis
Radical characterization, evolutionary explanatory models, 28
Reacting factors, in Vogt's theory, 51
Reaction norms
 biological phenomena affecting, 366–367
 genetic correlation, 367–368
 phenotypic plasticity, 364–365
Reconstructability analysis
 complexity and constraint, 248–250
 cyclicity, 247–248
 decomposition losses, 250–252
 definition, 237
 identification, 252–254
 information-theoretic, 239–241
 lattice of relations, 241–242
 lattice of structures, 243–246
 main formalisms, 238–239
 reconstruction, 252–254
 relation definition, 242–243
 set-theoretic, 239–241
 variables, 237–238
Reconstruction, in RA, 252–254
Relation, 242–243
Relaxed selection, 182
Rel pathway, 442
Repackaging, and character replication, 100
Repetition, in copy generation, 89
Replicating molecules, 89
Replication
 character, *see* Character replication
 DNA, 87
Representation, 218–219
Reproduction, *see also* Replication
 animal, Aristotle's theory, 59

in *Arabidopsis,* 120
 organisms, 114–116, 146–147
Reproductive success of organisms, 92–93
Rooting, 326–328, 333–335

scabrous, 403
Seeds, preangiophytes, 496–497
Segmentation
 arthropods, 456–459, 478–481
 Drosophila limb primordia, 462–464
 and epigenesis, 564
Selection
 context-dependent units, 149–153
 gradients, 124
 internal selection, 186, 190–191
 natural selection, 24, 149, 205–206
 on offspring phenotypes, 118
 in quantitative genetics, 125
 relaxed selection, 182
 stabilizing selection, 596
 on traits, 118
Self-maintenance, organisms, 146–147
Sensory bristle, in *Drosophila*
 as model system, 390–391
 QTL, 392–398
Serial homology
 evolution, 524–526
 individuation, 526–527
Sesamoids, phenotypic innovation, 570
Set-theoretic framework
 decomposition losses, 250
 in RA, 238–241
Sexual recombination, 103
Sexual reproduction
 and Buffon, Georges, 66–67
 QTL effects, 397
Sexual taxa, structural levels, 102
Shape, differences in parts, 274
Similarity, from character, 15–16
Skeletal elements, phenotypic innovation, 570
Skulls, as mechanical units, 180–182
Solanum hirtum, 505
Sonic hedgehog, 543–544
Soul, in *Hydra viridis* regeneration, 65
Southern puffer, *see Sphoeroides nephalus*
Spatial heterogeneity, as material process,
 1–2
Spatial relationships, parts, 271–274
Species
 Aristotle's origination theory, 58–59
 arrangement, 17

bryozoan parts, 278–279
delineation, 17
key innovations, 597–599
tree representation, 96–97, 100–101
Specificity, and brain pathological variations, 50
Specimens, character sets, 20
Spermatophytes
ovules and seeds, 496–497
phylogeny, 495–496
Sphenoid, mouse, morphological organization, 104
Sphoeroides nephalus, 289
Spontaneous generation, 66–67
Spotty, 353, 355–356
Stabilizing selection, 596
Stamen, angiosperms, 499
Starvation resistance, *Drosophila melanogaster,* 125
State models, 247–248
Static homology, 304–307, 310–313
Sternopleural bristle number, 393
Stochastic algorithms, 29
Stratigraphy, 325
Striped burrfish, *see Chilomycterus schoepfi*
Structural elements
duplication, 588–590
flowers, 494–495
Structural homology, 28
Structural identification, 39–40
Structural integration, organisms, 147–148
Structural key innovations
changes, 586–588
complexity, 593
duplications, 588–590
inhibiting factors, 594–596
promoting factors, 596
Structural units, as functional units, 177–179
Suborganismal objects, in individual organism, 144
Symmetrical structures, individuation, 523–524
Symmetry breaking, phase transitions, 234
Symmetry systems
evolution, 525
hierarchy, 522
nymphalid groundplan, 519
origin, 515–521
Sympetaly, angiosperms, 500–502
Synapomorphy, and homology, 310, 330
Syncarpy, angiosperms, 500
Syntagma philosophicum, 61
Systematics
character numbers, 15

character properties, 286
character usage, 18
and evolutionary theory, 32
historial treatment, 16–17
part correlation, 20

Taxonomy
as classificatory system, 201–202
developmental considerations, 212–213
developmental data, 213
free taxonomy, 204–205
Homo sapiens, 209–210
morphological characters, 389–390
Teeth, mammalian, 185
Templates, character replication copies, 85–86, 89
Temporal heterogeneity, as material process, 1–2
Temporal scale, for parts, 268–269
Tenuinucellar ovules, angiosperms, 503
Tests
allelic association tests, 402–403
association tests, 403–406
character testing, 584
quantitative complementation tests, 400–401
Tetradontiform fish
adductor mandibulae, 289–293
burst duration, 296–297
coughing, 293–294
motor pattern variability, 296
water blowing, 293–294
Tetrapods
limbs
basic plan, 544
studies, 548–549
in terrestrial vertebrates, 532
water-land transition, 590
Thecal organization, angiosperm stamen, 499
Tinman, 438
Tissues
arthropod limb morphogenesis, 477
decoupling, 591–593
and epigenesis, 563–564
multilayering and epigenesis, 564
Topistic unit
definition, 47
in Vogt's theory, 51
Trade-offs
in life history, 116, 128–129
in quantitative genetics, 124
in variation, 121

Traits
  genetic correlation, 412
  homology, 376
  life history, 126–130
  overall value, 366–367
  pleiotropic relationship, 130–131
  QTL
    Drosophila bristle number, 392–398
    to gene, 398–406
    in mandibular morphology, 426–429
    mapping, 132–134
    pleiotropic patterns, 420–424
  in quantitative genetics, 117–119
Transformed mouse, 129–130
Transplantation, in eyespot experiments, 353
Tree reasoning, for character replication,
    95–104
Tree representation
  fertilization, 99
  nucleotides, 99, 101–102
  populations and species, 96–97
Trembley, Abraham, 65
Trichoptera symmetry systems, 525

Uncoupling, eyespots, 356–357
Unitegmic ovules, angiosperms, 503

van Leeuwenhoek, Antoni, 63
Variables
  character variables, 215, 264
  latent variables, 247–248
  in RA, 237–238
Variational properties, functional units,
    192–193
Vertebral column, key innovation, 588–589
Vertebrates
  appendage outgrowth, 533
  leg, deep homology, 448–449
  limbs, 544–546, 568
  terrestrial, from tetrapod, 532
Vital processes, in Vogt's theory, 51
Vogt, Oskar and Cécile
  basic elements, 51–52
  biological characters, 38–39, 46–48, 50–52

brain architectures, 43
brain mapping, 42–43
brain pathological variations, 50
early work, 41
genetics, 45–46, 48–49
geographic variation, 43, 48–50
institution building, 46
neo-cortex characterization, 38
and pathology, 42
psychiatric disease, 43–44, 47
scientific publications, 41–42
von Baer, Karl Ernst, 70–71
von Haller, Albrecht, 64

Water blowing, in tetradontiform fish,
    293–294
Water-land transition, tetrapods, 590
Wernicke's area, in neo-cortex, 40
Wings
  butterfly
    color evolution, 359
    elements as characters, 514–515
    serial homology, 524–526
    structure, 512–513
  dimorphism, 377
  Drosophila, imaginal disc, 474–475
  polymorphisms, 129
Wnt pathways, 440

Xenopus
  amphibian organizer, 442–443
  BMP pathway, 448

Y model
  in constraint detection, 121
  for life history, 113–114

Zebra, 516
Zebrafish, *see Danio rerio*
Zone of polarizing activity, 534–535
Zoological systematics, 44–45
ZPA, *see* Zone of polarizing activity

Printed and bound by CPI Group (UK) Ltd, Croydon, CR0 4YY

08/05/2025

01864994-0001